Archaeology of Logic

Andrew Schumann
Head of the Department of Cognitive Science and Mathematical Modelling
University of Information Technology and Management in Rzeszow
Rzeszow, Poland

CRC Press
Taylor & Francis Group
Boca Raton London New York

CRC Press is an imprint of the
Taylor & Francis Group, an **informa** business
A SCIENCE PUBLISHERS BOOK

First edition published 2023
by CRC Press
6000 Broken Sound Parkway NW, Suite 300, Boca Raton, FL 33487-2742

and by CRC Press
4 Park Square, Milton Park, Abingdon, Oxon, OX14 4RN

CRC Press is an imprint of Taylor & Francis Group, LLC

Library of Congress Cataloging-in-Publication Data (applied for)

ISBN: 978-1-032-32734-1 (hbk)
ISBN: 978-1-032-32737-2 (pbk)
ISBN: 978-1-003-31646-6 (ebk)

DOI: 10.1201/9781003316466

Typeset in Times New Roman
by Radiant Productions

Preface

Around the third millennium B.C. in the Fertile Crescent, a world-system with a single world-economy, covering very vast regions, began to form. Mesopotamia became the center of this world-system. This was possible due to the development of common commercial law and logical competence there. The expansion of the world-economy during the Silk Road period from the 4th century B.C. to the early 5th century A.D. across various countries of Eurasia was accompanied by the spread of logical competence, first formed in Mesopotamia, as a mechanism of legal hermeneutics to draw logical conclusions without fallacies. This competence was simultaneously comprehended in different cultures connected by the Great Silk Road—in ancient Greek logic (4th–2nd centuries B.C.); ancient Chinese proto-logic (5th–2nd centuries B.C.); Judaic logical hermeneutics (1st–2nd centuries A.D.); and in Indian-Buddhist logic (2nd–6th centuries A.D.).

The book analyzes the emergence of logic and its spread and early forms of its reflection. Consequently, logical competence is seen not as an innate ability, but as a social practice first established in Mesopotamia. Logic as a science became possible only after the development of logical competence as an accepted social practice. On the other hand, this view is a non-Marxist assessment of the early form of the world-system, centered on international law and logical competence, which made the world-economy and international trade then possible.

Contents

Introduction

In the context of the information revolution associated with the advent of personal computers and the Internet, the amount of information transmitted in modern societies has grown significantly. In a sense, this has noticeably changed the status of socio-philosophical knowledge in the body of theories about man and society. The fact is that philosophical knowledge itself has been noticeably relativized in the context of the growth of opportunities for the creative self-expression, including the self-expression that is philosophical or pseudo-philosophical in nature. There has been a fragmentation of philosophical traditions and teachings.

In this regard, the standard ways of maintaining the philosophical tradition through the adoption of a certain set of values only further fragment the philosophy itself. Moreover, there are too many traditions now, nurtured in heterogeneous cultures, and in the context of globalization, there are opportunities for their mixing and hybridization, which entails an even greater fragmentation of philosophy.

For example, in the history of logic there was a certain set of philosophical values that substantiated the history of logic itself. This set of values included Eurocentrism, i.e., it was believed that logic appeared in Ancient Greece, starting with Aristotle, and its greatest achievements happened within the framework of European culture—during the period of philosophy of Ancient Greece and Rome, then medieval scholasticism, followed by the philosophy of modern times, and so on until the emergence of mathematical logic at the end of the 19th century. Accordingly, definite examples of European thinking were understood as a doctrine of logic.

However, it turned out that there existed and prevail philosophical traditions of logic outside of European culture. Such alternative traditions include the Indian-Buddhist logic, such as the schools of Yogācāra and Nyāya (Schumann, 2018a, 2018b, 2019c, 2020b, 2021c), and the Judaic logical hermeneutics (Schumann, 2008, 2010a, 2010b, 2011a, 2011b, 2012a, 2012b, 2012c, 2013a, 2013b, 2013c, 2015, 2016, 2017a, 2019b; Schumann and Kuznetsov, 2017). These logical doctrines were created in a completely different philosophical context than the ancient Greek logic. This forces us to abandon the standard approach, when a philosophical doctrine is substantiated through the very fact of an established tradition (for example, the history of logic is constituted through the fact of the existing tradition of European logic, starting with Aristotle). The problem arises of combining heterogeneous traditions and an adequate analysis of their achievements in conditions of the fact that it is no longer possible to choose only one tradition as a privileged one as there are too many traditions and they are not very compatible with each other.

To avoid any possible speculations when analyzing diverse traditions and their achievements in the context of different philosophical values, the work appeals to the structuralist analysis proposed by Vladimir I͡Akovlevich Propp (1968) and Claude Lévi-Strauss (1955) in relation to mythologies of heterogeneous traditions. This analysis allows us to focus not on some values of mythological traditions, but on some semantic structures within the mythological text—on stable syntagmatic and paradigmatic connections, in the language of Lévi-Strauss. This approach allows us to compare the identified structures of different myths of various peoples, and not the actual content of mythological narratives (Schumann, 2021d, 2022; Schumann and Arif, 2021; Schumann and Sazonov, 2021a, 2021c). This analysis has been revised in this book so that it could be applied to the reconstruction of the logical teachings of incompatible philosophical and even religious-philosophical schools. So, as a new way of substantiating the history of logic and philosophy, a new methodology for analyzing fragments of texts for various purposes is offered for the presence of logical competence and structural forms of its implementation in the form of some sets of logically correct reasoning schemes. This methodology can be called "archaeology of logic" (Schumann, 2017b, 2019a, 2020a, 2021a, 2021b) in the meaning of Michel Foucault (2005). According to this methodology, we are interested not so much in the living traditions of logic, but we focus on the recreation of the latent tradition of logical competence—when and where it first appeared, in what forms (structures) it was established, and how it was subsequently inherited. Let us note that the genealogical analysis of Polish logic was carried out by Jan Woleński (1999, 2011, 2013).

In this case, logic itself is understood as mathematical logic in all its diversity from classical to non-classical systems—i.e., as a complex of logical rules of inference, which allows us to draw conclusions from some judgments purely mechanically (Church, 1996; Copi et al., 2014; Hilbert and Ackermann, 1949; Hilbert and Bernays, 1968).

As a result of the structuralist analysis of the forms of logical competence in various cultural traditions, it turned out that the texts in the Sumerian and Akkadian languages (from the 22nd to the 5th centuries B.C.) contain the most ancient layer of fragments, the authors of which possessed the perfect logical competence in the sense of modern logic (Schumann, 2017b, 2019a, 2020a, 2021a, 2021b; Schumann and Sazonov, 2021b). These fragments are divided into the following three types of texts, in which conditional statements "If *A*, then *B*" are often encountered with subsequent conclusions from them on the basis of *modus ponens*, *modus tollens*, and a number of other inference rules: (1) hermeneutic texts ("If a sign/word *A* occurs, then we should understand it as *B*"); (2) predictions ("If in nature we observe a sign/omen *A*, then we should predict the occurrence of event *B*"); (3) legal codes and trial judgments ("If someone conducts an act *A*, then the punishment *B* will follow"). In all these three types of texts, a complex algebraic system for constructing a scheme of purely mechanical inference is easily reconstructed, as in modern logic.

On the other hand, if we analyze, for example, the logical fragments of the *Pāli Canon* (many texts that demonstrate logical reasoning), then we already find the lack of developed logical competence. Along with correct conclusions, sophisms immediately arise, and we see that the authors do not distinguish between them. The only text of the canon with impeccable logical inferences is presented by the *Milinda's Questions* (*Milindapañha*), created in Gandhāra (now Pakistan) during the reign of the Hellenes or the Hellenized Indo-Scythians in this region (from the 2nd century B.C. to the 2nd century A.D.).

Hence, the "archaeology of logic" makes it possible to identify the centers of logical cultures and the waves of their spread and diffusion through tracking the dynamics of logical competence: (1) the most ancient logical culture is found in the Sumerian and Akkadian texts; (2) this culture, even purely terminologically, is inherited in the logical fragments of the Jerusalem and Babylonian Talmuds; (3) this culture is also inherited in Stoic and Epicurean logical fragments (for example, in Cicero's *Topica*); (4) the Stoic and Epicurean fragments agree well with the propositional rules of the Buddhist schools of Madhyamaka and Yogācāra, as well as with the rules of the Hindu school of Nyāya, so that either their common source with Stoicism and Epicureanism or a separate influence of the Hellenistic logic on the Indian-Buddhist logic can be assumed. It shows a kind of linear development of propositional logic with the joint Sumerian-Akkadian root. But this tradition is reconstructed exclusively through tracing the logical competence which is an established family of inference techniques.

In the structuralist analysis of myths, there is a program proposed by Yuri E. Berezkin (b. 1946), see Thuillard et al. (2018), according to which the myth structure in the sense of V. I͡A. Propp (1968) and C. Lévi-Strauss (1955) is extremely stable and persists for many generations, but over time it can undergo minor mutations. The narrative in the myth can and does change, but the structure remains inert. Therefore, the myths of different peoples with an identical or almost identical structure say that there was a borrowing of the myth by one of the peoples from each other or by both from a third people. Thus, it is believed that the structure of the myth is created only once. Based on this assumption, the program of Yu.E. Berezkin makes it possible to apply the genealogical analysis to the structures of myth and to identify the waves of their spread among different peoples.

A similar hypothesis is basic for the study of the history of logic in my work. It is assumed that the structures of logical competence are extremely stable and inert, although they can undergo individual mutations, too. These mutations can be used to trace the dynamics of the spread and diffusion of these structures in different philosophical traditions.

This key research hypothesis is supported by the latest evidence in experimental psychology and cognitive science. According to these data, the logical competence is not an innate cognitive ability of a person and is transmitted only in the process of direct learning how to use it in practice. Even scientists may be deprived of this competence, according to some experimental studies (Kern et al., 1983; Sjoberg, 2017), although they are not just people with higher education, but also with some notable achievements in their scientific work. Ordinary people without a special preparation, as a rule, have no logical competence at all.

In this case, in accordance with the latest data of cognitive science, it can be argued that the logical competence did not exist among the tools of human cognition from the very beginning. It arose once and gradually spread along with the development of urban civilization within the framework of definite social practices through a direct training of this competence (Schumann, 2021e). Such practices were, first, legal practices of protection and guarantees of property rights for the legal support of trade. These practices have led to the creation of a judicial system in Mesopotamia for the effective resolution of any dispute, including the economic one, based on logical conclusions from legal documents. The judicial system of Mesopotamia became especially effective during the period of the III dynasty of Ur, who united the whole of Mesopotamia under their rule and reigned from about 2112 B.C. to 2004 B.C.

The legal practice of the III dynasty of Ur was preserved and continued by subsequent dynasties of Mesopotamia up to the New Babylonian Kingdom, which existed in the period 626–539 B.C. Legal documents, which include legal codes, judicial acts, business correspondence and contracts, acts of inheritance or donation, etc., were drawn up in Mesopotamia in dialects of the Akkadian language, and in culturally close states (such as the Hittite Kingdom) in their languages, but modeled on Akkadian documents. At the same time, Akkadian served as the language of international trade and diplomacy.

The Achaemenid Empire, which at the peak of its power, encompassed vast regions from Asia Minor, Egypt, Syria, and Palestine to present-day Afghanistan and Pakistan and ruled from about 550 to 330 B.C., fully inherited the legal culture of the New Babylonian Kingdom. Several documents continued to be written in Akkadian, but the most legal ones were already written in Aramaic. At this stage, it is this language that became the dialect of international trade and diplomacy.

The Kingdom of Alexander the Great, which existed from 336 to 323 B.C. (during his reign), eventually began to cover the borders of the Achaemenid Empire, including many new areas, primarily in the West. In this kingdom, a Hellenistic legal culture arose, which continued the heritage of the Achaemenids. From now on, the ancient Greek language became the main language of international trade and diplomacy. At this time, the Eurasian trade network was laid, which over time was called the Silk Road. The leading languages of this network were Aramaic, Ancient Greek, Bactrian, Sogdian, Tamil, Gāndhārī. Within the framework of this network, a high legal culture, and high lawful standards of trade, including the logical competence, were maintained—everything what first appeared in Mesopotamia as early as the 3rd millennium B.C. Therefore, it is not surprising that logic received a reflective comprehension among the peoples covered by the Silk Road trade network—with the main languages: Ancient Greek (Aristotle, Stoics), Aramaic (the tradition was preserved in the form of the Talmudic logic), Bactrian and Gāndhārī (the tradition was preserved in the form of the logic of the Madhyamaka, Yogācāra, and Nyāya schools).

As we can see, the "archaeology of logic" allows us to build the history of logic not based on the living philosophical traditions within the framework of traditional philosophical understanding, in particular within Eurocentrism, but on the basis of the structuralist reconstruction of different historical forms of logical competence, which can be traced back due to archaeology data—i.e., to some fragments of texts and inscriptions with well-fixed dating. Its oldest forms are found in the Sumerian-Akkadian texts.

So, my work gives a definition of logical competence—as some social ways of implementing logical knowledge. Thus, the forms of expression of rational thinking in speech practices are investigated. The study was carried out on the texts of early states to clarify the origin and genesis of logical competence from the earliest forms to the appearance of the very first treatises on logic. The oldest examples of developed logical competence are found in the texts of Mesopotamia, starting with judicial acts and business correspondence of the Akkadian period—i.e., from the 24th to the 20th centuries B.C. (Schumann, 2017b, 2019a, 2020a, 2021a, 2021b; Schumann and Sazonov, 2021b). By the time of the appearance of Aristotle's treatises on logic in the 4th century B.C., there was already a long history of the development of logical competence. The analysis of various forms of manifestation of this proficiency allows us to find out the nature of logical knowledge in its practical expression.

As a practical application of the proposed methodology, the prospects of using logic in the design of e-justice are analyzed within the framework of the general paradigm of e-government and e-democracy.

The Purpose and Objectives of the Study

The aim of this work is to identify specific and general patterns of development of logical competence at the early stages in the context of the structuralist approach.

To achieve this goal, the following tasks are formulated:

- to reveal the philosophical essence of the concept of logical competence;
- to develop the structuralist methodology called the "archaeology of logic" to identify the logical competence in its early historical forms;
- to identify the main mechanisms of manifestation of logical competence in social practices;
- to determine the conditions and reasons for the emergence of logical competence in the early states of the Middle East;
- to explicate the forms of manifestation of logical competence during the Hellenistic and then Roman periods, which became basic for humanity in the future;
- to identify specific patterns of the dynamics of logical competence at the early stage of formation—i.e., in the Sumerian-Akkadian (Babylonian) and Hellenistic or Hellenized cultures;
- to identify general patterns of functioning as well as patterns of development of logical competence in the historical period of the ancient world;
- to show the prospects for the practical application of the theory of logical competence in the modern legal culture.

The **object** of the research is presented by the logical competence as a necessary condition for reasoning that is correct from the standpoint of (modern or mathematical) logic.

The **subject** of the research is presented by the dynamics of logical competence at the early stages of development, i.e., by its specific and general patterns.

The choice of the subject of research is made due to current trends

- in modern scientific knowledge, consisting in rethinking the nature of logical competence,
- in the search for new philosophical foundations of logical knowledge,
- in the development of new strategies for the structuralist research. This choice is also done due to the need to comprehend the dynamics of logical knowledge without mistakes made in the past by traditional Eurocentrism, and to search for prospects for the development of logical competence in modern culture.

The scientific novelty and significance of the results of the research is determined by the fact that in it

- an integral definition of logical competence is given;
- a new structuralist methodology has been developed, which allows one to compare philosophical constructs within the semantic structures of different philosophical traditions and find similarities and differences in these structures;
- within the framework of the new methodology, a project of the so-called "archaeology of logic" has been proposed, which makes it possible to identify the logical competence in its historical forms;
- the dynamics of logical competence has been constituted as an independent scientific direction of logical and philosophical research;
- the mechanisms of functioning of logical competence in social practices have been disclosed, namely, the role and place of logical competence in the legal support of administrative and economic issues within the framework of the developed legal system has been shown; thus, the connection between the logical competence and the special legal culture of Mesopotamia has been established;
- the early historical forms of logical competence, which became basic for humanity in the future, have been revealed;
- the theory of logical argumentation in the Sumerian and Akkadian languages has been reconstructed; the logical rules of inference implicitly used in this theory have been explicated; it has been shown that propositional logic of the early Stoics became a reflective comprehension of the already existing tradition of legal argumentation of Mesopotamia;
- the ideas of logical determinism, formulated for the first time in the texts on divination in the Akkadian language, have been explicated; these ideas formed the basis of the system of modal logic of the Stoics; then these ideas became primary in the logical studies of schools of Indian philosophy: Madhyamaka, Yogācāra, and Nyāya;
- specific patterns of the dynamics of logical competence have been established, associated with its inherent effective way of regulating litigation and finding intersubjective grounds for resolving conflicts; with the growth of international

trade, starting with the developed states of Mesopotamia and, especially, during the Hellenistic and Roman periods, when the Eurasian transnational trade route, called the Silk Road arose, there was an urgent need for a unified legal support for trade, which would include the logical competence as an integral part of legal hermeneutics to ensure intersubjectivity;

- the author's theoretical model of the formation of logical competence in the historical period of the ancient world made it possible to establish the following general patterns of the functioning and development of logical competence in the context of different philosophical traditions: (a) the logical competence in the process of its application improves and becomes more complicated over time; (b) the logical competence in the process of reflection on it acquires more and more sophisticated forms; (c) the logical research eventually becomes self-sufficient and leaves its original cultural context, represented by legal practice and divination; this process of acquiring self-sufficiency begins at the stage of the emergence of three new literary genres of philosophical reflection: (1) the philosophical prose in Ancient Greece (from the 4th century B.C.); (2) the collections of Talmudic thematic dialogues (from the 1st century A.D.); (3) the thematic dialogues or philosophical *sūtra*s in Ancient India (from the 2nd century B.C.); within the framework of these three new genres, the first logical works appeared: (i) in the 4th–3rd centuries B.C. in Ancient Greece, (ii) in the 1st–2nd centuries A.D. in Ancient Judea; (iii) in the 2nd–4th centuries A.D. in Ancient India;
- some recommendations have been developed for the practical application of the theory of logical competence in the modern legal culture.

The formulation of the main objectives of this research is due to the practical needs of today's state policy in many countries on the use of e-government, in the systems of which a high logical competence is assumed.

The results of the study are of theoretical and methodological significance for the development of logic and theory of cognitions and for the practical application of the theory of logical competence in the evaluation of state programs of legal and socio-economic development, as well as in strategies and mechanisms for the implementation of relevant state and commercial projects in the field of e-justice and electronic government.

Basic Ideas of the Work

1. Logical competence is an ability to draw logically correct conclusions from the standpoint of modern symbolic logic. This involves the following two principles in reasoning: (1) premises in reasoning must be only true; (2) inferences from them should be based on special inference rules (such as *modus ponens* and *modus tollens*), the application of which to the true premises does not give contradictory consequences. This ability is not innate either for the animal world or for humans, and in its natural manifestations is inevitably accompanied by so-called cognitive distortions and biases in the form of the impact of emotional states, primarily stress or satisfaction, on decision-making. The logical

competence of a person arises as a social mechanism of cognition, transmitted through learning, which can be cleared of natural cognitive distortions. Animals, even swarms (i.e., pseudo-social), do not get this. Only in a purified form is the logical competence capable of giving logically correct reasoning. Such a pure form of competence is in demand in certain social practices, in which the achievement of an automatic consensus plays an important role, based on a system of conventions from which only consistent consequences can be drawn. One of the earliest forms of such a social practice, using the logical competence, was the legal system of Mesopotamia.

2. An implementation of the structuralist approach "archaeology of logic", proposed in the work, allows us to reveal the patterns of dynamics of logical competence as a new subject of scientific research, to constitute the study of this dynamics as an independent scientific direction of philosophy and history of logic, to explicate the diversity and unity of historical manifestations of logical competence.

   The essence of the "archaeology of logic" is to identify the logical competence in its historical forms and compare its manifestations in different cultures, tracing the possible influence of one tradition of using competence on another tradition. To discover the logical competence, a formalization by means of modern symbolic logic of logeme structures is used where the logemes are elementary steps of logically correct reasoning. The logeme structure *X* of one tradition may or may not have a similarity to the logeme structure *Y* of another tradition. Thus, a structure *X* is like a structure *Y* if and only if there is an injective mapping from *X* into *Y* with a preservation of the syntagmatic and paradigmatic orders of logemes. In other words, when *X* is a logical fragment of *Y*; for example, when in judicial acts of one period the correct use of the inference rules such as *modus ponens* and *modus tollens* occurs, and in judicial acts of the next period there is a correct use of not only the inference rules such as *modus ponens* and *modus tollens*, but also the introduction of conjunction and the introduction of disjunction.

   Let the logeme structures *X* and *Y* be found in cultures between which there were facts of influence or continuity. Then the mapping from *X* into *Y*, if *X* existed historically before *Y*, indicates that the tradition *Y* became a continuation and development of tradition *X*, and if *Y* existed historically before *X*, then it indicates that the tradition *X* took over the tradition *Y* in a weakened form.

   The application of the "archaeology of logic" to the materials of documents of different cultures of the ancient world of the historical period (from the Sumerian-Akkadian to the Ancient Indian one) has shown that the logical competence appeared in Mesopotamia no later than during the III dynasty of Ur and then developed within the framework of the spread and diffusion of legal standards of international trade until the emergence of transnational networks of the Silk Road two thousand years later.

3. The logical competence has its own historical forms and develops from elementary methods of logical reasoning to complex reasoning. As a social

mechanism of cognition, the logical competence is incorporated into relevant social practices within the legal system, in which consensus is guaranteed through the application of proficiency of drawing correct conclusions to legal norms. Thus, a connection is revealed between the logical competence and the special legal culture, which was first formed in Mesopotamia. The spread of the legal culture of Mesopotamia throughout the Middle East, and then through the Achaemenid Empire and the Hellenistic or Hellenized states across the countries of Eurasia, also influenced the spread and development of logical competence. In its rise, this competence had two phases: (a) pre-reflective, when it developed only in the context of relevant social practices (such as the judicial system); (b) reflective, when it received a philosophical reflection within the framework of four independent traditions that have survived to this day: (i) the ancient Greek logic (from the 4th to the 3rd centuries B.C.); (ii) the Judaic logical hermeneutics (from the 1st to the 2nd centuries A.D.); (iii) the Indian-Buddhist logic (from the 2nd to the 5th centuries A.D.), (iv) the Chinese legal hermeneutics, founded by Mozi as a proto-logic (4th century B.C.). Particularly, propositional logic, created by the philosophers of the Ancient Stoa in the 3rd century B.C., became a reflective comprehension of the tradition of legal argumentation, which was formed for the first time in Mesopotamia. In China, the Mohist pre-logic arose in the 5th–3rd centuries B.C., however, it was established outside of the Mesopotamian legal culture, but within its own legal tradition, where a critical discussion among conflicting parties was not assumed. Consequently, this pre-logic did not become a true logic and did not develop.

4. In the pre-reflective phase of the development of logical competence in Mesopotamia, the intuition of logical determinism appeared for the first time (the assumption that the judgment "Either *A* is necessary, or not-*A* is necessary" is always true). This determinism manifests itself in the idea that for any future event *Y* there is a set of omens *X* such that the implication "If *X*, then *Y*" is true. Meanwhile, the complete lists of the omens of *X*, which were compiled in the Assyrian and Babylonian books on divination, completely determined our expectations about subsequent events. This understanding of divination with the intuition of logical determinism influenced the development of the modal logic of the early Stoics, who were the first to formulate the logical determinism in an explicit form (in the reflective phase). Consequently, the idea of logical determinism also became primary in the logical studies of the schools of Indian philosophy: Madhyamaka, Yogācāra, and Nyāya.
5. Specific regularities of the functioning of logical competence in the early stages of development are determined by: (a) a social practice, which requires forms of reaching a consensus through an intersubjective decision-making procedure (cleared of possible cognitive biases) from existing conventions; (b) an illocutionary structure for resolving disputes with a minimization of cognitive biases; (c) an internal variety of the use of logically correct reasoning.

   Specific patterns of the development of logical competence are as follows: (i) the evolutionary nature of changes in the use of logical competence and the consolidation of successful forms of application of logical competence; (ii) the high adaptability to historical changes (a) due to a permanent improvement of the configuration of successful forms of application of logical competence and (b) due to the demand for social practices of finding a consensus, cleared of cognitive distortions; (iii) the ability to syncretize with new social practices due to the awareness of the importance of using a decision-making mechanism for these practices, cleared of cognitive distortions; (iv) the long-term historical accumulation of logically correct forms of reasoning.

6. In the context of the structuralist approach to studying the dynamics of logical competence, the following general patterns of functioning and development can be distinguished.

   I. *Patterns of functioning*: (i) the logical competence (as a social mechanism of human cognition) is universal in the sense that it is devoid of cognitive distortions and biases by definition, therefore any historical example of the application of logical competence lends itself to a consistent formalization by means of modern symbolic logic and can always be expanded to a more detailed reasoning; (ii) the effective ways of applying the logical competence are formed and exist due to the processes of self-organization, which consist in the deployment of more and more complex forms of building consistent inferences; (iii) the stability in the cultural tradition of successful ways of applying the logical competence in social practices is determined by the internal variety of possible logically correct reasoning.

   II. *Patterns of development*: (i) the ways of applying the logical competence improve and become more complex over time; (ii) there is a mechanism of reflection regarding the logical competence, presented in logical studies, which allows for an unrestricted complication of this proficiency in its implementations up to the creation of artificial intelligence systems; (iii) any logical research is self-sufficient and can be implemented outside the cultural context in which the logical competence has traditionally been used.

7. In the early stages of development, the logical competence was applied in legal discourse. Currently, this vector of practical applicability of logical competence in the legal system can be strengthened within the framework of the creation of e-justice and e-lawyer systems. These systems use artificial intelligence techniques to automate legal decisions, including judicial decisions. In such systems, logical fallacies are excluded, and hence the very possibility of judicial errors based on a lack of logical competence.

References

Church, A. 1996. *Introduction to Mathematical Logic*. Princeton, New Jersey: Princeton University Press.

Copi, I.M., Cohen, C. and McMahon, K. 2014. *Introduction to Logic*. 14th edn. Pearson Education Limited.

Foucault, M. 2005. *The Order of Things. An Archaeology of the Human Sciences*. Taylor and Francis e-Library.

Kern, L.H., Mirels, H.L. and Hinshaw, V.G. 1983. Scientists' understanding of propositional logic: An experimental investigation. *Social Studies of Science* 13(1): 131–146.

Lévi-Strauss, C. 1955. The structural study of myth. *The Journal of American Folklore* 68(270): 428–444.

Propp, V. IA͡. 1968. *Morphology of the Folktale*. 2nd edn.University of Texas Press.
Schumann, A. 2008. Non-well-foundedness in Judaic logic. *Studies in Logic, Grammar, and Rhetoric* 13(26): 41–60.
Schumann, A. (ed.). 2010a. *Judaic Logic*. Gorgias Press.
Schumann, A. 2010b. Preface, Introduction. pp. 7–8, 11–33. *In*: Schumann, A. (ed.). *Judaic Logic*. Book series: Judaism in Context, 8. Gorgias Press.
Schumann, A. 2010c. Judaic Syllogistics: The Baba Qama from the logical point of view. pp. 239–254. *In*: Schumann, A. (ed.). *Judaic Logic*. Book series: Judaism in Context, 8. Gorgias Press.
Schumann, A. 2010d. Preface: Religious logic as a part of philosophical logic. pp. 10–19. *In*: Schumann, A. (ed.). *Logic in Religious Discourse*. Ontos Verlag.
Schumann, A. 2011a. Preface. *History and Philosophy of Logic* 32(1): 1–8.
Schumann, A. 2011b. Qal wa-Homer and Theory of Massive-Parallel Proofs. *History and Philosophy of Logic* 32(1): 71–83.
Schumann, A. 2012a. *Talmudic Logic*. London: College Publications.
Schumann, A. 2012b. Kabbalistic-Leibnizian Automata for Simulating the Universe. pp. 259–280. *In*: Evgeny Katz (ed.). *Molecular and Supramolecular Information Processing. From Molecular Switches to Logic Systems*. Weinheim: Wiley-VCH.
Schumann, A. 2012c. Cellular automaton of the Lord. *Polish Journal of Philosophy* 6(2): 103–110.
Schumann A. 2013a. Logical cornerstones of Judaic Argumentation Theory. *Argumentation* 27(3): 305–326.
Schumann, A. (ed.). 2013b. *Pragmatic Studies in Judaism*. Gorgias Press.
Schumann, A. 2013c. Preface, Introduction. pp. 11–12, 15–24. *In*: Andrew Schumann (ed.). *Pragmatic Studies in Judaism*. Book series: Judaism in Context, 14. USA: Gorgias Press.
Schumann, A. 2013d. Pragmatic Limits of Halakhic Logic. pp. 67–98. *In*: Andrew Schumann (ed.). *Pragmatic Studies in Judaism*. Book series: Judaism in Context, 14. USA: Gorgias Press.
Schumann, A. 2015. *Logika Talmudu*. Kraków: Copernicus Center Press.
Schumann, A. 2016. Physarum Polycephalum syllogistic L-systems and Judaic roots of unconventional computing. *Studies in Logic, Grammar, and Rhetoric* 44(57): 181–201.
Schumann, A. (ed.). 2017a. *Philosophy and History of Talmudic Logic*. London: College Publications.
Schumann, A. 2017b. On the Babylonian origin of Symbolic Logic. *Studia Humana* 6(2): 126–154.
Schumann, A. and Kuznetsov, A.V. 2017. Talmudic foundations of mathematics. pp. 67–74. *BICT 2017 – 10th EAI International Conference on Bio-Inspired Information and Communications Technologies*.
Schumann, A. 2018a. On a logical reconstruction of the Kathavatthu. pp. 1509–1512. *In*: *Proceedings of the 3rd International Conference on Contemporary Education, Social Sciences and Humanities* (ICCESSH 2018).
Schumann, A. 2018b. Did Nāgārjuna know modal logic? *Логико-философские штудии* 16(1–2): 95.
Schumann, A. 2019a. Did the Neo-Babylonians construct a symbolic logic for legal proceedings? *Journal of Applied Logics – IfCoLog Journal of Logics and Their Applications* 6(1): 31–82.
Schumann, A. 2019b. Do ant colonies obey the Talmud? pp. 261–272. *In*: Adamatzky, A., Akl, S. and Sirakoulis, G.Ch. (eds.). *From Parallel to Emergent Computing*, CRC Press.
Schumann, A. 2019c. On the origin of Indian logic from the viewpoint of the Pāli Canon. *Logica Universalis* 13: 347–393.
Schumann, A. 2020a. Legal argumentation in Mesopotamia since Ur III. *Journal of Argumentation in Context* 9(2): 126–154.
Schumann, A. 2020b. The Milindapañha in the context of History of Indian Civilization. *RUDN Journal of Philosophy* 24 (4): 544–569.
Schumann, A. 2021a. Logical reasoning for forecasting in Mesopotamia. *Studies in Logic, Grammar, and Rhetoric* 66(3): 721–746.
Schumann, A. 2021b. On the origin of logical determinism in Babylonia. *Logica Universalis* 15: 331–357.
Schumann, A. 2021c. Nāgārjunian-Yogācārian modal logic versus Aristotelian modal logic. *Journal of Indian Philosophy* 49(3): 467–498.
Schumann, A. 2021d. Plato and Pure Land Buddhism. *Platonic Investigations* 15(2): 110–141.
Schumann, A. 2021e. Towards the Definition of Logical Competence. *Cogency* 13(2): 7–40.
Schumann, A. and Arif, S. 2021. Proto-Hindu Buddhist deities Maheśvara and Hārītī in Gandhāra. *Journal Asiatique* 309(2): 231–248.

Schumann, A. and Sazonov, V. 2021a. From Mesopotamian Nanāia to Indian Durgā. *Nouvelles Assyriologiques Bréves et Utilitaires* (NABU) 1: 63–67.

Schumann, A. and Sazonov, V. 2021b. Omens as Logic. *Nouvelles Assyriologiques Brèves et Utilitaires* (NABU) 3: 192–195.

Schumann, A. and Sazonov, V. 2021c. The long path of Nanāia from Mesopotamia to Central and South Asia. *Studia Antiqua et Archaeologica* 27(2): 279–299.

Schumann, A. 2022. Christian-Buddhist intercommunication of late antiquity. *Scrinium*.

Sjoberg, E. 2017. Logical fallacies in animal model research. *Behavioral and Brain Functions* 13(1): 3.

Thuillard, M., d'Huy, J., Berezkin, Y. and Le Quellec, J.-L. 2018. A Large-scale study of world myths. *Trames Journal of the Humanities and Social Sciences* 22(4): 407–424.

Woleński, J. 1999. *Essays in the History of Logic and Logical Philosophy*. Cracov: Jagiellonian University Press.

Woleński, J. 2011. *Essays on Logic and Its Applications in Philosophy*. Frankfurt am Main: Peter Lang GmbH, Internationaler Verlag der Wissenschaften.

Woleński, J. 2013. *Historico-Philosophical Essays*, Vol. 1, Kraków: Copernicus Center Press.

Chapter 1

Philosophical and Methodological Foundations for Studying the History of Logic of Various Religious and Philosophical Schools

Logical treatises have been written in different cultures and languages since ancient times. In this chapter, a methodology is substantiated which allows these texts to be considered simultaneously to identify general and specific patterns of the development of logical knowledge.

1.1 Structuralist Approach to the History of Logical Knowledge and the Archaeology of Logic

1.1.1 Problem Setting

There are two extreme approaches to the study of the history of philosophy authored by: (i) Diogenes Laërtius (ca. 3rd century A.D.) who wrote the book *Lives and Opinions of Eminent Philosophers* (Βίοι καὶ γνῶμαι τῶν ἐν φιλοσοφίᾳ εὐδοκιμησάντων; *Vitae Philosophorum*) and (ii) Georg Wilhelm Friedrich Hegel (1770–1831) who is an author of the following three books on this subject: *Vorlesungen über die Philosophie der Geschichte* (1837), *Vorlesungen über die Philosophie der Religion* (1832), and *Vorlesungen über die Geschichte der Philosophie* (1833–1836). The first approach is focused on differences among philosophers and their concepts. The second approach accepts some general features and joint viewpoints in philosophies to reconstruct a joint history of philosophy of all nations as a linear development.

The methodology for the history of philosophy reflected by Hegel is based on two principles, used by many philosophy historians so far: (1) the philosophical idea is considered given only as the history of this idea (each philosophical system has a genealogy and does not arise without the influence of previous systems, on the one hand, and competing systems, on the other hand); (2) the philosophical idea develops from its abstract forms to more concrete ones (after development, the philosophical system becomes more complex, and there is an increase in its philosophical reflection).

Hence, according to Hegel, each philosophical idea is defined by its genealogy in the retrospective view or by its history in the perspective view. In Hegel's terminology, each idea is a development and transition from the state *an sich* (in itself) to the state *für sich* (for itself) and it can be revealed only genealogically from the end of the transition process or historically from the beginning of the transition process.

For example, Brahman from the *Upaniṣad*s as the supreme existence and absolute reality was regarded by Hegel as "a supreme being, but one that merely thinks itself, or is merely at home with itself, outside which all other content and configuration still lies" (Hegel, 1984: 331). In this feature, He is close to "the God of Judaism". Both are "an abstraction, God in the spirit but not yet God as spirit" (*Ibid.*). As a consequence, Brahman of the *Upaniṣad*s as well as the God of Judaism is the God *an sich* (in Himself), i.e., He is just a beginning in the theological reflections, where Jesus Christ should become the God *für sich* (for Himself) as the end of theological reflections.

Hegel claims that the movement of the human spirit, including any philosophical reflection, has proceeded from the east to the west. It means that philosophy begins in the east and ends in the west. To the same extent, there were only three principles in religion proceeding from east to west: (i) the God in Himself as He is presented in the world of Far Eastern religions (Mongolian, Chinese, Indian); (ii) the God out of Himself as He is presented in the Islamic world; (iii) the God for Himself as He is presented in the Christian world:

> For if we cast our eyes around the world, we can discern three main principles in the older continents: the Far East (i.e., Mongolian, Chinese, or Indian) principle, which is also the first to appear in history; the Mohammedan world, in which the principle of the abstract spirit, of monotheism, is already present, although it is coupled with unrestrained arbitrariness; and the Christian, Western European world, in which the highest principle of all, the spirit's recognition of itself and its own profundity, is realised. This universal series has been described here as existing perennially; but in world history we encounter it as a sequence of successive stages (Hegel, 1975: 128–129).

Now, only Christian peoples play a significant role in the world history:

> The whole eastern part of Asia is remote from the current of world history and plays no part in it; the same applies to the north of Europe (Hegel, 1975: 172).

In this section, some strict limits in the Hegelian approach to the history of philosophy are shown. So, I am going to discuss that the deep problem of the approach founded by Hegel is that all the substantial differences among cultures and philosophies of different times and geographic locations are ignored so that a reconstructed philosophical tradition is examined as hermetic and self-sufficient—as a linear development from *an sich* to *für sich.* In Subsection 1.1.2, the traditional periodization of Indian philosophy is examined as made in accordance with the Hegelian approach and there its main problems are shown. In Subsection 1.1.3, I consider the criticism of the Hegelian approach proposed by Friedrich Wilhelm Joseph von Schelling (1775–1854). In Subsection 1.1.4, a mathematical centrism in the history of logic is regarded as a Hegelian paradigmatic mistake.

1.1.2 Periods of the Indian Philosophy According to the Hegelian Approach

A good illustration of applying Hegel's methodology can be presented by the following periodization of the Indian philosophy. This periodization is intended to reflect the linear development of Indian thought, starting from the period of the compilation of the Vedas: (H1) the four Vedas (*vedaḥ*): *Ṛgveda, Yajurveda, Sāmaveda,* and *Atharvaveda*, and the Vedic period as such (developing an abstract ideal picture of the world through organizing a complex religious ritual with reciting hymns); (H2) *Brāhmaṇa*s, *Āraṇyaka*s, and *Upaniṣads*—the first proto-philosophical books comprehending H1 (the birth of Indian proto-philosophy as a beginning of reflection carried out by the Brahmins in respect to the Vedic texts as well as Vedic rituals); (H3) *Sūtra*s belonging to Āstika (classical schools)—the first philosophical books in the strict sense as treating the texts of H2 (creating *ṣaḍdarśana* or six philosophical schools of Āstika: Sāṃkhya, Yoga, Nyāya, Vaiśeṣika, Mīmāṃsā, and Vedānta); (H4) *Sūtra*s belonging to Nāstika (non-classical schools, first of all, Buddhism and Jainism).

We see a linear development from H1 and through H2 to H3, and then H4 appears as a critical revaluation of previous periods. It is a step-by-step revelation of Vedic thought from its state *an sich* (H1) to its state *für sich* (H3) and then it comes to its negation (H4). This periodization was well formulated by Friedrich Max Müller (1823–1900), one of the founders of Indology, see (Müller, 1878): (1) *Khanda* period, earlier than 1000 B.C.—composing hymns of the Vedas and forming the Vedic religion; (2) *Mantra* period, from 1000 to 800 B.C.—collecting hymns into the four Vedas; (3) *Brāhmaṇa* period, from 800 to 600 B.C.—composing the texts of *Brāhmaṇa*s, *Āraṇyaka*s, and *Upaniṣads*; (4) *Sūtra* period, since 500 B.C.—first of all, the *Śulbasūtra* (considering the fire-altar construction) and the text of Pāṇini about the Sanskrit grammar—the *Aṣṭādhyāyī.*

Müller pays attention that the same periods are repeated as appropriate classes in the traditional study of *Ṛgveda*:

> A student of a Rig-Veda-*s*âkhâ (a recension of the Rig-Veda), if sharp and assiduous, takes about eight years to learn the Dasagranthas, the ten books, which consist of (1) The Sa*m*hitâ, or the hymns. (2) The Brâhma*n*a, the prose treatise on sacrifices, etc. (3) The Âra*n*yaka, the forest-book. (4) The G*r*ihya-sûtras, the rules on domestic ceremonies. (5–10) The six Angas, treatises on *S*ikshâ, pronunciation, *G*yotisha, astronomy, Kalpa, ceremonial, Vyâkara*n*a, grammar, Nigha*n*tu and Nirukta, etymology, K*h*andas, metre (Müller, 1878: 161).

In the meanwhile, Müller understands that Buddhism is out of this scheme and explains this fact based on archaeological data. According to these data, there were the Northern conquerors of India from the 1st century B.C. to the 4th century A.D. who were not believers in the Vedas, but they follow Buddhism with some own religious traditions such as Mazdeism and other Iranian worships. These conquerors were Indo-Scythians (Sanskrit: *Śaka*), i.e., they are one of the Iranian-speaking tribes from Tūrān (the region of today's Turkmenistan, Tajikistan, Kyrgyzstan, Uzbekistan, and the north-eastern parts of Afghanistan and Pakistan):

> The Northern conquerors, whatever their religion may have been, were certainly not believers in the Veda. They seem to have made a kind of compromise with Buddhism, and it is probably due to that compromise, or to an amalgamation of *S*aka legends with Buddhist doctrines, that we owe the so-called Mahâyâna form of Buddhism—and more particularly the Amitâbha worship—which was finally settled at the Council under Kanishka, one of the Turanian rulers of India in the first century A.D.
>
> If then we divide the whole of Sanskrit literature into these two periods, the one anterior to the great Turanian invasion, the other posterior to it, we may call the literature of the former period *ancient* and *natural*, that of the latter *modern* and *artificial*.
>
> Of the former period we possess, *first*, what has been called the *Veda*, i.e., Knowledge, in the widest sense of the word—a considerable mass of literature, yet evidently a wreck only, saved out of a general deluge; *secondly*, the works collected in the Buddhist Tripi*t*aka, now known to us chiefly in what is called the Pâli dialect, the Gâthâ dialects, and Sanskrit, and probably much added to in later times (Müller, 1883).

Thus, according to Müller, H4 differs from H1, H2, and H3 due to some external influences of the *Śaka*s (Indo-Scythians) on the Indo-Aryans. Without their invasion of North India up to some central parts, we would have a pure hermetic philosophical tradition from H1 and through H2 to H3, but after their invasion, Mahāyāna as a part of H4 appeared.

The periodization close to Müller on the basis of the Hegelian approach was also proposed by Sarvepalli Radhakrishnan (1888–1975), the Indian philosopher (Radhakrishnan, 1948: 57–59): (1) the Vedic period (1500 B.C.–600 B.C.) which covers the spread of the Aryan culture in India and "it was the time which witnessed the rise of the forest universities, where were evolved the beginnings of the sublime idealism of India" (Radhakrishnan, 1948: 57); (2) the epic period (600 B.C.–200 A.D.)—developing the early *Upaniṣad*s and the *ṣaḍdarśana*s, composing the two Indian great epics: the *Rāmāyaṇa* and the *Mahābhārata*, forming and expanding Buddhism, Jainism, Śaivism, Vaiṣṇavism; (3) the *sūtra* period (from 200 A.D.)—founding philosophy in the narrow sense in India; (4) the scholastic period (from 200 A.D.)—founding the tradition of philosophical commentaries.

Müller and other indologists have continued the Hegelian approach to the periodization of Indian philosophy as a hermetic and self-sufficient tradition.

For instance, Erich Frauwallner (1898–1974) proposed the following general periodization:

> First, the continuation of the Vedic thought-world and the beginning of the *Vedānta* up to the time of the system built by *Śaṅkara*. Secondly, the systems built by the *Śaiva*s. Thirdly, the decline of Buddhism and the rise of the Tantric Schools. Fourthly, the *Vedānta* system of the *Vaiṣṇava* and the other *Viṣṇuistic* Schools. Finally, is dealt the continuance still of the systems of the older period, so far as they continue in this period. A sub-division of the period of the modern Indian Philosophy renders itself to be unnecessary as it embraces only an entirely small compass of time. Thus, is given an organization of Indian Philosophy which, in my view, largely docs justice to the course of historical development and also simultaneously summarizes in clarity the phenomena belonging together, in well-arranged groups (Frauwallne, 1973).

Nevertheless, there is a great deal of textual evidence which refutes this insularity of Vedic tradition from H1 to H3 assumed in advance. First, many of the earliest philosophical *sūtra*s of Āstika contain quotations from Madhyamaka and Yogācāra—two early schools of Mahāyāna from North India. So, the *Gauḍapādīyakārikā*, on the one hand, represents the earliest available record of an uncompromising non-dualistic doctrine (*advaita-vāda*)—the central and principal concept of Vedānta school, and, on the other hand, shows that its author(s) had a good knowledge of Madhyamaka and Yogācāra texts (King, 1995). The *Nyāyasūtra*, the basic logical treatise of Āstika, also contains some direct quotations from Madhyamaka and Yogācāra books and was written surely after the Buddhist logical treatise *Milindapañha* (Schumann, 2019a).

The Pāli Canon was composed from the 1st century A.D. to the 4th century A.D. It is one of the earliest hermetic corpus of Indian texts with effective dating due to some inscriptions and cross-cultural textual analysis. It is quite surprising that many times there are mentioned not the four Vedas, as it can be expected, but only three Vedas (*Ṛk*, *Yajur*, and *Sāma*), for instance:

> *tena kho pana samayena brāhmaṇassa pokkharasātissa ambaṭṭho nāma māṇavo antevāsī hoti ajjhāyako mantadharo* ***tiṇṇaṃ vedānaṃ*** *pāragū sanighaṇḍukeṭubhānaṃ sākkharappabhedānaṃ itihāsapañcamānaṃ padako veyyākaraṇo lokāyatamahāpurisalakkhaṇesu* (*Ambaṭṭhasutta* 1, 3; Rhys Davids and Carpenter, 1890: 88).

> At that time Pokkharasāti had a student named Ambaṭṭha. He was one who recited and knew the hymns (*manta*) by heart, and was an expert in the **three Vedas** (*tiṇṇaṃ vedānaṃ pāragū*), together with their vocabularies (*nighaṇḍa*), ritual (*keṭubha*), phonology (*akkhara*) and etymology (*pabheda*), and the stories (*itihāsa*) as fifth. He knew philology (*pada*) and grammar (*veyyākaraṇa*) and was well versed in cosmology (*lokāyata*) and the [32] marks of a great man (*mahāpurisa*).

Why three? It is possible to explain by that the *Atharvaveda* was not a canonical book at least in the 1st century A.D. Another critical point in respect to the assumption of linear development from H1 to H3 is that at the time of Buddha and his disciples for a few hundred years early Sanskrit or Vedic was associated only to the Vedas and was not popular as a language of philosophy or other literature:

> *tena kho pana samayena yameḷakekuṭā nāma bhikkhū dve bhātikā honti brāhmaṇajātikā kalyāṇavācā kalyāṇavākkaraṇā. te yena bhagavā tenupasaṅkamiṃsu, upasaṅkamitvā bhagavantaṃ abhivādetvā ekamantaṃ nisīdiṃsu. ekamantaṃ nisinnā kho te bhikkhū bhagavantaṃ etadavocuṃ—"etarahi, bhante, bhikkhū nānānāmā nānāgottā nānājaccā nānākulā pabbajitā. te sakāya niruttiyā buddhavacanaṃ dūsenti. handa mayaṃ, bhante, buddhavacanaṃ chandaso āropemā"ti. vigarahi buddho bhagavā ... pe ... kathañhi nāma tumhe, moghapurisā, evaṃ vakkhatha—"handa mayaṃ, bhante, buddhavacanaṃ chandaso āropemā"ti. netaṃ, moghapurisā, appasannānaṃ vā pasādāya ... pe ... vigarahitvā ... pe ... dhammiṃ kathaṃ katvā bhikkhū āmantesi—"na, bhikkhave, buddhavacanaṃ chandaso āropetabbaṃ. yo āropeyya, āpatti dukkaṭassa. anujānāmi, bhikkhave, sakāya niruttiyā buddhavacanaṃ pariyāpuṇitun" ti* (*Khuddakavatthukkhandhaka, Cūḷavagga* 5, 33; Oldenberg, 1880: 139).

> At that time, Yameḷa and Kekuṭa were the names of two monks who were brothers, *brāhmaṇa*s (Brahmins) by birth, with beautiful voices, with excellent enunciation. They went to the Lord; and after arrival they greeted the Lord and sat down at a respectful distance. As they were sitting down at a respectful distance, these monks spoke the following to the Lord: "Recently, Lord, monks of various names (*nāma*), various clans (*gotta*; Sanskrit: *gotra*), various births (*jacca*; Sanskrit: *jāti*) have gone forth from various families (*kula*); these corrupt the words of the Buddha in his own dialect (*sakāya niruttiyā*). Now we, Lord, transform the words of the Buddha into the metrical form (*chandaso āropemā*) [of Vedic]." The Buddha, the Lord rebuked them, saying:
>
> "How can you, foolish men, speak thus: 'Now we, Lord, give the speech of the Buddha in the metrical form (*chandaso āropemā*) [of Vedic]'? It is not, foolish men, for pleasing those who are not pleased ..." And after rebuking them, he gave a reasoned talk—he talked to the monks the following words: "Monks, the speech of the Buddha should not be given in the metrical form (*chandaso āropemā*) [of Vedic]. Whoever should give it, there is an offence of wrong-doing. I allow you, monks, to learn the speech of the Buddha according to his own dialect (*sakāya niruttiyā*)."

The hypothesis that the expression *chandaso āropemā* means early Sanskrit or Vedic was put forward by Thomas William Rhys Davids (1843–1922), see his translation (1899–1921): (i) as the antithesis to "his own dialect"; (ii) because of using the word *chandasi* in the *Aṣṭādhyāyī* of Pāṇini with the meaning "the Veda-dialect"; (iii) since this change of sermon language was proposed by "Brahmins by birth"; (iv) within the traditional commentaries to this verse by Theravada scholars—so, Buddhaghosa (5th century A.D.) comments: *chandaso āropemā ti vedaṃ viya sakkaṭa-bhāsāya vācanā-maggaṃ āropema*, where *sakkaṭa* means *saṃskṛta*.

In the Pāli Canon, we do not find quotations from the Indian epics, only some references to epics as an especial genre (*itihāsa*). Furthermore, we do not find some contexts of phrases showing that their authors knew the Vedas or *Upaniṣad*s. The critique against the notion of *ātman* (the idea of non-self; Pāli: *anattā*; Sanskrit: *anātman*) is an attempt to develop a Buddhist type of reflection on all the cognitive and emotional states to distinguish them from ourselves. Initially, *ātman* (*attā*) is a singular reflective pronoun for all three persons and all three genders in Sanskrit (Pāli). The Buddha criticizes different idols of the mind and using the pronoun *ātman* is regarded by him as a sign of uncriticism in general. Hence, *anātman* (*anattā*) in the Pāli Canon is not directly connected to a critique against the *ātman* from the *Upaniṣad*s. It is a Buddhist critique against non-reflection and nothing more.

Nevertheless, in the Mahāyāna *sūtra*s we can find some ideas of *ātman* which are close to the *Upaniṣad*s. For instance, in the *Mahāparinirvāṇa-sūtra* (大般涅槃經; *Dàbān níhuán jīng*, T. 12, No. 376), the first Chinese translation of which appeared in 417 A.D., it is maintained that every separate mental state (*dharma*) [切法; *qiè fǎ*], according to its nature [其性; *qí xìng*], does not have itself (*anātman*) [無我; *wú wǒ*]. But it does not mean that the *ātman* does not exist. It is *dé* (*puṇya*) [我者是德; *wǒ zhě shì dé*] and it is obtaining mastery (*vaśitā*) [我者自在; *wǒ zhě zìzài*]. Thus, the *ātman* is the Mahāyāna path as such:

> 切眾生承如來言展轉相教皆說無我, 此是如來知時方便濟眾生故, 說一切法其性無我, 非如世間所受吾我, 故說一切法其性無我。時復說我, 如彼良醫明乳藥法, 當知我者是實, 我者常住非變易法非磨滅法, 我者是德, 我者自在, 如善乳藥醫, 如來亦然為諸眾生說真實法, 一切四眾當如是學

> *qiè zhòngshēng chéng rúlái yán zhǎnzhuǎn xiāng jiào jiē shuō wú wǒ, cǐ shì rúlái zhī shí fāngbiàn jì zhòngshēng gù, shuō yīqiè fǎ qí xìng wú wǒ, fēi rúshìjiān suǒ shòu wú wǒ, gù shuō yīqiè fǎ qí xìng wú wǒ. shí fù shuō wǒ, rú bǐ liáng yī míng rǔ yào fǎ, dāng zhī wǒ zhě shì shí, wǒ zhě chángzhù fēi biànyì fǎ fēi mómiè fǎ, wǒ zhě shì dé, wǒ zhě zìzài, rú shàn rǔ yào yī, rúlái yì rán wèi zhū zhòngshēng shuō zhēnshí fǎ, yīqiè sì zhòng dāng rú shì xué*
> (*Taishō Tripiṭaka* 1988, T. 12, No. 376, 0863a09–0863a16)

> All sentient beings who inherit the Tathāgata's words, change their cognitions and all say that there is no *ātman*. This is because the Tathāgata knows that it is convenient for all living beings. It is said that the nature of all *dharma*s has no ego [*ātman*], and it is not the same as the world accepts itself [*ātman*]. This is as in the case of the great doctor who well understands the *dharma* [*fǎ*] for the milk medicine, you should know that the *ātman* is true [*shí*], the *ātman* is permanent [*chánhzhù*], it is a non-changeable [*fēi biànyì fǎ*] and non-erasing *dharma* [*fēi mómiè fǎ*]. The *ātman* is virtue [*dé*, *puṇya*], the *ātman* is obtaining mastery [*zìzài*, *vaśitā*], like a good milk medicine doctor, and the Tathāgata is also the same who teaches all sentient beings about the true *dharma*, and all the four groups should learn it like this.

Hence, instead of a linear development of the Buddhist teaching from a Brahminical context to a more independent doctrine we encounter some Brahminical ideas such as the concept of *ātman* not in early Buddhist texts, but, on the contrary, only in later ones, i.e., dated from the 2nd century A.D. This is explained by the fact that Buddhism and Brahmanism developed in parallel for some time. This fact is well confirmed archaeologically, too. For example, refuting the linear development of the Vedic thought from H1 to H3 is confirmed by the fact that the earliest Sanskrit inscriptions are dated strongly from the 1st century B.C. to the 1st century A.D., not earlier (Salomon, 1998). And for a long time, we can observe a smooth change from Prakrits such as Gāndhārī into pure Sanskrit through some hybrid forms from the 2nd century A.D. to the beginning of the 5th century A.D. Only since the 4th–5th century A.D. there have been many long phrases in pure Sanskrit, although the earliest Prakrit inscriptions are dated to the 4th century B.C.

It is worth noting that the earliest Vaiṣṇava inscriptions are dated to the early 2nd to the late 1st century B.C., while all the early Śaiva objects and inscriptions are found exclusively at Buddhist sites for a long time within a syncretic Buddhist-Śaiva culture and only since the early 5th century A.D. Śaivism has been completely emancipated from Buddhism (Schumann and Arif, 2021).

Thus, the Hegelian approach to historically consider each philosophy as a linear development from the state *an sich* to the state *für sich* is not validated by the textual analysis and there is no archaeological evidence at least on the example of Indian philosophy. We face a mixture of various concurrent movements presenting H2, H3, and H4 until about the 2nd–4th centuries A.D., when the philosophical discourse in India had been finally formed.

The dating of the life of the Buddha is a decisive moment for the dating of the post-Vedic period (i.e., the period after H1). And there are two approaches to this: long and short chronology. According to the long chronology, Śākyamuni Buddha lived from ca. 566 to ca. 486 B.C. (i.e., Buddha's *parinirvāṇa* dates to 218 years before Aśoka's coronation). According to the short chronology, he lived from ca. 448 to ca. 368 B.C. (i.e., Buddha's *parinirvāṇa* dates to 100 years before Aśoka's first regnal year).

The short chronology was substantiated by Heinz Bechert (1989, 1991) who showed, based on the references to *Dīpavaṃsa* 1.24–26 and 5.55–59, that the long chronology of 218 years was a later development (Bechert, 1989: 104 ff.; 1991: 329–343). The short chronology is acknowledged by the following quotation from a 1st century A.D. Kharoṣṭhī manuscript (British Library fragment 4.6 recto): Aśoka was "a century after the Blessed Buddha achieved *parinirvāṇa* (*vasaśada pariṇurvude budhe bhagavade*)" (Neelis, 2011: 68).

According to the Gilgit manuscript of the *Bhaiṣajyavastu* (Dutt, 1984) written in Sanskrit and dated to the 8th century A.D., the short chronology may be even much shorter, namely Buddha's *parinirvāṇa* goes back to 400 years before the Kaniṣka *stūpa* (erected ca. 130 A.D.). It indicates the years of Śākyamuni's life from ca. 350 to ca. 270 B.C.:

> *bhagavān kharjūrikām neuropath | khajūrikāyāṃ bāladārakān pāṃsustūpakaiḥ krīḍato 'drākṣīt* | bhagavān bāladārakān pāṃsustūpakaiḥ krīḍato dṛṣṭvā ca punar vajrapāṇiṃ yakṣam āmantrayate | paśyasi tvaṃ vajrapāṇe bāladārakān pāṃsustūpakaiḥ krīḍataḥ | evaṃ bhadanta | eṣa caturvarṣaśataparinirvṛtasya mama vajrapāṇe kuśanavaṃśyaḥ kaniṣko nāma rājā bhaviṣyati | so 'smin pradeśe stūpaṃ pratiṣṭhāpayati | tasya kaniṣkastūpa iti saṃjñā bhaviṣyati | mayi ca parinirvṛte buddhakāryaṃ kariṣyati* (*Mūlasarvāstivādavinaya* 1: 2–3; Dutt, 1984).

> Bhagavān [Buddha] reached Kharjūrikā where he saw boys playing with a heap of earth. Seeing the boys playing with the mud heaps, he then turned to the *yakṣa* Vajrapāṇi, "Do you see, Vajrapāṇi, how the boys are playing with the mud heaps?" "Yes, sir". "Four hundred years after I have completely liberated, Vajrapāṇi, there will be a king named Kaniṣka of the Kuṣāṇa lineage. He shall set up a *stūpa* on this very spot, and it shall be called the Kaniṣka *stūpa*. Since I have been completely ceased, it will be he who will carry out the duty of the Buddha.

Evidently that the shorter the chronology is, the better we may explain various facts of the non-linear development from H2 to H4. The point is that in the short post-Vedic period until the 2nd–4th centuries C.E., Brahminical and Buddhist stages of development really coexisted. Furthermore, the shortest chronology with the dating from ca. 350 to ca. 270 B.C. agrees well with the facts of the beginning of the *śramaṇa* movement from ca. 400 A.D. in the context of the first large growth of Indo-Aryan urbanization in the Ganges Valley at that time.

1.1.3 Schelling versus Hegel

An alternative methodology for the history of philosophy was proposed by Friedrich Wilhelm Joseph von Schelling (1775–1854) in his late works such as *die Philosophie der Offenbarung* (Vorlesung, 1841–1842) and *die Philosophie der Mythologie* (Vorlesung, 1842). He was the first who grounded that the historic time (*die geschichtliche Zeit*) as a history of philosophical idea from *an sich* to *für sich* is just an official imagined history, i.e., it is a history within the current dominance of our certain ideology. To reveal the true genealogy of the given idea, we need to turn to its prehistoric time (*die vorgeschichtliche Zeit*)—we must refute the official ideology, where this idea is presented now within a linear imagined or made-up history of *geschichtliche Zeit*. It means we should go beyond a unified hermeneutics for one corpus of studied texts. Thus, the "Hegelian" periods from H1 to H4 correspond to the existed (*Śaiva* as well as *Vaiṣṇava*) traditions of today's Hinduism, for example, to the classes of studying the *Ṛgveda* (Müller, 1878: 161) mentioned above as corresponding to the periods from H1 to H3.

Schelling maintains that *die vorgeschichtliche Zeit* means to be before the historic process as such (to be *vorhistorische*) at the stage, where our consciousness did not yet reconstruct a linear development in the meaning of Hegel. At this stage we can observe religions and mythological ideas in their pure forms, i.e., without our imaginations and one-sided (historic) interpretations:

> Der wahre Inhalt der vorgeschichtlichen Zeit ist die Entstehung der formell und materiell verschiedenen Götterlehren, also der Mythologie überhaupt, welche in der geschichtlichen Zeit schon ein Fertiges und Vorhandenes, also geschichtlich ein Vergangenes ist (Schelling, 1907: 588).

> The true content of prehistoric time is presented by the emergence of formally and materially different doctrines of gods, therefore [by the emergence], of mythology in general, which is already given as something finished and available in the historic time, therefore, as something past historically.

Hence, according to Schelling, historic and prehistoric times are two different approaches to the history of philosophy and our thinking as such:

> Demgemäß sind die geschichtliche und die vorgeschichtliche Zeit nicht mehr bloß relative Unterschiede einer und derselben Zeit, sie sind zwei wesentlich verschiedene und voneinander abgesetzte, sich gegenseitig ausschließende, aber eben darum auch begrenzende Zeiten. Denn es ist zwischen beiden der wesentliche Unterschied, daß in der vorgeschichtlichen das Bewußtsein der Menschheit einer innern Notwendigkeit, einem Prozeß unterworfen ist, der sie der äußeren wirklichen Welt gleichsam entrückt, während jedes Volk, das durch innere Entscheidung zum Volk geworden, durch dieselbe Krisis auch aus dem Prozeß als solchem gesetzt und frei von ihm nun jener Folge von Taten und Handlungen sich überläßt, deren mehr äußerer, weltlicher und profaner Charakter sie zu historischen macht (Schelling, 1907: 588–589).

> Accordingly, historic and prehistoric times are no longer merely relative differences of one and the same time; they are two essentially different and separated, mutually exclusive, but just, therefore, also limiting times. So, there is an essential difference between the two so that in the prehistoric one the consciousness of humanity is subject to an inner necessity, to a process which, as it were, removes it [time] from the external real world, while every people, which has become a people through an internal decision, is also composed of the process as such through the same crisis, and free from it, it is now left to this sequence of deeds and actions for which a more external, worldly and profane character makes it [time] historical.

In line with the Schellingian idea of prehistoric time, Paul Michel Foucault (1926–1984) views genealogy in a new way (not Hegelian)—as a necessary method of philosophical analysis, in which we should get out of the isolation of one hermeneutic tradition with a one-sided historical reconstruction—in other words, we must abandon the Hegelian principle of ascent from the abstract to the concrete, while preserving the idea of the historicity of philosophical knowledge. According to Foucault, each cultural or social phenomenon can be philosophically investigated through a genealogical reconstruction of epistemic frameworks. To this end, he began to distinguish between the epistemological level of knowledge, representing what is now, and the genealogical reconstruction of existences. He called the genealogical reconstruction "the archaeological level of knowledge" or "the archaeology of knowledge." According to Foucault, philosophy also has its archaeology. It is one of the core objectives of philosophy:

> (…) archaeology, addressing itself to the general space of knowledge, to its configurations, and to the mode of being of the things that appear in it, defines systems of simultaneity, as well as the series of mutations necessary and sufficient to circumscribe the threshold of a new positivity (Foucault, 2005: xxv).

As we see, the archaeology of philosophy in the sense of Foucault should replace the history of philosophy in the sense of Hegel, if we would like to consider philosophical texts outside of only one closed philosophical tradition. The periodization of Indian philosophy from H1 to H4 reflects only one philosophical tradition with one hermeneutics—the Hindu philosophy of existing *Śaiva* as well as *Vaiṣṇava* religious traditions.

Within the framework of Foucault's approach, structuralist methods of text analysis are used. These methods are easily enhanced by the methodology of other humanities dealing with the history of knowledge: (a) historical reconstruction based on both archaeological data and other methods of studying material culture (description, interpretation, cataloging); (b) reconstruction presented in historical sociology, which studies societies in their historical dynamics; (c) methods of comparative textology and hermeneutics, which study different corpora of texts of the same or different traditions. All these methods are used in my research within the framework of a general approach called "the archaeology of logic", which goes back to the principles of Foucault's structuralist analysis and adapted specifically to study the dynamics of logical competence in the early stages of development.

1.1.4 Mathematical Centrism in the History of Logic

Traditionally, logic was viewed as a purely European tradition, founded in Ancient Greece by Aristotle, and then developed in Catholic and Byzantine scholasticism as well as in Islamic and Judaic Aristotelianism. This point of view may be called Eurocentrism in the study of the history of logic. The fact is that, as it is currently known, there were at least two ancient non-European logical traditions that lie outside of and parallel to the logic of Aristotle: The Judaic logical hermeneutics (from the 1st century A.D.) and the logic of *nyāya* and *yogācāra* schools in Indian philosophy (from the 2nd century A.D.). Both Asiatic logical traditions (Judaic as well as Indian) have survived to this day. Therefore, the Eurocentrism in the history of logic may be criticized for these reasons. Nevertheless, there exists another kind of 'centrism' in the history of logic which has not yet been criticized. According to this 'centrism', the notion of logic should be understood within the framework of philosophical tradition founded thanks to the Fregean revolution in logic and well explicated later in Hilbert's program. So, in this way, logic is significantly identified with some methods of set theory and the logical foundations of mathematics: "The scope and method of logic as we know it today eminently reflect the groundbreaking developments of set theory and the logical foundations of mathematics" (Lapointe, 2019: i). Hence, the history of logic would mean the linear development of symbolic tools for inferences in the direction of the earliest axiomatizations of various logical systems for mathematical purposes.

This new 'centrism' in the history of logic, which might be called *mathematical centrism* focusing only on set theory and the logical foundations of mathematics as ultimate goals of logic, is a controversial issue and it is completely consonant with the Hegelian concept of the development from *an sich* to *für sich*, but now in the history of logic. The main problem with this point of view is that Hilbert's program rather failed because it turned out that mathematics is not reducible to logical formalisms and recently logic has been developing to a greater extent not within the framework of mathematics, but within the framework of informatics and computer science, e.g., in artificial intelligence.

Let us consider the book edited by Sandra Lapointe (2019) as a symptomatic survey of the development of logic in the works of 19th-century philosophy. In her Introduction she highlights that there are two opposing standpoints in respect to the logical works of the 19th century: (i) a positive estimation of the logical ideas of Immanuel Kant and the 19th-century analytic philosophy after him, for example, agreeing with Wilhelm Windelband's (1912) positive attitudes toward them; (ii) a negative estimation of all purely philosophical works on the logic of the 19th century, for example, if we accept the following statements by Peter King and Stewart Shapiro: "when writers refer to 'traditional logic' they usually have this degenerate textbook tradition in mind. Since the beginning of the modern era most of the contributions to logic have been made by mathematicians" (King and Shapiro, 1995: 3) and by the statements of William Kneale and Martha Kneale: "very few works that contain anything at once new and good" (Kneale and Kneale, 1962: 298). Consequently, in the treatment of philosophical logic of the 19th century, we find the mathematical centrism defined above, considering logic only as a mathematical subdiscipline (King and Shapiro, 1995; Kneale and Kneale, 1962), and only very few historical works concentrate on some of the significant philosophical implications of logic of the 19th century with their positive evaluation (Windelband, 1912).

Evidently, Kant's logical ideas did nothing for mathematical logic, namely, as Sandra Lapointe notes, they were "at best defective" (Lapointe, 2019: 2) from the point of view of this logic. But Kant's ideas may be analyzed by answering the question: "How did Kant's views on the scope, method and place of logic impact the development of the discipline?" (Lapointe, 2019: 12, cf. Lapointe and Pincock, 2017). In this way, we can distinguish between what Kant himself calls "pure general logic" (formal logic in the modern sense) and "transcendental logic". While formal logic is a study of the 'form' of thinking, transcendental logic is an *a priori* study of the conditions under which the thinking of objects of experience takes place. Thus, while formal logic is the limit of various meaningful combinations of signs in inference, which might be studied mathematically, transcendental logic is a process of thinking in inference, which cannot be reduced to formalisms at all. In the historical study of formal logic, we can turn to the mathematical centrism, representing a kind of linear dynamics of logical ideas, consonant with the Hegelian approach to philosophy history as an ascent from the abstract to the concrete. But it is impossible for logic in its wider meaning.

Kant's transcendental logic, of course, did not become mainstream in the 20th-century analytic philosophy (during the period of the dominance of mathematical logic), but there were some traditions that continued Kant's logic, for example, we can remember content-genetic logic in the USSR (Ilyenkov, 1974; Shchedrovitsky, 1964; Zinoviev, 1957), treating logic as a study of the origins of knowledge (not as a study of ready-made knowledge by means of signs), that is, as a method in which the knowledge was obtained. Nevertheless, after the emergence of cognitive science, it became clear in the 21st century that human thinking is not free from cognitive biases and in its natural form does not proceed in the framework of formal logic. Therefore, Kant's transcendental logic now acquires a certain meaning as a logical study of the real processes of thinking. Furthermore, there is the so-called semantic inferentialism (Brandom, 2000), according to which the meaning of an expression is identified more with its contextual use in reasoning, than with denotations. This semantic approach is close to Kant's transcendental logic, too.

The book edited by Sandra Lapointe (2019) consists of two groups of contributions: (i) six chapters are focusing on the linear development of formal logic from Kant to Russell and Hilbert; (ii) and five chapters are showing some interesting impacts of Kantian ideas on the 19th-century logic that recently are outside of mathematical logic.

(i) The first group includes, first, the chapter *Kant and Formalism: Hilbert, Russell and Whitehead* written by Nicholas Griffin, where it is traced how the Kantian approach to formal logic may be regarded as a very preliminary step towards Hilbert's approach to the axiomatization of mathematics, presented in his *Foundations of Geometry* (1899), shared by Russell and Whitehead. In the next chapter of the first group, entitled *Sigwart, Russell and the Emergence of Scientific Philosophy* by Sean Morris, it is shown what role Christoph Sigwart's *Logic* (English translation 1895) played in these dynamics of Kantian ideas towards Russell with his book *Our Knowledge of the External World* (1914). In the chapter *Platonism in Lotze and Frege*, Nicholas F. Stang considers the mathematical Platonism of Frege, although any Platonism is sharply criticized by Kant and his followers including Hegel. In the chapter, *The Logic in Dedekind's Logicism*, Erich H. Reck shows that Dedekind understands arithmetic as a "part of logic" in his booklet *Was sind und was sollen die Zahlen?* (1888) that was consonant later with Hilbert's logicism. The chapter *Laws of Thought and Laws of Logic after Kant* by Lydia Patton is concentrated on drawing an explicit connection between logic and algebra, made by Boole in his *The Mathematical Analysis of Logic* (1847) and *An Investigation of the Laws of Thought* (1854). In the chapter *Bolzano on Logic in Mathematics and Beyond*, Sandra Lapointe mainly analyses Bolzano's *Wissenschaftslehre* (1837) from the point of view of modern logic.

(ii) The second group of contributions is to show a philosophical context of logic in the 19th century. This part is more non-standard for the history of logic. In the chapter, *The Logicians of Kant's School*, Jeremy Heis analyses some major works of the Kantian school such as Schultz's *Prüfung der Kantischen Critik der reinen*

Vernunft (1789), Jakob's *Grundriss der allgemeinen Logik* (1791), Kiesewetter's *Grundriss einer allgemeinen Logik nach Kantischen Grundsätzen* (1791), Hoffbauer's *Analytik der Urtheile und Schlüße* (1792), Maass' *Grundriss der Logik* (1793), Krug's *Denklehre oder Logik* (1806), Herbart's *Lehrbuch zur Einleitung in die Philosophie. Zweiter Abschnitt. Die Logik* (1813), Mansel's *Prolegomena Logica: An Inquiry into the Psychological Character of Logical Processes* (1851). In the chapter *Kant's Excessive Tenderness for Things in the World, and Hegel's Diatheism*, Graham Priest traces Hegel's dialectical logic (i.e., his dialetheism) back from Kant's account of the antinomies that certain noumenal objects have always contradictory properties. The chapter *Hegel's Conception of Thinking in His Logics* by Clinton Tolley represents some basic features of Hegel's dialectical logic such as claiming that thinking is an absolute self-mediation. In the chapter *Demystifying Cohen's Logik*, Frederick Beiser shows that Hermann Cohen understands logic as an epistemological and non-formal discipline, since logic deals with knowledge along with its content (to the same extent as it will be treated in content-genetic logic). In the chapter, *What Russell Meant When He Called Moore a Logician*, Consuelo Preti pays attention to Kant's formulation of judgment that was applied in Moore's preparation for his moral sciences studies.

This survey of 19th-century logicians is not complete, especially for the second group of contributions. There are a lot of Kantian, neo-Kantian or post-Kantian philosophers who would be worth remembering. For example, some relevant logical impacts in the works of Arthur Schopenhauer (1788–1860) are considered in Lemanski (2020). Johann Friedrich Herbart (1776–1841), the Kantian philosopher, in his fundamental works *Lehrbuch der Psychologie* (1816) and *Psychologie als Wissenschaft, neu gegründet auf Erfahrung, Metaphysik und Mathematik* (1824–1825) has proposed a mathematical model of thinking as a concurrent process which may be consistent with process calculus that was developed and implemented in business process management and business process mapping. Heinrich John Rickert (1863–1936), the prominent representative of the Southwest or Baden school of Neo-Kantianism, in his books *Die Grenzen der naturwissenschaftlichen Begriffsbildung* (1896) and *Kulturwissenschaft und Naturwissenschaft* (1899) has shown that basic concepts in natural science may be formalized by means of mathematics and formal logic, while basic concepts in historical science cannot be examined in this way.

Rickert's idea of distinguishing between two logical natures of physical and historical concepts can be defined in terms of modern logic to show that it is really promising. Let us take the notions of 'mass' and 'energy' from physics. They are general and abstract. It means that they could be defined as primary concepts by means of non-logical axioms $A_1, \ldots, A_n$ of a formal theory $\mathbf{T}$. Let $\mathbf{M}[A_1, \ldots, A_n]$ be a class of models for the theory $\mathbf{T}[A_1, \ldots, A_n]$. Now, let us take a new general and abstract notion such as 'entropy'. To define it as a primary concept of physics we should add more non-logical axioms $A_{n+1}, \ldots, A_m$ to the formal theory $\mathbf{T}[A_1, \ldots, A_n]$. As a result, we obtain a new formal theory $\mathbf{T}[A_1, \ldots, A_n, A_{n+1}, \ldots, A_m]$. All the theorems proved in $\mathbf{T}[A_1, \ldots, A_n]$ are provable in $\mathbf{T}[A_1, \ldots, A_n, A_{n+1}, \ldots, A_m]$, too, but not vice versa. In other words, $\mathbf{T}[A_1, \ldots, A_n] \subset \mathbf{T}[A_1, \ldots, A_n, A_{n+1}, \ldots, A_m]$.

Let $\mathbf{M}[A_1, \ldots, A_n, A_{n+1}, \ldots, A_m]$ be a class of models for the theory $\mathbf{T}[A_1, \ldots, A_n, A_{n+1}, \ldots, A_m]$. According to the Galois connection, $\mathbf{M}[A_1, \ldots, A_n, A_{n+1}, \ldots, A_m] \subset \mathbf{M}[A_1, \ldots, A_n]$. So, the broader the theory is, the smaller its model class is.

Now, let us take the notion of 'Indo-Scythian' from history. It is general, but not abstract. For example, we cannot consider 'Indo-Scythian' as a well-defined attribute, according to some properties. So, according to the property of language, to be 'Indo-Scythian' means rather to be a native of one of the Eastern Middle Iranian languages (Bactrian, Sogdian, etc.). But there was a dynasty of the Western Kṣatrapas who identified themselves with Śakas (Indo-Scythians), but in their inscriptions they used only Prakrits (Middle Indo-Aryan languages). Further, according to the property of self-identification, there was the Indo-Scythian dynasty of Kuṣāṇas with the self-designation κoþανo (*kušano*) in Bactrian, while almost the same self-designation *kuśiññe* was used by the Tocharians in their Tocharian B texts, although this tribe was not Indo-Scythian. We can continue these attempts to find well-defined criteria for representing 'Indo-Scythian' as a well-defined attribute, but each criterion shall have some exceptions. Hence, 'Indo-Scythian' as well as all other general notions from history cannot be presented as primary concepts such as 'mass' or 'energy'.

Let $D_1, \ldots, D_n$ be some descriptions of each case covered by the notion of 'Indo-Scythian' and assume that $\mathbf{T}[D_1, \ldots, D_n]$ is a descriptive theory. Let $\mathbf{M}[D_1, \ldots, D_n]$ be a class of models for the theory $\mathbf{T}[D_1, \ldots, D_n]$, for example, represented by some archaeological data. Now, let us take a new general notion of history such as 'Buddhist' fixed by new descriptions $D_{n+1}, \ldots, D_m$. Thus, $\mathbf{T}[D_1, \ldots, D_n] \subset \mathbf{T}[D_1, \ldots, D_n, D_{n+1}, \ldots, D_m]$. At the same time, we do not have the Galois connection here, namely $\mathbf{M}[D_1, \ldots, D_n] \subset \mathbf{M}[D_1, \ldots, D_n, D_{n+1}, \ldots, D_m]$. So, the broader the theory in history is, the larger its model class is.

I have specifically considered the Rickertian idea of distinguishing between the two ways of forming concepts in natural science and history in terms of mathematical logic to show that Rickert's 19th-century logical approach is very relevant. He showed excellently that a formal theory is not possible in history, at least by means of a classical model theory, as we can add. This demonstrates that a lot of logical ideas of the 19th-century philosophers, which were beyond the linear development from Kant to Hilbert, may be very relevant nowadays. Hence, the Hegelian idea of linear development of philosophical (logical) thought from the state *an sich* to the state *für sich* is not correct and this fact may be readily proved even on the material of the 19th-century logic. Consequently, the mathematical centrism in the history of logic may be considered a Hegelian paradigmatic mistake.

1.1.5 Conclusions

To sum up, the Hegelian approach to consider the history of philosophy and logic with one hermeneutics causes the one-sided presentation with many abstractions and imaginations. The Schellingian criticism against Hegel is based on the idea to go out of one hermeneutics within one "historic time". This idea was well developed by Foucault who showed that we can address to a whole arsenal of historical methods in our historical investigations of philosophical ideas. This approach is called the **archaeology of philosophy**. Here we focus on a cross-textual analysis and archaeological data. It can be used also for the logic history as the **archaeology of logic** by studying different historic forms of logical competence.

1.2 Main Logical Traditions within the Framework of Archaeology of Logic

1.2.1 Problem Setting

The earliest logical treatises were as follows: (1) in Ancient Greece, Aristotle's *Organon* and some Stoic and Epicurean logical works, such as *On Phenomena and Inferences* written by Philodemus of Gadara (Φιλόδημος ὁ Γαδαρεύς; ca. 110–ca. 40 B.C.); (2) in Ancient India, the logical works, first of all, of the Nyāya and Yogācāra schools (from the 2nd to the 6th century A.D.); (3) in Judea, the *B̲āraytā' Dərabbî Yišmā'ē'l* (1st century A.D.); (4) in Ancient China, the *Dialectical Chapters* from the *Mohist Canon* (from the 5th to the 3rd century B.C.) that is rather a proto-logical text. In all these treatises, there is an explicit formulation of the distinction between the particular and the general as well as the understanding that the general follows from the particular of the same genus. It is a cornerstone of any logic. But we can try to find within the framework of "archaeology of logic" where and when this understanding first arose.

Among all the ancient documents, this understanding is first traced in the legal discourse of Mesopotamia dating back to at least the 21st century B.C. So, in Mesopotamian law codes we see conditional formulations of articles ("if a crime, then a punishment") allowing the Sumerians and Akkadians to distinguish general notions from specific cases and to use a naïve set theory. The analysis of Old-Babylonian and New-Babylonian business correspondence, trial records, and other legal documents shows us many examples of difficult logical schemata as results of applications of some inference rules to law codes. The main idea of the Babylonian courts was that any trial judgment should be final in solving problems, and its verdict should be complete and be inferred from the list of arguments (facts and documents): "If facts and documents, then a trial verdict." In case of an incomplete set of arguments for making a final decision, the court issues a conditional verdict: "If facts and documents, then if an additional document that is missing, then a trial verdict" (that is logically equivalent to the following sentence: "If facts and documents and an additional document that is missing, then a trial verdict"). For instance: "Five branded sheep were seen in the flock of Kīnaya. Zēriya testifies against Kīnaya, proving that Kīnaya stole three of the sheep. The assembly decrees that Kīnaya must repay those sheep thirtyfold. Kīnaya claims that the remaining two sheep were given to him by a shepherd. Kīnaya must present the shepherd to the administrators of the Eanna. If he does not present the shepherd, then Kīnaya must repay the Eanna thirtyfold for those two sheep, as well [29 October, 547 B.C.]" (Holtz, 2014: 52).

Thus, we can claim that the earliest logically correct reasoning first arose in Mesopotamia and this tradition was grounded in the Sumerian and Akkadian jurisprudence. The only known codification of the Greeks is the *Law Code of Gortyn* (Crete, the 5th century B.C.). It was compiled using many elements of the Babylonian legal tradition—there are many articles that are very similar to the corresponding articles from the Torah or the *Laws of Hammurabi*. In this way we can assume that the Greek logic reflected the existing forms of correct inferences within the framework of legal hermeneutics, preserved from Mesopotamia.

The specific (representing a case or species) and the general (representing a genus) are connected by implication: if a particular A is regarded as a case of an appropriate general B, then "If A, then B" holds true. For instance, the following conditional statement is always true: "If it is a silver fir (A), then it is a tree (B)" ("Each silver fir (A) is a tree (B)"), where "silver fir" is a particular and 'tree' is a general.

In the Sumerian and Akkadian legal codes, for the first time there were introduced some general notions as generalizations of particulars. The word to denote a generalization is *mimma* or *mimma šumšu* in Akkadian ('whatever'), e.g., in the *Laws of Hammurabi* §113, §116: "Whatever he originally gave as the loan [*mimma mala iddinu ītelli*]" (Roth, 1995). Implicitly, it means that suitable Hammurabi laws §113 and §116 concerning all the items given as the loan cover all the cases: "If there is whatever he originally gave as the loan, then rules §113 and §116 should take place." Let us assume that somebody gave an ox as the loan. Then we could apply the following composite implication: "If he gave an ox as the loan (A), then it is the case of whatever he originally gave as the loan (B)." From this it follows that rule §113 and §116 of the *Law Code of Hammurabi* should be applied for giving this ox as the loan (C). Formally: $((A \Rightarrow B) \& (B \Rightarrow C)) \Rightarrow (A \Rightarrow C)$. Hence, this *mimma* ('whatever') assumes a logically correct construction of conditional propositions (implications) with a logical rule of transitivity of implication.

The expressions "a man who…" (*awīlum ša…*) and "a woman" (*sinništum*) from the *Laws of Hammurabi* are related to all free citizens according to their gender. Both expressions are another form to denote a generalization. Each actor is examined as a particular case of *awīlum* or *sinništum* covered by appropriate rules of the *Laws of Hammurabi*. Also, there are many other expressions in Akkadian denoting 'whoever' such as *mannummê* and *attamannu*.

Thus, the Sumerian-Akkadian codes allow us to appeal to general notions if they cover all the precise instances. As a result, in these codes, for the first time the implication as a kind of logical proposition with a correct semantics was introduced. Each article of the code, i.e., each rule, is formulated in the form of implication: "If/when (Akkadian: *šumma*) this or that occurs, this or that must be done as a trial judgment." The Semitic legal tradition (including the Talmud) is a continuation of the Sumerian and Akkadian law formulations. So, in the Bible we can find out the following three ways of law formulations which are also typical for Aramaic legal texts:

1) '*Casuistic*': "If/when (Aramaic: *hn* or *ʾm*) this or that occurs, this or that action must be undertaken or this or that punishment must be inflicted." Hebrew: *wəʾim* or *ʾim*. Akkadian: *šumma*. An example from the Bible is as follows:

   > And if [*wəʾim*] he be not redeemed in these years, then he shall go out in the year of jubile, both he, and his children with him (*Leviticus* 25: 54).

2) '*Apodictic*': "Thou shall not... (Aramaic: prohibitions in the second person singular of the imperfect, sometimes by using the negative particle *l'*)." A Biblical example:

Thou shalt not [*lō'*] go up and down as a talebearer among thy people: neither shalt [*lō'*] thou stand against the blood of thy neighbour; I am the Lord (*Leviticus* 19: 16).

3) '*Relative*': "The man who… (Aramaic: *'īš zī* or *gəbar zī* or *'enāš zī*)" or "Whoever … (Aramaic: *zī* or *mn*)." This *zī* or *mn* corresponds to *awīlum* in Akkadian. Some examples from the Bible:

And the Lord said unto Moses, whosoever [*mî*] hath sinned against me, him will I blot out of my book (*Exodus* 32: 33).

If [*kî*] a man [*'îš*] have a stubborn and rebellious son, which will not obey the voice of his father, or the voice of his mother, and that, when they have chastened him, will not hearken unto them (*Deuteronomy* 21: 18).

And whoever soul [*wə**k̲ol**-hannepeš*] it be that doeth any work in that same day, the same soul will I destroy from among his people (*Leviticus* 23: 30).

It is worth noting that in the Torah, the word *kol* (Hebrew: 'all') is often used for expressing the notions 'whoever' and 'whatever'. In Akkadian: *kala, kali, kaluma* ('everything', 'everyone', 'everybody').

Hence, the Sumerian and Akkadian legal styles were integrated in the broader context of the Near Eastern juridical terminology, including the Aramaic and Hebrew legal terms.

The only Greek law code preserved until now is the *Law Code of Gortyn* (Willetts, 1967) written in the Dorian dialect and dated to the first half of the 5th century B.C. It was a codification of the civil law of the ancient Greek city-state of Gortyn located in southern Crete. It is a type of stone inscription on the wall of a public civic building in the agora of Gortyn. Its script style is called boustrophedon, in which alternate lines must be read in opposite directions rather than from left to right, or right to left. This code contains the articles formulated in the form of implications αἰ δέ… (the Dorian dialect of Old-Greek: "and if…"; the same in Hebrew: *wə'im*) in the way of Sumerian and Akkadian legal tradition. This text includes expressions for general notions such as ὅς κ'… ("whosever…") and ὅτι δέ τίς κ'… ("and whatever anyone…"), etc.

In the Stoic propositional logic established by Chrysippus of Soli (Χρύσιππος ὁ Σολεύς; ca. 279–ca. 206 B.C.), the general notions are expressed in the spirit of the legal tradition by the terms 'whatever' and 'whoever', as was done first in the *Code of Ur-Nammu* and the *Laws of Hammurabi*, and then in the Greek statutes, such as the *Law Code of Gortyn*.

For instance, the well-known proposition "Man is a rational, mortal animal" (ἄνθρωπός ἐστι ζῷον λογικὸν θνητόν) was reformulated in the following manner: "**Whatever** thing is man, that thing is a rational, mortal animal" (εἴ τί ἐστιν ἄνθρωπος, ἐκεῖνο ζῷόν ἐστι λογικὸν θνητόν) (Sextus Empiricus, *Against the Ethicists* 9). Another example of Stoic universal affirmative propositions: "**Whosoever** are men, they are either Greeks or barbarians" (τῶν ἀνθρώπων οἱ μέν εἰσιν Ἕλληνές οἱ δὲ βάρβαροι) (Sextus Empiricus, *Against the Ethicists* 11).

The main difference of the Stoic formulation of universal affirmative propositions from the Aristotelian one is that the Stoic formula is immediately interpreted as an implication: "Whatever is *A*, it is *B*" is understood as "If *A*, then *B*." All the same as it took place in the Sumerian and Akkadian legal culture.

Summing up, each law code in Sumerian or Akkadian consists of implications $A \Rightarrow B$ which are examined as true forever. Each court had to consider the actual case *C* of claim that was verified by testimonies or signed documents, and then the court had to find out an appropriate general *A* for this *C*. After that the court issued a verdict *B* by *modus ponens* applied twice:

$$\frac{A \Rightarrow B;\ C \Rightarrow A;\ C}{B.}$$

The latter sentence is a verdict what should be done (which punishment *B* should be chosen) according to the rule $A \Rightarrow B$ from the legal code.

If the situation *C* of claim was not suitable for the legal rule $A \Rightarrow B$ from the code, *modus tollens* was applied one time:

$$\frac{C \Rightarrow A;\ A \Rightarrow B;\ \neg A}{\neg C.}$$

Hence, the task of any court is to examine all the facts, such as testimonies or signed documents, for verifying the claim *C* or falsifying its general case *A* within an appropriate law $A \Rightarrow B$. Therefore, "making a decision by a court" was called *dīna parāsu* in Akkadian, where *dīnu* (or *dīnum*) means 'law' and *parāsu* means "to separate," "to divide," or "to get a particular". In other words, the direct meaning of *dīna parāsu* is to deduce something from a law or to make a particular case within the framework of law. Thus, *dīna parāsu* became the first word denoting a deduction as such in human languages. "Examining the facts" was called *warkata parāsu* in Akkadian. It is worth pointing out that "giving a decision by an omen" was called *dīna šakānu*—"to establish a law". Omens in Akkadian were formulated by conditional propositions "If …, then …", too, and *dīna šakānu* meant to put forward a conditional law.

In the Talmud, the general term is called *kəlāl* and the particular term is called *pərāṭ*. The *kəlāl* corresponds to the Akkadian word *kalû* ('all' or 'totality'), in Hebrew *kol*, while the *pərāṭ* with the meaning *paraṭ* ("to separate," "to divide," or "to get a particular") corresponds to the Akkadian word *parāsu*. In the Babylonian Talmud, it is stated that the judges should focus on particulars in their trial judgments:

אָמַר רַבִּי יוֹחָנָן אֵין לְמֵדִין מִן הַכְּלָלוֹת וַאֲפִילּוּ בִּמְקוֹם שֶׁנֶּאֱמַר בּוֹ חוּץ

ʾāmar rabbî yôḥānān ʾên ləmēḏîn min hakkəlālôṯ waʾăp̄îlû bimqôm šenneʾĕmar bô ḥûṣ.

> Rabbi Yoḥanan says: One does not draw general conclusions (***hakkəlālôṯ***), but as if there is a place where it says: except [to exclude something] (*Qîdûšîn* 34a: 4)

In the Babylonian legal tradition, the law code was to be published (widely announced) as the main source of legality. This edition was commemorated by the erection of a stele with official inscriptions, or by the engraving of these inscriptions on the wall. One of the most famous steles with such inscriptions belongs to Hammurabi and represents his laws. Another example is the inscriptions on the wall in the agora of Gortyn representing the Greek laws.

The legal code was considered a set of non-logical axioms announced for all due to its publications on the stone. To the same extent, the Tablets of the Law (Hebrew: *luḥōṯ habbərîṯ*, לוחות הברית), inscribed, according to the Torah, by God Himself, were the official inscriptions of the Judaic law code written on stones:

> And he gave unto Moses, when he had made an end of communing with him upon mount Sinai, two tables of testimony, tables of stone, written with the finger of God (*Exodus* 31: 18).

In the *Samaritan Pentateuch* it is stated more explicitly that the Israelites had to write down their code on stones:

> 14a And when Shehmaa your Eloowwem will bring you to the land of the Kaanannee which you are going to inherit it. 14b You shall set yourself up great stones and lime them with lime. And you shall write on them all the words of this law. 14c And when you have passed over the Yaardaan you shall set up these stones, which I command you today, in Aargaareezem. 14d And there you shall build an altar to Shehmaa your Eloowwem, an altar of stones, you shall lift up no iron on them (*Exodus* 20: 14, the *Samaritan Pentateuch*, translated by Benyamim Tsedaka).

This fragment is quite significant because it does not appear in the Hebrew Bible. The Tablets of the Law were considered here to be made by people and inserted into the wall in the temple on Mount Gerizim ('Aargaarizem'). It was common among other Semitic peoples to write the basic rules on stones for the public announcement of the legal code. For instance, we have no fragments of the Old Assyrian law code, but a publication of this code on stones is cited many times in some judicial records and letters, containing official verdicts, by references to "the words of the stele" (*awāt naruāim*), see Veenhof (1995). In Akkadian, a stele with laws was called *narûm* (Akkadian: "inscribed stone"). In Old Babylonian texts, "a stele" is mentioned, for instance, in the following manner: "[F]or the shortfall which occurs one will treat him in accordance with the text of the **stele** (*kīma pī narīm*)" (Veenhof, 1995: 1721). In Old Assyrian fragments, an appropriate mention is as follows: "To swear him with/by the three words of (variant: which are written on) the **stele**" (Veenhof, 1995: 1721) and "The creditors of Šukubum, from whatever Šukubum possesses, in accordance with the words of the **stele**, when it is confirmed by witnesses (each), will take his silver in/at/from/by means of his ..." (Veenhof, 1995: 1729).

Thus, the "archaeology of logic" may reveal the genesis of logical terminology and schemes of correct reasoning, which became the prerequisites and intuitive foundations for the logical treatises in Ancient Greece, Ancient India, and Judea, as well as for the proto-logical treatises in Ancient China. The legal discourse is a good context for reconstructing this genesis. Greek and Judaic logical treatises had a very similar legal context, mentioned above, that is, they had a quite similar logical terminology with distinguishing between the particular and the general, coming from jurisprudence. The trial in Mesopotamia was presented as a critical discussion between the claimant and the defendant with involving only logical argumentation. This tradition was continued in Judea and Ancient Greece. Nevertheless, in Ancient China and India the trial was never a critical discussion. So, in China the law (法; *fǎ*) was understood as a system of punishments (刑; *xíng*) carried out by officials without a critical discussion, but considering only facts and their correspondence to the legal statutes or codes (律; *lǜ*). In India, on the one hand, "the judge demands from the suitor a document written against the accused person … and in the document the well-established proof of the justice of his suit. In case there is no written document, the contest is settled by means of witnesses without a written document. The witnesses must not be less than four" (Sachau, 1910: 158). On the other hand, if there are no arguments, then there are different oaths, e.g., "the accused person is invited to drink the *bîsh* (*visha*?) called *brahmaṇa* (?). It is one of the worst kinds; but if he speaks the truth, the drink does not do him any harm" (Sachau, 1910: 159). Consequently, in India the judicial process was mere traditional, without statutes or legal codes, but sometimes with some purely religious rituals for verifying the truth.

As we see, in Judea and Greece there was a continuation of the same Mesopotamian legal system. It means that there were some law codes, and the trial always represented a critical discussion between the claimant and the defendant. In China there were statutes as instructions for the officials on how to govern and punish. Therefore, there was no need for critical discussions in trials. In India, within the framework of Hindu traditions, there were neither legal codes nor critical discussions in courts. The *Manusmṛiti* was composed as a code with the conditional formulation of some articles, but this text dates from the 5th century A.D. and was not used as a foundation for judicial decisions in Medieval India. This text declared the Buddhist royal dynasties of North India outcasts (*Manusmṛiti* 10: 44), but this only became possible after ca. 400 A.D., when they all fell.

The absence of legal codes in India is well explained by another understanding of law. It was treated as *dharma* (Pāli: *dhamma*)—a natural order of things that cannot be formalized through codification. The term *dharma* was mentioned in the *Edict* of Aśoka, the Indian emperor of the Maurya Dynasty who ruled almost the whole Indian subcontinent from ca. 268 to 232 B.C. It is the first known legal document, issued in India. It is worth noting that it was written down on the stone wall or stele, too—in the Babylonian (Sumerian or Akkadian) style that was obviously borrowed. This text was written in a Prakrit close to Pāli. But several translations of the *Edict* into some other languages were published simultaneously with the main text. The Greek and Aramaic versions of this *Edict* excavated in Kandahar are unique because they can be regarded as an outer cultural commentary to the concept of *dhamma* of

Aśoka, proposed by a Greek scholar and by an Aramean legalist who compiled these translations. Let us consider first the Greek text of the *Edict*:

1. δέκα ἐτῶν πληρη[....]ων βασι[λ]εὺς
2. Πιοδασσης **εὐσέβεια[ν** ἔδ]ε[ι]ξεν τοῖς ἀν-
3. θρώποις, καὶ ἀπὸ τούτου εὐσεβεστέρους
4. τοὺς ἀνθρώπους ἐποίησεν καὶ πάντα
5. εὐθηνεῖ κατὰ πᾶσαν γῆν• καὶ ἀπέχεται
6. βασιλεὺς τῶν ἐμψύχων καὶ οἱ λοιποὶ δὲ
7. εἴ τινες ἀκρατεῖς πέπαυνται τῆς ἀκρα-
8. σίας κατὰ δύναμιν, καὶ ἐνήκοοι πατρὶ
9. καὶ μητρὶ καὶ τῶν πρεσβυτέρων παρὰ
10. τὰ πρότερον καὶ τοῦ λοιποῦ λῷιον
11. καὶ ἄμεινον κατὰ πάντα ταῦτα
12. ποιοῦντες διάξουσιν.

Ten years being completed king Piyadassi [*A.Sch.*: Aśoka] showed piety (i.e., *dhamma*) [*A.Sch.*: εὐσέβεια] to men. And from that time [onwards] he made men more pious. And all things prosper throughout the whole world. And the king refrains from [eating] living beings, and indeed other men and whosoever [were] the king's huntsmen and fishermen have ceased from hunting, and those who were without control [over themselves] have ceased as far as possible from their lack of [self-] control, and [have become] obedient to father and mother and to elders, such as was not the case before. And in future, doing all these things, they will live more agreeably and better than before (Thapar, 1997: 260–261).

The Greek scholar translated the *dhamma* as εὐσέβεια that means piety. So, establishing *dhamma* in the Empire by Aśoka meant, for the Greeks, establishing a priority of religious customs in the everyday life.

The Aramean author was more rigorous in his commentary than the Greek:

1. שנן 10 פתיתו עביד זי מראן פרידארש מלכא קשיטא מהקשט
2. מן אדין זעיר מרעא לכלהם אנשן וכלהם אדושיא הובד
3. ובכל ארקא ראם שתי ואף זי זנה כמאכלא למראן מלכא זעיר
4. קטלן זנה למחזה כלהם אנשן אתהחסינן אזי נוניא אחדן
5. אלך אנשן פתיזבת כנם זי פרבסת הוין אלך אתהחסינן מן
6. פרבסתי והופתיסתי לאמוהי ולאבוהי ולמזישתיא אנסן
7. איך אסרהי חלקותא ולא איתי דינא לכלהם אנשיא חסין
8. זנה הותיר לכלהם אנשן ואוסף יהותה.

1. Our Lord, Prīyad ʾarš [*pryd ʾrš*], the king, has repented for ten years, showing [*mhqšṭ*] a straight way [*qšyṭ ʾ*].
2. Since that time evil has decreased for all men [*klhm ʾnšn*] and he has destroyed all the confrontation.
3. And a foundation [*štī*] arose on the whole earth [*wbkl ʾrq ʾ*]. And besides, this concerns food: for Our Lord, the King, little
4. is slaughtered. Seeing this, all men have stopped [to do it]. And those men who [*zī*] were catching living beings [*zwny ʾ*],
5. have been forbidden [to do it]. So, who were bound, those have ceased to
6. be bound. And good obedience [is observed] to his mother and to his father and to the elder men
7. as fate imposed upon him. And the law [*dīn ʾa*] does not exist regarding anyone who is strong.
8. This has benefited all men and will benefit all of them.

Hence, the Aramean legalist translated the *dhamma* of Aśoka in the following two manners: (i) the *dhamma* is a simple/straight way proposed by King Prīyad ʾarš (פרידארש מלכא קשיטא מהקשט); (ii) this way is not a law [*dīn ʾa*]; as a result, if we follow the *dhamma*, we do not need any legal proceeding [the same term *dīn ʾa*] at all (חסין ולא איתי דינא לכלהם אנשיא). It means that the *dhamma* is above any law and cannot be formalized as the Akkadian *dīnum* (i.e., as the Aramaic *dīn ʾa*). In the Talmud, there is a concept of "fundamental ethics" called "road of the earth" (*dere<u>k</u>- ʾereṣ*, דרך ארץ) which denotes the ethics before the Judaic legality that is called *dîn tôrāh* (Hebrew: דִּין תּוֹרָה, "the law of the Torah"). Respectively, this Hebrew concept of the "road of the earth" is close to the *dhamma* as it was understood by the author of the Aramaic version of the *Edict*.

In the Hebrew Qabbalah, the Aramaic word *dīn ʾa* denoting a law has negative connotations linking legality just to prohibitions, too. The matter is that since the 1st century A.D. the post-Akkadian approach to legality with an emancipation from the strong *dīn ʾa* has taken place in the Aramaic world including Judea of that time; according to this approach, ethics began to be considered more important than any law. This post-Akkadian approach was very well expressed in Christianity, first Syriac (Aramaic).

Thus, the *dharma* cannot be formalized by a series of implications, as was the case with *dīn ʾa*. This feature was also evident to the author of the Aramaic version of Aśoka's *Edict*. The matter is that the *dharma* is a natural order of things. Heraclitus of Ephesus (Ἡράκλειτος; ca. 535–ca. 475 B.C.) called this order *logos* (λόγος). For the Buddhists, there are many *dharma*s presenting different potencies in realizing the order of things. According to the Stoics, each of these potencies is called a seminal *logos* (σπερματικός λόγος). So, between *dharma*s of the Buddhists and seminal *logoi* of the Stoics there is a similarity in some connotations.

In India within the framework of Hindu traditions, there was no critical discussion in courts, but the tradition of critical discussion about *dharma* or *dharma*s between different schools of thought was rooted to the same extent that the tradition of philosophical dispute about *logos* was rooted in Ancient Greece but relatively earlier (approximately from one to two or even to three centuries earlier).

In this manner, we see that the phenomenon of critical discussion was presented in Ancient India within the framework of debates between different philosophical schools, according to some strict argumentation standards. We can add that in the Talmud one can also observe peculiar philosophical disputes in the form of a series of debates between sages about the Judaic law (*hălāḵāh*). However, the phenomenon of critical discussion in the narrow sense (meaning with higher logical norms) was not present in Ancient China either in legal proceedings or in philosophical debates, although legal proceedings were highly normalized and standardized, and in the beginning, there were many philosophical schools.

In this way, we can summarize some data collected within the framework of the archaeology of logic in respect to the genesis of logical discourse in Greece, Judea, India, and China, see Table 1.2.1. We detect that the tradition of critical discussion is the main mark that was necessary and sufficient for the appearance of the first logical treatises. In the following three subsections, I consider some of the essential features of logical discourse from the logical treatises written by the earliest Greek, Judaic, Indian, and Chinese logicians and I explain why logic in the narrow sense did not develop in China, although there were some predispositions for it.

Table 1.2.1. Countries with legal codes or some forms of critical discussions.

Countries	**Judea**	**Greece**	**India**	**China**
Legal codes	The Tablets of the Law (*luḥōṯ habbərîṯ*) and the Torah in general.	The *Law Code of Gortyn* as the only preserved example of Greek codes.	No, the articles from the *Manusmṛiti* and other *dharmaśāstra*s were not used as premises in courts of Medieval India.	The statutes (律; *lǜ*) of the Qin and Han dynasties.
Trial as a critical discussion between the claimant and the defendant	Legal documents of South Egypt in Aramaic of the Achaemenid period.	Legal documents of South Egypt in Old Greek of the Ptolemaic period.	No in general for Hindu traditions, but there are legal documents of Buddhist communities from North India and Central Asia in Gāndhārī, Sogdian and Bactrian with the proof of the existence of critical discussions in the courts of North India from the 1st to the 7th century A.D.	No, although there are trial records from the *Book of Submitted Doubtful Cases* dated to the time of the early Han dynasty, but they do not contain a critical discussion in the sense of Babylonian jurisprudence.
Phiosophical dispute as a critical discussion	Legal debates from the Talmud about *hălāḵāh*.	Philosophical debates about λόγος, first of all the debates between the Stoics and the Epicureans about sign-inferences.	Philosophical debates about *dharma*, first of all the debates between the Nyāya and Yogācāra schools about *pramāṇa* (sources of knowledge).	No.

Subsection 1.2.2 is to examine the basic logical inference rules of Judaic hermeneutics. In Subsection 1.2.3, the analogical syllogisms of Mozi's hermeneutics are reconstructed. In Subsection 1.2.4, the logical discourses of Greek and Indian philosophers about the natural order of things are compared. Then the next chapters will explore the genesis and cultural contexts of these discourses with focusing on the development of Greek and Indian logical systems.

Table 1.2.1 shows in which countries there were legal codes and some forms of critical discussions. In this manner, in Late Antiquity, there were simultaneously critical discussions of both kinds in Judea, Greece and North India (but rather within the framework of Buddhist communities). However, there was no form of critical discussions in China. From this it follows that the critical discussions of both types provide a sign sufficient and necessary for the appearance of the first logical treatises.

1.2.2 Rabbi Ishmael's Thirteen Hermeneutic Rules as a Kind of Logic

The Judaic hermeneutics contains some logical inference rules called *middôṯ* in Hebrew. It is difficult to date them, but it is known that Hillel the Elder (Hīllēl; ca. 110 B.C.–10 A.D.) first presented seven *middôṯ* for the interpretation of the Torah before the wise men from the Bathyra family. So, in the *ʾĀḇôṯ Dərabbî Nāṯān* (אָבוֹת דְּרַבִּי נָתָן) 35: 10, we find the following reference to these rules listed by Hillel:

ז' מדות דרש הילל הזקן לפני בני בתירה אלו הן ק"ו וגזרה שוה בנין אב מכלל ופרט ומפרט וכלל כיוצא בו במקום אחר דבר הלמד מענינו אלו שבע מדות שדרש הילל הזקן לפני בני בתירה

Hillel the Elder presented seven hermeneutic rules before the House of Bathyra: (a) an *a fortiori* argument [*qal wāḥōmer*], (b) an analogy [*gəzērāh šāwāh*], (c) a prototype [*binyan ʾāḇ*], (d) a general statement and a specific statement [*kəlāl ûp̱ərāṭ*], (e) a specific statement and a general statement [*pərāṭ ûḵəlāl*], (f) a similar case in another place [*kayyôṣēʾ bô bəmāqôm ʾaḥēr*], (g) a principle learned out from context [*dāḇār hallāmēḏ mēʿinyānô*]; these are the seven hermeneutic rules that Hillel the Elder presented before the House of Bathyra.

Rabbi Ishmael (Rabbî Yišmāʿēʾl; ca. 70–ca. 135 A.D.) developed this corpus of rules and added six more rules to them. He only changed the names of rules (c) and (f)—according to him, (c) is called *binyan ʾāḇ mikkāṯûḇ ʾeḥāḏ* ("a prototype based on one passage"), and (f) is called *binyan ʾāḇ miššənê kəṯûḇîm* ("a prototype based on two passages").

The purpose of this subsection is to determine whether the thirteen hermeneutic rules of Rabbi Ishmael may be understood as a kind of logic in the modern sense.

We know that Judaic authorities have evaluated the *middôṯ* as a logic since the early medieval period. The Karaite scholars, such as Yaʿqūb al-Qirqisānī (first half of the 10th century), who wrote the *Kitāb al-Anwār wa-al-Marāqib* (*Book of Luminaries and Observatories*), stated that instead of the *middôṯ*, the Aristotelian syllogisms should be involved in inferring laws from the Pentateuch. For instance, in the *Leviticus* 3: 17 there is the following general statement:

> It shall be a perpetual statute for your generations throughout all your dwellings, that ye eat neither fat nor blood.

It is the major premise, consisting of the universal affirmative proposition. Then Yaʿqūb al-Qirqisānī noted in his book that the verse about the entire tail fat in the *Leviticus* 3: 9 is the minor premise, also consisting of the universal affirmative proposition. Consequently, we deal with the Aristotelian syllogism, called *Barbara*, and conclude that the entire tail fat is prohibited, too (Ravitsky, 2018: 151–152).

In the dispute about the logic of hermeneutics, Maimonides (Mōšeh ben Maimôn; 1135–1204), the greatest Judaic philosopher and scholar of the medieval time, in his *Treatise on the Art of Logic* (*Maqālah fī ṣināʿ at al-mantiq*), defended the *middô<u>t</u>* of Judaic hermeneutics and called them "dialectical arguments" (*maqāyis jadaliyyah*), see Ravitsky (2011: 32), also see Ravitsky (2004).

Different Judaic commentaries to Rabbi Ishmael's thirteen *middô<u>t</u>* are collected in Freimann (1917), also see Ravitsky (2009). In the 19th and early 20th centuries, many Judaic scholars argued that the *middô<u>t</u>* is a logic that may even be formalized (Hirschfeld, 1840, 1847; Schwarz, 1897, 1901, 1909, 1913, 1916, 1921). A more modern philosophical interpretation of the *middô<u>t</u>* is proposed in Jacobs (1961) and Neusner (2002).

In the 21st century, the book series *Studies in Talmudic Logic* (College Publications, London) was organized in which the authors aim to formalize different fragments of Judaic reasoning including the *middô<u>t</u>*, from the point of view of modern symbolic logic, e.g., see Abraham et al. (2010a, 2010b, 2013) about a formalization of some of the hermeneutic rules.

The main subject of the subsection is to prove that the thirteen hermeneutic rules provided by Rabbi Ishmael can be formalized as a consistent logical system. It is a good argument that the project of the book series *Studies in Talmudic Logic* really makes sense not only from the standpoint of modern logic, but also from the standpoint of historical studies. The matter is that this means that even the founders of Judaic logical hermeneutics—Hillel and Rabbi Ishmael—explicitly formulated the principles of inference to avoid any contradictions in commentary reasoning and did not make logical fallacies thereby.

The earliest logical treatise in Hebrew, dated to the 2nd century A.D., is presented by the *<u>B</u>āraytāʾ Dərabbî Yišmāʿēʾl* (בָּרַיְתָא דְּרַבִּי יִשְׁמָעֵאל) that constitutes the Introduction to the *Si<u>p</u>rāʾ*—that is, to the *mi<u>d</u>rāš* to the Book of Leviticus (Weis, 1962). It contains a compendium of thirteen hermeneutic rules, collected (but not invented) by Rabbi Ishmael to infer new laws from the Torah just logically. This short treatise begins as follows:

> רבי ישמעאל אומ':
> בשלש עשרה מדות התורה נדרשת: (א) מקל וחומר; (ב) מגזירה שוה; (ג) מבנין אב מכתוב אחד, מבנין אב משני כתובים; (ד) מכלל ופרט; (ה) מפרט וכלל; (ו) מכלל ופרט וכלל – אי אתה דן אלא כעין הפרט; (ז) מכלל שהוא צריך לפרט ומפרט שהוא צריך לכלל;(ח) כל דבר שהיה בכלל ויצא מן הכלל ללמד, לא ללמד על עצמו יצא, אלא ללמד על הכלל כולו יצא. (ט) כל דבר שהיה בכלל, ויצא מן הכלל ליטען טען אחר שהוא כענינו – יצא להקל ולא להחמיר. (י) כל דבר שהיה

בכלל, ויצא מן הכלל ליטען טען אחר שלא כענינו יצא להקל ולהחמיר. (יא) כל דבר שהיה בכלל,
ויצא מן הכלל לידון בדבר חדש, אי אתה יכול להחזירו לכללו עד שיחזירנו הכתוב לכללו בפירוש.
(יב) דבר הלמד מעניינו, ודבר הלמד מסופו; (יג) וכן שני כתובין המכחישין זה את זה – עד שיבא
הכתוב השלישי ויכריע ביניהן

rabbî yišmāʿēʾl ʾômēr: bišlōš ʿešōrēh middôṯ hattôrāh niḏrāšeṯ. miqqal wāḥōmer. ûmiggəzērāh šāwāh. mibbinyan ʾāḇ mikkāṯûḇ ʾeḥāḏ, ûmibbinyan ʾāḇ miššənê ḵəṯûḇîm. mikkəlāl ûp̱ərāṭ. ûmippərāṭ ûḵəlāl. kəlāl ûp̱ərāṭ ûḵəlāl, ʾî ʾattāh ḏān ʾellāʾ kəʿên happərāṭ. mikkəlāl šehûʾ ṣārîḵ lip̱rāṭ, ûmippərāṭ šehûʾ ṣārîḵ liḵlāl. kol dāḇār šehāyāh biḵlāl wəyāṣāʾ min hakkəlāl ləlammēḏ, lōʾ ləlammēḏ ʿal ʿaṣmô yāṣāʾ, ʾellāʾ ləlammēḏ ʿal hakkəlāl kullô yāṣāʾ. kol dāḇār šehāyāh biḵlāl, wəyāṣāʾ liṭʿôn ṭōʿan ʾaḥēr šehûʾ ḵəʿinyānô, yāṣāʾ ləhāqēl wəlōʾ ləhaḥămîr. kol dāḇār šehāyāh biḵlāl wəyāṣāʾ liṭʿôn ṭōʿan ʾaḥēr šellōʾ ḵəʿinyānô, yāṣāʾ ləhāqēl ûləhaḥămîr. kol dāḇār šehāyāh biḵlāl wəyāṣāʾ liddôn baddāḇār heḥāḏāš, ʾî ʾattāh yāḵôl ləhaḥăzîrô liḵlālô, ʿaḏ šeyyaḥăzîrennû hakkāṯûḇ liḵlālô bəp̱ērûš. dāḇār hallāmēḏ mēʿinyānô, wəḏāḇār hallāmēḏ missôp̱ô. wəḵēn šənê ḵəṯûḇîm hammaḵḥîšîn zeh ʾeṯ zeh, ʿaḏ šeyyāḇōʾ hakkāṯûḇ haššəlîšî wəyaḵrîaʿ bênêhen.

Rabbi Ishmael says: By thirteen methods the Torah is interpreted. [It is interpreted] (i) by means of *a fortiori* argument [*miqqal wāḥōmer*]; (ii) by means of an analogy [*ûmiggəzērāh šāwāh*]; (iii) by means of a prototype based on one passage [*mibbinyan ʾāḇ mikkāṯûḇ ʾeḥāḏ*] and by means of a prototype based on two passages [*ûmibbinyan ʾāḇ miššənê ḵəṯûḇîm*]; (iv) by means of a general statement and a specific statement [*mikkəlāl ûp̱ərāṭ*]; (v) by means of a specific statement and a general statement [*ûmippərāṭ ûḵəlāl*]; (vi) by means of a general statement and a specific statement and a general statement [*kəlāl ûp̱ərāṭ ûḵəlāl*]—you decide only according to the subject of the specific statement [*ʾî ʾattāh ḏān ʾellāʾ kəʿên apparatṣ*]; (vii) by means of a general statement which requires the specific statement, and by means of a specific statement which requires the general statement [*mikkəlāl šehûʾ ṣārîḵ lip̱rāṭ, ûmippərāṭ šehûʾ ṣārîḵ liḵlāl*]; (viii) anything which is included in the general statement and which is specified in order to teach [something], teaches not only about itself but also teaches about everything included in the general statement [*kol dāḇār šehāyāh biḵlāl wəyāṣāʾ min hakkəlāl ləlammēḏ, lōʾ ləlammēḏ ʿal ʿaṣmô yāṣāʾ, ʾellāʾ ləlammēḏ ʿal hakkəlāl kullô yāṣāʾ*]; (ix) anything that is included in the general statement and which is specified as a requirement concerning another requirement which is inkeeping with the general statement, is specified in order to make [the second requirement] less stringent and not more stringent [*kol dāḇār šehāyāh biḵlāl, wəyāṣāʾ liṭʿôn ṭōʿan ʾaḥēr šehûʾ ḵəʿinyānô, yāṣāʾ ləhāqēl wəlōʾ ləhaḥămîr*]; (x) anything that is included in the general statement and which is specified as a requirement in the general statement and which is specified as a requirement concerning another requirement which is not inkeeping with the general statement, is specified either to make less or more stringent [*kol dāḇār šehāyāh biḵlāl wəyāṣā ʾ liṭ ʿôn ṭō ʿan ʾaḥēr šellō ʾ ḵə ʿinyānô, yāṣā ʾ ləhāqēl ûləhaḥămîr*]; (xi) anything that is included in the general statement and which is excepted from it by an entirely new [provision], you may not return it to [the provision] of its [original] statement unless Scripture expressly indicates that you may do so [*kol dāḇār šehāyāh biḵlāl wəyāṣā ʾ liddôn baddāḇār heḥāḏāš, ʾî ʾattāh yāḵôl ləhaḥăzîrô liḵlālô, ʿaḏ šeyyaḥăzîrennû hakkāṯûḇ liḵlālô bəp̄ērûš*]; (xii) a thing is to be explained from its context, and a thing is to be explained from what follows it [*dāḇār hallāmēḏ mē ʿinyānô, wədāḇār hallāmēḏ missôp̄ô*]; (xiii) and thus two passages which contradict each other [cannot be reconciled] unless a third passage comes and decides between them [*wəḵēn šənê ḵəṯûḇîm hammaḵḥîšîn zeh ʾeṯ zeh, ʿaḏ šeyyāḇō ʾ hakkāṯûḇ haššəlîšî wəyaḵrîa ʿ bênêhen*] (Porton, 1977: 63–64).

Hence, we face an enumeration of the hermeneutic rules/methods (*middôṯ*) for treating the Torah without demonstrating a criterion of their logical division. But then the author properly defines these rules and exemplifies each of them by showing how to interpret different passages from the Pentateuch in terms of each method. These examples allow us to understand to what extent the thirteen *middôṯ* may be evaluated by us as a logical system in the narrow sense.

We know that logical systems of antiquity, such as those of Aristotle and Chrysippus, were based on the idea that we should introduce some inference rules to avoid any inconsistency in our reasoning. So, the logical system is an inference process without possible contradictions. Meanwhile, there were no ideas of axiomatizations and subsequent verification of their completeness, as is the case of modern logic. The consistency, well defined and postulated for reasoning, was sufficient. And we see that Rabbi Ishmael also introduced the principle of non-contradiction as his thirteenth hermeneutic method: *wəḵēn šənê ḵəṯûḇîm hammaḵḥîšîn zeh ʾeṯ zeh, ʿaḏ šeyyāḇō ʾ hakkāṯûḇ haššəlîšî wəyaḵrîa ʿ bênêhen.*

This rule is illustrated by two examples in the *Ḇāraytā ʾ Dərabbî Yišmā ʿē ʾl*. Let us assume that we find two contradictory passages in the Bible: '*A*', 'not-*A*'. According to the rule of introducing conjunction, we may then infer that "*A* and not-*A*", that is, obtain a contradiction. And Rabbi Ishmael proposes two ways to reject this contradiction. First, we can find a new passage *B* from the Bible such that we have: "If *A*, then *B*"; "If not-*A*, then not-*B*". Then, instead of two statements '*A*', 'not-*A*', we formulate the following one axiom which is a tautology in propositional logic: "If *A*, then *B* **or** if not-*A*, then not-*B*." Rabbi Ishmael's own example is as follows. One verse (*Exodus* 19: 20) states: '*A*'—"And the Lord came down upon Mount Sinai, on the top of the mount." Another verse (*Exodus* 20: 22): 'not-*A*'—"I have talked with you from heaven." So, the Lord spoke "from the top of the mount" ('*A*') and He spoke "from heaven" ('not-*A*'). A third verse resolves the contradiction (*Deuteronomy* 4: 36): "Out of heaven he made thee to hear his voice, that he might instruct thee: and upon earth he shewed thee his great fire; and thou heardest his words out of the midst of the fire." Thus, the voice is either not from fire if from heaven or from fire if it is from the top of the mount ("If not-*A*, then not-*B*

or if *A*, then *B*", where '*B*' is fire). The later statement is a tautology of propositional logic, where 'or' is a standard disjunction.

The second way to reject the contradiction of '*A*' and 'not-*A*' is as follows. We should find '*B*' such that "If *B*, then not-*A* **or** if not-*B*, then *A*". Rabbi Ishmael takes the following example to first show a contradiction: '*A*'—"And when Moses was gone into the tabernacle of the congregation to speak with him" (*Numbers* 7: 89), and 'not-*A*'—"And Moses was not able to enter into the tent of the congregation" (*Exodus* 40: 35). This is resolved by the new verse (*Exodus* 40: 35): '*B*'—"the cloud abode thereon, and the glory of the Lord filled the tabernacle". Indeed, if the cloud was there, Moses did not enter ("If *B*, then not-*A*"). When the cloud departed, he entered and spoke with Him ("If not-*B*, then *A*"). Consequently, we have the tautology: "If *B*, then not-*A* **or** if not-*B*, then *A*."

Hence, if we face a contradiction "*A* and not-*A*" in the Torah, we replace it by one of the two axioms:

(xiii') "*A* and not-*A*" → "If not-*A*, then not-*B* **or** if *A*, then *B*;"
(xiii") "*A* and not-*A*" → "If *B*, then not-*A* **or** if not-*B*, then *A*."

Here the sign → means a replacement of one passage by another, according to an appropriate hermeneutic rule.

To sum up, Rabbi Ishmael considers the *middôṯ* on the assumption that in the Torah there are no contradictions at all, therefore some logical inference rules are always applicable to this text.

According to the twelfth rule (*wəḏāḇār hallāmēḏ missôpô*), each obscure statement of the Pentateuch is clarified later. Let *A* be a statement which seems to be unclear, then, instead of *A*, we should take two strings simultaneously: *AB*, where *B* is a later clarification of *A*. In this way, the verse *A* from the *Leviticus* 14: 34 is clarified by the verse *B* from the *Leviticus* 14: 45. Therefore, both verses *AB* should be regarded together, instead of only one *A*:

(xii') $A \to AB$,

where *B* is a later statement clarifying *A*.

Let us assume that we have two strings in the Torah, occurring together: *AB*. In accordance with the second part of the twelfth rule (*dāḇār hallāmēḏ mēʿinyānô*), *B* may be considered through its context as a particular case of *B* restricted to the situation of *A*. Formally: *p*(*B*/*A*), where *p*(*B*/*A*) = "*A* and *p*(*B*)." Then, instead of *AB*, we deal with *Ap*(*B*/*A*), i.e., we replace *B* by *p*(*B*/*A*). For instance, in the *Leviticus* 13: 40, "he is clean" is not understood as "clean of all plague-spot uncleanliness", but it concerns scalls alone:

(xii") $AB \to Ap(B/A)$.

Several rules from the *middôṯ* are grounded on the difference between particular and general notions. Let *A* be a statement. Then *p*(*A*) means a particularization of *A* and g(*A*) means a generalization of *A*. Subsequently, the following implications hold true: "If *p*(*A*), then *A*"; "If *A*, then g(*A*)"; "If *p*(*A*), then g(*A*)". By the seventh rule (*mikkəlāl šehûʾ ṣārîḵ liprāṭ, ûmippərāṭ šehûʾ ṣārîḵ liḵlāl*), some particulars are obtained by a set-theoretic intersection of their generalizations. Let us examine a string $g(A)g(B)g(C)$ consisting of three general notions: $g(A)$—"all the firstborn"; $g(B)$—"whatsoever openeth the womb"; $g(C)$—"all the firstling males". They are taken from the following two verses: "Sanctify unto me all the firstborn, whatsoever openeth the womb among the children of Israel, both of man and of beast: it is mine" (*Exodus* 13: 2); "All the firstling males that come of thy herd and of thy flock thou shalt sanctify unto the Lord thy God" (*Deuteronomy* 15: 19). Each member (atomic string) from $g(A)g(B)g(C)$ is a particular case for another and, at the same time, it's a general case. This means that we have an intersection "$g(A)$ and $g(B)$ and $g(C)$". Therefore, we should replace the string $g(A)g(B)g(C)$ by this intersection. It gives all the firstborn who are male and who did not have the Caesarian section (i.e., "openeth the womb"):

(vii) $g(A)g(B)g(C) \rightarrow$ "$g(A)$ and $g(B)$ and $g(C)$".

According to the fourth, fifth, and sixth rules, the order of particular and general notions of the same genus in one and the same complex string changes its meaning. Assume that we find a specific statement after a general statement (*kəlāl ûpərāṭ*) of the same genus: $g(A)p(A)$. For example, "If any man of you bring an offering unto the Lord, ye shall bring your offering of the beasts, of the cattle, and of the sheep" (*Leviticus* 1: 2, the modified translation of KJV according to the Judaic commentaries). Here "of the beasts" is general (that is, all animals) and "of the cattle and of the sheep" is specific (that is, domesticated animals). Then, in accordance with the fourth rule, we consider only the specific (i.e., only the domesticated animals):

(iv) $g(A)p(A) \rightarrow p(A)$.

Now, let us suppose that we see a general statement after a specific statement (*pərāṭ ûḵəlāl*). For instance, in the *Exodus* 22: 9, "for ox, for ass, for sheep, for raiment" is specific (let it be denoted by $p(A)$) and immediately after that we find the phrase "for any manner of lost thing which another challengeth to be his" that is general (let it be denoted by $g(A)$). Then, according to the fifth rule, we accept only the general:

(v) $p(A)g(A) \rightarrow g(A)$.

According to the sixth rule (*kəlāl ûpərāṭ ûḵəlāl*), if we find the triple of general, specific, and general: $g'(A)p(A)g''(A)$, then we should accept the latter general as a restricted to the previous specific $g'''(p(A))$. For instance, "and thou shalt bestow that money for whatsoever thy soul lusteth after, for oxen, or for sheep, or for wine, or for strong drink, or for whatsoever thy soul desireth" (*Deuteronomy* 14: 26). Here "thou shalt bestow that money for whatsoever thy soul lusteth after" is general. "For oxen, or for sheep, or for wine, or for strong drink" is specific. "For whatsoever thy soul desireth" is again general, but it is a generalization of the previous specific. In this generalization, we accept something that is one nature from another (e.g., wine from grapes) to exclude mushrooms and truffles (which, although they are fruits, do not come from another fruit):

(vi) $g'(A)p(A)g''(A) \rightarrow g'''(p(A))$,

where $g'''(p(A))$ is a subset of $g''(A)$, but $g'''(p(A))$ is not equal to $g''(A)$.

According to the eighth rule (*kol dāḇār šehāyāh biḵlāl wəyāṣā' min hakkəlāl ləlammēḏ, lō' ləlammēḏ 'al 'aṣmô yāṣā', 'ellā' ləlammēḏ 'al hakkəlāl kullô yāṣā'*), if a particular case $p(A)$ already covered in a generalization $g(A)$, but it is nevertheless treated separately: $p(A)B$, then the same particularized treatment $p(A)B$ is applied to all other cases which are covered in that generalization: $g(A)B$. For example, in the following verse of the *Leviticus* 7: 20, we find $p(A)B$: "But the soul that eateth of the flesh of the sacrifice of peace offerings, that pertain unto the Lord, having his uncleanness upon him, even that soul shall be cut off from his people". So, we have a specific case presented by "the sacrifice of peace offerings" (i.e., $p(A)$) in order to teach about "cutting off from his people" (i.e., about B). Let the category of sacrifice, generalizing burnt-offering, meal-offering, sin-offering, guilt-offering, offering of investiture, and peace-offering, be denoted by $g(A)$. Then the following passage is to teach about "cutting off from his people" (i.e., about B) on the basis of all offerings (i.e., of $g(A)$): "Whosoever he be of all your seed among your generations, that goeth unto the holy things, which the children of Israel hallow unto the Lord, having his uncleanness upon him, that soul shall be cut off from my presence" (*Leviticus* 22: 3). Symbolically:

(viii) $p(A)B \rightarrow g(A)B$.

Quite close to this rule are also the ninth and tenth *middôṯ*. So, according to the ninth method (*kol dāḇār šehāyāh biḵlāl, wəyāṣā' liṭ'ôn ṭō'an 'aḥēr šehû' ḵə'inyānô, yāṣā' ləhāqēl wəlō' ləhaḥămîr*), if a specific statement $p(A)$ is subsumed in a general category $g(A)$ and departed from that category for treating B, then it is departed for leniency and not for stringency—let us denote it by $p^l(B)$. For instance, in the verse: "Or if there be any flesh, in the skin whereof there is a hot burning, and the quick flesh that burneth have a white bright spot, somewhat reddish, or white" (*Leviticus* 13: 24), we see "a white bright spot", i.e., $p(A)$, departed from the category of all plague-spots (*Leviticus* 13: 2), to teach about quarantine, i.e., B. Then it is departed for leniency and not for stringency: $p^l(B)$. Therefore, one week of quarantine suffices as opposed to the two-week requirement of the general category:

(ix) $p(A)B \rightarrow p(A)p^l(B)$.

By the tenth method (*kol dāḇār šehāyāh biḵlāl wəyāṣā' liṭ'ôn ṭō'an 'aḥēr šellō' ḵə'inyānô, yāṣā' ləhāqēl ûləhaḥămîr*), if a specific statement $p(A)$ is subsumed in a general category $g(A)$ and departed from that category for treating B, then it is departed both for leniency and for stringency—let us denote it by $p^{ls}(B)$. For example, "If a man or woman have a plague upon the head or the beard" (*Leviticus* 3: 29), this verse separately mentions "head and beard", which are in the general category of skin and flesh, to show that they are not affected by white hair (leniency) and that they are affected by yellow hair (stringency)—$p^{ls}(B)$:

(x) $p(A)B \rightarrow p(A)p^{ls}(B)$.

According to the eleventh rule (*kol dāḇār šehāyāh biḵlāl wəyāṣāʾ liddôn baddāḇār heḥāḏāš, ʾî ʾattāh yāḵôl ləhaḥăzîrô liḵlālô, ʿaḏ šeyyaḥăzîrennû hakkāṯûḇ liḵlālô bəp̄ērûš*), if something specific $p(A)$ is subsumed in a general category $g(A)$, and departed from that category for a new learning (i.e., $p(A)B$), but after that the Holy Scripture shows another learning from the general category (i.e., $g(A)C$), then we accept it as a learning from this specific category, too (i.e., $p(A)C$). For example, "for as the sin offering is the priest's…" (*Leviticus* 14: 13). This is departed from the category for a new learning about the placing of the blood on the thumb of priest's right hand and of his right foot and on his right ear—$p(A)B$. Then the Scripture explicitly restores it to its general category $g(A)$ to tell us that just as a sin-offering requires the placing of blood on the altar, so does this guilt-offering require it—$g(A)C$. Then we accept this learning for the specific category, too:

(xi) $p(A)Bg(A)C \rightarrow p(A)Bp(A)C$.

We see that the *middôṯ* from (viii) to (x) are competing—based on the same premise they draw different conclusions. This demonstrates the specifics of logical inference in the hermeneutic tradition of Rabbi Ishmael: in one logical conclusion, the inference rules numbered by (iv), (v), (vi), (vii), (viii), (ix), (x), (xi), (xii'), (xii''), (xiii'), (xiii''), may only be applied once. So, these inference rules assume only one-step entailing. And if we see a premise $p(A)B$, then we can obtain either $g(A)B$ (by (viii)) or $p(A)p^{l}(B)$ (according to (ix)) or $p(A)p^{ls}(B)$ (by (x)), and we cannot continue inferring any further.

The first and second rules listed by Rabbi Ishmael are two types of inference by analogy. The first one (*qal wāḥōmer*) is perhaps one of the oldest inference rules in the West Semitic culture. Its early occurrences are found in the Amarna archive dated between ca. 1360–1332 B.C. This archive, mainly written in the Canaanite dialect of Akkadian, contains diplomatic correspondence between the Egyptian administration and its representatives in Canaan and Amurru (today's Israel and Syria). And we find many expressions containing the word *appūnamma* with the meaning 'even', 'moreover', 'furthermore' in a clause relating to previous statements to draw a conclusion by analogy corresponding to the *qal wāḥōmer*. This *appūnamma* (other spellings: *appūna, appunnāma, appunāna*) is close to the Hebrew *ʾap̄* ('even'). For example, in the following conditional statement, its author claims that "if the king should come forth and all the lands are hostile to him", then moreover (*ap-pu-na-ma*) this king will even more destroy the author who is more hostile to him:

ù šum-ma ap-pu-na-ma yu-ṣa-na šàr-ru/ù ka-li KUR.KUR.KI *nu-kúr-tu*$_4$ *a-na ša-šu/ù mi-na yi-pu-šu a-na ia-ši-nu.*

But if, moreover, the king should come forth and all the lands are hostile to him, then what can he do to us? (EA, 74: 39–41; Rainey, 1996: 230).

In the next example, it is stated that if the pharaoh does not speak, then the author "will abandon the city and depart", but the author will even more do so it if the pharaoh even does not respond:

> *šum-ma ki-a-ma la-a ti-iq-˹bu˺* (?)/*ù i-te*$_9$*-zi-ib* UR[U] *ù/pa-aṭ-ra-ti ša-ni-tam šum-ma la-a/tu-te-ru-na a-wa-ta*$_5$ *a-na ia-ši/ù i-te*$_9$*-zi-ib* URU *ù/pa-aṭ-ra-ti qa-du* LÚ.MEŠ/*ša i-ra-a-mu-ni.*

> If thus you do not speak, then I will abandon the ci[ty] and depart; moreover, if you don't send word back to me, then I will abandon the city and depart (EA, 83: 45–51; Rainey, 1996: 360).

To the same extent as the Akkadian *appūnamma*, the Hebrew *ʾap̱* in the Torah is sometimes used to express the inference by analogy, called *qal wāḥōmer*. For instance, in the following verse this conclusion is formulated by means of the particles *wəʾap̱ kî* ("so then how much more"):

> *kî ʾānōḵî yāḏaʿtî ʾeṯ-meryəḵā wəʾeṯ-ʿorpəḵā haqqāšeh hēn bəʿōwḏennî ḥai ʿimmāḵem hayyōwm mamrîm hĕyiṯem ʿim-ʾăḏōnāi* ***wəʾap̱ kî****-ʾaḥărê mōwṯî.*

> For I know thy rebellion, and thy stiff neck: behold, while I am yet alive with you this day, ye have been rebellious against the Lord; and how much more after my death? (*Deuteronomy* 31: 27).

It means that if they are rebellious against the Lord when the author is alive, then moreover they will be even more rebellious after his death.

Other Hebrew particles to express the *qal wāḥōmer* (analogy through *a fortiori*) in the Pentateuch is presented by *wəʾêk* ("then how"):

> *hēn kesep̱ ʾăšer māṣāʾnû bəp̱î ʾamtəḥōṯênû hĕšîḇōnû ʾēlêḵā mēʾereṣ kənāʿan* ***wəʾêḵ*** *nignōḇ mibbêṯ ʾăḏōnêḵā kesep̱ ʾōw zāhāḇ.*

> Behold, the money, which we found in our sacks' mouths, we brought again unto thee out of the land of Canaan: how then should we steal out of thy lord's house silver or gold? (*Genesis* 44: 8).

According to this passage, if they brought the money back, then they cannot steal out silver or gold from the lord's house (they will even bring this silver or gold back).

Since the *qal wāḥōmer* was very popular manner of thinking in the West Semitic culture, we also find many examples of this reasoning in the New Testament. First, in Jesus' sermons, we see particles *kəmā hāḵīl* ("how much more"), used to demonstrate the *qal wāḥōmer* in Aramaic (Syriac), for instance:

> *ᵓeṯbaqaw bənaᶜḇe dəlā zārᶜīn wəlā ḥāṣdīn wəlayt ləhon tawāne wəḇēṯ qəpāse wălāhā məṯarse ləhon* ***kəmā hāḵīl*** *ᵓatton yattīrīn ᵓətton men pārḥāṯā.*

> Observe the ravens; for they do not sow nor reap, and they have no storerooms and barns; and yet God feeds them; how much more important are you than the fowls? (*Peshitta, Luke* 12: 24; translation by George Lamsa).

According to this reasoning, if the Lord feeds the ravens, then moreover He will even more feed the humans.

The conclusion by *qal wāḥōmer* was often drawn by apostles, too. Let us see the following inference composed by Paul the Apostle to underline the high importance of restoration:

> *wen tūqlaṯhon həwāṯ ᶜūṯrā ləᶜālmā wəḥayyāḇūṯhon ᶜūṯrā ləᶜamme* ***kəmā hāḵīl*** *šūmlāyhon.*

> Now if their stumbling has resulted in riches to the world, and their condemnation in riches to the Gentiles; how much more is their restoration? (*Peshitta, Romans* 11: 12; translation by George Lamsa).

Hence, the *qal wāḥōmer* rule was pointed out by Rabbi Ishmael as the first hermeneutic method, due to its extreme popularity in Hebrew and Aramaic discourses, which inherited some thought patterns of reasoning by analogy that took place in Akkadian using *appūnamma* (then in Hebrew *ʾap̱*). Rabbi Ishmael's own example of *qal wāḥōmer* refers to the following reasoning from the Pentateuch:

> And the Lord said unto Moses, if her father had but spit in her face, should she not be ashamed seven days? (*Numbers* 12: 14).

According to this verse, if Miriam's father had spat in her face (this means, according to Judaic commentaries, that he was just offended with her), then she would be in shame for seven days. It is a premise from the Torah. Then we know from the Torah that the *Šəḵînāh* (שְׁכִינָה; the divine presence of God) rebukes her. We can then assume that she ought to be put to shame even for fourteen days. Nevertheless, according to the principle of sufficiency (*dayyô*), we cannot sentence more than what is stated in the premise ("seven days"). Then the ultimate sentence for Miriam is as follows:

> Let her be shut out from the camp seven days, and after that let her be received in again (*Numbers* 12: 14).

Rabbi Ishmael explains the *qal wāḥōmer* as containing only two premises, namely let $p'(A)$ and $p''(A)$ be two particulars of the same genus $g(A)$. Suppose that one $p'(A)$ teaches for B, i.e., $p'(A)B$. But we do not know what is taught by $p''(A)$. Then by the *dayyô*, we infer $p''(A)B$ after $p'(A)B$:

(i) $p'(A)Bp''(A) \rightarrow p'(A)Bp''(A)B$,

where either "If $p'(A)$, then $p''(A)$" or "If $p''(A)$, then $p'(A)$" holds true. In Rabbi Ishmael's example, "If her father rebukes her, then the Lord [any the more] rebukes her" holds true. It is a kind of *a fortiori* argument—we apply 'moreover' (*appūnamma*, *ʾap̱*), connecting both particulars $p'(A)$ and $p''(A)$ through an implication in one of the two possible directions. It allows us to extrapolate a property ('teaching') B from $p'(A)$ to $p''(A)$.

More complicated examples discussing the *qal wāḥōmer* with more than two premises are given, for example, in the *Bāḇāʾ Qammāʾ* 25a of the Babylonian Talmud (as well as in the *T̲ôsāp̲ôt̲* to this folio). A logical formalization of these complex patterns of *qal wāḥōmer* is suggested in (Abraham et al., 2009).

The second rule by analogy is called *gəzērāh šāwāh*. Let us look at Rabbi Ishmael's own example. We find in the *Exodus* 22: 11 in respect to a hired watchman that an oath shall be between him and his neighbor that the watchman did not put his hand against the deposit of the neighbor:

> Then shall an oath of the Lord be between them both, that he hath not put his hand unto his neighbour's goods; and the owner of it shall accept thereof, and he shall not make it good (*Exodus* 22: 11; *Šəmôt̲* 22: 10).

Let it be *p'*(*A*)*BC*, where *p'*(*A*) is the hired watchman, *B* is his oath that he did not put his hand, and *C* is "between both", that is, except for the watchman's heirs and the owner's heirs. Then we read about someone who watches gratis (*Exodus* 22: 8; *Šəmôt̲* 22: 7): "[oath], whether he have put his hand unto his neighbour's goods". Let it be *p''*(*A*)*B*, where *p''*(*A*) is the watchman for gratis and *B* is his oath that he did not put his hand. Then we conclude that in this oath his heirs as well as the owner's heirs are also excluded. Hence, we add "between both" (*C*) to him:

(ii) $p'(A)BCp''(A)B \rightarrow p'(A)BCp''(A)BC$.

The third hermeneutic rule of Rabbi Ishmael is called *binyan ʾāb̲* ('prototype'). First, it may proceed by means of a prototype based on one passage (*binyan ʾāb̲ mikkāt̲ûb̲ ʾeḥād̲*). For instance, *miškāb̲* ("what is lain upon"), *A*, and *môšāb̲* ("what is sat upon"), *B*, are different categories. We know that *zāb̲* ("one whose body flows") defiles them—*ADBD*, where *D* is defiling through *zāb̲*. This *zāb̲* also defiles clothing—*CD*, where *C* is clothing. Their possible joint general attribute "*g*(*A*) and *g*(*B*) and *g*(*C*)" is that they are articles only for our bodily comfort. Then, by induction, we conclude that all articles which are designed for the bodily comfort (that is, "*g*(*A*) and *g*(*B*) and *g*(*C*)") are defiled by *zāb̲*:

(iii') $ADBDCD \rightarrow$ "$g(A)$ and $g(B)$ and $g(C)$"D.

Second, it may proceed by means of a prototype based on two passages (*binyan ʾāb̲ miššənê k̲ət̲ûb̲îm*). The example provided by Rabbi Ishmael is as follows. The subject of the lamps on the *mənôrāh* (let it be *A*) is introduced through the notion *ṣaw* ('command')—let it be *C*, that is, we have *CA*:

> *ṣaw ʾet̲-bənê yiśrāʾēl wəyiqḥû ʾēlêk̲ā šemen zayit̲ zāk̲ kāt̲ît̲ lammāʾōwr ləhaʿălōt̲ nēr tāmîd̲.*

> Command [*ṣaw*] the children of Israel, that they bring unto thee pure oil olive beaten for the light, to cause the lamps to burn continually (*Leviticus* 24: 2).

The subject of sending the unclean outside of the encampment (*B*) is introduced by the notion *ṣaw* (*C*), too, that is, *CB*:

> *ṣaw ʾet̲-bənê yiśrāʾēl wîšalləḥû min-hammaḥăneh kol-ṣārûaʿ wək̲ol-zāb̲ wək̲ōl ṭāmēʾ lānāp̲eš.*

> Command [*ṣaw*] the children of Israel, that they put out of the camp every leper, and every one that hath an issue, and whosoever is defiled by the dead (*Numbers* 5: 2).

Then we know that *CA* and *CB* should be applied both immediately (*D*) and for future generations (*E*). The immediate (*D*) performing of the lamps as a command (*CA*) is discussed in the *Leviticus* 8: 2 and this (*CA*) for future generations (*E*) in the *Leviticus* 24: 3. So, we have *CADE*. The *Numbers* 5:4 speaks of the immediate (*D*) departure of the unclean outside the camp as a command (*CB*) and the *Numbers* 5: 4 speaks of that (*CB*) for future generations (*E*). Consequently, *CBDE*. From this reasoning, we conclude by induction that all commandments introduced by *ṣaw* are applicable both immediately and for future generations: *C*"*g*(*A*) or *g*(*B*)"*DE*.

(iii'') *CADECBDE* → *C*"*g*(*A*) or *g*(*B*)"*DE*.

We see that Rabbi Ishmael presents hermeneutic rules (i), (ii), (iii'), (iii''), (iv), (v), (vi), (vii), (viii), (ix), (x), (xi), (xii'), (xii''), (xiii'), (xiii'') rather as *topoi* (τόποι, *loci*) from argumentation theory (evidently, he knew nothing about the Old Greek argumentation theory formed since Aristotle). This means that these argumentation schemes can be applied to the same premises only once, unlike syllogisms, which can be repeatedly applied to the same set of premises, building a chain of inference.

Nevertheless, in the book *Zəḇāḥîm* (49b–51a) of the Babylonian Talmud, there is a hot discussion about chains of conclusions, applying several inference rules one after another. So, some limits of inference chains are considered for the following four rules: *qal wāḥōmer*, *gəzērāh šāwāh*, *binyan ʾāḇ*, and *heqqēš*. Let us note that *heqqēš* is not listed among Rabbi Ishmael's thirteen hermeneutic rules, but it means a kind of inference by analogy. Some limits discussed in the treatise are as follows:

- what is 'learnt' (inferred) through a *heqqēš* does not in turn 'teach' through a *heqqēš* (i.e., this *heqqēš* cannot be applied as next inference rule);
- what is 'learnt' by a *heqqēš* does not in turn 'teach' through a *gəzērāh šāwāh*;
- what is 'learnt' through a *heqqēš* cannot in turn 'teach' through a *binyan ʾāḇ*;
- what is 'learnt' through a *heqqēš* 'teaches' in turn by a *qal wāḥōmer*.

Some attempts of a logical formalization of these limits in drawing inference chains are proposed in Franks (2012) and Schumann (2013b).

The thirteen hermeneutic rules (i), (ii), (iii'), (iii''), (iv), (v), (vi), (vii), (viii), (ix), (x), (xi), (xii'), (xii''), (xiii'), (xiii'') have been symbolically written in the following formal system: **F** = (**K**, **V**, **P**, **I**, **L**), where **K** = {*A*, *B*, *C*,... } is an alphabet consisting of signs, **V** = {*X*, *Y*, *Z*,... } is a set of variables, **P** = {*p*, *g*, *p*', *g*', *p*'', *g*'',...} is a set of one-place predicates defined on signs or variables (where *p* is read as "a particular of" and *g* as "a general of" so that for any $x \in \mathbf{K} \cup \mathbf{V}$, "If *p*(*x*), then *x*" and "If *x*, then *g*(*x*)" hold true), **I** consists of comma (,) and the symbol of replacement (→), **L** is a set of axioms defining strings and their replacements. Each sign, variable, or expression *h*(*x*), where $x \in \mathbf{K} \cup \mathbf{V}$ and $h \in \mathbf{P}$, is called an atomic string. In **F**, we also construct the second-order predicates of the form *h*'(*h*''(*x*)), where $x \in \mathbf{K} \cup \mathbf{V}$ and *h*', *h*'' ∈ **P**. Then *h*'(*h*''(*x*)) is an atomic string, too. Let us suppose that atomic strings are closed under the set of logical operations. It means that an atomic string can also be a compound proposition "$f(x, y, \ldots, z)$", where f is a propositional function of two-valued logic (built by using some logical connectives such as negation, conjunction, disjunction, or implication) and defined on atomic strings $x, y, \ldots, z$. The molecular string is of the form $x_1 \ldots x_n$, where $x_1, \ldots, x_n \in \mathbf{K} \cup \mathbf{V}$ or for some i, $x_i := h(y)$ or $h'(h''(y))$, where $y \in \mathbf{K} \cup \mathbf{V}$ and $h, h', h'' \in \mathbf{P}$, or for some i, $x_i :=$ "$f(y, \ldots, z)$", where f is a propositional function defined on atomic strings $y, \ldots, z$. It is a concatenation of atomic strings. Thus, each atomic or molecular string is a string. Our intuition is that each string $x_1 \ldots x_n$ means that the sequence of symbols $x_1 \ldots x_n$ expresses one of the Judaic laws, contained in or derived from the Pentateuch. For any two strings x, y, we may construct their replacement: $x \rightarrow y$. Some strings and their replacements are collected as axioms in **L**.

This system **F** is to generate strings over signs, variables, predicates, and compound propositions, using axioms and the following two inference rules: (i) the *substitution rule*—we can replace the same variable by the same sequence of signs; (ii) *modus ponens*—if there are axioms x and $x \rightarrow y$, then we deduce a string y, where x and y are strings over **F**.

Proposition 1.1.1. Let **L**' be a set of axioms of the first-order logic without quantifiers, formulated in the language of **F**, and **S**' be a logical system closed under applications of substitution rule and *modus ponens* to **L**'. Then **S**' is consistent and complete.

Proof. It is trivial. The concatenation operation $xy \ldots z$, coupling strings $x, y, \ldots, z$ from **F**, can be treated as the implication "If x, then if y, then … z". The Judaic inference $x \rightarrow y$ for replacing the string x by the string y can be treated as the implication "If x, then y", too. In this way, we can use any axioms **L**' of propositional logic, formulated in **F**. Then **S**' is the standard first-order logic without quantifiers that is consistent and complete, as we well know.

All the laws of the Pentateuch along with the thirteen *middôṯ* of Rabbi Ishmael (i), (ii), (iii'), (iii''), (iv), (v), (vi), (vii), (viii), (ix), (x), (xi), (xii'), (xii''), (xiii'), (xiii'') may be understood as some non-logical axioms **L**'' added to **L**', where **L**' is taken from Proposition 1.1.1. Then the inference, based on (i), (ii), (iii'), (iii''), (iv), (v), (vi), (vii), (viii), (ix), (x), (xi), (xii'), (xii''), (xiii'), (xiii''), is directly drawn by *modus ponens* from **L**''. For example, let us take (iv). If we find the string $g(A)p(A)$ in the Torah (it means that this string is contained in **L**''), then from this string and (iv) (but this (iv) is also contained in **L**''), we infer by *modus ponens* that $p(A)$.

Let **S**'' be a logical system with axioms **L**'' ∪ **L**'. We know that the inference rules of Rabbi Ishmael's hermeneutics from (viii) to (x) compete. To avoid possible contradictions because of drawing two competing propositions 'A' and 'not-A', applying these rules simultaneously, we must forbid the use of the substitution rule. In other words, we must delete the set **V** from **F**. In this way, we obtain **F**'' = (**K**, **P**, **I**, **L**'' ∪ **L**')—the formal system for the logic **S**''. It is natural since we only deal with signs and not with variables in the Pentateuch. Furthermore, we suppose that in the *middôṯ* of (i), (ii), (iii'), (iii''), (iv), (v), (vi), (vii), (viii), (ix), (x), (xi), (xii'), (xii''), (xiii'), (xiii''), all the signs are different. In this manner, we cannot apply these *middôṯ* more than once. Each application of them is fixed as an appropriate non-logical axiom (i), (ii), (iii'), (iii''), (iv), (v), (vi), (vii), (viii), (ix), (x), (xi), (xii’), (xii”), (xiii’), (xiii”) from **L**”. This immediately proves the following statement:

Proposition 1.1.2. The logic **S**” is consistent.

Thus, the thirteen *middôṯ* collected by Rabbi Ishmael should be considered as axioms of **L**” and not as inference rules in the narrow sense. It is a logic but in its wide meaning—a system of consistent reasoning. It is sufficient to be called logic.

1.2.3 Some Proto-Logical Aspects of Mozi’s Hermeneutics

In classical Chinese literature from the 4th century B.C., we can find patterns of logical or quasi-logical reasoning that can be analyzed in terms of modern logic. There are some attempts to discover some correct conclusions from the point of view of modern propositional logic with great disregard for their context, e.g., see Chmielewski (2009). The author of this book simply found many examples of correct conclusions from various texts of different times and collected them. Nevertheless, such an attempt does not make sense. First, we should not follow the mathematical centrism in the history of logic, according to which only classical propositional logic should be reconstructed based on ancient documents. The fact is that sometimes we meet some complex reasoning by analogy, namely, the same case of Chinese or Judaic hermeneutics. Therefore, we must restore their own logical meaning—how conclusions are drawn by analogy. Second, a search in different books for different examples of correct conclusions cannot give something for us. We must reconstruct the own logical tradition of Chinese philosophy, which was recognized and philosophically reflected by Chinese scholars themselves and consciously applied in their practice within the framework of the relevant norms of argumentation. If we randomly found correct patterns of reasoning, it does nothing to understand the Chinese logic or proto-logic.

The purpose of this subsection is to analyze the earliest Chinese logical fragments of the following two types: (1) a kind of philosophical reflection on correct logical inferences, first presented in the work of Mozi (from the 5th to the 4th century B.C.); (2) a social practice involving some correct reasoning that first appeared in the legal discourse of the Qin and Han empires (from 221 B.C. to 220 A.D.), in order to find some strict correlations between logical conclusions of both types of texts.

Mozi became the founder of logical discourse in China; he first applied many correct inference rules in his hermeneutics. For instance, let us take the following proposition from his treatise: “Stopping: If another, by enumerating this is so, just considers everything is so, then I will ask him by enumerating that is not so” (止 彼舉然者 以為此其然也 則舉不然者而問之; *zhǐ: bǐ jǔ rán zhě, yǐwéi cǐ qí rán yě, zé jǔ bùrán zhě ér wèn zhī*; *Mohist Canon* 40: 99; Hansen, 1983: 165). Then we may formalize it as follows (Hansen, 1983: 166): (i) Some S_i are P; (ii) Therefore, all S are P. Refutation: (i) Some S_j are not P; (ii) S_i and S_j are of the same kind as that for all S. (iii) Therefore, not all S are P.

Nevertheless, he did not construct a logical system, but he founded a hermeneutics containing some inference rules by analogy. The main influence of his hermeneutics on the Chinese classical philosophy was that Mozi introduced a kind of situation semantics close to the modern-day situation theory (Cooper et al., 1990):

> A name (名; *míng*) is used for defining objects (實; *shí*), a proposition (辭; *cí*) is used for expressing a meaning (意; *yì*), an explanation (說; *shuō*) is used for exploring a reason, a class [of situations] is used for a generalization, and a class [of situations] is used for a detalization (以名舉實 以辭抒意 以說出故 以類取 以類予; *yǐ míng jǔ shí, yǐ cí shū yì, yǐ shuō chū gù, yǐ lèi qǔ, yǐ lèi yǔ*; *Mohist Canon* 45: 1).

This semantics played the role of logical discourse in Ancient China and, in the absence of pure logic there, it was sufficient to organize the Chinese legal system. In this subsection, I will examine the logical nature of the *Mohist Canon*, then I will find its schemes of reasoning in the legal discourse of the Qin and Han dynasties, and I will trace a general critique of logical consistency against Mozi from the early School of Names.

The earliest proto-logical reasoning in China belonged to different scholars of the Warring States period (戰國; *Zhànguó*; from ca. 481 to 221 B.C.). For example, Zhuang Zhou (莊子; *Zhuāngzi*; around the 4th century B.C.) reasoned as follows, comparing positive and negative attributes/'pointings' (指; *zhǐ*) with each other in respect to 'horse' (馬; *mǎ*):

> 以指喻指之非指，不若以非指喻指之非指也；以馬喻馬之非馬，不若以非馬喻馬之非馬也。天地，一指也；萬物，一馬也

> *yǐ zhǐ yù zhǐ zhī fēi zhǐ, bù ruò yǐ fēi zhǐ yù zhǐ zhī fēi zhǐ yě; yǐ mǎ yù mǎ zhī fēi mǎ, bù ruò yǐ fēi mǎ yù mǎ zhī fēi mǎ yě. tiāndì, yī zhǐ yě; wànwù, yī mǎ yě* (*Zhuāngzi* 2: 6).

> To use an attribute to show that attributes are not attributes is not as good as using a nonattribute to show that attributes are not attributes. To use a horse to show that a horse is not a horse is not as good as using a non-horse to show that a horse is not a horse, Heaven and earth are one attribute; the ten thousand things are one horse (Watson, 2003: 35).

Meanwhile, another Chinese scholar of the Warring States era, Mozi (墨翟; *Mò Dí*; 墨子; *Mòzi*; 5th–4th century B.C.), can be named the founder of Chinese proto-logic in the full sense of the word. He first involved some basic logical notions: 'term' or 'name' (名; *míng*); 'premise' (说; *shuō*); "logical conclusion" (实意故; *shíyì gù*); 'checking'—one of the four types of inference by similarities and differences (效; *xiào*), etc., (see Liang, 1999: 3186–3187). Some other basic terms used by him: 'reasoning' (辩; *biàn*); 'proposition' or 'judgment' (辞; *cí*); 'assumption' (假; *jiǎ*), etc. (Hu, 2003: 253–258).

He wrote the *Mohist Canon* (墨經; *Mòjīng*; or 墨子; *Mòzi*), see Johnston (2010), which became the fundamental text for the school named Mohism after him. This book contains the so-called *Dialectical Chapters* (墨辯; *Mòbiàn*), namely the chapters from 40 to 45, devoted to different semantic and logical aspects of reasoning and names. Meantime, the 45th chapter among them may be called the oldest proto-logical treatise in Chinese. It is entitled the *Minor Illustrations* (小取; *Xiǎoqǔ*).

Let us consider its main impact to the Chinese traditional hermeneutics—how Mozi's principles for understanding reasoning and names became basic in Chinese traditional culture.

The first principle of the Mohist hermeneutics, with which the *Minor Illustrations* begins, is called a principle of distinguishing which "is used to make clear the distinction between right and wrong" (夫辯者 將以明是非之分; *fū biàn zhě, jiāng yǐ míng shìfēi zhī fēn*; *Mohist Canon* 45: 1). Here the goal of reasoning (辯; *biàn*) is defined to clearly distinguish between what is so and what is not so (是非; *shìfēi*, i.e., right or wrong, true or false).

For instance, this principle is applied in defining differences between "periods of trial and chaos, to clarify decisions on similarities and differences" (審治亂之紀 明同異之處; *shěn zhì luàn zhī jì, míng tóng yì zhī chù*; *Mohist Canon* 45: 1). The second principle speaks of the strong limitation in applying similarities: "being some does not mean being all and supposing something does not mean that it exists now" (或也者 不盡也 假者 今不然也; *huò yě zhě, bù jìn yě. jiǎ zhě, jīn bùrán yě*; *Mohist Canon* 45: 2). There are the following four inference rules by similarities with finding a checking (效; *xiào*) to establish the standard/law (法; *fǎ*) through it (*Mohist Canon* 45: 2): (i) analogy (辟; *pì*) to substantiate something on the basis of other; (ii) similarity/concurrency (侔; *móu*), when two phenomena occur together; (iii) leading/authority (援; *yuán*), when someone follows the leader's opinion; (iv) extending (推; *tuī*) from something found to another.

According to the third principle of hermeneutics, each phenomenon cannot be considered only from a single perspective: "there are many methods [such as analogy, similarity, leading, extending], different classes [of situations], various reasons/causes, so you cannot apply them separately from a single perspective" (言多方 殊類 異故 則不可偏觀也; *yán duōfāng, shū lèi, yì gù, zé bùkě piān guān yě*; *Mohist Canon* 45: 3).

Then in this chapter, we see four examples of the following analogical syllogism: "If the particular is the general, then the use of/the act with the particular is the use of/the act with the general" ("If *B* follows from *A*, then using/acting with *B* follows from using/acting with *A*"). For instance, "A white horse is a horse; riding a white horse is riding a horse" (白馬 馬也 乘白馬 乘馬也; *báimǎ, mǎ yě; chéng báimǎ, chéng mǎ yě*; *Mohist Canon* 45: 4); "Zāng is a man; loving Zāng is loving a man" (臧 人也 愛臧 愛人也; *zāng, rén yě; ài zāng, àirén yě*; *Mohist Canon* 45: 4). These inferences belong to the case of positive premise and positive conclusion (是而然者也; *shì ér rán zhě yě*; *Mohist Canon* 45: 4).

Meanwhile, we see examples of the following analogical syllogism, contradictory to the previous one: "If the particular is the general, then the use of/the act with the particular is **not** the use of/the act with the general" ("If *B* follows from *A*, then using/acting with *B* does not follow from using/acting with *A*"). For example, "His younger brother is handsome; loving his younger brother is not loving handsome people" (其弟 美人也 愛弟 非愛美人也; *qí dì, měirén yě; ài dì, fēi ài měirén yě*; *Mohist Canon* 45: 5); "A boat is wood; boarding a boat is not boarding wood" (船 木也 入船 非入木也; *chuán, mù yě; rùchuán, fēi rù mù yě*; *Mohist Canon* 45: 5). These inferences belong to the case of positive premise and negative conclusion (此乃是而不然者也; *cǐ nǎi shì ér bùrán zhě yě*; *Mohist Canon* 45: 5).

Hence, we face two contradicting inference rules, according to their schemes: "If *B* follows from *A*, then using/acting with *B* follows from using/acting with *A*" and "If *B* follows from *A*, then using/acting with *B* does not follow from using/acting with *A*". From the same first premise "*B* follows from *A*" and according to these inference rules, we obtain a contradiction: "using/acting with *B* follows from using/acting with *A*" and "using/acting with *B* does not follow from using/acting with *A*". This situation is to illustrate the third Mohist hermeneutic principle, according to which only one method, that is only one scheme of inference, is not sufficient for analyzing phenomena. We should always combine complementary methods, which may be even contradictory by their schemes. For instance, "acting with" may be transitive for the implication: "If $A \Rightarrow B$, then "acting with *A*" $\Rightarrow$ "acting with *B*"", and may not: "If $A \Rightarrow B$, then it is false that "acting with *A*" $\Rightarrow$ "acting with *B*"".

Then we see some more examples of analogical syllogisms: "Acting with *A* is not *A*, liking acting with *A* is liking *A*". For instance, "Reading books is not books; liking reading books is liking books" (且夫讀書 非 書也 好讀書 好書也; *qiě fū dúshū, fēi shū yě; hǎo dúshū, hǎo shū yě*; *Mohist Canon* 45: 6); "Being about acting with *A* is not acting with *A*, stopping about to acting with *A* is stopping acting with *A*". For example, "Being about to fall into a well is not falling into a well; stopping someone about to fall into a well is stopping someone from falling into a well" (且入井 非入井也 止且入井 止入井也; *qiě rùjǐng, fēi rùjǐng yě; zhǐ qiě rùjǐng, zhǐ rùjǐng yě*; *Mohist Canon* 45: 6).

In these examples, Mozi demonstrates that if there is no generalization from *A* to *B* (that is, we do not have a true implication $A \Rightarrow B$), but 'liking' or 'stopping' can, on the contrary, be transitive: "liking/stopping *A*" $\Rightarrow$ "liking/stopping *B*". All these inferences with this scheme belong to the case of negative premise and positive conclusion (此乃是而然者也; *cǐ nǎi shì ér rán zhě yě*; *Mohist Canon* 45: 6).

After that the author tries to define a universal quantifier 'all' ('universally') through the following analogical syllogism: "If acting with *A*, then acting with all *A*; as a consequence, if not acting with *A*, then it is not true that acting with not all *A*". More easily: "If *A*, then all *A*; consequently, if not *A*, then it is not true that not all *A*". For instance, "If loving people, then universally loving people to count as loving people; if not loving people, then it is not true that universally not loving people. Not universally [not loving people], because not loving people" (愛人 待周愛人而後為愛人 不愛人 不待周不愛人 不周愛 因為不愛人矣; *àirén, dài zhōu àirén érhòu wèi àirén. bù àirén, bùdài zhōu bù àirén; bù zhōu ài, yīn wéi bù àirén yǐ*; *Mohist Canon* 45: 7). The more literary translation: "If you love people, then you love all people, including loving everyone; if you do not love people, it does not mean that you love not all people. If you do not love people, then [you do not love] not all people". In other words, if you love people (i.e., you are kind), then you love all people (e.g., including strangers); if you do not love people (i.e., you are not kind), then you love only some people (e.g., excluding strangers)—you do not love not all people. Semantically, it is correct. In terms of categorical statements: "If *A* is *B*, then *A* is for all *B*; therefore, if *A* is not *B*, then it is not true that *A* is for not all *B*. If *A* is not *B*, then *A* is not for not all *B*".

Here we see a logical fallacy which is readily detected in constructing a contrapositive statement based on an appropriate conditional statement. Instead of the correct scheme: "it is *A*" ⇒ "it is for all *A*"; therefore, "it is not for all *A*" ⇒ "it is not *A*", we see the following fallacy: "it is *A*" ⇒ "it is for all *A*"; therefore, "it is not *A*" ⇒ "it is for **some** *A*". Let us note that "do not love (not *P*) not all people (not all *A*)" is logically equivalent to "love (*P*) some people (some *A*)" in the modern predicate logic: "not all *A* such as not *P*(*A*)" is equivalent to "some *A* such as *P*(*A*)", where *P* is a predicate true on *A*. This rule was also known to Aristotle, e.g., see the 10th chapter of his *De Interpretatione*.

Thus, in the *Mohist Canon* 45: 7, we read the following conclusion: "If not [不; *bù*] loving people [愛人; *àirén*], it does not mean [不; *bù*] that loving [愛人; *àirén*] not all people [不待周; *bùdài zhōu*]. If not loving people [不愛人矣; *bù àirén yǐ*], then [not loving] not all people [不周愛; *bùdài zhōu*]". So, according to the contrapositive statement, the conclusion should be: "If not loving all people [不待周愛人; *bùdài zhōu àirén*], then not loving people [不愛人矣; *bù àirén yǐ*]".

Meanwhile, according to Mozi, there is possible another analogical syllogism: "If acting with *A*, then acting with not all *A*; as a consequence, if not acting with *A*, then not acting with all *A*". More easily: "If *A*, then not all *A*; consequently, if not *A*, then it is not true that all *A*". For example, "If riding horses, then not universally riding horses to count as riding horses; if riding horses, then having ridden upon a horse. By contrast, if not riding horses, then universally not riding horses to count as not riding horses" (乘馬 不待周乘馬然後為乘馬也 有乘於馬 因為乘馬矣 逮至不乘馬 待周不乘馬而後不乘馬; *chéng mǎ, bùdài zhōu chéng mǎ ránhòu wèi chéng mǎ yě; yǒu chéng yú mǎ, yīnwèi chéng mǎ yǐ. dǎi zhì bù chéng mǎ, dài zhōu bù chéng mǎ érhòu bù chéng mǎ*; *Mohist Canon* 45: 7). The more literary translation: "If riding horses, then riding not all horses to count as riding horses; if riding horses, then having ridden upon a horse. By contrast, if not riding horses, then not riding all horses to count as not riding horses". In other words, if you ride a horse, then you ride not all horses (it is only one); therefore, if you do not ride a horse, then you do not ride all horses. Semantically, it is correct. However, in this case the propositional formula corresponding to this inference: "it is *A*" ⇒ "it is for not all *A*"; therefore, "it is not *A*" ⇒ "it is not for all *A*", is not a tautology, too. It means that the conclusion is not correct from the point of view of classical predicate logic—it is a logical fallacy.

So, we have two complementary syllogisms again: "If it is *A*, then it is for all *A*; consequently, if it is not *A*, then it is for **some** *A*" and "If it is *A*, then it is for not all *A*; consequently, if it is not *A*, then it is not for all *A*". This means that we should apply one of both syllogisms according to our intuition and there is no general scheme with the universal quantifier for all phenomena. Both syllogisms belong to the case called "one requires a universal quantifier, and one does not" (此一周而一不周者也; *cǐ yīzhōu ér yī bù zhōu zhě yě*; *Mohist Canon* 45: 7).

As we have just shown, both ways of defining universal and existential quantifiers in the *Mohist Canon* 45: 7 are not correct from the standpoint of modern predicate logic. Nevertheless, they correspond to the non-Aristotelian square of opposition considered in Schumann (2013a):

“All *A* such as *P*(*A*)” “Some *A* such as *P*(*A*)”

“All *A* such as not *P*(*A*)” “Some *A* such as not *P*(*A*)”

In this new square, “all *A* such as *P*(*A*)” [“universally loving people”/“universally riding horse”] and “some *A* such as *P*(*A*)” [“not universally not loving people”/“not universally not riding horse”] are contrary (i.e., they can be false together, but they cannot be true together), where *P* is a predicate, true on *A*. “All *A* such as not *P*(*A*)” [“universally not loving people”/“universally not riding horse”] and “some *A* such as not *P*(*A*)” [“not universally loving people”/“not universally riding horse”] are subcontrary (i.e., they can be true together, but they cannot be false together). The following two pairs are contradictory (i.e., they cannot be true or false together): (i) “all *A* such as *P*(*A*)” [“universally loving people”/“universally riding horse”] and “some *A* such as not *P*(*A*)” [“not universally loving people”/“not universally riding horse”]; (ii) “some *A* such as *P*(*A*)” [“not universally not loving people”/“not universally not riding horse”] and “all *A* such as not *P*(*A*)” [“universally not loving people”/“universally not riding horse”]. “All *A* such as not *P*(*A*)” follows from “all *A* such as *P*(*A*)” [“universally not riding horses”/“riding a horse” semantically follows from “universally riding horses”], and “some *A* such as not *P*(*A*)” follows from “some *A* such as *P*(*A*)” [“not universally loving people”/“not loving some people” follows from “not universally not loving people”/“loving some people”].

Indeed, “loving people” semantically implies “loving all people”. “Not loving people” semantically implies “not loving not all people”, that is, “loving some people”. This means that “loving all people” and “loving some people” are considered as opposing each other by Mozi—in the same way as it is done in the non-Aristotelian square of opposition, namely they are contrary. Then “riding horses” semantically implies “riding not all horses”, that is, “not riding some horses”. “Not riding horses” semantically implies “not riding all horses”. Consequently, “not riding some horses” and “not riding all horses” are considered as opposing each other by Mozi, too—they are also opposing in the non-Aristotelian square of opposition, namely they are subcontrary.

Hence, the universal and existential quantifiers are correctly treated by Mozi only from the point of view of non-Aristotelian approach to them. Aristotle treated the existential quantifier as “some but maybe all”, while Mozi treated this quantifier as “some but surely not all”. As a result, “all *A* such as *P*(*A*)” and “some *A* such as *P*(*A*)” cannot be true together, but they can be false together, according to Mozi.

Then the author enumerates some examples of syllogisms with a contradictory structure for particular and general cases: “If a property of *A* is not *A*, then a property of particular case of *A* is a particular case of *A*”. For example, “A person’s ghost is not a person; one’s elder brother’s ghost is one’s elder brother” (人之鬼 非人也 兄之鬼 兄也; *rén zhī guǐ, fēi rén yě; xiōng zhī guǐ, xiōng yě*; *Mohist Canon* 45: 8).

"If acting with a property of *A* is not acting with *A*, then acting with a property of particular case of *A* is acting with a particular case of *A*". For example, "Sacrificing to a person's ghost is not sacrificing to a person; sacrificing to one's elder brother's ghost is sacrificing to one's elder brother" (人之鬼 非祭人也 祭兄之鬼 乃祭兄也; *rén zhī guǐ, fēi jì rén yě; jì xiōng zhī guǐ, nǎi jì xiōng yě*; *Mohist Canon* 45: 8).

In the *Minor Illustrations*, there are many other examples of syllogisms, but almost all of them have no logical meaning—most of them are not logical tautologies (always true propositions) in classical logic. The following example is to show a generalization: "One horse is a horse. Two horses are a horse" (馬 馬也 二馬 馬也; *mǎ, mǎ yě; èr mǎ, mǎ yě*). "If a horse is white, then two horses are white; not a singular horse is white" (馬或白者 二馬而或白也 非一馬而或白; *mǎ huò bái zhě, èr mǎ ér huò bái yě, fēi yī mǎ ér huò bái*; *Mohist Canon* 45: 8).

All the instances from the *Mohist Canon* 45: 8 are called "one positive and one negative" (此乃一是而一非者也; *cǐ nǎi yī shì ér yī fēi zhě yě*; *Mohist Canon* 45: 8).

As we see, the syllogisms presented in this short treatise are mainly inferences by analogy and cannot be understood as a logic in the narrow sense, because their schemes are sometimes contradictory, sometimes are logically erroneous for classical logic, and in any case, they are treated by Mozi himself as complementary to each other and as not sufficient for all conclusions. Nevertheless, some Chinese scholars found some strong logical ideas in this text. So, they believe that in all illustrations of analogical syllogisms of the 45th chapter of the *Mohist Canon*, we may reconstruct the possibility of four types of distributions of affirmative and negative propositions in premise (彼; *bǐ*) and conclusion (此; *cǐ*): (a) "positive in premise and positive in conclusion" (是而然; *shì ér rán*); (b) "positive in premise and negative in conclusion" (是而不然; *shì ér bùrán*); (c) "negative in premise and positive in conclusion" (不是而然; *bùshì ér rán*); (d) "negative in premise and negative in conclusion" (不是而不然; *bùshì ér bùrán*) (see Fung, 2020: 44–45). According to (Mo, 1981: 433–434), the scheme of (d) is another. It is presented by "one is adequate, but another is not" (一周而 一不周; *yīzhōu ér yī bù zhōu*) and "one is correct, but another is not" (一是而一非; *yī shì ér yī fēi*). Thereby, Shaokui Mo (1981: 433–436) calls scheme (a) (that is, *shì ér rán*) a "normal phenomenon", which is presented by the formula, according to him: "$A = B$, and $CA = CB$", where C is an additional attribute. Other schemes from (b) to (d) are called by him "abnormal phenomena", so that (b) has the following formula: $A = B$ but $CA \neq CB$; (c) has the formula: $A \neq B$ but $CA = CB$; and (d) has the formula: $f(A) = g(A)$ but $f(B) \neq g(B)$. In any case, even in such a formalization, it is not a logic, since at least (a) and (b) give a contradiction.

Fung (2020: 44–45) and Mo (1981: 433–436) have generalized the following enumeration of possible conclusions provided by Mozi himself: "Items are sometimes positive and positive, sometimes positive and negative, sometimes one requires a universal quantifier and one does not, sometimes one positive and one negative" (夫物或乃是而然 或或一是而一非也是而不然 或一周而一不周; *fū wù huò nǎi shì ér rán, huò shì ér bùrán, huò yīzhōu ér yī bù zhōu, huò yī shì ér yī fēi yě*; *Mohist Canon* 45: 2).

Hence, the Mohist *Dialectical Chapters* (墨辯; *Mòbiàn*), and especially its 45th chapter, are not a kind of logic, but a kind of proto-logic. Its goal is to show semantic relationships between various terms (names, *míng*) to find "similarities and differences" (同異; *tóng yì*) among them. But semantic relations are much wider than logical ones. For example, Mozi's examples of universal and existential quantifiers have some semantic validations, although these quantifiers are beyond the classical predicate logic.

In the *Mohist Canon*, there is a logically correct understanding that any general (genus) follows from an appropriate particular (species), e.g., 'horse' follows from "white horse". However, no inference rules such as *modus ponens* or *Barbara* have been proposed to use this relationship in inferring. For finding "similarities and differences" there have been shown the four ways of reasoning: (i) analogy (辟; *pì*); (ii) similarity/concurrency (侔; *móu*); (iii) leading/authority (援; *yuán*); (iv) extending (推; *tuī*). Each of these ways is to build a suitable *topos* (τόπος, *locus*) or dialectical argument in the Aristotelian meaning. But they are not logical inference rules in the narrow sense. And Mozi provided many *topoi* in the *Minor Illustrations* to show some of the limitations in finding "similarities and differences".

Let us remind that he put forward the following three hermeneutical principles, which may be reconstructed in his text: (1) we can always observe the distinction between right and wrong (the principle of non-contradiction) (夫辯者 將以明是非之分; *fū biàn zhě, jiāng yǐ míng shìfēi zhī fēn*; *Mohist Canon* 45: 1); (2) there is a strong limitation in applying similarities (或也者 不盡也 假者 今不然也; *huò yě zhě, bù jìn yě. jiǎ zhě, jīn bùrán yě*; *Mohist Canon* 45: 2); (3) "there are many methods [such as analogy, similarity, leading, extending], different kinds, various reasons, so you can't apply them separately from a single perspective" (言多方 殊類異故 則不可偏觀也; *yán duōfāng, shū lèi, yì gù, zé bùkě piān guān yě*; *Mohist Canon* 45: 3)—all rhetorical *topoi* are complementary to each other, there is no single way in reasoning. Thus, the aim of these *topoi* is to approach as many semantic details as possible in defining items. In fact, according to Mozi, we deal with only three logical processes: generalization (達; *dá*), detalization (私; *sī*), and finding similar cases (類; *lèi*). Therefore, in his opinion, logical inference rules, which are universally held, are completely impossible. We can only check the validity of conflicting assertions by their semantic meanings to determine one of the two assertions as 'acceptable' (可; *kě*).

Thus, Mozi's opinion that there are no universal logical schemes is consonant with his doctrinal claim against fate (非命; *fēimìng*). The 35th, 36th, and 37th chapters of his *Mohist Canon* are dedicated to this subject – according to Mozi, everything is in our hands, that is, there is no place for logical determinism (命; *mìng*). The law (法; *fǎ*) should be organized by ourselves through "similarities and differences" (明同異之處; *míng tóng yì zhī chù*; *Mohist Canon* 45: 1).

In the 44th chapter, called the *Major Illustrations* (大取; *Dàqǔ*), from the *Dialectical Chapters* of the *Mohist Canon*, Mozi defined the situation semantics for propositions, which is close to the modern-day situation theory (Cooper et al., 1990), considering not facts but classes of situations as meanings of propositions: "A proposition (辭; *cí*) is made on the basis of a relevant condition (故; *gù*), developed in accordance with a relevant idea (理; *lǐ*) and held in accordance with an appropriate class [of situations] (類; *lèi*)" (夫辭以故生 以理長 以類行也者; *fū cí yǐ gù shēng, yǐ lǐ zhǎng, yǐ lèi xíng yě zhě*; *Mohist Canon* 44: 25).

To sum up, Mozi's hermeneutics cannot be considered a logic in the narrow sense, but it does contain the situation semantics as a useful tool for justifying the meanings of words and propositions. Chapters 40–44 of the *Mohist Canon* contain a list of definitions of different terms. Sometimes, these definitions are based on logical reasoning, e.g.:

> **罪**, 犯禁也。
> **罪**: 不在禁, 惟害無**罪**, 殆姑

> *zuì, fànjìn yě.*
> *zuì: bùzài jìn, wéi hài wúzuì, dài gū.*

> Committing a crime is transgressing a prohibition.
> Committing a crime: If no prohibition exists, even if there is harm, there is no crime (*Mohist Canon* 40: 37; Johnston, 2010).

Here the crime is understood as a particular case of an appropriate prohibition: "If a crime of class *A*, then a prohibition of the same class *A*". Therefore, by *modus tollens* we can conclude: "If no prohibition of class *A*, then no crime of the same class *A*". This conclusion is logically correct.

For each term, according to Mozi, we should find its general cases (達; *dá*), its class of situations (類; *lèi*), and its particular cases (私; *sī*):

> 名 達 類 私
> 名 物 達也 有實必待之名也 命之馬 類也 若實也者 必以是名也 命之臧 私也 是名也止於是實也 聲出口 俱有名 若姓字灑

> *míng, dá, lèi, sī.*
> *míng: wù, dáyě. yǒu shí bì dài zhī míng yě. mìng zhī mǎ, lèi yě. ruò shí yě zhě, bì yǐ shì míng yě. mìng zhī zāng, sī yě. shì míng yě zhǐ yúshì shí yě. shēng chūkǒu, jù yǒumíng, ruò xìng zì sǎ.*

> A name [may be] generalized (*dá*), attributed to the class of similar items (*lèi*), and detalized (*sī*).
> 'Thing' is a generalization. If there is an entity, it necessarily gets this name. Its class of situations is 'horse'. If there is an entity like this, it is necessarily named by this [word of 'horse']. Naming someone Zāng is a [final] detalization. This name stops at this entity. The sounds issued from the mouth all have names—like the pairing of surname and style (*Mohist Canon* 40: 79).

The Mohist hermeneutics with its situation semantics was also incorporated into the Confucian hermeneutics of Xunzi (荀況; *Xún Kuàng*; 荀子; *Xúnzi*; 3rd century B.C.). He wrote a Confucian canon named *Xúnzi* after him, which includes the 22nd chapter entitled the *Rectification of Names* (正名; *Zhèngmíng*) about hermeneutics grounded on the principles and ideas of Mozi. It is about the correct use of 'names' (名; *míng*) and how to justify their semantic relations to 'stuff', 'objects', 'events', or 'situations' (實; *shí*). The Legalist canon *Hánfēizi* (韓非子) attributed to Han Feizi

(韓非子; *Hán Fēizi*; 3rd century B.C.) also contains the Mohist way of treating "names and reasoning" (名辯; *míng biàn*).

Mozi's proto-logical hermeneutics is well traced in the twenty-two trial acts from the *Book of Submitted Doubtful Cases* or the *Collection of Court Requests Zòu and Court Requests Yàn* (奏讞書; *Zòuyàn shū*), dated to 246–196 B.C. (Barbieri-Low and Yates, 2015: 95–96). It is a compendium of judicial inquiries written by officials to higher authorities on legal cases in which the officials have some doubts about the issuance of court decisions. These legal documents show the fundamental difference between the Chinese legal system, on the one hand, and the Mesopotamian system inherited much later by the Greeks and Romans, on the other hand. In Ancient China, the law (法; *fǎ*) is a standard for decisions of officials that does not involve any discussion between plaintiffs and defendants. There is only one party in courts, represented by the officials who also play the role of judges, and if they have any doubts, then they ask their higher authorities about the judgment. The lawsuit in Mesopotamia at least from the 3rd dynasty of Ur proceeded as *di-tila*—a critical discussion between a plaintiff and a defendant over a claim with applying only arguments based on logical inference rules, while the role of judges was to test arguments and make a final decision grounded only on valid documents and verified facts, grouped logically (Culbertson, 2009, 2015). The lawsuit in Ancient China took place as a criminal investigation and clarification of all details of the case. It did not assume any critical discussion between a plaintiff and a defendant over a claim. Meanwhile, in this criminal investigation the following three logical procedures, introduced by Mozi, were adopted by the officials: detalization (私; *sī*), generalization (達; *dá*), and finding similar cases (類; *lèi*).

Let us exemplify the detalization (私; *sī*) by considering the case of hair and grass in the Lord's food solved by Shǐ Yóu from the *Book of Submitted Doubtful Cases* (Korolkov, 2013: 84–86; Barbieri-Low and Yates, 2015: 1362–1367). This criminal case was triggered by finding hairs, three *cùn* in length (approx. 6.93 cm), on the roast, served by Shuō for the Lord of Wèi, and some wild grass, half a *cùn* in length (approx. 1.16 cm), in the food, served by Mèi for the Lady of Wèi. As a result, "the Lord and the Lady both became angry and made an official accusation" (Barbieri-Low and Yates, 2015: 1363). According to the Wèi Law (衛 法; *wèi fǎ*), "not being careful in preparing food for the Lord or the Lady is a crime [punished with] death" (為君∠、夫人治食不謹, 罪死; *wèi jūn ∠, fūrén zhì shí bù jǐn, zuì sǐ*; Barbieri-Low and Yates, 2015: 1363). Hence, in accordance with this law and the fact that the hairs and grass were found in the food for the Lord and his Lady, the senior chef Shuō and the food-serving slave-maid Mèi must both be sentenced to death by *modus ponens*. Nevertheless, Shǐ Yóu's judgment was different: "Shuō is not guilty. Mèi matches receiving a bestowal of clothing" (說毋罪, 媚當賜衣; *shuō wú zuì, mèi dāng yī*; Barbieri-Low and Yates, 2015: 1363). In the additional judicial inquiry requested by the Lord himself there was a detalization of the case made again by Shǐ Yóu before the Lord. Consequently, it was clarified that Shuō cannot be responsible for the hairs on the roast. First, his knife that used to cut meat is "newly forged and extremely sharp" (Barbieri-Low and Yates, 2015: 1363). He cuts fatty beef into the cubes which are "not larger than one *cùn* (approx. 2.31 cm)"

(Barbieri-Low and Yates, 2015: 1363) so that the three-*cùn*-long hairs had to be cut, too. But they were not cut. Furthermore, "the meat is well scorched using [the high quality] mulberry charcoal and the sturdy tongs" (Barbieri-Low and Yates, 2015: 1365). Therefore, the hairs also had to be scorched, but they were not. Second, after the investigative experiment it was found that these hairs are like the hairs of the Lord which take flight after fanning him. In this way, they can get onto the roast. Hence, the chef Shuō is not guilty, as Shǐ Yóu concluded.

Also, Shǐ Yóu examined the dining hall of the Lady and found, on the one hand, that "the plaster on the ceiling is very carefully applied, and the spreading out of the canopies and curtains is well arranged, and in the dining hall there is no wild grass" (Barbieri-Low and Yates, 2015: 1365). On the other hand, he inspected Mèi's rush sleeping-mat, which is worn out, while "the warp is broken, and the rush is fragmented" (Barbieri-Low and Yates, 2015: 1365). Her clothes are worn out, too. Then Shǐ Yóu found six half-*cùn*-long pieces of grass in the worn-out mat. They can be attached to the floss of Mèi's clothes, and they are identical with the wild grass in the cooked grain. Thus, Mèi is not guilty, as well, in Shǐ Yóu's opinion.

After this investigation, the Lord declared to release Shuō and to bestow new clothing upon Mèi, in accordance with Shǐ Yóu's decision.

It is an example of detalization (私; *sī*) as a logical procedure in making legal decisions to check whether the details of the case are consistent with the legal norm. According to Shǐ Yóu's investigation, the details were not consistent with the norm "not being careful in preparing food for the Lord or the Lady is a crime [punished with] death" (Barbieri-Low and Yates, 2015: 1363), therefore the legal decision based on an appropriate norm cannot be made. Then both Shuō and Mèi are free.

However, there are legal acts from the *Book of Submitted Doubtful Cases* with the details which are contradictory to the norm, but the legal decision was made on the norm and did not change after reconsideration by the higher authorities. For example, in the judicial case concerning the absconding indigenous conscript that was submitted to higher authorities for a final decision and dated September 19, 196 B.C. (Korolkov, 2013: 41–42; Barbieri-Low and Yates, 2015: 1174–1177), the adult male Wúyōu was declared as a garrison conscript of the chief commandant, although he is a Barbarian *mányí* (蠻夷) and he paid fifty-six cash annually to match his government service, including a special poll tax in order to avoid being drafted into the garrison—this special tax is according to the translation of Korolkov (2013: 41–42). Hence, based on the statute, he cannot be drafted to the army, but he was. Then "he left and absconded" (Barbieri-Low and Yates, 2015: 1175). The officials decided that even though it was not fitting that he may be made a conscript, he was mobilized and became a garrison conscript. Furthermore, there is no prohibition to order service in garrisons, even in the case of a special poll tax paid for draft evasion. Hence, even after reconsidering the case, Wúyōu was sentenced to death as deserter: "He matches being cut in two at the waist" (當腰斬; *dāng yāo zhǎn*; Barbieri-Low and Yates, 2015: 1177).

In this legal act, the administrative system is more important than each singular norm (statute, law), and in all legal collisions, preference is given to specific orders of a particular official, if they do not contradict the orders of higher authorities. Thus, strict logic in inferences may be insufficient and insignificant for judicial decisions. According to the law, Wúyōu cannot be drafted to the army, but he was drafted. And he is guilty of desertion, because he left, being mobilized. The law (法; *fǎ*) is a submission to superiors, not a critical discussion about the conformity of facts to existing legal norms, as it was in Mesopotamia. It also shows that judicial decisions in Ancient China were grounded not on deductive logic as in Ur III, but on analogical syllogisms. Therefore, legal collisions were always possible and to avoid them a class of similar cases (類; *lèi*) was further extended. For instance, leaving the garrison after the mobilization is sufficient to be deserter, even if the mobilization was wrong.

The legal act concerning the sword fight between a runaway 'slave' and a thief catcher is the next example of extending an appropriate class of similar cases (類; *lèi*) after a legal collision. This act, taken from the *Book of Submitted Doubtful Cases* (Korolkov, 2013: 50–52; Barbieri-Low and Yates, 2015: 1218–1221), is dated August 20, 197 B.C. It is stated that the "member of the rank and file" Jūn made the denunciation to Constable Chí that Jūn's adult male slave Wǔ absconded, but he was seen west of the police station. Based on this denunciation, Chí sent the thief catcher Shì to pursue and arrest Wǔ. During the arrest, a sword fight broke out between Shì and Wǔ with injuring each other. Nevertheless, as it turned out, the denunciation was wrong, since Wǔ is no longer Jūn's slave. Therefore, the arrest of Wǔ was illegal. But Wǔ did not obey the submission to his superiors, injuring the thief catcher Shì. The class of similar cases (類; *lèi*) for this crime is "maliciously injuring another person". The final decision was as follows: "Wǔ matches undergoing tattooing and being made a wall-builder. Remove Shì's [liability for a crime, although his denunciation was false]" (黥武為城旦∠。除視; *qíng wǔ wèi chéng dàn ∠. chú shì*; Barbieri-Low and Yates, 2015: 1221). But in the *Statute of Denunciations* (告律; *Gàolǜ*), which was found along with the *Book of Submitted Doubtful Cases*, it is stated about a direct liability for a false denunciation: "Making an intentionally false denunciation of another person for a crime [that matches] the death [penalty]: tattoo [the denouncer] and make [him or her] a wall-builder or grain-pounder; for each other [case], reverse the crime" (誣告人以死**罪**: 黥為城旦舂; 它各反其**罪**; *wúgào rén yǐ sǐzuì: qíng wèi chéng dàn chōng; tā gè fǎn qí zuì*; Barbieri-Low and Yates, 2015: 547); "for a crime [that matches] undergoing tattooing and being made a wall-builder or grain-pounder: leave [the denouncer or criminal] intact and make [him or her] a wall-builder or grain-pounder" ((黥為城旦舂)**罪**: 完=為=城=旦=舂=; (*qíng wèi chéng dàn chōng*) *zuì*: *wán =wèi =chéng =dàn =chōng* =; Barbieri-Low and Yates, 2015: 547). But this statute was not obeyed in the court decision.

As we see on the examples from the *Book of Submitted Doubtful Cases*, legal norms as fundamentals for the court decisions are not used as major premises for deductive inference rules, such as *modus ponens*, *modus tollens*, introducing conjunction, deleting conjunction, and so on. These rules as well as some of the foundations of propositional logic in the modern sense are unknown in Ancient China. Therefore, some legal collisions/conflicts were always possible, as mentioned above. Legal cases are assumed to be subject to norms through generalization (達; *dá*). But a generalization may establish a new rule that goes beyond existing legal norms. Let us illustrate this by the act from the *Book of Submitted Doubtful Cases* concerning an assistant scribe robbing grain (Korolkov, 2013: 87–88; Barbieri-Low and Yates, 2015: 1370–1371).

In accordance with the norm of Lǔ (魯灋; *lǔ fǎ*): "Robbing [items valued at] 1 cash up to 20 [cash]: fine one *liǎng* (approx. 15.5 g) of gold. More than 20 and up to 100 [cash]: fine two *liǎng* (approx. 31 g) of gold. More than 100 and up to 200 [cash]: make [the criminal] a white convict. More than 200 and up to 1,000 [cash]: leave [the criminal] intact and make [him or her] a singer" (盜一錢到廿: 罰金一兩。過廿到 百: 罰金二兩。過百到二百: 為日（白）2徒。過二百到千: 完為倡; *dào yī qián dào niàn: fájīn yī liǎng. guò niàn dào bǎi: fájīn èr liǎng. guò bǎi dào èrbǎi: wèi rì (bái) tú. guò èrbǎi dào qiān: wán wèi chàng*; Barbieri-Low and Yates, 2015: 1371). Dīng who serves as an assistant scribe robbed one *dǒu* (approx. 2 l) of untreated grain, valued at 3 cash. Then by deleting conjunction and *modus ponens* he should be sentenced to paying one *liǎng* (approx. 15.5 g) of gold. However, he was maximally sentenced to the punishment of being made a singer as if he robbed items valued at more than 200 and up to 1000 cash. Why?

The matter is that the generalization (達; *dá*) for finding an appropriate norm to define the crime may be done based on analogical syllogisms. Dīng is "wearing a kingfisher cap"—it means that "he could manage the rites and wear Confucian clothing". This is a sign that he is a gentleman and obeys the Confucian principles. Stealing the grain contradicts this clothing. Another norm of Lǔ is as follows: "For all those who deceive their superiors about government affairs: sentence them according to the categorical principle on white convicts. For one who is guilty of two crimes [matching] the punishment of being made a white convict: add one degree to his or her crime" (諸以縣官事訑其上者: 以日（白）徒灋（法）論之。有日（白）徒罪二者, 駕（加）其罪一等; *zhū yǐ xiàn guān shì yí qí shàng zhě: yǐ rì* (*bái*) *tú fǎ* (*fǎ*) *lùn zhī. yǒu rì* (*bái*) *tú zuì èr zhě, jià* (*jiā*) *qí zuì yī děng*; Barbieri-Low and Yates, 2015: 1371). Dīng is guilty of the following two crimes: he "robbed the principles of the gentleman. And, furthermore, he has robbed the learning of a gentleman in order to advance his merit" (今丁有宵（小）人心, 盜君子節。有（又）盜君子學以上功; *jīn dīng yǒuxiāo* (*xiǎo*) *rénxīn, dào jūnzǐ jié. yǒu* (*yòu*) *dào jūnzǐ xué yǐshàng gōng*; Barbieri-Low and Yates, 2015: 1371). Hence, the two details together that Dīng is "wearing a kingfisher cap" and he is guilty of robbing the grain should be generalized up to the norm of "deceiving his superiors about government affairs", committed by him twice. With the help of analogical syllogisms, the generalization established here the new rule regarding robbers in Confucian clothing, according to which they should be sentenced to the maximum punishment.

Summing up, we can say that all the acts from the *Book of Submitted Doubtful Cases* are a good illustration of what kinds of reasoning were used in real court judgments. We do not find out some inference rules of propositional logic, such as *modus ponens*, *modus tollens*, deleting conjunction, etc., but we do see some modes of reasoning, including analogical syllogisms, proposed earlier by Mozi.

The *Book of Submitted Doubtful Cases* was found in Zhangjiashan tomb no. 247 along with the *Laws and Decrees of the Second Year* (二年律令; *Èrnián lǜlìng*) that includes 27 laws and one collection of orders. All these documents from tomb no. 247 represent early legal texts belonging to the Han Empire (漢朝; *hàncháo*), founded by the second imperial dynasty of China (202 B.C.–220 A.D.). But the practice of submitting doubtful cases to higher authorities have existed since the time of the Qin Empire (秦朝; *qíncháo*), founded by the very first imperial dynasty (221–207 B.C.). The matter is that all the legal institutions of the Qin dynasty, such as written statutory laws (codes), "were adopted by subsequent dynasties as the primary means for maintaining administrative and social control" (Caldwell, 2018: 1).

Many legal documents of the Qin dynasty including different statutes were excavated in Shuihudi tomb no. 11. They contain the earliest law codes found in China, among which there are the following texts: the *Eighteen Varieties of Qin Statutes* (秦律十八種; *Qínlǜ shíbā zhǒng*), the *Statute on Checking* (效律; *Xiàolǜ*), the *Miscellaneous Transcriptions of Qin Statutes* (秦律雜抄; *Qínlǜ zá chāo*), the *Questions and Answers on Laws and Statutes* (法律答問; *Fǎlǜ dá wè*) (see Caldwell, 2018: 5). For instance, the *Statute on Checking* (or the *Xiàolǜ*) consists of 30 individual articles "covering administrative auditing of county-level offices and the government accounts and provisions these offices oversee" (Caldwell, 2018: 6). Most of these articles are composed as conditional statements, that is, in the way of the earliest Mesopotamian statute—the *Code of Ur-Nammu* dated to ca. 2100–2050 B.C. This means that in the antecedent (protasis) some proscribed actions are listed, while in the consequent (apodosis) they are followed by appropriate penal or administrative sanctions. This logical structure of articles in the form of implications allows judges to apply the inference rule called *modus ponens* in making decisions, as it was done in Mesopotamia: "If an act *A* is committed, then a sanction *B* is realized. The act *A* was committed. As a consequence, the sanction *B* is fulfilled." However, as we see on the examples from the *Book of Submitted Doubtful Cases*, instead of the rule of *modus ponens*, there was applied another logical or proto-logical technique: either a generalization up to the norm (達; *dá*), if there are no doubts of higher authorities, or finding another class of similar situations (類; *lèi*), if there are any doubts.

Statutes or legal codes were published on special bamboo slips (扁書; *biǎnshū*) in the Qin and Han Empires and then widely distributed from the center to the distant periphery to announce appropriate legal norms in the state. For example, we observe the mention of this fact in the texts dated to the Western Han dynasty (206 B.C.–25 A.D.):

> 明白大扁書鄉亭市里門外謁舍顯見處**令**百姓盡知之
>
> *míngbái dà biǎnshū xiāng tíng shì lǐ ménwài yè shě xiǎnjiàn chù lìng bǎixìng jǐn zhīzhī.*
>
> Prominently display [the edict/law] on a large *biǎnshū* in conspicuous locations at townships, stations, markets, outer gates, and hostels causing the hundred names to fully comprehend it (57 2000ES7S: 4A; Caldwell, 2018: 35).

> 明白扁書亭隧顯見處**令**吏卒盡知之
>
> *míngbái biǎnshū tíng suì xiǎnjiàn chù lìng lì zú jǐn zhīzhī.*
>
> Prominently display [the edict/law] on *biǎnshū* at stations and outposts in conspicuous locations, causing officials and troops to fully comprehend it (2000ES9SF4: 3; Caldwell, 2018: 35).

As we see, the legal norms of the Qin and Han dynasties were distributed among "officials and troops", and not among all citizens, as was the case in Mesopotamia or Ancient Greece, although the norms were preserved at public places to the same extent.

The idea of unifying all the laws (punishments) and standards (norms) of all provinces in the state first belonged to the Qin dynasty. The principle of the unification was called *tiānxià* (天下)—"[all]-under-Heaven". It was well formulated in the *Twenty-sixth Year Edict* (廿六年詔; *Niànliù nián zhào*; 二十六年詔; *Èrshíliù nián zhào*):

> 廿六年皇帝盡併天下諸侯黔首大安立號為皇帝乃詔丞相狀綰灋度量則不壹歉疑者皆明壹之
>
> *niàn liù nián huángdì jǐn bìng tiānxià zhūhóu qiánshǒu dà ānlì hào wèi huángdì nǎi zhào chéngxiàng zhuàng wǎn fǎ dùliàng zé bù yī qiàn yí zhě jiē míng yī zhī.*
>
> In the twenty-sixth year (221 B.C.), the August Thearch completed the annexation of all-under-Heaven. The various lords and commoners were greatly pacified. [He] inaugurated the title of August Thearch. Thereupon, [he] instructed the grand councilors Zhuàng and Wǎn: "Those laws and measurements that are not standard or are ambiguous should all be distinguished and standardized" (Caldwell, 2018: 62, Wèizǔ and Gǔfù, 1997: 4–19).

In the Qin edicts and statutes, the law (法; *fǎ*) was treated rather as a unified and standardized system of punishment (刑; *xíng*). It is worth noting that some books of punishments (刑書; *xíng shū*) were mentioned before the Qin dynasty, e.g., in sources of the 6th century B.C. (Caldwell, 2018: 30–33). But since the time of the Qin Empire the punishment has already been understood as a unified system for the state, and the legal norms have been considered instructions for the officials on how to govern and punish.

The legal reforms of the Qin dynasty and their role in the history of China may be compared only with the legal reforms of the Ur III dynasty and their role in the history of Mesopotamia. Both dynasties have unified and standardized legal procedures and norms. But the established systems have become very different.

After the first reforms in Ur III, the court decision (*di-tila*) began to be made not only by the governor, but also by representatives of the city elite, who were additional judges from zero to four (Culbertson, 2009, 2015). Then, at the time of Shu-Sin, the fourth king of Ur III, in addition to the governor, there were between one and six representatives of the city elite, also between one and two case reporters ("case witnesses") as well as from zero to nine high-ranking jurors ("men who stood at the case"), see Culbertson (2009, 2015). As we see, after the reforms in Ur III, the court decision was made not only by the governor, but also by the representatives of the whole city community. Meanwhile, the procedure of the court was organized as a critical discussion between a proponent (plaintiff) and an opponent (defendant). In Ur III, this *di-tila* as a trial judgment was understood as a final decision.

After the legal reforms of the Qin dynasty, accepted also by the Han dynasty, the court decision was made only by the officials of the following ranks in turn (Korolkov, 2013: 31–32): (a) officers of the government located in the counties (縣官; *xiànguān*); (b) district chiefs, i.e., senior officials of the relevant districts (縣令; *xiànlìng*, or 縣長; *xiànzhǎng*); (c) county governors (郡守; *jùnshǒu*). In case of doubts, legal acts were submitted to higher authorities. Hence, the judiciary was not independent of the administration—the officials carried out the trials. Therefore, the court was not a critical discussion between a plaintiff and a defendant, since the court was a decision solely about a punishment in accordance with instructions, such as books on punishments and administration (扁書; *biǎnshū*). So, the legal norms were instructions for the officials on how to manage and punish. This system did not need propositional logic for organizing a critical discussion. It was enough to have several analogical syllogisms to justify the following three logical processes in judgments of the officials: detalization (私; *sī*), generalization (達; *dá*), and finding similar cases (類; *lèi*).

Mozi put forward the principle of logical consistency, allowing him to introduce the situation semantics which became very useful in the legal discourse. However, this principle was criticized by the founders of the School of Names (名家; *míngjiā*, literally "experts on names" or sophists), e.g., by Hui Shi (惠施; *Huìshī*; 4th century B.C.)—they formulated different paradoxes such as "Heaven and earth are alongside each other" or "Eggs have feathers" (Fraser, 2020). Some logical paradoxes were proposed by Gongsun Long (公孫龍; *Gōngsūn Lóng*; ca. 320–250 B.C.)—one of the most famous 'dialecticians' or 'sophists' (辯者; *biàn zhě*) of the early School of Names. It is difficult to say whether he was a real character. His name was used mostly in the context of rhetorical tropes "with a strong negative moral connotation" (Indraccolo, 2017: 2). In his text *Gōngsūn Lóngzi* (公孫龍子), he tries to show that any reasoning (辯; *biàn*) is mainly inconsistent. This treatise is quite short and consists of six chapters, among which the following two are of logical importance (Chén, 1937): the 1st chapter, entitled the *Dialogue on the White Horse* (白馬論; *Báimǎ lùn*), and the 2nd chapter, entitled the *Dialogue on the Pointed Things* (指物論; *Zhǐwù lùn*).

According to Angus C. Graham, the *Dialogue on the White Horse* existed from at least as early as the 1st century A.D. (Graham, 1986). The main statement of this chapter is as follows: "A white horse is not a horse" (白馬非馬; *báimǎ fēi mǎ*). This claim is a direct criticism against Mozi's thesis "A white horse is a horse" (白馬 馬也; *báimǎ, mǎ yě*; *Mohist Canon* 45: 4). But Gongsun Long's criticism seems to be an evident logical fallacy, since the genus ('horse') always follows from the corresponding species ("white horse").

In the secondary literature about the *Dialogue on the White Horse*, there were many attempts to understand this illogical phrase that a white horse is not a horse (*báimǎ fēi mǎ*). For instance, Fung (1948: 88) proposed to treat 'horse' (*mǎ*) as "the universal horseness" and "white horse" (*báimǎ*) as "the universal white-horseness", then the horseness and the white-horseness are two different notions, although the word 'horse' may also be used to refer to individual horses. A similar view belongs to Chad Hansen (1983: 30–49) who argued that *mǎ* in Chinese is a mass noun referring to stuff, whereas count nouns, such as 'horse' in English, are to designate each of the individuals with an appropriate characteristic. Consequently, the phrase means that the stuff of the white horses is not the stuff of the horses. Nevertheless, *mǎ* in Chinese is well combined with numericals, see also (Byeong-uk, 2014), therefore it is a count noun, e.g.: "Two horses are white" (二馬而或白也; *èr mǎ ér huò bái yě*; *Mohist Canon* 45: 8).

To prove the thesis that a white horse is not a horse, Gongsun Long turned to the following three arguments:

1) "Horses are named for their shape; whites are named for their color. The named for color is not the named for shape" (馬者, 所以命形也; 白者, 所以命色也。命色者非命形也; *mǎ zhě, suǒyǐ mìng xíng yě; bái zhě, suǒyǐ mìng sè yě. mìng sè zhě fēimìng xíng yě*; *Dialogue on the White Horse* 4). Hence, we have two different processes of generalization: the first ends with the shape and the second with the color. These two generalizations are different. Then the two notions are different, too.
2) "If you request a horse, then a brown or a black horse may come; if you request a white horse, then a brown or a black horse will not come" (求馬, 黃、黑馬皆可致; 求白馬, 黃、黑馬不可致; *qiú mǎ, huáng, hēimǎ jiē kě zhì; qiú báimǎ, huáng, hēimǎ bùkě zhì*; *Dialogue on the White Horse* 6). "Requesting any horse" and "requesting only white horses" are two contrary propositions to the same manner and to the same extent as "requesting all horses" and "requesting some horses" are two contrary propositions in Mozi's hermeneutics. Both propositions cannot be true together, but they can be false together. Thus, 'horses' in the meaning of all horses and "white horses" in the meaning of some horses are contrary, therefore they are different. However, this reasoning is a logical fallacy from the point of view of classical predicate logic. It supposes the non-Aristotelian square of opposition.
3) "A horse [as an abstract notion] is not [given], but the white is a horse; and a white [as an abstract notion] is not [given], but a horse is white. Combining the horse with the white is named a white horse. This is the name of not being together as a combined thing, but it is impossible" (馬未與白為馬, 白未與馬為白。合馬與白, 復名白馬。是相與以不相與為名, 未可; *mǎ wèi yǔ bái wéi mǎ, bái wèi yǔ mǎ wèi bái. hémǎ yǔ bái, fù míng báimǎ. shì xiāng yǔ yǐ bù xiāng yǔ wéi míng, wèi kě*; *Dialogue on the White Horse* 9). In this reasoning, Gongsun Long assumes that there exist only individuals (denotations). Any name, such as "white horses", is an association of different individuals together, although these combinations are a speculation and they do not exist as real objects.

As we can see, Gongsun Long appeals to Mozi's logical ideas, using his generalization (達; *dá*) as well as detalization (私; *sī*), to criticize his thesis that a white horse is a horse. But the main assumption of Gongsun Long is specific. He argues that there are only individuals (物; *wù*) and abstract notions do not exist beyond these individuals. This assumption is well expressed in the next chapter of his treatise—*Dialogue on the Pointed Things*, see also Cheng (1997). In Gongsun Long's opinion, the white horses taken together are not the same things as the horses taken together. Any abstract notion is nothing but 'pointing' (物指; *wùzhǐ*) as an actualization of appropriate denotations. Things under our consideration are "pointed things" (指物; *zhǐwù*).

Hence, Gongsun Long accepted some modes of reasoning proposed by Mozi, including generalization (達; *dá*) and detalization (私; *sī*), and he shared Mozi's understanding of existential and universal quantifiers. But he did not accept the existence of the class of similar items (*lèi*) postulated by Mozi. If we assume their existence, then we can define "white horses" as a particular class of situations and 'horses' as a general class of situations such that 'horses' follow from "white horses". In this case, the statement that a white horse is not a horse is obviously false.

It is known that logic in China is well traced from the 19th century A.D. through a European influence (Kurtz, 2011). Nevertheless, some Buddhist logical treatises were translated into Chinese much earlier and were adopted within the Buddhist discipline of "knowledge of reasons" (因明; *yīnmíng*). But this logic did not arouse the interest of Chinese scholars outside of Buddhist monasteries. As a result, the Buddhist logic existed only for Buddhist religious matters.

Mozi founded the situation semantics, in which there may exist an inclusion relation between different classes of situations, e.g., the class of white horses is included in the class of horses. This situation semantics played the role of logic, in which instead of deductive inference rules there were analogical syllogisms, and all propositional relations were tested semantically, but not logically. Due to this semantics, a developed legal discourse was expanded, and an efficient legal system was organized, starting with the Qin and Han dynasties.

Trial judgments in Ancient China did not represent a critical discussion between plaintiffs and defendants, as was the case in Mesopotamia and then in the Achaemenid Empire and Ancient Greece. The point is that any court issued an appropriate administrative decision within the framework of a unified system of punishments, so that the only actors in these criminal or administrative investigations were officials. In these proceedings, there was no need to invoke deductive inference rules, since it was sufficient for the officials to apply the following three logical procedures: detalization (私; *sī*), generalization (達; *dá*), and finding similar cases (類; *lèi*), which were first formulated by Mozi.

Thus, logic is not an innate ability to be found in any developed discourse. We can have some very effective mechanisms of societal regulations without deductive logic and Ancient China is a good proof of this.

1.2.4 Sign-inferences in the Greek and Buddhist Logic

The emergence in Ancient India of its own logical tradition, distinct from the Greek, has been recognized by European scholars since the time of early Indologists such as Max Müller (1823–1900):

> The sciences of Logic and of Grammar were, as far as history allows us to judge, invented or originally conceived by two nations only, by Hindus and Greeks. All other nations, if they ever cultivated these sciences, received the first impulse from without. The Romans from the Greeks, the Germans

> from the Romans, the Arabs from the Greeks, the Jews from the Arabs. That the two most highly gifted nations of the world, the Hindus, and the Greeks, should both have been led, each in its own way, to a study of the laws of thought and the laws of language, seems in itself perfectly natural (Müller, 2001: 59).

This idea that the Indian logic is unique and has nothing in common with the Greek logic was widely shared by other Indologists, e.g., see Schayer (2001). Meanwhile, the Indian logic in these assessments is compared only with Aristotle, although in Ancient Greece there were also the Stoic and Epicurean systems, but they are ignored in this comparison. Initially, the origin of logic in Ancient India dates to the 7th century B.C., i.e., to the possible time of composition of the *Nyāyasūtra*:

> Indian Logic … [is divided] into three periods: Ancient (650 B.C.–100 A.D.), Medieval (up to 1200 A.D.) and Modern (from 900 A.D.). The standard texts for each of these periods were *Nyāyasūtra* by Akṣapāda, *Pramāṇa-samuccaya* by Dignāga, and *Tattvachintāmaṇi* by Gaṅgeśa Upādhyāya respectively (Vidyabhusana, 1921: xiii).

In this manner, it was stated that the Indian logic appeared much earlier than Aristotle lived. Nevertheless, this dating of the *Nyāyasūtra* and the corresponding periodization of Indian logic is now unacceptable, as there are many arguments in favor of the fact that the *Nyāyasūtra* cannot be dated earlier than the 2nd century A.D. (Schumann, 2019a). Therefore, the Indian logic in the narrow sense began, most probably, only then.

The more adequate periodization of Indian logic is as follows: (1) the pre-Dignāga period from the earliest times "which cannot be definitely fixed" to the 5th century A.D. with the following subperiods in their chronological order: (i) the *Vaiśeṣikasūtra*, (ii) the *Nyāyasūtra*, (iii) Vātsyāyana's commentary on the *Nyāyasūtra*, and (iv) Praśastapāda's commentary on the *Vaiśeṣikasūtra* (Shastri, 1976: 11); (2) the period of conflict with the Yogācāra school of Buddhism (mainly, with the disciples of Dignāga) from the 5th to the 11th century A.D., resulted in the development of the Nyāya-Vaiśeṣika realism (Shastri, 1976: 12); (3) the period of decay or the post-Buddhist period, the emergence of the Navya-nyāya school (Shastri, 1976: 12).

In general, it is difficult to date the starting point of Indian logic, but some roots of logical thinking have been detected in Ancient India since ca. 400 B.C., when the *śramaṇa* movement became widespread as a reaction to the new wave of urbanization in the Ganga Valley at that time (this urbanization gave rise to the two earliest scripts for the Indo-Aryan language, Brāhmī and Kharoṣṭhī, as well as to the earliest states of the Indo-Aryans in the narrow sense, although chiefdoms existed there before). The ascetic movement of the *śramaṇa*s consisted of various philosophical and religious sects, among which Jainism and, especially, Buddhism was formed over time as new influential social networks that widely covered the post-Vedic society of India, mainly in cities. Philosophical disputes about *dharma* (the natural order of things) and other religious issues between different sects became very popular.

In time, there was an understanding what critical discussion (*kathā*) is, namely, that there are two opposing parties: the proponent (*vādin*) and the opponent (*prativādin*), who may come to a consensus. In debates (*jalpa* of the Naiyāyikas and *vāda* of the Yogācārins), both parties try to justify their own positions against each other and have different points of their defeat (*nigrahasthāna*), see Gokhale (1993: xv–xvi).

Hence, in both Ancient India and Ancient Greece, logic originated as the main tool for composing arguments in critical discussion. And it is significant that we observe the phenomenon of critical discussion and a philosophical reflection on this issue in these two regions almost simultaneously. In both cultures, the natural order of things (λόγος for Heraclitus and *dharma* for the Buddhists) became an essential subject of philosophical reflection and philosophical debates.

According to Aristotle, reasoning in argumentation is called λόγος that can be scientific, and it is mainly used in deliberations and courts as two forms of critical discussion, that is, often outside of philosophical disputes:

> In both deliberation and in court the opening speaker should state his own premises first, then should meet those of his opponent by disproving and tearing them to pieces before he can make them. But if the opposition has many good points to make, put the refutations first, as Callistratus did in the Messenian assembly; for first removing the objections they were going to voice, he then spoke his own case. But if you speak second, you should reply first to the opposing speech, refuting and offering opposed syllogisms, especially if what was said seems to have met with approval (*Rhetorica* III: 17, 1418b; Kennedy, 2007: 245).

Meanwhile, the *logos* gives a scientific knowledge only by means of syllogisms, studied in the Aristotelian *Prior Analytics* (Ἀναλυτικὰ Πρότερα/*Analytica Priora*) and his *Posterior Analytics* (Ἀναλυτικὰ Ὕστερα/*Analytica Posteriora*). Some other Greek philosophers, such as Philodemus of Gadara, who was an Epicurean, argued that the *logos* in the critical discussion is more the result of learning and exercise, so there is neither science nor art for reasoning:

> Another point: even if it is possible to win with exercise and not with art, it must follow that some people [apply themselves] with more care and experience, while others with less, and because of this some people turn out superior, while others (turn out) inferior. And those who leave the subject matter of sophistic to a child, and have common sense, train people skilled in legal matters and [produce] what they have observed in assemblies and courts as much as they are able (Philodemus, *Rhetorica*, Col. XIV Longo, 71; Chandler, 2006: 28).

Philodemus compared the Stoic logical handbooks with the omen books of the Chaldaeans (Babylonians), suggesting that they represent the same logical technique based on implications, *modus ponens*, and *modus tollens* that is equally fabulous:

> And by publishing technical manuals they are able to publish dreams for the purpose of deceiving those who read them, as the Chaldaeans and diviners do, and perhaps they too are in error themselves. And it follows, if anything, that sophistic is an art, but that it is also an art of politics, [not even] they themselves urgue. And no one, surely, intended to derive those who speak with ability and intelligence in assemblies and courts as experts [from that source].
>
> And yet [it was proposed] with regard to rhetoric [whether] certain people [seem] to conduct themselves technically [in assemblies] and courts. But those who are trained in the schools do not share even one of the virtues of those men (Philodemus, *Rhetorica*, Col. XVIII, Longo, 79; Chandler, 2006: 30).

As we see, logical practice was presented in Ancient Greece within the framework of critical discussion, most often in assemblies and courts. Therefore, there were many handbooks in Greek devoted to rhetoric and argumentation, based on logical inferences. Among these texts, there are the following books, written by Aristotle: The *Topics* (Τοπικά/*Topica*), the *On Sophistical Refutations* (Σοφιστικοὶ Ἔλεγχοι/*De Sophisticis Elenchis*), the *Rhetoric* (Ῥητορική/*Rhetorica*), as well as books of some other authors, such as the *Topica* of Marcus Tullius Cicero (106–43 B.C.).

In Ancient India, these forms of critical discussion can also be reconstructed based on some excavated legal records in Gāndhārī (Burrow, 1940) and Bactrian (Sims-Williams, 2000, 2007), including legal acts, contracts, and business correspondence. In Bactrian, these texts are compiled according to the standards and structure of Greek legal documents. However, in logical treatises devoted to the critical discussion, such as in the *Vādanyāya* of Dharmakīrti (7th century A.D.), who was a famous representative of the Yogācāra school, we find some examples taken from philosophical debates and not from legal proceedings. This means that the philosophical reflection of critical discussion in Ancient India was rather focused on disputes between various religious sects.

Meanwhile, there was the epistemic context of logical studies in Ancient Greece and Ancient India that was very similar. In both regions, inferences from signs (*liṅga*, σημεῖον) to hidden phenomena were analyzed, and both investigations were equally involved into natural studies, including medicine. So, in both Ancient Greece and in Ancient India, sign-inferences were applied in medical diagnoses and prognoses. Hippocrates of Kos (Ἱπποκράτης ὁ Κῷος; ca. 460–ca. 370 B.C.) understood signs as medical symptoms in his *Corpus Hippocraticum* (Adams, 1868) and in this way his diagnoses and prognoses were drawn as logical conclusions from signs. For example, he examined "deadly symptoms/signs" (σημεῖα θανατώδεα) in his *De morbis popularibus* (Περὶ Ἐπιδημιῶν) VI: 4, 22, and 'tokens' (τεκμήρια) in his *De diaeta* IV (Περὶ Διαίτης Δ'): 86, as some premises for entailing. To the same extent, the *Caraka-saṃhitā* (Sharma, 2014), one of the earliest medical treatises in India, was based on signs and *pramāṇa* (the logical study of true sources of knowledge). The authorship of this text was traditionally attributed to Caraka (from the 1st to the 2nd century A.D.) who, according to the *Saṃyukta-ratna-piṭaka-sūtra* (Levi, 1896), was the medical attendant of Kaniṣka the Great and, hence, lived in Gandhāra (present-day Pakistan) during the time of this Kuṣāṇa ruler.

The purpose of this subsection is to show that the doctrine of sign-inferences taught in the Yogācāra school is consonant partly with the doctrine of the Stoics, partly with the doctrine of the Epicureans, and partly with the doctrine of Aristotle. Therefore, the Yogācāra system can be understood as a specific synthetic generalization of Hellenistic logic of different schools. In any case, the logical discourse of the Yogācārins bears a strong structural and functional resemblance to the logical discourses of the Stoics, the Epicureans, and Aristotle simultaneously. First, I discuss some of the fundamentals of sign-inferences in the Stoic and Epicurean schools. Then I trace the structural and functional similarities between the Hellenistic logic of different schools and the Yogācāra logic.

Logical inference, as it is understood in the Indian logic, is based on a sign (*liṅga*) as something visible like 'smoke' p to conclude about something hidden like 'fire' q, that is, we have the following implication: $p \Rightarrow q$, which holds true, while the Aristotelian syllogistic is based on the middle term (species) M, connecting a visible phenomenon S to its genus P, that is, we have the transition of two implications: "If $M \Rightarrow P$ and $S \Rightarrow M$, then $S \Rightarrow P$" (*modus Barbara*). Hence, the Indian and Aristotelian systems seem to be different. The middle term M of Aristotle as a basis for the inference is taken, being a general characteristic for the minor term (visible phenomenon) S, to conclude about possessing the more general characteristics P. Thus, in the Indian inference we go from the visible sign p to other phenomena q, while as in the Aristotelian syllogism we go from species (middle terms) M to genera (major terms) P.

However, in Ancient Greece, a theory of 'sign-inference' (σημείωσις) was also developed, considering a conclusion from a sign p which may be perceived to its meaning q which may be hidden: $p \Rightarrow q$ (Burnyeat, 1982: 218–224). The Stoics presented this conclusion as *modus ponens*: $p \Rightarrow q$; p is observed; then q should be also. To draw the 'sign-inference' we first observe a 'sign' (σημεῖον, *signum*) represented as 'token' (τεκμήριον) for other phenomena that allows us to infer. But it is worth noting that Aristotle (Ἀριστοτέλης; 384–322 B.C.) was also one of the first Greek philosophers who proposed this sign-inference along with his syllogistic.

According to Aristotle, each syllogism is a *logos* (λόγος), "in which, certain things having been laid down, something other than these things necessarily results through them" (*Topica* I: 1, 100A25–7; Smith, 1997: 273). Hence, a syllogism is an automatic transgression from one statement, called a premise, to another, called a conclusion. Meanwhile, the premise can contain a sign (σημεῖον). In this case, there are possible three relationships between the premise and the conclusion: (1) the premise, containing the sign, is more particular than it is needed in the conclusion, for example, "If Socrates was both wise and just, then all wise men are just" (ἔστιν δὲ τῶν σημείων τὸ μὲν ὡς τὸ καθ' ἕκαστον πρὸς τὸ καθόλου ὧδε, οἷον εἴ τις εἴπειεν σημεῖον εἶναι ὅτι οἱ σοφοὶ δίκαιοι, Σωκράτης γὰρ σοφὸς ἦν καὶ δίκαιος; *Rhetorica* I:

2, 18,1357b; Ross, 1959), but this reasoning is invalid, because it cannot be reduced to a syllogistic form; indeed, from both premises that "Socrates was wise" and "Socrates was just" we cannot draw a conclusion; (2) the premise, containing the sign, is more general than it is necessary for the conclusion, for example, "If this man breathes hard, then he has a fever", but it is possible for a man to breathe hard without having a fever (τὸ δὲ ὡς τὸ καθόλου πρὸς τὸ κατὰ μέρος ἔχον, οἷον εἴ τις εἴπειεν ὅτι πυρέττει σημεῖον εἶναι, πυκνὸν γὰρ ἀναπνεῖ. λυτὸν δὲ καὶ τοῦτο, κἂν ἀληθὲς ᾖ: ἐνδέχεται γὰρ καὶ μὴ πυρέττοντα πνευστιᾶν; *Rhetorica* I: 2, 18,1357b; Ross, 1959); in this case, we have the following logical fallacy: "All those with fever breathe roughly; this man breathes roughly; then he has a fever" ($p \Rightarrow q$; q, then p); the correct reasoning would be as follows: "All those with fever breathe roughly; this man has a fever; then he breathes roughly" ($p \Rightarrow q$; p, then q); (3) the premise, containing the sign, is a particular from which the conclusion follows as its general, for example, "If a man has a fever, then he is ill" or "If a woman has milk, then she has had a child" (οἷον εἴ τις εἴπειεν σημεῖον ὅτι νοσεῖ, πυρέττει γάρ, ἢ τέτοκεν, ὅτι γάλα ἔχει, ἀναγκαῖον; *Rhetorica* I, 2, 18,1357b; Ross, 1959); these are examples of *modus ponens*: "All those with fever are ill; this man has fever; then he is ill" and "All those with milk has had child; this woman has milk; then she has had child" ($p \Rightarrow q$; p, then q). Only the third type of conclusion, by Aristotle's statement, is irrefutable if the fact is true (μόνον γάρ, ἂν ἀληθὲς ᾖ, ἄλυτόν ἐστιν; *Rhetorica* I: 2, 18,1357b; Ross, 1959) and can be called the sign-inference based on a token (τεκμήριον). The first and the second types are non-syllogistic (ἀσυλλόγιστον) and their signs are not tokens for drawing conclusions. As we see, Aristotle argued that the token-inferences assuming *modus ponens* are logically correct/syllogistic, cf. (Allen, 2001: 30):

> τούτων δὲ τὸ μὲν ἀναγκαῖον τεκμήριον, τὸ δὲ μὴ ἀναγκαῖον ἀνώνυμόν ἐστι κατὰ τὴν διαφοράν. ἀναγκαῖα μὲν οὖν λέγω ἐξ ὧν γίνεται συλλογισμός: διὸ καὶ τεκμήριον τὸ τοιοῦτον τῶν σημείων ἐστίν (*Rhetorica* I: 1, 16–17, 1357b; Ross, 1959).

> Necessary signs are called tokens; those which are not necessary, have no proper name. I call those necessary signs from which a syllogism can be constructed, wherefore such a sign is called token.

The token-inferences belong to the class of enthymemes or "rhetorical syllogisms". Together with the example (παράδειγμα) as a "rhetorical induction", they constitute the whole logical arsenal of rhetoric:

> καλῶ δ' ἐνθύμημα μὲν ῥητορικὸν συλλογισμόν, παράδειγμα δὲ ἐπαγωγὴν ῥητορικήν (*Rhetorica* I: 2, 8, 1356b).

> So, I call an enthymeme a rhetorical syllogism, and an example rhetorical induction.

Thus, the token-inferences are assumed to be a kind of deduction/syllogism. However, the best form of syllogism is called 'demonstration' that is studied in the Aristotelian syllogistic (see his *Prior Analytics* and the *Posterior Analytics*):

> By demonstration [ἀπόδειξις] I mean a syllogism which produces scientific knowledge [συλλογισμὸς ἐπιστημονικός], in other words one which enables us to know by the mere fact that we grasp it (*Analytica Posteriora* I: 2, 71b; Tredennick and Forster, 1960: 31).

Sextus Empiricus (Σέξτος Ἐμπειρικός; 2nd century A.D.), retelling the Stoic concepts, considered inferences as a transition (μετάβασις) from the evident (φαινόμενα, πρόδηλα, ἐναργῆ), what is accessible to our direct perceptual observations, to something other including non-evident matters (ἄδηλα), see his *Outlines of Pyrrhonism* (Πυρρώνειοι ὑποτυπώσεις/*Pyrrhoniae Hypotyposes* II: 97–133;) as well as the *Against the Logicians* (Πρὸς λογικούς/*Adversus Logicos* II: 141–299). He, following the Stoics, more explicitly than Aristotle maintained that there are only two logical ways of drawing conclusions: signs (σημεῖα) and demonstrations (ἀποδείξεις), see (*Pyrrhoniae Hypotyposes* II: 96; *Adversus Logicos* I: 24–26; II: 140, 319). One of the main examples of sign-inferences, most likely borrowed from the Stoics and provided by Sextus Empiricus, is as follows: "If there is smoke, then there is fire (although it is unseen)", e.g.:

> ἐπεὶ καπνὸν ἰδών τις σημειοῦται πῦρ καὶ οὐλὴν θεασάμενος τραῦμα γεγενῆσθαι λέγει (*Pyrrhoniae Hypotyposes* II: 102).

> If someone sees smoke, then he diagnoses fire and if he observes a scar, then he says that a wound was inflicted. Also see (*Adversus Logicos* II: 152, 157).

According to Sextus Empiricus, there are two types of signs (*Pyrrhoniae Hypotyposes* II: 100–101; *Adversus Logicos* II: 152–154): (a) the commemorative or recollective signs (ὑπομνηστικὸν σημεῖον) which are obtained through the recollection (ὑπόμνησις) of what was co-observed with the signified matter (e.g., 'smoke' for 'fire'); (b) the indicative signs (ἐνδεικτικὸν σημεῖον) which are obtained through the abstraction from the own nature and constitution of the signified (ἐκ τῆς ἰδίας φύσεως καὶ κατασκευῆς).

Let p be a sign, for example 'night', and q be its derivative, for example 'dark'. Then, according to the Stoics, as Sextus Empiricus claimed, the conditional statement "If p, then q" is valid if and only if p is true. In other words, if and only if the conjunction "p and if p, then q" holds true (*Adversus Logicos* II: 415). For instance, now is not night, but day. Hence, the conjunction "It is night and if it is night, then it is dark" is false (*Adversus Logicos* II: 416) and we cannot draw a conclusion from the sign 'night' to its derivative 'dark'. In classical logic: $p \,\&\, (p \Rightarrow q)$. Let p be false. Then $p \,\&\, (p \Rightarrow q)$ is false.

The sign as an antecedent of true implication is called προκαθηγούμενον in the *Outlines of Pyrrhonism* and καθηγούμενον in the *Against the Logicians*. If it is true in the conjunction, it is called ἡγούμενον (*Pyrrhoniae Hypotyposes* II: 106). So, from the sign, its derivative directly follows, but the sign can be either true or false, then we may or may not draw a conclusion.

Between two opposing signs, such as 'day' and 'night', there is a contrary relationship. This means that both cannot be true together, but both can be false together. From each of opposing signs, an appropriate signified is derived and between these signifieds, such as 'light' and 'dark', there is a contradictory relationship. This means that both cannot be false together, and both cannot be true together. Obviously, that between 'day' and 'dark' (also, between 'night' and 'light') there is the same contradictory relationship. Hence, we obtain the following basic inference rules, which can be treated as the Stoic square of opposition for the signs, see Fig. 1.2.1:

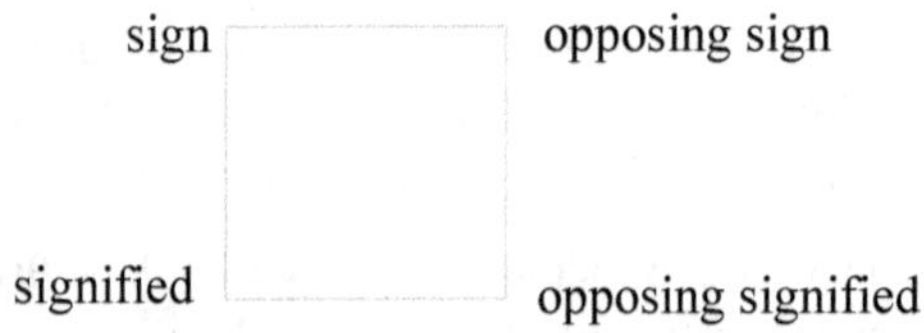

Figure 1.2.1. The Stoic square of opposition for opposing two signs 'day', 'night' and two opposing signifieds 'light', 'dark'. The difference from the Aristotelian square is that the bottom side of the square represents not a subcontrary relation, but a contradictory one.

i) *Modus ponendo ponens* (or shortly *modus ponens*): "If it is day, then it is light; it is day now; therefore, it is light" (Sextus Empiricus, *Adversus Logicos* II: 224, 431; *Pyrrhoniae Hypotyposes* II: 157); formally: $p \Rightarrow q$; p; then q.

ii) *Modus tollendo tollens* (or shortly *modus tollens*): "If it is day, then it is light; it is not light; therefore, it is not day" (Sextus Empiricus, *Adversus Logicos* II: 225, 433; *Pyrrhoniae Hypotyposes* II: 157); formally: $p \Rightarrow q$; $\neg q$; then $\neg p$.

iii) *Modus ponendo tollens*: "It cannot take place that it is both day and night; it is day now; therefore, it is not night" (Sextus Empiricus, *Adversus Logicos* II: 225; *Pyrrhoniae Hypotyposes* II: 158); formally: $\neg(p \;\&\; q)$; p; then $\neg q$.

iv) *Modus tollendo ponens* I: "It is either light or dark; but it is light; hence it is not dark" (Sextus Empiricus, *Adversus Logicos* II: 225; *Pyrrhoniae Hypotyposes* II: 158); formally: $p \oplus q$; p; therefore, $\neg q$.

v) *Modus tollendo ponens* II: "It is either light or dark; but it is not dark; therefore, it is light" (Sextus Empiricus, *Pyrrhoniae Hypotyposes* II: 158); formally: $p \oplus q$; $\neg q$; therefore, p.

These inference rules are primary (atomic) for making various compound inferences, such as:

> τὰ παρ' ἡμῖν ὅ[σ]α βραδείας ποιεῖται τὰς παρὰ τὰ ἐπιπροσθετοῦντ' ἐκ[φ]άσεις ἤτοι [π]αρὰ τὸ βραδέω[ς] κịνεῖσθαι [π]άσχει τοῦτο ἢ παρὰ τὸ μεγάλα σφοδρ' ἔχειν τὰ [μ]εγ[έ]θη· καὶ τὸν ἥλιον οὖ\ν/ἀναγκαῖ[ον], ἐπεὶ βραδέως ἐκφαίνεται, τ[ῶ]ν δυεῖν πεπονθένα[ι] θ̣ἄτερον· [κ]α̣ὶ βραδέ[ω]ς μὲν οὐ κ̣[ι]νεῖται [δή]που, κα[θὼ]ς ἐν δώ[δεχ'] ὥραις ̣τ[ὴν] ἀπ' ἀνα̣τολῆ[ς εἰς δύ]σιν [ὁδὸν ἀ]νύει, [δι-]άστημ[α μέ]γα τ[οῦτο συ]ṿεχῶ̣[ς].

> All objects in our experience that reappear slowly from behind objects that eclipse them have this character either because they move slowly, or because they are very large. Since the sun reappears slowly, it must of necessity have one of these two characteristics. But it does not move slowly, since it completes the path from sunrise to sunset in twelve hours, passing through a very great distance. Therefore, it must be very large [*A.Sch.*: and only hence the sun reappears slowly from behind objects] (Philodemus, *De Signis*, Col. X: 4–16; De Lacy and De Lacy, 1941: 45).

Formally:

[((“Objects move slowly” ⇒ “objects reappear slowly from behind objects”) ⊕ (“objects are very large” ⇒ “objects reappear slowly from behind objects”)) & (“the sun reappears slowly” & (“it completes the path from sunrise to sunset in twelve hours, passing through a very great distance” ⇒ “it does not move slowly”))] ⇒ (“the sun is very large” ⇒ “the sun reappears slowly”).

The four axioms:

1) (“Objects move slowly” ⇒ “objects reappear slowly from behind objects”) ⊕ (“objects are very large” ⇒ “objects reappear slowly from behind objects”).
2) “The sun reappears slowly”.
3) “It completes the path from sunrise to sunset in twelve hours, passing through a very great distance” ⇒ “it does not move slowly”.
4) “It completes the path from sunrise to sunset in twelve hours, passing through a very great distance”.

   After applying the substitution rule in (1), we obtain:

5) (“The sun moves slowly” ⇒ “the sun reappears slowly from behind objects”) ⊕ (“the sun is very large” ⇒ “the sun reappears slowly from behind objects”).

By *modus ponens* from (3) and (4), we have:

6) "The sun does not move slowly".

By introducing conjuction, from (2) and (6) it follows that

7) "The sun does not move slowly" & "the sun reappears slowly".

But (7) is a negation of the statement: "the sun moves slowly" ⇒ "the sun reappears slowly from behind objects". Then we apply *modus tollendo ponens* II to (5) and (7) and obtain:

8) "The sun is very large" ⇒ "the sun reappears slowly from behind objects".

Thus, the cause of reappearing slowly is that the sun is very large.

To sum up, the Stoic sign-inferences were regarded as a logical system that can be treated as nowadays classical propositional logic. Another view of the signs was proposed by Epicurus (Ἐπίκουρος; 341–270 B.C.) and his disciples. The Epicurean conception of sign-inferences, as compared with the corresponding Stoic conception, is well represented in a work written by Philodemus of Gadara in Syria (Φιλόδημος ὁ Γαδαρεύς; ca. 110–ca. 40 B.C.) and entitled the *On Methods of Inference* (Περὶ σημειώσεων, *De Signis*) or the *On Phenomena and Inferences* (Περὶ φαινομένων καὶ σημειώσεων), PHerc. 1065.

The main difference between the Epicurean and the Stoic approach to the sign-inferences is grounded on the difference between deductive and inductive logic. Thus, according to the Stoics, the sign-inferences are demonstrations in the Aristotelian meaning, i.e., completely scientific, and are constructed from the evident (perceived) through the consequence (ἀκολουθία), when p is a subset of q in the implication $p \Rightarrow q$ (e.g., "If day, then light"), and through the exclusion (μάχη), when two signifieds q and q' are exclusive (e.g., "light and dark"). In contrast, Epicurus appeals to the couple of similarity (συμφωνία) and exclusion (μάχη), instead of the couple of consequence and exclusion (Allen, 2001: 204). In the sign-inferences, the Stoics use the "elimination method" (ὁ κατ' ἀνασκευὴν τρόπος), or just 'elimination' (ἀνασκευή), while the Epicureans use the "similarity method" (ὁ καθ' ὁμοιότητα τρόπος), see (Asmis, 1996).

The elimination means that "If p, then q" holds true if and only if not-q is incompatible with p. The same equivalence takes place in classical logic: $p \Rightarrow q$ if and only if $\neg(p \,\&\, \neg q)$. Also, "If p, then q" holds true if and only if we have that "If not-q, then not-p". The same equivalence is valid in classical logic: $p \Rightarrow q$ if and only if $\neg q \Rightarrow \neg p$.

The sign relations established through the elimination are given *a priori*, whereas the Epicureans maintained that signs are based on a sense perception and its extrapolation through induction. The similarity method of the Epicureans means that the sign p and its signified q should be always true together in our experience: "p and q" is true.

> We say that anyone who thinks that a certain man is good because he is rich is using an unsound and common sign, since some men who are rich prove to be bad and some good. Therefore, if the particular and cogent sign cannot exist except along with the unperceived object which, we said, is

> by necessity connected with it, and of which it is the sign, it would never be possible to reveal that which is unperceived by means of the analogical method of inference (Philodemus, *De Signis*, Col. I: 5–15; De Lacy and De Lacy, 1941: 25).

Hence, the difference between the Epicureans and the Stoics may be presented by Table 1.2.2.

Table 1.2.2. The relationship between the sign and the signified is defined as *a priori* by the Stoics and as *a posteriori* by the Epicureans.

Philosophical School	Empirical Relationship between the Sign *p* and its Signified *q*.	Logical Relationship between the Sign *p* and its Signified *q*.
Stoics	not-*q* is always incompatible with *p* (or from not-*q* it follows that not-*p*).	If *p*, then *q*.
Epicureans	*p* and *q* are always true together ("*p* cannot exist except along with *q*").	If *p*, then *q*.

As we know, there are things that cannot be observed (τὰ ἄδηλα) and for them we have sign-inferences (σημεῖα). The Stoics argued that the sign-inferences represent an additional source of true knowledge, which is *a priori*, while the Epicureans opined that sign-inferences are only conclusions by analogy (ἡ κατὰ τὸ ὅμοιον μετάβασις) with a generalization of some cases after observing many repeating examples and, therefore, the sign-inferences are a continuation of our perception and do not go beyond our experience—they are *a posteriori*. For example, the conclusion that all humans are mortal is inferred by the analogy that the human beings we met are mortal:

> ἐπεὶ οἱ παρ' ἡμῖν ἄ̣ν[θρω]ποι θνητοί εἰσι, καὶ τοὺς π[άντας, τ]ὸ διὰ τῆς ὁμοιότ[η]τọ[ς ὑγιές <ἐστιν> εἰ] κατὰ πάντα.

> Since men in our experience are mortal, all men are (Philodemus, *De Signis*, Col. II: 26–29; De Lacy and De Lacy, 1941: 27).

This conclusion is grounded on the analogy that "men who are in unperceived places are similar to men in our experience in all respects, so that they will be similar also in respect to mortality; without this assumption it will not be valid" (Philodemus, *De Signis*, Col. II; De Lacy and De Lacy, 1941: 29). Hence, instead of the inference from one sign 'man' that "If he is a man, then he is mortal", we have two premises for the conclusion:

> ἐ[πε]ὶ οἱ παρ' ἡμῖν ἄνθ[ρωποι θνητ]οί ẹἰσιν, κα[ὶ εἴ] που κατ' ἄλλους τόπους εἰσὶν ἄνθρωποι τοῖς παρ' ἡμῖν ὡμοιωμένοι κατά τε τἆλλα καὶ κατὰ τὸ θνητοὶ εἶναι, θνητοὶ ἂν εἴησαν.

> [*A.Sch.*: **the first premise**:] Since the men in our experience are mortal, [*A.Sch.*: **the second premise**:] if there are men in other places who are similar to men in our experience in all respects, including mortality, they are mortal (Philodemus, *De Signis*, Col. II: 37–38–Col. III: 1–4; De Lacy and De Lacy, 1941: 29).

Another formulation of the same conclusion with two premises:

> ἐπεὶ οἱ παρ' ἡμῖν [θν]ητοί εἰσιν ἄνθρωποι, κ̣αὶ εἴ πού [εἰσ]ιν θνητοὶ ἄ[ν-]θρω[ποι, θν]ητοί [εἰσι]ν.

> [*A.Sch.*: **the first premise**:] Since the men in our experience are mortal, [*A.Sch.*: **the second premise**:] if there are mortal men anywhere, they are mortal (Philodemus, *De Signis*, Col. III: 10–12; De Lacy and De Lacy, 1941: 29).

According to the Stoics, if there is a relationship between a sign *p* and its signified *q* in the form of the following implication: "If *p*, then *q*", then we can construct the inference by *modus ponens*. So, the *Stoic inference* has two premises and one conclusion:

(S_1) If *p*, then *q*;
(S_2) there is *p*;
(C) therefore, there is *q*.

According to the Epicureans, instead of *modus ponens* with two premises, we should have the same *modus ponens* but with one more premise demonstrating the empirical validation for the concurrent occurrence of the sign and its signified. Thus, the *Epicurean inference* by *modus ponens* has the following three premises and one conclusion:

(E_1) *p* and *q* occur together in our experience;
(E_2) if *p*, then *q*;
(E_3) there is *p*;
(C) therefore, there is *q*.

In the *De Signis*, Philodemus insisted that in our reasoning we cannot go beyond our experience, therefore we should address to similarity or analogy in the first premises (i.e., in (E_1)) and do it even for reasoning based on contrapasition (that is, in *modus ponens* and *modus tollens*):

> [δεῖ δὲ διὰ τῆς καθ' ὁμοιότητα σημειώ]σεως ἢ δι[ὰ τῆς ἀνα]λογίας συλλογίζε[σθαι] μήτ' οὐδὲ ταύτης <... > οὔσης ἀφίστασθαι τῆς μόνης ἐκ τῶν πραγμάτων διδομένης ἀναλογίας. ἔτι δ[ὲ τ]ὰ μὲν τοῦ εἶναί τι[ν' ἐ]ν τοῖς ἀ[δήλ]οις τιθώμεθα ση[μεῖ]α τὰ δὲ [τ]ο̣ῦ το̣ιαῦτ' εἶναι τὰ [δ]ὲ τοῦ μ[ὴ ὑπ]άρχειν τὰ δὲ τ[οῦ] μὴ τοιαῦ[θ' ὑ]πάρχειν· καὶ μὴ διὰ τῶν ἐνη[λλ]αγμένως σημαιν[όν]των ποι[ώμ]εθα τοὺς συλλογισμούς. ὡσα[ύ]τως δὲ καὶ τὰ κατ' ἀ[νασκευὴν τατ]τ̣όμενα σημεῖα [οὐκ ἐπάγομεν, εἰ] \μὴ/ [κ]ατὰ γεννή-

> [ματα τῶν φύσει ὄντ]ων ἀδήλ[ων],
> [ἐφ' ὧν ἐκ τῶν φανερῶν ποι]ούμε-
> [θα τὴν σημείωσιν].

> We ought to reason through inference according to similarity or through analogy; and this being so, we should not depart from the analogy that is furnished by empirical facts. Further, let us posit some things as signs of the existence of unperceived objects, some of a determinate existence, some of non-existence, some of the absence of a determinate existence; and let us not reason from things whose signification varies. We do not infer from the signs tested by contraposition, except in the case of things unperceived by nature about which we make inferences from the apparent (Philodemus, *De Signis*, Frag. III; De Lacy and De Lacy, 1941: 115).

The matter is that any evidence for our reasoning comes from the testimonies based on our observations through "perception, anticipations, mental perceptions, and feelings":

> [..]A[..] [ἡ]
> [τῶν ση]μειώσεω[ν σ]ύνθεσις [οὐ-]
> [κ ἀνασ]κευῆι τοῦ εἰ τόδε το[ῦτ',]
> [ἀλλὰ κατ]ελήφθη διὰ τῶν φαν-
> [τασιῶν] τῶν τὰς χρείας ἀποδι-
> [δουσῶν]. ὁ δέ γε ἀπορῶν πῶς
> [κριθήσε]ται [ἡ] τῆς διανοίας φαν-
> [τασία συ]νθετ[έ]ον τὰς σημει-
> [ώσεις οἴ]εται, ἐὰν ἐπιμαρτυ[ρῆ-]
> [ται διὰ τῆ]ς ὄψεως μήτε τοῖς πᾶ-
> [σιν,] [ἃ κριτ]ήρια λέγεται τῶν ἀδή-
> [λων κα]τὰ τὰς αἰσθήσεις καὶ
> [προλήψεις κ]αὶ τὰς [φα]ντασικὰς
> [ἐπιβολὰς τῆς διανοί]ας καὶ τ[ὰ]
> [πάθη,] [ἀντιπίπτηι].

> The construction of inferences is not established by contraposition of the argument "in so far as this is such", but by appearances which give the necessary evidence. Indeed, even if one does not know how mental perception will be judged, he thinks that inferences from signs should be constructed if they are verified by observation and do not conflict with present appearances, which are called the criteria of the unperceived: namely, perception, anticipations, mental perceptions, and feelings (Philodemus, *De Signis*, Frag. I; De Lacy and De Lacy, 1941: 113).

However, the Stoics argued that we do not need the first premise of the Epicurean inference scheme denoted by (E_1) to indicate the empirical basis for the *logos*. Indeed, let us take the Epicurean first premise, showing a generalization from our perception: "If men in our experience are vulnerable and mortal, men everywhere are vulnerable and mortal", then its contraposition is as follows: "If men in unperceived places do not have all the differences that we mention, neither do those in our experience" (Philodemus, *De Signis*, Col. XXXVIII: 1–10; De Lacy and De Lacy, 1941: 111). But now this contraposition is *a priori* and has nothing empirical. Logically, both the implication and its contraposition are identical, then we can avoid the first premise of the Epicureans—we can conclude without the empirical validation for the relationship between the sign and its signified. But, according to Philodemus, inconceivability (ἀδιανοησία) as a criterion of inference should be distinct from contraposition (Philodemus, *De Signis*, Col. XXVIII: 15–25). For the implication "If there is motion, then there is void", its appropriate contraposition as its empirical validation means that "for when void is denied by hypothesis, by the mere denial of it, motion will also be denied" (Philodemus, *De Signis*, Col. XII; De Lacy and De Lacy, 1941: 51), i.e., "If there is no void, then there is no motion". But sometimes we deal with conjunctions, such as "Plato is a man, and Socrates is a man", which cannot be completely verified in this way by contraposition. On the one hand, "if this is true, it is true also that, 'If Socrates is not a man, neither is Plato a man'" (Philodemus, *De Signis*, Col. XII; De Lacy and De Lacy, 1941: 51). Formally, it looks like that

("Plato is a man" & "Socrates is a man") ⇒ ("Socrates is *not* a man" ⇒ "Plato is *not* a man").

$(p \,\&\, q) \Rightarrow (\neg q \Rightarrow \neg p)$.

This propositional formula is a tautology (always true proposition) in classical logic. But on the other hand, we do not have the equivalence between $(p \,\&\, q)$ and $(\neg q \Rightarrow \neg p)$ as in the case between $(p \Rightarrow q)$ and $(\neg q \Rightarrow \neg p)$. Therefore, we should go beyond the contraposition and address here to the procedure when "by the denial of Socrates Plato is denied along with him, but because it is inconceivable for Socrates not to be a man and Plato to be a man. But this inference is derived from the method of analogy" (Philodemus, *De Signis*, Col. XII; De Lacy and De Lacy, 1941: 51).

In the *De Signis* Philodemus noted that if the sign is more general than it is necessary and sufficient for drawing conclusions (that is, more general than the signified), then it is called "common sign" (κοινόν σημεῖον). If it is more particular than the signified, then we can draw conclusions and it is called "particular sign" (ἴδιο σημεῖον). The Stoics applied the following composite proposition:

> The common sign exists whether the unperceived object exists or not, whereas the particular sign exists only when the unperceived object exists and does not exist when the unperceived object is non-existent (Philodemus, *De Signis*, Col. XIV; De Lacy and De Lacy, 1941: 55).

(("The unperceived object exists" ⊕ "the unperceived object does *not* exist") ⇒ "the common sign exists") & ("the unperceived object exists" ⇒ "the particular sign exists") & ("the unperceived object is *non*-existent" ⇒ "the particular sign does *not* exist").

From this reasoning it follows that "every particular sign is an instance of the method of contraposition" (Philodemus, *De Signis*, Col. XIV; De Lacy and De Lacy, 1941: 55). Indeed, by contraposition, we know that if the common sign does not exist, then the unperceived object exists and not and if the specific sign does not exist, then the unperceived object is non-existent, while if the particular sign exists, then the unperceived object exists.

As a model example of the Stoic contraposition, Philodemus referred to the relationship between 'fire' and 'smoke': "If there is smoke, then there is fire" if and only if "If there is no fire, then there is no smoke". It means that 'fire' is a cause of 'smoke':

> καὶ ἵνα πυρὸς μὴ ὄντος ἢ γεγονότος ὁ καπνὸς ἀνασκευασθῆι τῶι πάντως καὶ ἐπὶ πάντων καπνὸν ἐκ πυρὸς ἐκκρινόμενον τεθεωρῆσθαι διατεινόμεθα.

> And in order to make the contraposition, "If there is or has been no fire, there is no smoke," we contend that always in all cases smoke has been observed to be given off by fire (Philodemus, *De Signis*, Col. XXXVI: 1–7; De Lacy and De Lacy, 1941: 107).

Hence, the Stoics did not use similarity in inferences (Philodemus, *De Signis*, Col. XXVIII: 25–29) and they did not recognize the empirical validation as the first premise for conclusions (Philodemus, *De Signis*, Col. XXVIII: 37–XXIX: 4). As a result, they denied the *a posteriority* for the *logos* (Philodemus, *De Signis*, Col. XXIX: 4–15). The Epicureans maintained that we always need either a verification (ἐτημαρτύρησις) as a direct perception that the object of judgment is present, cf. (Sextus Empiricus, *Adversus Logicos* II: 324), or a falsification (οὐκ ἀντιμαρτύρησίς) as the criterion to be false, cf. (Sextus Empiricus, *Adversus Logicos* I: 216). In any case, we always start from the evidence that is given (φανερόν). Therefore, our inferences are always inductive (ἐπιλογισμός).

Both the Stoics and the Epicureans accepted that there are things which are immediately perceived and, therefore, they do not need inferences for knowing them. Also, there are unperceived notions (ἄδηλα). Among them there are notions which can never be perceived or known through inferences, such as the either odd or even number of the stars or the exact number of grains of sand in Libya. Meanwhile there are notions which cannot be perceived but known through inferences. For instance, atoms and void for the Epicureans. Also, there are notions which can be perceived, but not present now and known through inferences. This Stoic-Epicurean classification of sources of true knowledge is given by Sextus Empiricus in the *Adversus Logicos* (II: 145–147, 316–320) and the *Pyrrhoniae Hypotyposes* (II: 97–99).

Now, let us compare the Stoic and Epicurean conception of sign-inference with the Indian logic.

From the very beginning of Buddhist epistemology and ontology, a unique doctrine of *abhidharma* arose, in which all our mental states (*dharma*s) are classified. Within the framework of this classification, *dharma* as a cogitative state or an intelligible entity was distinguished from *dharmin* as something what is believed to be the bearer of this entity or marked by it. Later, both concepts of *dharma* and *dharmin* became basic for the whole Indian logic within the framework of the common doctrine *dharma-dharmi-bheda* (essential differentiation between the substratum and its properties). In logic, the term *dharma* denotes a property that can be known, and the term *dharmin* denotes a possessor of this property. For example, let the fire be located on a mountain, then 'fire' is to represent the *dharma* and 'mountain' is to show the *dharmin* as an appropriate locus of 'fire'.

According to the *Nyāyasūtra*, one of the earliest logical texts in India that is dated, most probably, to the 2nd century A.D., there are only four sources (*pramāṇa*) for knowing *dharma*s and their relationships with *dharmin*s:

> *pratyakṣānumānopamānaśabdāḥ pramāṇāni* (*Nyāyasūtra* I: 1, 3; Tarkatirtha and Tarkatirtha, 1982: 85).

> The sources of true knowledge (*pramāṇa*) [about the relationships between *dharma*s and *dharmin*s] are as follows: perception (*pratyakṣa*), inference (*anumāna*), comparison (*upamāna*), and authoritative word (*śabda*).

The *pratyakṣa* (the first source) was understood as the Stoic and Epicurean αἴσθησις ('perception'). The *anumāna* (the second source) was treaded as the Stoic and Epicurean σημείωσις ('sign-inference'). The *upamāna* (the third source) corresponded to the Epicurean καθ' ομοιότητα μετάβασις ("inference through similarity") or αναλογία ('analogy'). The *śabda* (the fourth source) was not presented in the Greek logic at all. It meant that the holy texts (*śruti*, first, Vedas) are a separate source of true knowledge.

In the *Nyāyasūtra*, each syllogism (*parārthānumāna*) linking *dharma*s to *dharmin*s consists of five members/steps:

> *pratijñāhetūdāharaṇopanayanigamanāny avayavāḥ* (*Nyāyasūtra* I: 1, 32; Tarkatirtha and Tarkatirtha, 1982: 269).

> The members of a syllogism are as follows: thesis (*pratijña*), reason (*hetu*), illustrative example (*udāharaṇa*), application (*upanaya*), and conclusion (*nigamanāna*).

By this definition, the *modus ponens* has the following form (it is called *the Nyāya conclusion*):

(N_1) *q* (*pratijñā*);
(N_2) *p* (*hetu*), such that if *p*, then *q*;
(N_3) if *p*, then *q*, since there is an example that *p* and *q* occur together (*udāharaṇa*);
(N_4) *p* is observed (*upanaya*);
(C) *q* is inferred (*nigamana*).

For instance:

(N'_1) There is fire on a mountain (*q*), and this fire is understood within the framework of *agnihotra*—the Vedic fire ritual;
(N'_2) there is smoke (*p*), such that if smoke, then fire;
(N'_3) if there is smoke, then there is fire; for example, in a kitchen (if *p*, then *q*, because there is an example that *p* and *q* occur together);
(N'_4) the mountain has smoke (*p* is observed), and 'mountain' is a *dharmin* for 'smoke';
(C') the mountain has fire (*q* is inferred), and 'fire' is a *dharma* for 'mountain'.

In each syllogism (*parārthānumāna*), all the members participating in the conclusion from (N_1) to (N_4) are based on a suitable source of true knowledge: (N_1) the thesis (*pratijña*) is received from the authoritative word (*śabda*); (N_2) the reason (*hetu*) appears from the inference (*anumāna*); (N_3) the illustrative example (*udāharaṇa*) is obtained through the comparison (*upamāna*); (N_4) the application (*upanaya*) follows from the perception (*pratyakṣa*). For example, (N'_1) the criterion of *śabda*: 'fire' (*agni*) is a word about the Vedic ritual of *agnihotra*; (N'_2) the criterion of *anumāna*: 'smoke' is connected to 'fire' as the reason (*hetu*) through the inference that if there is smoke, then there is fire; (N'_3) the criterion of *upamāna*: from the observation that smoke and fire occur always together, for example, in a kitchen (*udāharaṇa*), we generalize these facts for all situations; (N'_4) the criterion of *pratyakṣa*: we feel smoke and, as a result, we can apply (*upanaya*) the syllogism.

Hence, in the Hellenistic terms, we have the following presentation of *modus ponens* in Nyāya:

> (N_1) *q* (authoritative word)—it is absent in the Hellenistic logic as an unnecessary element for our reasoning;
> (N_2) *p* (σημεῖον, 'sign'), such that if there is a sign, then there is its signified (if *p*, then *q*);
> (N_3) if *p*, then *q*, since there is an example that *p* and *q* occur together (καθ' ομοιότητα μετάβασις, "inference through similarity")—the Epicurean inference with the following two premises: (E_1) *p* and *q* occur together in our experience; (E_2) if *p*, then *q*;
> (N_4) *p* is observed (αἴσθησις, 'perception');
> (C) *q* is inferred.

Thus, in Nyāya we have four premises in *modus ponens* from (N_1) to (N_4) instead of two for the Stoics from (S_1) to (S_2) and three for the Epicureans from (E_1) to (E_3).

The reason (*hetu*) in sign-inferences is understood in the *Nyāyasūtra* as a token (*liṅga*). In the commentary on the *Nyāyasūtra* III: 2, 41, Vātsyāyana (from the 4th to the 5th century A.D.) defined 'token' as follows:

> *liṅgaṃ punaḥ saṃyogi samavāyy ekārthasamavāyi virodhi ceti. | yathā dhūmo 'gneḥ, gor viṣāṇam, pāṇiḥ pādasya, rūpaṃ sparśasya abhūtaṃ bhūtasyeti* (Tarkatirtha and Tarkatirtha, 1982: 878).

> The token is [of four types]: by connection (*saṃyoga*), by inherency (*samavāya*), by co-inherency in one substratum (*ekārthasamavāy*), and by contradiction (*virodha*). Examples: "smoke [as a connective token] of fire", "horns [as an inherent token] of a cow", "hands [as a co-inherent in one substratum token] of legs or color of tangibility", "non-existent [as a contradictory token] of existing".

The Nyāya notion of token (*liṅga*) corresponds to the Aristotelian τεκμήριον ('token') and to the Stoic-Epicurean ἴδιο σημεῖον ("particular sign").

Along with sign-inferences, there are also analytic inferences from species to genera and, instead of tokens, the specific difference (*lakṣaṇa*) is used here, defined by Vātsyāyana in his commentary on the *Nyāyasūtra* III: 2, 41, in the following manner:

> *lakṣaṇaṃ paśvavayavasthaṃ gotrasya smṛtihetuḥ, vidānām idaṃ gargāṇām idam iti* (Tarkatirtha and Tarkatirtha, 1982: 878).

> The specific difference is a reason for remembering the genus, associated with the representatives of the body of the animal, for example: "This is [a specific difference of] the genus Vida, and that is [of] the genus Ganga".

The Nyāya notion of specific difference (*lakṣaṇa*) has the close meaning to the Aristotelian *differentia specifica* (διαφορά ειδοποιός), used in his syllogistic.

Summing up, in the Nyāya doctrine, there are the following four criteria of true knowledge: perception (*pratyakṣa*), inference (*anumāna*), comparison (*upamāna*), and authoritative word (*śabda*). We remember that, according to both the Stoics and the Epicureans, there are only two criteria of true knowledge: perception (αἴσθησις) and inference (σημείωσις), see (*Adversus Logicos* II: 145–147, 316–320; *Pyrrhoniae Hypotyposes* II: 97–99). The same number of criteria is proposed in the Buddhist logic. First, we find only perception and inference among possible criteria of true knowledge in the *Milindapañha*—the Buddhist logical treatise written in Gandhāra from the 1st century B.C. to the 2nd century A.D., but surely earlier than the *Nyāyasūtra* was compiled, for some arguments see Schumann (2019a). Second, only two sources of right knowledge are enumerated in the Yogācāra logic, founded in the same region of Gandhāra:

> *dvividhaṃ samyag-jñānam. pratyakṣam anumānaṃ ca* (Dharmakīrti, *Nyāyabindu* I: 2).

> Right knowledge is of two kinds—perception (*pratyakṣa*) and inference (*anumāna*) (Gangopadhyaya, 1971: 86).

See also *Pramāṇasamuccaya* by Dignāga (Hattori, 1968: 24).

It is worth noting that both criteria of right knowledge (perception and inference) are separately mentioned in the *Nyāyasūtra* as the main criteria of acceptance:

> *dṛṣṭānumitānāṃ hi niyogapratiṣedhānupapattiḥ* (*Nyāyasūtra* III: 1, 51; Tarkatirtha and Tarkatirtha, 1982: 779).

> It is wrong to question or deny things that are perceived and inferred.

By Dignāga (5th century A.D.), we perceive the objects through the direct sense perception (*pratyakṣa* or *grahaṇa*), while we infer new objects through our intellect (*vikalpa*) or thought (*adhyavasāya*). In the *Pramāṇa-samuccaya* compiled by him, it is stated that we can perceive only "unique characteristics" (*sva-lakṣaṇa*), but they cannot be inferred (Hattori, 1968: 24), see also:

> *tasya viṣayaḥ svalakṣaṇam* (Dharmakīrti, *Nyāyabindu* I: 12; Malvania, 1955).

> Its object [of perception] is the unique characteristics (*svalakṣaṇa*).

These underlying things are "devoid of any attribution (*āropa*)" (Gangopadhyaya, 1971: 111) and represent the ultimate reals (*paramārtha-sat*). Therefore, they are excluded from similar objects (*sajātīya-vyāvṛtta*) as well as from dissimilar objects (*vijātīya-vyāvṛtta*). This term of *svalakṣaṇa* has the same logical meaning as the Aristotelian ὑποκείμενον/*subiectum* ("underlying thing")—something which cannot be a predicate of others, but can be predicated by others (Aristotle, Κατηγορίαι/*Praedicamenta* 1a20). So, each *svalakṣaṇa* can be predicated by other *lakṣaṇa*s.

Instead of underlying things (*svalakṣaṇa*s), we can only deduce "general characteristics" (*sāmānyalakṣaṇa*), but these cannot be perceived and are not real (Hattori, 1968: 24). See also:

> *anyat sāmānyalakṣaṇam | so anumānasya viṣayaḥ* (Dharmakīrti, *Nyāyabindu* I: 16–17; Malvania, 1955).

> The general characteristics is another kind [of intelligible object]. It is the object of inference.

We see the same criterion for distinguishing between perception and inference in the logic of the Yogācārins as in the logic of the Stoic-Epicureans. If something is *apratyakṣa* or ἄδηλα and it can be intelligible, then it can be inferred, see (*Adversus Logicos* II: 145–147, 316–320; *Pyrrhoniae Hypotyposes* II: 97–99).

In the Nyāya scheme of syllogisms with the steps from (N_1) to (N_5), it is assumed that all four sources of knowledge (*pramāṇa*s) are intermixed and participate equally in drawing conclusions. This Nyāya doctrine is called *promāṇa-samplava* ("the theory of intermixture of *pramāṇa*s"). In accordance with the Yogācārins, the only two sources, i.e., perception and inference, give two different types of knowledge, as we have just seen, and they are not mixed. This Buddhist theory is called *pramāṇa-vjavasthā* ("the theory of restriction of *pramāṇa*s"), see (Shastri, 1976: 36). This view is consistent with the Stoic logic, in which perception and inference are also concurrent.

According to Nyāya-Vaiśeṣika, the external objects are real (*bāhyārtha-vāda*), while in the Yogācāra school only knowledge exists (*vijñāna-vāda*) and the real entities (*svalakṣaṇa*) are *Dinge an sich* in the Kantian meaning, see (Tachikawa, 1981; Shastri, 1976). For example, Vinītadeva (8th century), in his commentary on Dharmakīrti's *Nyāyabindu*, explained one of the properties of *pratyakṣa*, being "not illusory" (*abhrānta*), in the following way. It is treated by him as "in respect of [the knowledge] that makes one reach the object" (*prāpaka-viṣaya*) and not "in respect of the object supporting a cognition" (*ālambana-viṣaya*) (Gangopadhyaya, 1971: 94).

In Yogācāra, inference (*anumāna*) is a combination of two types of conclusions that took place in Ancient Greece: (a) inferences through signs (σημεῖα) or tokens (τεκμήρια) of the Stoics and Epicureans, based in cause-and-effect relationships; (b) inferences through demonstrations (ἀποδείξεις) of Aristotle, based on species-and-genus relationships. In the inference, by means of signs (tokens), we go from signs to hidden phenomena (*kārya-anumāna*), and in the inference, by means of demonstrations, we go from more particular characteristics to more general ones (*svabhāva-anumāna*). The main difference between the Yogācāra logic and the Hellenistic logic is that all syllogisms containing negation (*anupalabdhi*) in premises are considered by the Yogācārins separately as an additional group of conclusions. In this way, the Yogācārins have the following three types of conclusions: (a) inferences through signs or tokens of the Stoics and Epicureans (*kārya-anumāna*) but without negation (that is, only *modus ponens*); (b) inferences through demonstrations (*svabhāva-anumāna*) but without negation (that is, only *modus Barbara*); (c) inferences through signs or tokens with negation (*anupalabdhi-anumāna* with *kārya*) (that is, *modus tollens*, *modus tollendo ponens*, *modus ponendo tollens*) as well as inferences through demonstrations with negation (*anupalabdhi-anumāna* with *svabhāva*) (that is, *modus Camenes*).

An inferential token that is used in all inferences is called *liṅga*. So, there are three types of *liṅga* in accordance with the three groups of conclusions mentioned above: (a) through tokens of causes (*kārya*)—the causal tokens; (b) through demonstrations or analyticity (*svabhāva*)—the analytic tokens; (c) through negation or non-perception (*anupalabdhi*)—the non-perceptual tokens, please see:

> *anupalabdhiḥ svabhāvaḥ kāryaṃ ca iti* (Dharmakīrti, *Nyāyabindu* II: 11; Malvania, 1955).

> [The three tokens are] non-perception (*anupalabdhi*), analyticity (*svabhāva*), and causality (*kārya*).

> *trividhameva hi* ***liṅga****mapratyakṣasya siddheraṅgam -* ***svabhāvaḥ, kāryam, anupalambhaśca*** *| tasya samarthanaṃ* ***sādhyena*** *vyāptiṃ prasādhya* ***dharmiṇi*** *bhāvasādhanam | yathā 'yat sat kṛtakaṃ vā, tat sarvamanityam, yathā ghaṭādiḥ, san kṛtako vā śabdaḥ'; iti* (Dharmakīrti, *Vādanyāya* 2).

> Because three kinds of **tokens** (= antecedents of implications) constitute the proof of an unperceived object: **analyticity**, **causality**, and **non-perception**, Its justification (i.e., justification of antecedents) means nothing but proving its existence in the **dharmin** (= subject of the thesis), after proving its pervasion with the **dharma** (= predicate of the thesis). For example – "All that is real or is a product is impermanent, like the things such as pot. And sound is real or a product" (Gokhale, 1993: 4; the translation is slightly modified).

Analyticity

The demonstration (ἀπόδειξις) or analyticity (*svabhāva*), according to Dharmakīrti, is grounded on identity (*tādātmya*) between two essential characteristics, *svabhāva*s, of the same thing (Gangopadhyaya, 1971: 92). The implication "If there is a token (*liṅga*) *p*, then *q*" is analytic (*svabhāva*) if and only if *p* is a species and *q* is its genus (*jāti*). The Yogācāra model example for this type of conclusion is as follows: "If it is 'being a *Dalbergia sisso*' (*śiṃśapātva*), then it is 'being a tree' (*vṛkṣatva*)". This reasoning can be presented as the following *modus ponens* of the Stoics: "If it is a *Dalbergia sisso* (*liṅga*, 'token'), then it is a tree (*dharma*, "inferential property"). It (*dharmin*, "substratum of reasoning") is a *Dalbergia sisso*. Therefore, it is a tree". To the same extent, it can be presented as the following *modus Barbara* (Figure I of syllogism) of the Aristotelian syllogistic: "All the cases of *Dalbergia sisso* is a tree. This thing is a case of *Dalbergia sisso*. Therefore, it is a tree". In this syllogism, "a case of *Dalbergia sisso*" is a middle term or *liṅga*, "this thing" is a minor term or *dharmin* (the subject of the conclusion), 'tree' is a major term or *dharma* (the predicate of the conclusion). As we see, in the minor premise, the Yogācārins find a property for an individual (*svalakṣaṇa*, ὑποκείμενον, *subiectum*), but we can apply only individual statements in all minor premises, since they always are statements about individuals. These statements cannot be particular and only have the meaning of universal statements. Logically, "this one is…" and "all this is…" are identical, e.g., "Socrates is…" is equal to "All Socrates is…". Therefore, if both the major and minor premises are positive (affirmative), then we have only one *modus* of the Aristotelian syllogistic, called *Barbara*, in which both premises are universal and affirmative at the same time. Consequently, the Yogācārins have only this *modus* of the Aristotelian syllogistic with both affirmative premises.

Causality

The sign-inference (σημείωσις) or causality (*kārya*), according to Dharmakīrti, is expressed through cause-and-effect relationships (*tadutpatti*) (Gangopadhyaya, 1971: 92). The implication "If there is a token (*liṅga*) *p*, then *q*" is causal (*kārya*) if and only if *p* is an effect and *q* is its cause. The Yogācāra model example: "If there is smoke, then there is fire". Please see:

> *kāryaṃ yathā vahnir atra dhūmād iti* (Dharmakīrti, *Nyāyabindu* II: 17; Malvania, 1955).

> The reason based on causation (*kārya*) is, for example, if there is smoke, then there is fire.

This reasoning is expressed by *modus ponens*: "If there is smoke (*liṅga*, 'token'), then there is fire. On this locus (*dharmin*, "substratum of reasoning") there is smoke. Therefore, on this locus there is fire (*dharma*, 'property')".

Non-perception

The group of Yogācāra syllogisms containing negative premises (*anupalabdhi-anumāna*) consists of eleven types of non-perception schemes, enumerated in the *Nyāyabindu* of Dharmakīrti:

1) "The non-perception of the analytic token; for example, if there is the non-perception of what satisfies the conditions of knowledge [i.e., there is no analytic token for smoke], then there is no smoke (*svabhāva-anupalabdhir yathā, na atra dhūma upalabdhi-lakṣaṇa-prāptasya anupalabdhir iti*)" (Dharmakīrti, *Nyāyabindu* II: 31; Malvania, 1955). "If there is no analytic token for smoke (its general properties do not exist, not-*q*), then there is no smoke (not-*p*)". Formally it is *modus tollens*: $p \Rightarrow q$; $\neg q$; then $\neg p$, where the implication expresses the relationship of analyticity between the effect *p* and its analytic token (general property) *q*.
2) "The non-perception of the causal token; for example, if there is no smoke, then there is no efficient cause of it (*kārya-anupalabdhir yathā, na iha apratibaddha-sāmarthyāni dhūma-kāraṇāni santi dhūma-abhāvād iti*)" (Dharmakīrti, *Nyāyabindu* II: 32; Malvania, 1955). "If there is no smoke (not-*q*),

then there is no fire (not-p)". Formally it is *modus tollens*: $p \Rightarrow q$; $\neg q$; then $\neg p$, where the implication expresses the relationship of causality between the cause p and its effect q.

3) "The non-perception of the genus; for example, if it is not a tree, then it is not a *Dalbergia sisso* (*vyāpaka-anupalabdhir yathā, na atra śiṃśapā, vṛkṣa-abhāvād iti*)" (Dharmakīrti, *Nyāyabindu* II: 33; Malvania, 1955). "If it is not a tree (not-q), then it is not a *Dalbergia sisso* (not-p)". Formally it can be expressed as *modus tollens*: $p \Rightarrow q$; $\neg q$; then $\neg p$, where the implication expresses the relationship of analyticity between the species p and its genus q. But to the same extent, this reasoning can be expressed as *modus Camenes* (Figure IV of syllogism): "All *Dalbergia sisso* are trees; no trees are this substratum; therefore, this substratum is not a *Dalbergia sisso*". In this syllogism, "*Dalbergia sisso*" is a major term or *dharma* (the predicate of the conclusion), "this substratum" is a minor term or *dharmin* (the subject of the conclusion), and 'tree' is a middle term or *liṅga*.
4) "The perception of the contradictory by nature; for example, if there is fire, then there is no sensation of cold (*svabhāva-viruddha-upalabdhir yathā, na atra śīta-sparśo vahner iti*)" (Dharmakīrti, *Nyāyabindu* II: 34; Malvania, 1955). "If there is a sign p ('fire'), then there is no contradictory ('cold') q' to its signified q ('warm')". Either q or q'; if p, then q; p; then not-q', where the implication expresses the relationship of sign-inference between the token p and its signified q. Formally it is a composite *modus tollendo ponens* I of the Stoics: "It is either cold (q') or warm (q); but it is warm (q), because there is fire; hence, it is not cold (not-q')". Indeed, $q \oplus q'$; q; therefore $\neg q'$.
5) "The perception of an effect produced by the opposite; for example, if there is smoke, then there is no sensation of cold (*viruddha-kārya-upalabdhir yathā, na atra śīta-sparśo dhūmād iti*)" (Dharmakīrti, *Nyāyabindu* II: 35; Malvania, 1955). "If there is a token (p) of a cause (r) and this cause (r) has a next effect (q), then there is no opposition of the latter effect (not-q')". Formally it is a composite *modus tollendo ponens* I of the Stoics: "It is either cold (q') or warm (q); but it is warm (q), because there is smoke (p) and if there is smoke (p), then there is fire (r) and if there is fire (r), then it is warm (q); hence, it is not cold (not-q')". Indeed, $q \oplus q'$; q; therefore $\neg q'$.
6) "The perception of the invariable concomitant of the opposite; for example, the destruction of even a created entity (*bhāva*) is not something permanent (*dhruvabhāvin*), because it depends on a further cause [which in turn is invariably associated with the property of impermanence] (*viruddha-vyāpta-upalabdhir yathā, na dhūma-bhāvī bhūtasya api bhāvasya vināśaḥ, hetv-antara-apekṣaṇād iti*)" (Dharmakīrti, *Nyāyabindu* II: 36; Malvania, 1955). "If there is an effect of a cause, then there is no opposition to the invariable concomitant of this effect". Formally it is a composite *modus ponendo tollens* of the Stoics: "It cannot take place that it is both depending on a further cause (q) and permanent (q'); it is depending on a further cause (q), because it is a destruction of even a created entity (p); then it is not permanent (not-q')". Indeed, $\neg(q \,\&\, q')$; q; therefore $\neg q'$.

7) "The perception of a contrary to the cause; for example, if there is fire, then there is no efficient cause of cold (*kārya-viruddha-upalabdhir yathā, na iha apratibaddha-sāmarthyāni śīta-kāraṇāni santi, vahner iti*)" (Dharmakīrti, *Nyāyabindu* II: 37; Malvania, 1955). "Both *p* or *p'* cannot take place together; there is *p*; therefore, there is no *p'*." Formally it is the *modus ponendo tollens* of the Stoics: "It cannot take place that it is both fire (*q*) and its opposition (the cause of cold, *q'*); there is fire (*q*); then there is no efficient cause of cold (not-*q'*)". Indeed, $\neg(q \,\&\, q')$; *q*; therefore, $\neg q'$.
8) "The perception of a contradictory to the similar cases of the signified; for example, if there is fire, then the touch inhering in ice does not exist here (*vyāpaka-viruddha-upalabdhir yathā, na atra tuṣāra-sparśo vahner iti*)" (Dharmakīrti, *Nyāyabindu* II: 38; Malvania, 1955). "Either *q* or *q'*; *q*, because there is *p*; therefore, there is no *q'*." Formally it is a composite *modus tollendo ponens* I of the Stoics: "Either it is warm (*q*) or it is a touch inhering in ice (*q'*); it is warm (*q*), because there is fire (*p*); then there is no touch inhering in ice (not-*q'*)". Indeed, $q \oplus q'$; *q*; therefore $\neg q'$. This reasoning is close to the following Stoic conclusion: "Fire which heats by nature appears to all as heating, and snow which chills by nature appears to all as chilling" (Sextus Empiricus, *Pyrrhoniae Hypotyposes* III: 179; Bury, 1976, vol. I: 447). Therefore, if there is fire, then 'chilling' is impossible.
9) "The non-perception of the cause; for example, if there is no fire, then there is no smoke (*kāraṇa-anupalabdhir yathā, na atra dhūmo vahny-abhāvād iti*)" (Dharmakīrti, *Nyāyabindu* II: 39; Malvania, 1955). Formally it is *modus tollens*: if *p*, then *q*; not-*q*; then not-*p*, where the implication expresses the relationship of sign-inference between the causal token *p* and its signified *q*.
10) "The perception of the contradictory to the effect; for example, if there is a specific form of fire, then this person does not feel any special symptom of cold, such as having erect hair (*kāraṇa-viruddha-upalabdhir yathā, na asya roma-harṣa-ādi-viśeṣāḥ, saṃnihita-dahana-viśeṣatvād iti*)" (Dharmakīrti, *Nyāyabindu* II: 40; Malvania, 1955). "If there is fire, then there is no effect of having erect hair." Formally it is a composite *modus tollendo ponens* II of the Stoics: "It is either cold (*q*) or warm (*q'*); but it is not cold (not-*q*) (this person does not feel any special symptom of cold, such as having erect hair), because there is fire (*p*); hence, it is warm". Indeed, $q \oplus q'$; $\neg q$; then *q'*. The corresponding Stoic reasoning: "When I feel warm through the approach of fire, I take my own condition as a sign that the external substance of fire is warm, and the same may be said of the other objects of sense" (Sextus Empiricus, *Adversus logicos* I: 365; Bury, 1976, vol. II: 193). As a result, symptoms of cold are then excluded.
11) "The perception of the signified produced by the contrary token; for example, if there is smoke, then this symptom is not typical for a person who has some special symptoms of cold, such as having erect hair (*kāraṇa-viruddha-kārya-upalabdhir yathā, na roma-harṣa-ādi-viśeṣa-yukta-puruṣavān ayaṃ pradeśaḥ, dhūmād iti*)" (Dharmakīrti, *Nyāyabindu* II: 41; Malvania, 1955). "If there is

smoke, then there is no causal token of having erect hair". Formally it is the *modus ponendo tollens* of the Stoics: "It cannot take place that it is both smoke (q) and its opposition (the token of cold, q'); it is smoke (q); then there is no token of cold (not-q')". Indeed, $\neg(q \& q')$; q; then $\neg q'$.

Thus, in the group of inferences containing the negative premises (*anupalabdhi*, 'non-perception'), we see the following atomic inference rules: *modus tollens* in (1), (2), (3), (9); *modus Camenes* in (3); *modus ponendo tollens* in (11). Other schemes contain premises which are based on other schemes, such as *modus ponens*. These inference rules are presented by their composite versions: *modus tollendo ponens* I in (4), (5), (8); *modus ponendo tollens* in (6), (7); *modus tollendo ponens* II in (10). This shows that in the Indian logic of ancient times, we do not detect the Aristotelian or Stoic idea that some inference rules are atomic, and all conclusions can simply be considered a composition of these atomic rules. The atomic rules of the Stoics: *modus ponens*, *modus tollens*, *modus ponendo tollens*, *modus tollendo ponens* I, *modus tollendo ponens* II. They form the Stoic square of opposition, see Fig.1.2.1. In Dharmakīrti's inference rules from (1) to (11), we can reconstruct at least one analogue to the Stoic square of opposition, see Fig. 1.2.2.

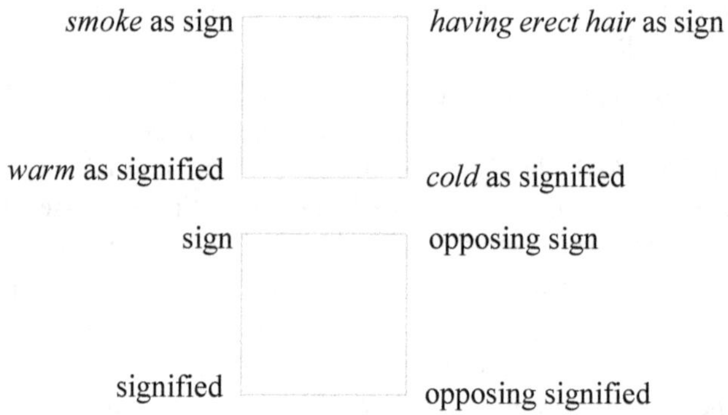

Figure 1.2.2. Dharmakīrti's square of opposition for opposing two signs 'smoke', "having erect hair" and two opposing signifieds 'warm', 'cold'. The difference from the Aristotelian square is that the bottom side of the square represents not a subcontrary relation, but a contradictory. This square is the same as the Stoic square, see Fig. 1.2.1.

Summing up, the Yogācāra logic represents the classical propositional logic in the Stoic version, but additionally the Yogācāra system has two inference rules from the Aristotelian syllogistic: *modus Barbara* and *modus Camenes*. Nevertheless, each syllogism, according to the Yogācārins, is different from the Stoic formulation, since it should contain one additional premise to show a generalization from our experience, as the Epicureans did. So, the *Yogācāra conclusion* of *modus ponens*, according to Dignāga, is as follows:

(C) q (*pakṣa*, *sādhya*, 'thesis'—what is inferred);
(Y_1) if p, then q (*hetu*, 'reason')
(Y_2) there is an example that p and q occur together (*dṛṣṭānta*, 'example').
(Y_3) p.

In the thesis, an inferential predicate (*ādhyadharma*) is attributed to an appropriate *dharmin*. According to the teachers of Dignāga, such as Vasubandhu (from the 4th to the 5th century A.D.) and Asaṅga (4th century A.D.), the resason (*hetu*) for inferring the inferential predicate (*ādhyadharma*) should satisfy three necessary conditions: (a) it must be a property of the inferential subject (*pakṣadharmatva*); (b) it must exist in all or some homogeneous instances for this subject (*sapakṣe sattvam*); (c) it must never exist in any heterogeneous instance for this subject (*vipakṣe 'sattvam eva*) (Hattori, 1968: 4). The method of *sapakṣe sattvam* has the meaning of the Epicurean similarity method (ὁ καθ' ὁμοιότητα τρόπος), and the method *vipakṣe 'sattvam eva* corresponds to the Stoic elimination method (ὁ κατ' ἀνασκευὴν τρόπος). In Dignāga's terminology, the similarity method was called *sādharmya* ('homogeneous'), and the elimination method was called *vaidharmya* ('heterogeneous'). The same two terms were used by Dharmakīrti. These two terms homogeneity and heterogeneity should be demonstrated in (Y_2) by means of some examples.

Let *p* be a token (*liṅga*) and *q* be an inferable object (*anumeya*). Then *p* must be present in the substratum (*dharmin*) of *q*, present in a similar case, and absent in a dissimilar case:

> *trairūpyaṃ punar liṅgasya anumeye sattvam eva, sapakṣa eva sattvam, asapakṣe ca asattvam eva niścitam | anumeyo atra jijñāṣita-viśeṣo dharmī* (Dharmakīrti, *Nyāyabindu* II: 5–6; Malvania, 1955).

> The three properties of a token, which must be satisfied, are its presence in the inferable object, its presence in the similar case (*sapakṣa*) and its absence in the dissimilar case (*asapakṣa*). An inferable object here is a substratum (*dharmin*) for which its property is aimed to be known (*jijñāsita*).

Dharmakīrti's formulation of the *Yogācāra conclusion* by *modus ponens*:

(Y_1) if *p*, then *q* (*pakṣadharmatā*, "setting a property of the property-bearer");
(Y_2) there is an example that *p* and *q* occur together (*vyāpti*, "invariable concomitance").
(Y_3) *p*.
(C) *q*.

This reasoning scheme is illustrated by Dharmakīrti in the following way:

> *kāryahetoḥ prayogaḥ: yatra dhūmas tatra agniḥ, yathā mahānasa-ādau. asti ca iha dhūma iti* (Dharmakīrti, *Nyāyabindu* III: 21; Malvania, 1955).

> The formulation of a reason based on causation [is as follows]: if there is smoke, then there is fire; for example, in the kitchen or any similar place. And smoke exists here.

It is worth pointing out that this structure of syllogism in India first appeared in the *Milindapañha*. In this text the syllogism consists of the following four propositions:

(M_1) If *p*, then *q* (*anumāna*);
(M_2) there is an example that *p* and *q* occur together (*opamma*);

Table 1.2.3. The injective mapping from the set of logical terminology of the Yogācāra doctrine into the set of logical terminology of the Stoics, the Epicureans, and Aristotle.

School of Philosophy	**Sources of True Knowledge**	**Real Entities**	**Logical System**	**Model Examples for Cause-and-effect Relationships**
Aristotle	Perception (αἴσθησις) and inference (ἀπόδειξις), and the inference is needed for unperceived or hidden phenomena (ἄδηλα).	Underlying thing (ὑποκείμενον/*subiectum*).	Aristotelian syllogistic, figures I, II, III, IV. We do not need to exemplify the major premises by similarities.	—
Stoics	Perception (αἴσθησις) and inference (σημείωσις), and the inference is needed for unperceived or hidden phenomena (ἄδηλα).	Pneuma or fire.	Classical propositional logic with the following atomic inference rules: *modus ponens*, *modus tollens*, *modus ponendo tollens*, *modus tollendo ponens* I, *modus tollendo ponens* II. We do not need to exemplify the major premises by similarities.	"If there is ['sign', σημεῖον, *signum*] smoke, then there is fire"; "If there is ['sign', σημεῖον, *signum*] fire, then it is warm"; "If there is ['sign', σημεῖον, *signum*] snow, then it is cold".
Epicureans	Perception (αἴσθησις) and inference (σημείωσις), and the inference is needed for unperceived or hidden phenomena (ἄδηλα).	Atoms.	Inductive logic. Some inference rules of the Stoics, such as *modus ponens*, but with a necessary exemplification of each major premise by means of similarities (ὁ καθ' ὁμοιότητα τρόπος).	—
Yogācārins	Perception (*pratyakṣa*) and inference (*anumāna*), and inference is needed for the unperceived (*apratyakṣa*).	Underlying thing (*sva-lakṣaṇa*).	*Modus Barbara* (figure I) and *modus Camenes* (figure IV) of the Aristotelian syllogistic, and all the inference rules of the Stoic logic: *modus ponens*, *modus tollens*, *modus ponendo tollens*, *modus tollendo ponens* I, *modus tollendo ponens* II, but without the understanding that there are atomic and composite inference rules. Each inference rule (its major premise) should be exemplified by similarities (*sapakṣa*).	"If there is ['sign', *liṅga*] smoke, then there is fire"; "If there is ['sign', *liṅga*] fire, then it is warm"; "If there is ['sign', *liṅga*] ice/having erect hair, then it is cold".
The Hellenistic Analogue for the Yogācārins	Aristotelian, Stoic, and Epicurean	Aristotelian	Aristotelian, Stoic, and Epicurean	Stoic

(M_3) *p* (observation);
(C) *q* (thesis and conclusion).

Thus, instead of four premises from (N_1) to (N_4) that are needed, according to the Naiyāyikas, for one conclusion, the Yogācārins applied only three premises (Y_1) to (Y_3), for drawing one conclusion. Meanwhile, (Y_1) expresses *hetu*, (Y_2) is an analogue for *udāharaṇa*, (Y_3) for *upanaya*, and (C), that is *pakṣanirdeśa*, is simultaneously the thesis of syllogism (analogue for the *pratijñā* of the Naiyāyikas) and its conclusion (analogue for the *nigamana* of the Naiyāyikas).

We can find some doctrinal differences between the teachings of the Stoics, the Epicureans, and the Yogācārins about sign-inferences (σημείωσις), but all three are related to the same logical theory, although with various aspects, and in the Yogācāra school we see a specific generalization of the Stoic and Epicurean logical teachings with small inclusions of Aristotelian ideas. These facts are summarized in Table 1.2.3.

To sum up, the Yogācāra doctrine on sign-inferences may be regarded as a synthetic conception combining not only the Stoic and Epicurean ideas, but also some Aristotelian ones.

1.2.5 Conclusions

So, in the Classical Antiquity we can find only four centers of philosophical reflection on logical conclusions based on signs (σημείωσις): Judaic, Chinese, Greek, and Indian. The Judaic and Chinese centers were exclusive and had no analogues. Meanwhile, the Judaic cultural-logical hearth developed in the form of Judaic logical hermeneutics, which in the Middle Ages was influenced by Aristotle, while the Chinese hearth did not become logic in the proper sense of the word. The Indian doctrine of logical signs (*liṅga*), together with the Greek doctrine (σημεῖον), cannot be considered exclusive, since there is a strong similarity between them, particularly, the Yogācāra system can be regarded as a kind of synthesis of the teachings of the Stoics, the Epicureans, and Aristotle. In subsequent chapters, it will be shown that for the first time the doctrine of logical signs was formed in Mesopotamia and refers to signs from the field of law and from the field of prediction. The Judaic logic also has its source in the legal culture of Mesopotamia.

1.3 Analysis and Classification of Approaches to the Definition of Logical Competence

1.3.1 Problem Setting

There may be two general approaches to an explication of logical knowledge: normative and phenomenological. In the normative approach, logical concepts and relationships are regarded as established through a creation of logical norms. This approach began with introducing some logical laws (by Aristotle and Gottfried Wilhelm Leibniz) such as *tertium non datur* and the principle of sufficient reason and then it was transferred into a qualitatively new level with creating the first axiomatic systems of logic (by Bertrand Russell, David Hilbert, and many others logicians of the beginning of the 20th century). The second approach emphasizes descriptions, identifications, and developments of logical structures in existing knowledge systems, for example, in jurisprudence. Nevertheless, even an explication of logic in the light of the phenomenological approach may have a normative character, too—such a way was implemented, for example, by E.V. Bulygin in his analysis of logical relations in legal discourse (Al'churron and Bulygin, 2013).

This section examines the nature of logical knowledge within the normative approach, namely, logic is presented as a *logical competence*—the normalized ability to build consistent conclusions, applying certain inference rules. As a result of such an analysis, it is possible to determine the very boundaries of the normalization of logical knowledge as such. Within the framework of this consideration, the logical competence is understood, first of all, as a social mechanism of cognition within the framework of common human ways of finding a consensus.

In the everyday life of people, the logical competence is detected only with some biased forms (Tverky and Kahneman, 1974). This raises problems how far logic or logical competence can be considered an innate ability of humans. In the 19th century, it was maintained (within the logical psychologism), see Heymans (1905), that logic gives some limits of our thinking and, therefore, is absolutely natural for us, because it is a part of our psychology. In this section, I'll show that logic is biased not only in our reasoning, but also in biological processes of transmitting signals from the external world, and it can be biased more or less, but its pure (unbiased) forms are almost impossible. So, logical competence is not natural and it arises as a special social practice of finding different forms of consensus among humans. In Subsection 1.3.2, the logical psychologism presenting logic as an innate human ability is sharply criticized. In Subsection 1.3.3, it is shown that biased and unbiased forms of reasoning are studied together in informal logic or argumentation theory. In Subsection 1.3.4, on the basis of intra- and inter-cellular effects caused by a lateral activation and a lateral inhibition, it is demonstrated that each cell receives signals most often through some 'biases'. Let us note that lateral activation can be treated as a state of satisfaction on the microlevel, while lateral inhibition as a state of stress. In Section 1.3.5, there are regarded some biased ways of intelligent reactions performed by different animals to show that these reactions repeat the lateral activation and the lateral inhibition, but now on the macrolevel. In Section 1.3.6, it is shown that logical competence without some biased manifestations can be realized only as a special social practice among humans and this practice was first formed within the Sumero-Akkadian legal tradition and only then it was reflected within philosophy.

1.3.2 Logical Psychologism and Its Criticism

Due to some experimental data of cognitive science (Tverky and Kahneman, 1974), it became known that logic is not an innate ability of humans and it can be cultivated only in the process of learning and practical applications. Accordingly, the use of logical competence in legal discourse or other human areas does not occur by itself, but it is possible only through a direct constitution of some norms.

In the 19th century, there was another general understanding of logical competence, according to which this competence was considered an integral part of human thinking and cognition, and logic itself with all the laws was reduced to psychology and its laws. This approach to logical competence (in the light of justifying the reducibility of logic to psychology) is called *logical psychologism*. According to it, logical competence as a whole is natural for human cognition (Heymans, 1905). The development of logic is just the development of reflection on the already existing forms of logical thinking.

The logical psychologism has its roots in traditional definitions of logic as a natural way of thinking, e.g., in this manner logic is defined by Antoine Arnauld (1612–1694) and Pierre Nicole (1625–1695):

> La logique est l'art de bien conduire sa raison dans la connaissance des choses, tant pour s'instruire soi-même que pour en instruire les autres.
> Cet art consiste dans les réflexions que les hommes ont faites sur les quatre principales opérations de leur esprit, *concevoir*, *juger*, *raisonner* et *ordonner* (Arnauld and Nicole, 1662/1992: 30).

> Logic is an art of leading one's reason well to the knowledge of things, both to educate oneself and to instruct others.
> This art consists of the reflections that men have made on the four main operations of their mind: *representation*, *judgement*, *entailment*, and *ordering*.

This understanding of logic was common for logic handbooks written from the 19th century to the early 20th century: logic is "a science of the laws of correct thinking" (Chelpanov, 1917: 1). The point is that logic gives general frameworks of our knowledge:

> Logic can do nothing but bring to our knowledge the conditions under which this subjective feeling of necessity arises, and express them in a general form (von Sigwart, 1895: 15).

Our cognitions are possible only due to logic:

> Logic is a science of the methods of thinking, with the help of which the human mind seeks to achieve the knowledge of existence (Povarnin, 1916: 3).
>
> Logic is a science that shows how thinking should be done in order for truth to be achieved (Chelpanov, 1917: 1).

There are several laws of our cognition (first of all, the principle of identity, the law of contradiction, and the law of excluded middle) and several forms of thinking (first of all, concept, judgement, and inference) which are studied in logic:

> Logic is the science of the laws and forms of correct constructions of thoughts (Vinogradov and Kuz'min, 1954: 3).

The logical psychologism was preserved in Russian handbooks of logic till now:

> The term 'logic' … is used both to designate a set of rules that manage the process of thinking, reflecting reality, and to designate the science of the rules of reasoning and the forms in which it is carried out. <…> It [*A.Sch.*: logic] explores thinking as a means of knowing the objective world, those forms and laws in which the world is reflected in the process of thinking (Getmanova, 2011: 11).

Friedrich Ludwig Gottlob Frege (1848–1925) was one of the first to offer sharp criticism of logical psychologism and to convincingly show that logic, nevertheless, is completely autonomous from psychology and represents an intersubjective knowledge obtained through the study of methods of compositional combination of meaningful sentences with achieving a conclusion (Frege, 1934). Logic, in the end, is a more general science than mathematics.

According to Frege, logic studies our thinking as drawing conclusions, but it does not study the external world and its representations:

> Die Gedanken sind weder Dinge der Außenwelt noch Vorstellungen. Ein drittes Reich muß anerkannt werden. Was zu diesem gehört, stimmt mit den Vorstellungen darin überein, daß es nicht mit den Sinnen wahrgenommen werden kann, mit den Dingen aber darin, daß es keines Trägers bedarf, zu dessen Bewußtseinsinhalte es gehört (Frege, 1918/1919: 69).

> Thoughts are neither things of the external world nor representations. A third empire must be recognized. What belongs to this agrees with the idea that it cannot be perceived by the senses, but with things that there is no need for an actor to whose consciousness it belongs.

Frege's approach to the logical competence is now called *logicism*. According to this approach, mathematics is reducible to logic, and the logical competence itself is understood as an ability to draw correct conclusions based on meaningful sentences chosen as axioms or hypotheses (assumptions). This ability is associated exclusively with a knowledge of inference rules, which give consistent results. These inference rules are formulated within constructing a logical system and can be different. In particular, in the 20th century the so-called non-classical logics arose, which have axioms and inference rules a lot different from those used by the creators of mathematical logic (Gottlob Frege, Bertrand Russell, and others), but they also provide consistent knowledge. For example, there are many-valued logics, where the law of excluded middle does not hold true, and paraconsistent logics, where the law of contradiction does not hold true, etc.

Thus, in logic there are no laws—it is just a study of inferences by their formalizations taking into account the fact that these formalizations can be different, because they are acceptable only due to their consistency:

> <...> ein neues Feld der Forschung eröffnet, indem die Formalisierung des logischen Schließens zu einer systematischen Beweistheorie verwertet wurde, welche die Frage nach der Tragweite der logischen Schlußweisen, die von der traditionellen Logik nur in einer sehr speziellen Form gestellt und gelöst wurde, in systematischer Allgemeinheit behandelt und durch deren Untersuchungsmethode die Probleme der Grundlagen der Mathematik mit den logischen Problemen in unmittelbaren Zusammenhang treten (Hilbert and Bernays, 1968: 44).

> <...> there was opened up a new field of research by utilizing the formalization of logical inference into a systematic theory of proof, which addresses the question of the scope of logical inference, which traditional logic posed and solved only in a very special way, dealt with in a systematic generality and whose investigation method brings the problems of the fundamentals of mathematics into direct connection with the logical problems.

Kurt Friedrich Gödel (1906–1978) showed, in turn, the boundaries of logicism and thus justified the autonomy of mathematics from logic (Gödel, 1931). As a consequence, logical competence is now understood as a unique ability that is not directly related to a mathematical competence. This is the ability, first of all, to automatically draw conclusions, avoiding contradictions. Investigations of this ability began to be called formal logic:

> Our subject is *logic*—or, as we may say more fully, in order to distinguish from certain topics and doctrines which have (unfortunately) been called by the same name, it is *formal logic*.
> Traditionally, (formal) logic is concerned with the analysis of sentences or of propositions and of proof with attention to the *form* in abstraction from the *matter* (Church, 1996: 1).

Other names of the study of inferences: theoretical logic (*die theoretische Logik*), mathematical logic (*die mathematische Logik*), and symbolic logic (*die symbolische Logik*), e.g., see Hilbert and Ackermann (1949: 1). Now logic as such is understood as this investigation: "Logic is a science that studies principles and methods of inference" (Mortari, 2001: 2); "Logic is the study of the methods and standards of inference" (Kelley, 2014: 2). On the other hand, logic also began to be studied in the context of cognitive science (Stenning and van Lambalgen, 2011, 2008).

1.3.3 Cognitive Biases and Informal Logic

From the new understanding of logical competence as an ability to mechanically draw consistent conclusions, some plans arose to create *artificial intelligence*—abstract machines (such as Alan Turing's machines), which themselves are able, based on the data received at the input of the system, to give new data (like conclusions) at the output from the system so that these new data contain already a new knowledge in the form of processing the received data according to the required algorithms. By the beginning of the 21st century, significant successes have been achieved in the creation of artificial intelligence systems, which have led to a series of information revolutions.

However, in the form of understanding logical competence in computer science, it is no longer a natural ability of thinking—neither of animals nor of humans. It is presented exclusively in the implementation of electronic devices, where there is information at the input as well as algorithms that transform it to obtain information at the output. Moreover, these algorithms are taken from classical mathematics with the assumption that all actions in them are sequential and deterministic.

At the beginning of the 21st century, in computer science, a new direction has emerged, called *unconventional computing*, which studies the possibility of representing any complex natural process (mechanical, chemical, biological) as a specific system of logical (or, more generally, arithmetic) functions (Adamatzky, 2017). For instance, it was shown that *Physarum polycephalum*, the unicellular organism behaving as a giant amoeba, can approximate some simple algorithms of Kolmogorov-Uspensky machines (Adamatzky, 2007) or algorithms of cellular automata (Shirakawa et al., 2015). This amoeba can be examined even as some biological logic gates (Adamatzky and Schubert, 2014; Jones and Adamatzky, 2010; Mayne and Adamatzky, 2015; Schumann, 2015; Whiting et al., 2014). Then these results were generalized to demonstrate that some proteins of amoeboid organisms are responsible for their intelligent behavior. Therefore, these proteins can be considered a medium implementing different logical and arithmetic functions. So, some networks of these proteins such as actin filament networks can be presented as a logic or computation medium, too (Adamatzky and Mayne, 2015; Alonso-Sanz and Adamatzky, 2016; Schumann, 2016, 2017; Siccardi and Adamatzky, 2015, 2016; Siccardi et al., 2016). Hence, these studies of actin filament networks and behaviors of unicellular organisms are being carried out in order to find out what specific logical and mathematical problems can be encoded in the biochemical processes of living organisms, thus presenting the organisms themselves in the form of computers—in other words, "the logical competence" of unicellular beings is studied (Schumann, 2018). One of the important results of these studies is that noticeable cognitive distortions were shown in the process of receiving signals by the single-celled organisms with subsequent reactions of the organisms to them. It turns out that cognitive distortions and violations of the rules of logic are a natural and integral part of cognition already at the level of behavior of unicellular organisms. Their "logical competence" is not perfect—the way computers 'think' is significantly different (Schumann, 2018, 2019c; Shirakawa et al., 2020).

We can show, therefore, that logic is implemented biologically only in some biased forms. A pure logic does not exist in nature. As we remember, in the philosophy of the 19th century, the idea that logic is a natural way of thinking for all humans was dominant within the logical psychologism. Nevertheless, in the 20th century it was proved experimentally that humans do not think logically in its strict sense surely—their logical competence is imperfect. They make gross blunders even at the primary level of logic. For example, the logical conjunction A & B is often understood not correctly—some referents do not know that its probability $P(A \& B)$ must be always smaller than the individual probability $P(A)$ or $P(B)$ (Tversky and Kahneman, 2008). It is an insensitive blunder in understanding conjunction as one of the basic logical operations.

Let $\vdash$ be a sign of logical deducibility. So, $A \vdash B$ means that from A we can deduce B, i.e., A is a premise for inferring B, and $\vdash B$ means that B is an axiom, i.e., it holds without premises. Then any inference of A from B in classical logic: $A \vdash B$, assumes that an appropriate implication $A \Rightarrow B$ holds. This statement is called the deduction theorem: if $A \vdash B$, then $\vdash A \Rightarrow B$.

In classical logic there are the following two basic inference rules:

- *modus ponens*: If $\vdash A \Rightarrow B$ and $\vdash A$, then $\vdash B$ (it means that from the true implication $A \Rightarrow B$ and the true statement A we can infer that B is true, also);
- *modus tollens*: If $\vdash A \Rightarrow B$ and $\vdash \neg B$, then $\vdash \neg A$ (it means that from the true implication $A \Rightarrow B$ and the true negation of B we can infer the true negation of A, as well).

These inference rules are always valid. There are similar deductions which are invalid in fact:

- *affirmation of the consequent*: If $\vdash A \Rightarrow B$ and $\vdash B$, then $\vdash A$ (but it is invalid that from the true implication $A \Rightarrow B$ and the confirmation of its consequent B we can infer that A is true);
- *denial of the antecedent*: If $\vdash A \Rightarrow B$ and $\vdash \neg A$, then $\vdash \neg B$ (but it is invalid that from the true implication $A \Rightarrow B$ and the denial of its antecedent A we can infer the denial of its consequent B).

Understanding that *modus ponens* and *modus tollens* are valid, while the affirmation of the consequent and the denial of the antecedent are invalid, constitute the nuts and bolts of elementary propositional logic. Without this understanding, any reasoning will contain fallacies. But it turned out that sometimes even the scientists cannot draw elementary correct conclusions from conditionals presented by true implications. So, seventy-two scientists including psychologists, biologists, and physicists from a large US university completed a questionnaire (Kern et al., 1983) to check their abilities to rate the validity of four forms of propositional inference: *modus ponens*, *modus tollens*, affirmation of the consequent, and denial of the antecedent. As a result, some of them made serious fallacies, see Table 1.3.1. In the meanwhile, there was detected the failure of nearly half of the scientists to recognize the logical validity of *modus tollens*, one of the two fundamental inference rules of propositional logic. This experiment shows that even the scientists do not think purely logically with avoiding unintentional fallacies.

Table 1.3.1 shows the gross fallacies scientists make in understanding *modus ponens*, *modus tollens*, affirming the consequent, denying the antecedent. Even the logic training for them is not so helpful.

As it was shown (Sjoberg, 2017), the fallacy of affirming the consequent is very common for biologists. Sometimes they recognize the following reasoning as logically valid: "If bipolar patients have reduced cognitive processing speed, then we will observe slower reaction time on a test. We observe a slower reaction time, and therefore bipolar patients have reduced cognitive processing speed". Furthermore, they often use the false analogy to extrapolate from sharing some characteristics between two groups of animals to sharing their other characteristics (Sjoberg, 2017).

Our reasoning with its biased forms is studied in two modern logical disciplines: critical thinking and informal logic (or argumentation theory). So, critical thinking is examined as an ability to detect cognitive biases to avoid them (Baker, 2020)

Table 1.3.1. Gross fallacies that scientists make in understanding *modus ponens*, *modus tollens*, affirming the consequent, denying the antecedent.

Referents	**Psychologists**	**Biologists**	**Physicists**	**Overall Percentage across Disciplines (no Logic Training)**	**Some Logic Training (n = 26)**
1. Recognizing the logical validity of *modus ponens*	91.5%	96%	96%	94.5%	100%
Abstract scheme:	83%	92%	92%	89%	100%
Concrete example:	100%	100%	100%	100%	100%
2. Recognizing the logical invalidity of denial of the antecedent	79%	66.5%	87.5%	77.5%	83.5%
Abstract scheme:	83%	58%	75%	72%	67%
Concrete example:	75%	75%	100%	83%	100%
3. Recognizing the logical invalidity of affirmation of the consequent	79%	62.5%	83.5%	72.5%	84.5%
Abstract scheme:	83%	75%	75%	73%	87%
Concrete example:	75%	50%	92%	72%	82%
4. Recognizing the logical validity of *modus tollens*	41.5%	58%	66.5%	55%	72%
Abstract scheme:	33%	33%	58%	41%	53%
Concrete example:	50%	83%	75%	69%	91%

Source: The data is taken from the work of L.H. Kern, H.L. Mirels and V.G. Hinshaw (1983).

and informal logic is defined as a science that studies both biased and unbiased inferences:

> Logic … is the branch of philosophy that studies the reasoning process and seeks to understand the differences between good and bad reasoning (Crews-Anderson, 2007: 8).

> Logic is the study of the methods and principles used to distinguish correct from incorrect reasoning (Copi et al., 2014: 2).

Thus, logic is not natural for humans and even for scientists and we study it to avoid some biased inferences. We then wonder if we can expect to see a pure logic in nature—how far biological processes of *Physarum polycephalum* or actin filament networks can implement the pure logic.

1.3.4 Lateral Inhibition and Lateral Activation as a Biological Mechanism of Logical Operations

We may observe the following fundamental duality in all active reactions of different organisms to external stimuli: each stimulus either attracts the organism or repels it.

So, each stimulus which changes the behavior of individuals, in fact can be considered either an attractant (such as food pieces or sex pheromone) or a repellent (such as predators or poisons). This duality is topological: the attractant can be occupied **from** different directions, but from the repellent each individual moves **in** different directions (Shirakawa et al., 2012; Fackler and Grosse, 2008; Pollard and Borisy, 2003; Van Haastert and Devreotes, 2004). On the basis of this topological duality, it is possible to present (i) the attractants $|p|, |q|, \ldots, |r|$, occupied by some individuals, as true propositional variables $p, q, \ldots, r$; and (ii) the repellents $|\dot{p}|, |\dot{q}|, \ldots, |\dot{r}|$, avoided by some individuals, as true negations $\neg p, \neg q, \ldots, \neg r$.

Evidently, different attractants (repellents) have a different intensity of their influence on individual behaviors: they can more or less attract (repel) organisms. This property can be expressed by a degree of truth (falsehood) presented by the interval of real numbers [0, 1], where it is possible to be closer to 0 (it means, do more to repel) or to 1 (do more to attract).

Now, let us consider several attractants $|p|, |q|, \ldots, |r|$ and several repellents $|\dot{p}|, |\dot{q}|, \ldots, |\dot{r}|$. Surely all the repellents $|\dot{p}|, |\dot{q}|, \ldots, |\dot{r}|$ must be avoided. Then this reaction of individuals can be treated as the following true conjunction: $\neg p \,\&\, \neg q \,\&\, \ldots \,\&\, \neg r$ (it means, $|\dot{p}|, |\dot{q}|, \ldots, |\dot{r}|$ are avoided simultaneously). However, in respect to the attractants $|p|, |q|, \ldots, |r|$, reactions of individuals can be very different. First of all, there are the following two typical reactions: (i) fusion as an attempt of focusing only on one attractant (or on a small number of attractants) to occupy it (them) first and foremost; (ii) splitting as an attempt of occupying several attractants at once, the more the better.

Both possible reactions to the attractants are explained by the following two complementary mechanisms in perceiving signals on the intra- and inter-cellular levels: (i) a lateral inhibition corresponding to a fusion, it expresses some stress conditions which increase the contrast of one signal to react more directly and efficiently and to be better focused; (ii) a lateral activation corresponding to a splitting, it shows some comfortable conditions which decrease the contrast of signals to perform concurrent actions (Jones, 2015; Schumann and Kuznetsov, 2018). These mechanisms in perceiving stimuli are scale-invariant. First, they occur during assembling and disassembling of actin filament waves to cause a fusion of protoplasmic tubes of *Physarum polycephalum* or splitting them. Second, they occur between groups of neurons within the visual cortex to process visual cues more or less focused. Third, they occur between active zones of swarms to react to their environments to occupy attractants more or less focused, too.

Repellents $|\dot{p}|, |\dot{q}|, \ldots, |\dot{r}|$ cause always a stress in individuals, therefore each individual tries to be concentrated on the repellents due to the lateral inhibition to avoid all of them in advance. But for attractants $|p|, |q|, \ldots, |r|$, each individual can be more laterally inhibited or more laterally activated. As a consequence, it starts to occupy all attractants or only some of them. The attractants in focus are connected by conjunction. The attractants which are not in focus are connected by disjunction. Hence, the lateral inhibition for signals $|p|, |q|, \ldots, |r|$ gives their increasing $p \,\&\, q \,\&\, \ldots \,\&\, r$, while the lateral activation for signals $|p|, |q|, \ldots, |r|$ gives their decreasing $p \vee q \vee \ldots \vee r$. The problem is that there is a smooth transformation between lateral inhibition and lateral activation. Therefore, there is the same smooth transformation between conjunction and disjunction which is difficult to be presented logically. In other words, conjunction as well as disjunction can be more or less biased (activated).

Many optical illusions occur, because our brain is trained to switch between lateral inhibition and lateral activation automatically under different conditions. For example, in the Müller-Lyer illusion, see Fig. 1.3.1, we perceive a length of a line depending on whether the line has the inward wings, see Fig. 1.3.1(a) (when we follow the lateral inhibition effect), or the outward wings, see Fig. 1.3.1(b) (the lateral activation effect). It is worth noting that T. Sakiyama and Y.-P. Gunji (2013) showed that the Müller-Lyer illusion holds for the group behavior of foraging ants. They explain this phenomenon by that each swarm of ants has the following two main logistic tasks: (i) to build a global route system connecting the nest with food sources to monopolize all reachable food sources (it corresponds to the lateral activation, that is, to the colony's ability to discover new food sources through exploration); (ii) to exploit effectively and efficiently each food source that is found (it corresponds to the lateral inhibition, i.e., to the colony's ability to concentrate on some food sources).

In the Ebbinghaus illusion (Fig. 1.3.2), as well as in the Delboeuf illusion (Fig. 1.3.3), perceiving the disc surrounded by bigger figures is laterally inhibited, therefore this disc seems to be smaller, while perceiving the disc surrounded by smaller figures is laterally activated, therefore this disc seems to be larger. It was shown (Howard et al., 2017) that a kind of Ebbinghaus or Delboeuf illusion can be observed in the behavioral patterns of honeybees. So, when the viewing distance was unrestricted, an appropriate illusion occurred. However, when the viewing distance was restricted, bees did not demonstrate perceiving either the Ebbinghaus or Delboeuf illusion.

Note that at least one Grey parrot (*Psittacus erithacus*) reports the Müller-Lyer illusion under certain conditions (Pepperberg et al., 2008). The Ebbinghaus and

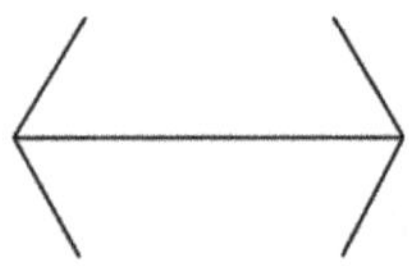

(a) Line with inward wings.

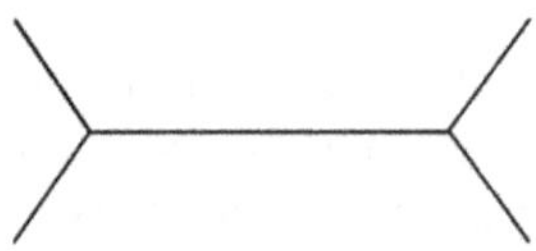

(b) Line with outward wings.

(c) Line without wings.

Figure 1.3.1. The Müller-Lyer illusion. The lines of (a), (b), (c) are of the same length. Nevertheless, it seems to us that the line with the inward wings (i.e., (a)) is shorter than line (b), and the line with the outward wings (i.e., (b)) is longer than line (a).

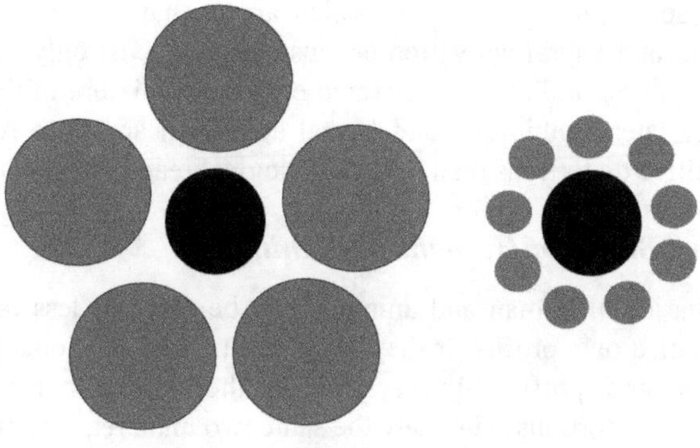

Figure 1.3.2. The Ebbinghaus illusion. The two discs in both centers are of the same size, but the right one seems to be larger than the left one because of the surrounding discs of different diameters.

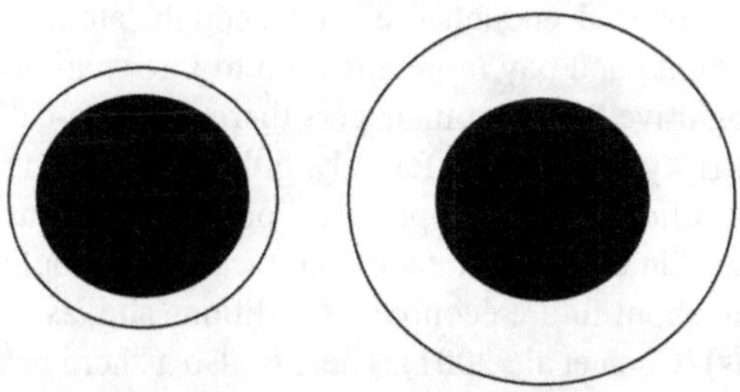

Figure 1.3.3. The Delboeuf illusion. The two discs in both centers are of identical size, but the left one seems to be larger than the right one because of the surrounding rings of different diameters.

Delboeuf illusions are well traced in perceptions of macaque monkeys and some other species of monkeys (Parrish et al., 2015; Huang et al., 2002). The point is that the perception of brightness in macaques and humans is quite similar because of common underlying neural computations in the two species. The same illusions under some conditions are well traced in perceptions of red-footed tortoises (*Chelonoidis carbonaria*) and bearded dragons (*Pogona vitticeps*) (Santacà et al., 2019). Nevertheless, they are not so well reported for domestic dogs (Byosiere et al., 2020). Thus, different mechanisms of lateral inhibition and activation may alter perceiving or not perceiving different illusions in a particular species. This data could improve our understanding of animal perception. In any case this perception is always biased. So, a reaction of any organism to stimuli can be more conjunctive (laterally inhibited) or more disjunctive (laterally activated) than it is in fact.

Hence, the negation as a logical operation has its biological meaning presented as appropriate responses to repellent stimuli. The conjunction as a logical operation may be biologically treated as laterally inhibited responses to several stimuli at once. The disjunction as a logical operation may be understood as laterally activated responses to several stimuli simultaneously. This allows us to implement some logic gates within the biological reactions (Whiting et al., 2014; Schumann and Pancerz, 2019). But the problem is that these implementations can be realized only with strict limits and strong approximations. The matter is that the negation, conjunction, and disjunction are some pure ideal forms which are unchangeable, while repelling, lateral inhibition, and lateral activation are instable and exist only in some hybrid forms with being more inhibited or activated or being a mixture of those reactions. And that is why lateral inhibition and lateral activation are more responsible for biases and cognitive distortions than for purely logical reactions.

1.3.5 Cognitive Biases of Humans and Animals

Behavioral scenarios of human and animals may be more or less rational—as an attempt to maximize own profits. If they are at least partly irrational (i.e., there are no forecasted maximal profits after carrying out these scenarios), they are called cognitive biases or distortions. There are the same two main reasons of many biases: stress as psychic reaction (analogous to the lateral inhibition with one neural network) and satisfaction (analogous to the lateral activation of neurons). For example, managers usually overestimate the probability of positive forecasting (Heaton, 2002)—they are not focused enough (i.e., not enough laterally inhibited—in the terms of neural networks) and pay more attention to information that supports their opinions. Among cognitive biases of managers there is the so-called framing (when they may frame the risky outcome differently, either considering the worst case or the best case) and confidence (their optimistic or pessimistic assessment of one's ability to perform, e.g., in a period of economic recession, people are less confident and more pessimistic about future economic conditions and, as a consequence, they avoid risky decisions) (Chen et al., 2001). There is also a "herd behavior", when they copy actions of a larger group due to the social pressure of conformity, e.g., some investors may follow others to buy the same stocks (Gilovich et al., 2002). The study of different cognitive biases and their influences on making financial decisions is presented in behavioral finance (Goldberg and von Nitzch, 2001), a new approach in finance studying some real irrational patterns of decision-making. So, managers are not logical enough and their decisions are at least partly always biased—through stress or satisfaction.

Now, cognitive biases and distortions are studied for animals, too, first of all for animals of zoo and aquaria to check their affective states to know whether their welfare is acceptable or it should be improved (Clegg, 2018). It is stated by Spruijt et al. (2001) that a balance of positive and negative affective states is always preferable for managing zoo or aquaria. For instance, cognitive biases are analysed for bottlenose dolphins (*Tursiops truncatus*) (Clegg et al., 2017) and western lowland gorillas (*Gorilla gorilla gorilla*) (Daoudi et al., 2017). It is learnt that under stress conditions animals follow more pessimistic judgments about ambiguous stimuli, pay more attention to negative stimuli, and remember more often negative memories (Clegg, 2018; Eysenck et al., 1991; Wright and Bower, 1992). As a consequence, it was detected (Robinson and Roiser, 2016) that, similar to humans, animals in their negative affective states exhibit negative affective biases on decision-making and memory tasks. Their emotional responses to a stimulus can be quick, but at the same time there are affective states which are longer (such as anxiety or depression) and they are not caused by a single stimulus but are the results of their experiences (Robinson and Roiser, 2016). There was developed a special system of judgment bias tasks (Roelofs et al., 2016) to assess emotional states of animals by a cognitive measure of their optimism and/or pessimism by recording behavioral responses to ambiguous stimuli.

Hence, cognitive biases are normal mechanisms of thinking not only of humans, but also of animals. People, like animals, do not think purely logically, which means that their logical competence, by definition, is far from perfect. At the same time, there are two main affects in decision-making (i.e., in building an inference): stress and pleasure. At the micro level, both affects are expressed by two mechanisms in the perception of signals: lateral inhibition (analogue of stress) and lateral activation (analogue of satisfaction).

Different forms of cognitive distortions in our reasoning are called *logical fallacies*. They are not directly related to stress and satisfaction as two essential moods of our emotions. Among them there are "motivational biases" which reflect a desire, involved in inferences, to achieve a result in discussions. Usually, these biases are committed unintentionally (Correia, 2011). The matter is that "people are more likely to arrive at conclusions that they want to arrive at, but their ability to do so is constrained by their ability to construct seemingly reasonable justifications for these conclusions" (Kunda, 1990: 480). In other words, people tend to seem rational—based on some arguments, but they do not know logical norms well. For example, the alcoholic can justify his or her excessive drinking by affirming: "I am not an alcoholic. I drink because…" (Correia, 2011: 119). Nevertheless, his or her further reasoning thereby will be imperfect logically. Also, there are some intentional fallacies such as the straw man fallacy presenting a strategy whereby the proponent misrepresents the opponent's position in a way that makes it easier to be refuted, and then refutes this made-up position instead of the real one (Talisse and Aikin, 2006; Johnson and Blair, 1983; Walton, 1989). In any case each logical fallacy has an appropriate belief that biases reasoning and gives an appropriate emotional distortion in accepting logically correct inferences (Cohen, 1992). Furthermore, logical fallacies are rooted in real discussion so that purely logical forms of inferring do not seem to be a real practice in the every-day argumentation (van Eemeren, 1986; Walton, 1989).

1.3.6 Logical Competence as a Kind of Social Practice

Within the framework of unconventional computing, the group or swarm behavior of animals is also being explored for their logical competence, namely, logical competence is discovered in the behavior of so-called 'social' animals (such as bees, ants or horses), which coordinate their behavior among themselves, distributing specific roles from a leader or queen to workers. As a result of these studies, a special direction of unconventional computing arose, called *swarm computing* (or *swarm intelligence*)—in it the logical competence is subjected to a logical and mathematical analysis in the form as it is presented in a particular swarm (for example, in an anthill, a nest of bees, a herd of horses, a flock of birds, a school of fish, etc.), see (Bonabeau et al., 1999; Schumann, 2020b). Swarms were found to be capable of solving complex logistic problems such as the traveling salesman problem (Dorigo, 1997) or the generalized assignment problem (Ozbakir et al., 2010).

But the swarm behavior is not social in the strict sense of the word—there are no pure conventions consciously shared among the members of the group. From the standpoint of behaviorism, any collective behavior is a complex system that is controlled by stimulating individuals. And in this regard, it would seem that the 'social' behavior of animals is just a simpler stimulation system than the behavior of humans. However, some inevitable cognitive distortions are also recorded in the group behavior of swarm animals, which shows the strict boundaries of behaviorism, its inapplicability in modeling complex forms of group behavior, in which noticeable cognitive distortions are simply inevitable, and, as a consequence, the same combination of stimuli can give different reactions, which contradicts the basic assumptions of behaviorism (Shirakawa et al., 2020; Schumann, 2019c).

In the theory of symbolic interactionism, the creators of which are George Herbert Mead (1863–1931) and Herbert Blumer (1900–1987), the social behavior is analyzed initially in the light of possible cognitive biases and distortions that are responsible for attributing a special symbolic meaning to some objects when these items get an added value. The founders of this theory noticed that people act in relation to things based on symbolic, and not direct, meanings of these things. In other words, some cognitive biases involved in the cognition of people are fixed in symbolic meanings shared by others, which, thereby, provide social interactions of people, i.e., interactions beyond the biological behavior. The social behavior is built, therefore, on symbolic interactions, which are simply impossible in the swarm behavior due to the absence of the symbolic meanings as such. It turns out that the social behavior is impossible without a material culture—without some objects that have acquired an additional symbolic meaning, consciously shared by this group of people.

At the same time, in order to stabilize societies, there is a need for a long-term consolidation of symbolic meanings and other conditions of symbolic interactions. This is how social institutions are born with legal support as long-term forms of social contact. As a result, the logical competence acquires a new meaning—it is a form of intersubjective decision-making without cognitive distortions from some existing conventions (social norms, provisions, assumptions), which, in turn, are built on symbolic meanings (cognitive biases involved in the formatting of social space).

Thus, the logical competence of people can be understood in different ways: (1) as an individual mechanism of cognition in combination with cognitive distortions; (2) as a swarm cognition combined with cognitive distortions; (3) as a social cognitive mechanism that can be cleared of natural cognitive biases. This research analyzes the origins and early genesis of logical competence in the third sense.

So, the logical competence of a person as a certain practice of thinking, devoid of logical fallacies, is not an innate ability, since in its natural form it is not separable from cognitive distortions, but is a special social practice of making logically correct decisions, which once arose and was subsequently developed and spread through a direct learning.

The logical competence without distortions is implementable within *artificial intelligence* that is presented as some useful techniques of human cognitions which are free of cognitive biases in advance. Artificial intelligence now is the study of designing expert systems, which help us to analyze the environments and produce actions which maximize success according to some algorithms. The idea of some machines with drawing conclusions in an automatic way which support us in decision-making and finding consensus is quite old and has been popular since Raymundus Lullius (ca. 1235–1315) and Gottfried Wilhelm von Leibniz (1646–1716). For instance, Leibniz claimed that his analytic language of reasoning based on arithmetic operations (*characteristica universalis*) offers the unbiased forms of conclusions and, therefore, it can be used for converting the Chinese empire to Christianity, because the *characteristica universalis* is the best way of finding any consensus including the religious one (Swetz, 2003).

Traditionally, Aristotle is considered the father of logic as a science of unbiased inferences. So, we can suppose that only after him we can trace back some samples of perfect logical competence. But it is not true. There is one tradition that is much older than the Aristotlian logic—it is a Sumerian-Akkadian legal hermeneutics established at least since Ur III (Culbertson, 2009; Schumann, 2019b, 2020a), i.e., about 4,000 years ago. Within this hermeneutics, laws were presented as law codes, where each article from the code was formulated as a conditional statement (implication): "If an act of *p* is conducted, then its actor is sentenced to *q*". As a consequence, each court should make a decision in accordance with *modus ponens*:

Law article from a code: "If *p*, then *q*";
Fact-checking at the trial: "There is a fact *p*";

Court decision: "There is a sentence *q*".

Many legal documents written in Sumerian and different dialects of Akkadian have been excavated: law codes, trial records, contracts, testaments, etc. And their authors possess such a high logical competence with demonstrating the good knowledge of various inference rules: *modus ponens*, *modus tollens*, introducing conjunction, introducing disjunction, etc. (Schumann, 2019b, 2020a). Furthermore, in these documents there are no logical fallacies and we can explicate ten or even more logical steps in inferring. This tradition is unique and has no analogues in other ancient cultures.

Hence, the Sumerian-Akkadian legal hermeneutics has its origin in Mesopotamia and has been expanded later through different empires: Neo-Assyrian, Neo-Babylonian, and Achaemenid, then through Hellenistic states, the Roman Empire, and some partly Hellenized states of Central and South Asia participating in trading networks of the Silk Road, such as the Kuṣāṇa Empire. Laws from state to state can have been different, but the legal hermeneutics, at the same time, was very similar, i.e., the legal matters were changeable, but the logical forms in making legal decisions on the basis of laws were constant. First this hermeneutics was mainly in Akkadian, then in Aramaic and Greek, and only then in some other official languages of the Silk Road such as Bactrian or Sogdian.

Let us show that marriage contracts composed in Akkadian, Aramaic, Greek, and Bactrian have had the same structure since the Old Babylonian period and, accordingly, they have belonged to the same joint legal hermeneutics indeed.

Furthermore, it is known that Aramaic legal documents were a continuation of legal tradition in Akkadian. Then Greek became the *lingua franca* of diplomacy and international trade with many clichés borrowed from the Babylonian jurisprudence.

Marriage contracts were needed for obtaining a legal status of wife and husband in an appropriate society according to the Old Babylonian law codes. For example, in the Code of Eshnunna, §27–28, we read:

> [**§27**] *šum-ma* lú dumu. SAL l[ú *b*]*a-lum ša-al a-bi-ša ù um-mi-ša i-ḫu-si-ma* ⌈*ù*⌉[2] [*k*]*ir-ra-am ù ri-ik-⟨sa⟩-tim a-na a-bi-ša ù um-mi-ša la i*[*š-ku-u*]*n* u_4*-mi* mu.l.kam *i-na é-šu li-ši-im-ma ú-ul* dam
> [**§28**] *šum-ma ul*[3] *ri-ik-sa-tim ù kir-ra-am a-na a-bi-ša ù um-mi-ša iš-ku-un-ma i-ḫu-us-si* dam u_4*-um*[4] *i-na su-un* lú ⌈*iṣ-ṣa-ab*⌉*-ba-tu i-ma-at ú-ul i-ba-al-lu-*⌈*uṭ*⌉.

> **§27** If a man took another man's daughter without asking her father and mother and did not arrange for (lit. establish) a libation and marriage contract with her father and mother, though she live in his house for a year, she is not a wife.
> **§28** If... he arranged for a marriage contract and libation with her father and mother and took her, she is a wife; the day she is caught with (another) man she shall die; she shall not live (Greengus, 1969: 505).

More formally:

> **§27** If *r* (a man and a woman live together) and not-*p* (there is no marriage contract), then not-*q* (she is not a wife).
> **§28** If *p* (there is a marriage contract), then *q* (she is a wife). If *q* (she is a wife) and *s* (she is caught with another man), then *t* (she gets death penalty).

The Code of Hammurabi, §128, has almost the same article as §27:

> [**§128**] *šum-ma a-wi-lum aš-ša-tam i-ḫu-uz-ma ri-ik-sa-ti-ša la iš-ku-un* SAL *ši-i ú-ul aš-ša-at.*

> **§128** If a man took a wife and did not arrange for her marriage contract, that woman is not a wife (Greengus, 1969: 505).

Let us assume that there are two facts: not-*r* and *p*, checked at the trial, i.e., a man and a woman do not live together, but they have a marriage contract. Then, according to §28 and *modus ponens*, the court logically infers that *q*, i.e., she is a wife. If we have *r* and *p*, the court again infers *q* due to §28. Hence, the articles are formulated in the way that allows us to apply inference rules of propositional logic to entail legal decisions automatically—for example, to prove that there is a lawful marriage in fact.

Something is treated as a fact in the Sumero-Akkadian hermeneutics only after checking at the trial. To be a fact any state of affairs should be either fixed in a well-composed legal document (past trial decision, contract, will, etc.) or testified by at least two witnesses. In legal documents, one main thesis as one declarative or a series of declarations (e.g., that the woman is a wife) should be well formulated. Then the declarations are supported by several witnesses who sign (i.e., put their seals on) an appropriate document containing this main thesis. Each additional declarative of the legal document is formulated in the form of implication: "If a state of affair *p* takes place, then a legal decision *q* shall be made" so that the next court can use this conditional declarative for making the next decision *q* just by applying *modus ponens* after checking the fact of *p*.

Let us quote the marriage contract composed at the time of Erra-Imittī, the ninth king of the first dynasty of Isin who ruled in ca. 1868–1861 B.C.:

Obv.
^{m}*reš-tum* dumu-munus ur-^{d}utu
^{md}*šu-i-lí-šu*
dumu ^{m}*èr-ra-mu-ba-lí-iṭ*
nam-dam-šè in-ni-tuku
mu-lugal-ka-ni in-pàd
túg-íb-lá nu-mu-un-zu-[a]-ni\in-du$_8$
é a-šà ù níg-gur$_{11}$-ra a-na [me-a-bi]
lú-lú-gin$_7$ al-dím
tukum-bi u$_4$-kúr-šè
^{m}*reš-tum* ^{md}*šu-i-lí-šu* dam-a-ni-éš
dam-ĝu$_{10}$ nu-me-en ba-na-an-du$_{11}$
½ ma-na kù-babbar ì-lá-e
ù tukum-bi ^{md}*šu-i-lí-šu*
^{m}*reš-tum* dam-a-ni-éš
Rev.
dam-ĝu$_{10}$ nu-me-en ba-⟨na⟩-an-du$_{11}$
½ ma-na kù-babbar ì-lá-e
mu-lugal-bi in-pàd-dè-eš
[igi] ^{m}*na-bi*-^{d}EN.ZU dumu ^{m}*na-bi-ì-lí-šu*
igi [^{m}*a*]-*ta-na-aḫ* dumu ^{m}*na-bi-ì-lí-šu*
igi *a-du-ba* dumu èr-^{d}en-líl-lá
igi ^{md}EN.ZU-*ki-ma-ì-lí-a*
dumu ^{md}EN.ZU-*mu-ba-lí-iṭ*
igi ^{m}*a-ad-da-wa-qar* [dumu xx]-*lí-wa-qar*
igi ^{m}*àm-ma-nu-um* [lú]-nar
igi ^{md}EN.ZU-*i-din-na-šu* DUMU *la-bi-šu*
iti giš-apin-du$_8$-a ud-15-kam
mu ^{d}*èr-ra-i-mi-ti* lugal
4 urudu ur-maḫ-gal-gal
é-AN-nir-ra mu-na-dím
Seal impression: ^{md}*šu-i-lí-šu*

Obv.
Rēštum daughter of Ur-Utu
Šū-ilīšu
son of Erra-muballiṭ
took her for wife
he swore in the name of the king

> he has loosened her “belt of virginity”
> the house, the field and the possessions,
> as much as [they may be]
> they shared like equals.
> **If** in future,
> Rēštum to Šū-ilīšu, her husband
> says: you are not my husband
> she will weigh out ½ mina of silver
> **and if** Šū-ilīšu
> to his wife
> **Rev.**
> says: you are not my wife
> he will weigh out ½ mina of silver
> They swore in the name of the king.
> Witnesses:
> **Nabi-Sîn son of Nabi-ilīšu**
> **Atanaḫson of Nabi-ilīšu**
> **Aduba son of Warad-Enlila**
> **Sîn-kīma-ilīya son of Sîn-muballiṭ**
> **Adda-waqar son of … -li-waqar**
> **Ammanum [the] singer**
> **Sîn-iddinaššu son of Labišu**
> Fifteenth day of November
> year king Erra-imittī
> 4 large copper lions
> he fashioned on the Ziggurat.
> Seal impression: Šū-ilīšu
> (Al-Shamari and Al-Jalili, 2020: 130).

This legal document is signed by several witnesses who thereby support the following two declarations: (1) Rēštum is a wife of Šū-ilīšu; (2) the property is owned by both spouses equally (50/50). Then there are two additional declarations formulated as conditionals: (1) **if** (tukum-bi) Rēštum declares that Šū-ilīšu is not her husband, **then** she pays ½ mina of silver; (2) **if** (tukum-bi) Šū-ilīšu declares that Rēštum is not his wife, **then** he pays ½ mina of silver.

Marriage contracts in Aramaic have almost the same logical structure as those documents in Akkadian (Yaron, 1958), which is not surprising taking into account the fact that it is the same Babylonian legal tradition. Aramaic began to supplant Akkadian as a new legal language from the 6th century B.C.

Many legal documents in Aramaic were excavated in Elephantine (an island on the Nile). At the same place were excavated legal documents in Greek also, but dated later than the Aramaic ones. It shows that Greek started to supplant Aramaic as a new legal language from the 4th century B.C.

Let us consider the marriage contract between Heraclides and Demetria, composed in 311/310 B.C.—at the time of Alexander, the satrape of Ptolemy:

Ἀλεξάνδρου τοῦ Ἀλεξάνδρου βασιλεύοντος ἔτει ἑβδόμωι, Πτολεμαίου σατραπεύοντος ἔτει τεσαρε-
σκαιδεκάτωι(*) μηνὸς Δίου. συγγραφὴ συνοικισίας Ἡρακλείδου καὶ Δημητρίας. λαμβάνει Ἡρακλείδης(*)
Δημητρίαν Κώιαν γυναῖκα γνησίαν παρὰ τοῦ πατρὸς Λεπτίνου Κώιου καὶ τῆς μητρὸς Φιλωτίδος ἐλεύθερος
ἐλευθέραν προσφερομένην εἱματισμὸν(*) καὶ κόσμον (δραχμῶν) Α, παρεχέτω δὲ Ἡρακλείδης Δημητρίαι
ὅσα προσήκει γυναικὶ ἐλευθέραι πάντα, εἶναι δὲ ἡμᾶς κατὰ ταυτὸ(*) ὅπου ἂν δοκῆι ἄριστον εἶναι βουλευομένοις κοινῆι
βουλῆι Λεπτίνηι καὶ Ἡρακλείδηι(*). εἰὰν(*) δέ τι κακοτεχνοῦσα ἁλίσκηται ⟦ἁλίσκηται⟧ ἐπὶ αἰσχύνηι τοῦ ἀνδρὸς Ἡρακλείδου Δημητρία,
στερέσθω ὧμ(*) προσηνέγκατο πάντων. ἐπιδειξάτω δὲ Ἡρακλείδης ὅτι ἂν ἐγκαλῆι Δημητρίαι ἐναντίον ἀνδρῶν τριῶν,
οὓς ἂν δοκιμάζωσιν ἀμφότεροι. μὴ ἐξέστω δὲ Ἡρακλείδηι γυναῖκα ἄλλην ἐπεισάγεσθαι ἐφ' ὕβρει Δημητρίας μηδὲ
τεκνοποιεῖσθαι ἐξ ἄλλης γυναικὸς μηδὲ κακοτεχνεῖν μηδὲν παρευρέσει μηδεμιᾶι Ἡρακλείδην εἰς Δημητρίαν·
εἰὰν(*) δέ τι ποῶν(*) τούτων ἁλίσκηται Ἡρακλείδης καὶ ἐπιδείξηι Δημητρία ἐναντίον ἀνδρῶν τριῶν, οὓς ἂν δοκιμάζωσιν(*)
ἀμφότεροι, ἀποδότω Ἡρακλείδης Δημητρίαι τὴμ(*) φερνὴν ἣν προσηνέγκατο (δραχμῶν) Α, καὶ προσαποτεισάτω ἀργυρί-
ου Ἀλεξανδρείου (δραχμὰς) Α. ἡ δὲ πρᾶξις ἔστω καθάπερ ἐγ(*) δίκης κατὰ νόμον τέλος ἐχούσης Δημητρίαι καὶ τοῖς μετὰ
Δημητρίας πράσσουσιν ἔκ τε αὐτοῦ Ἡρακλείδου καὶ τῶν(*) Ἡρακλείδου πάντων καὶ ἐγγαίων καὶ ναυτικῶν. ἡ δὲ συγγραφὴ
ἥδε κυρία ἔστω πάντηι πάντως ὡς ἐκεῖ τοῦ συναλλάγματος γεγενημένου, ὅπου ἂν ἐπεγφέρηι Ἡρακλείδης κατὰ
Δημητρίας(*) ἢ Δημητρία τε καὶ τοὶ(*) μετὰ Δημητρίας πράσσοντες ἐπεγφέρωσιν(*) κατὰ Ἡρακλείδου. κύριοι δὲ ἔστωσαν(*) Ἡρακλεί-
δης καὶ Δημητρία(*) καὶ τὰς συγγραφὰς αὐτοὶ τὰς αὑτῶν φυλάσσοντες καὶ ἐπεγφέροντες κατ' ἀλληλως(*). μαρτερες(*)
Κλέων Γελῷιος, Ἀντικράτης Τημνίτης, Λῦσις Τημνίτης, Διονύσιος Τημνίτης, Ἀριστόμαχος Κυρηναῖος, Ἀριστόδικος
Κῷιος.
(seal) (seal) (seal) (seal)

In the 7th year of the reign of Alexander, son of Alexander, the 14th year of the satrapship of Ptolemy, in the month Dius. Marriage contract of Heraclides and Demetria. Heraclides takes as his lawful wife Demetria, Coan, both being freeborn, from her father Leptines, Coan, and her mother Philotis, bringing clothing and ornaments to the value of 1000 drachmae, and Heraclides shall supply to Demetria all that is proper for a freeborn

wife, and we shall live together wherever it seems best to Leptines and Heraclides consulting in common.
If Demetria is discovered doing any evil to the shame of her husband Heraclides, she shall be deprived of all that she brought, but Heraclides shall prove whatever he alleges against Demetria before three men whom they both accept.
It **shall not be lawful** for Heraclides to bring home another wife in insult of Demetria nor to have children by another woman nor to do any evil against Demetria on any pretext. **If** Heraclides is discovered doing any of these things and Demetria proves it before three men whom they both accept, Heraclides shall give back to Demetria the dowry of 1000 drachmae which she brought and shall moreover forfeit 1000 drachmae of the silver coinage of Alexander. Demetria and those aiding Demetria to exact payment shall have the right of execution, **as if** derived from a legally decided action, upon the person of Heraclides and upon all the property of Heraclides both on land and on water.
This contract shall be valid in every respect, wherever Heraclides may produce it against Demetria, or Demetria and those aiding Demetria to exact payment may produce it against Heraclides, **as if** the agreement had been made in that place.
Heraclides and Demetria shall have the right to keep the contracts severally in their own custody and to produce them against each other.
Witnesses **Cleon**, **Gelan**; **Anticrates**, **Temnian**; **Lysis**, **Temnian**; **Dionysius**, **Temnian; Aristomachus**, **Cyrenaean**; **Aristodicus**, **Coan**.
(Hunt and Edgar, 1932: 2–3).

We see that several witnesses signed their confirmation that (1) Demetria is a wife of Heraclides, (2) Demetria has a dowry with its description, (3) the property obtained after the marriage is owned now by both spouses equally (50/50). The two additional declarations formulated as implications: (1) **if** Demetria is caught with another man **and** this fact is confirmed by at least three witnesses, **then** she shall be deprived of the dowry; (2) **if** Heraclides brings home another woman with the testimony of three witnesses **or** he has children by another woman with the testimony of three witnesses **or** he is caught with another woman with the testimony of three witnesses, **then** he shall give back to Demetria her dowry and shall moreover pay 1000 drachmae. To infer from the second implication, we should apply the inference rule of deleting disjunction and only then *modus ponens*.

Legal documents in Bactrian (one of the Eastern Middle Iranian languages) were a continuation of legal tradition in Aramaic and Greek. Furthermore, Bactrian was being written in the Greek script and it supplanted Greek as a new official language in ca. 127 A.D. in today's Afghanistan and Pakistan, according to the edict issued by Kaniṣka I, the emperor of Kuṣāṇas (Rabatack inscription).

Let us quote one of the Bactrian marriage contracts:

Document A

A1 χϸονο ρ’ ι’ Αυρηζνο μαο σαχτο Αβαμοχοινο ρωσο καλ̣δο νοβιχτο μο ολοβω-
A2 στογο μαλο αβο μο ρωβαγγο ϸαρο αβιιοανδαγο σιδο στηβο ριζδο αζδηβδο ειμουανο αστ-
A3 ηβιγ̣ανο αζαβοργανο οιγαλφανο μολροδηβγα̣νο κ̣ιδο ι ωνδο λιβο ζαν̣ινδο ταδο αβο μασκο̣
A4 πιδοναμγο νοβιχτιγινδο οδο κιδο ι ωνδο λιβο ναζανινδο ταδοιηνο αβο μο νιγο λιστοβαρο κ̣ι-
A5 ρδο πισ̣ο οινδοωρομοζδο κολαγανο ιανδαγοβιδο οδο πισο χοαδηολαδο ιαστονικανο οδο π̣ι-
A6 σο οαραζοωρομοζδο χοασραογανο οδο πισο αβδαβοκο παβογανο οδο πισο ασπαλομιρο
A7 ιωλικανο οδο πισο βαγοβαο ιαβιϸταδο οδο πισο μανδαρουανισο στηβιγαν̣ο αζαβοργανο̣
A8 οιγαλφανο μολροδηβγανο πιδοοασατο χοακαμο χοασινδο βαγοφαρνο ζαμωδο̣ πορο̣
A9 οδο βαβο οδο πιδοκο βαγοφαρνο ποραναν̣ο ζαουοζαδγανο κιδο ωσο νοβανδο νινδοκο ι ω-
A10 χϸοβαδογα̣νο πιδο αζαδο ασπασο ποροχηζαμο μισιδο ζιϸτο μα̣νο βαγοφαρνο ασαφαγο
A11 φαροοηϸο οδασαφαγο νωγοσανινδο ιαλωγδο ασονωυο ειο ζινο κιδο ραλικο να̣μο φαρι-
A12 μουανο ποραναν̣ο ολο φαρο βαβο οδο̣ φαρο πιδοκο ιθαυατανο ληρηιο αζο βαγοφαρνο̣
A13 αβο μασκο νοβιχτισο ζινο κιδο ραλικο ναμο ασνωυο ασνωυογωγγο αβο οισπο αστο σ̣ιδ̣ο̣μ-
A14 ηνο ωσο αστο οδο σιδανο ασο μαβρισο αλφανζαμηιο ιθασαγωνδο ασνωυο πορδο αβο
A15 λριγδο σαγωνδο αβο μο λαμο νοβανδανο οτο πιδοοασατο βαβο οδο πιδοκο αγγιτιδο αμαχ-
A16 ο μανο βαβο οδο πιδοκο αβο ραλικο ολο ολογωγγο αβο οισπο αστο σιδομηνο ωσο αστ-
A17 ο οδο σιδανο ασο μαβρισο αλφανζαμηιο φινζο φρομανζο ιθασαγωνδο αβο μο λαμο νοβα-
A18 νδανο οτανο μαπαδοαϸχανηιο αζο βαβο οδαζο πιδοκο ανδαρο ολο αβο κιρδο οδο μαζαδ̣ο πιδο-
A19 ρωφσο αβο λριγδο σιδανο ραλικο μαυοσινδηιο οδο καλδανο αζο βαβο αλαζο πιδοκο ανδαρο̣
A20 ολο κιρινδηιο αλο αζαδο πιδορωφσο ληρινδηιο σιδανο ραλικο μαυοσινδηιο ταδανο̣
A21 λαυινδηιο ταρ̣ανο αβο ϸαογανο γαζνο ζαροζιδγο κ› διναρο οδο δαφρηλο φαρο παδαρλο̣-

A22 γο [ο]το πιδοοασατο νινδοκο μοζδαοανινδο πορο οδο ιαμϸοβανδαγο οδο παπο οδο ιατο̣

A23 νινδοκο πορανανο κιδο καδγο ωχϸοβαδογανο ριζδο κοοαδο αμαχο αβο ραλικο χοαδο ζ-

A24 ιϸτο οτο πιδιδο ιειρο υοσινδαμο οτανο μαπαδοαϸχανινδηιο αζο νινδοκο οδαζο ιαμϸο̣-

A25 βανδαγο οδαζο παπο οδαζο ιατο φαρο ραλικο ιαοαργο οδο ταραζο αβο λαδο οδο μασιδανο α-

A26 σο̣ ραλικ̣ο αζιιηιο οδο σιδανο ασο ραλικο πορο αζιιηιο ταδανο πιδο χοβο χοαρηιο̣

A27 οτο πιδο υαργο κιρη̣ιο πιδο αζαδο ασπασο σαγωνδο νιιαγο οδο πιδο κιρδο οδο σιδα̣νο̣

A28 λ̣ογδο αζιιηιο τα̣κ̣οοαδανο πιδο οδο ιμαδο οδο γωυριγο υ̣οσινδηιο αβηγ̣ο αβο λαδο ταδ̣[ο]

A29 ιανο λανινδηιο οδο καλδανο αζο νινδοκο αλαζο ιαμϸοβανδαγο αλαζο παπο αλαζο ιατο οιζα̣-

A30 γγ[οδ]αρο κιρη̣ιο αλδανο φαρο ραλικο ιαοαργο αλο ταραζο λαυινδηιο οδαλδο ιθαχοανι̣ γ̣ δ̣-

A31 ηιο κοοαδομηνο αβο ραλικ̣ ο ζα̣δικανο βανζο αλο μαρηγο βουδο βηγο πιδο αζαδο ασπασο

A32 σαγωνδο νιιαγο οδο πιδο ταδο λαυινδηιο ταοανο αβο ϸαογανο γαζνο ζ[α]ροζιδγο κ' δι̇ναρο

A33 οδο δαφρηλο φαρο παδαρλογο οτομηνο ιθαησο χοανδο ο̣δ̣ο̣ ο̣ι̣ σ̣α̣ρο̣[ναοαο]ορο αβουδ̣ηιο

A34 ναμαγγο̣ κωσ̣οβο α' βαρζο α' οανοσ̣τ̣ ο̣σ̣ι̣ γο α' ραγαζο •'[...]δ̣ ομιγγο α' λ̣ ι-

A35 σ̣τ̣ οβα̣νδιγο δ' ϸκ̣ ομιγ ο β' καφϸινο ιογδιγο γ' ποσο̣ β' [...γανδο]μο μιδο̣ γ''

Document A

(It was) the year 110, the month Ahrezhn, the day Abamukhwin having elapsed, when this marriage contract was written [2]here in the borough of the city of Rob which is called Steb, with the cognizance of these [3]freemen of Asteb, the witnesses (and) those who have impressed (their) seals (upon it), who witness the present document and [4](whose) signatures are written hereupon—and those who do not witness the present document have put (their signatures) on the other manuscript—[5](namely) in the presence of **Wind-ormuzd Kulagan** the chief of the district, and in the presence of **Khwadew-lad Yastunikan**, and in the presence of **Waraz-ormuzd Khwas-rawgan**, and in the presence of **Abdabuk Pabugan**, and in the presence of **Aspal-mir** [7]**Yolikan**, and in the presence of **Bag-bandag** the master craftsman, and also in the presence of the other freemen of Steb, [8]the witnesses (and) those who have impressed (their) seals (upon the documents). (Then this) declaration was freely (and) willingly made (by me), Bag-farn, son of Zamod, [9]and (by us), Bab and Piduk the legitimate sons of Bag-farn, who now serve with Ninduk [10]Okhsh-badugan in free service.

Now: there has been requested by me, Bag-farn, from you, [11]Far-wesh, and from you, Nog-sanind, this woman whose name (is) Ralik (as) a fully privileged daughter-in-law, [12](as) a wife for these (my) sons, Bab and Piduk, so that I, Bag-farn, **may treat** [13]the woman described herein, whose name (is) Ralik, (as) a daughter-in-law (to be treated) like a daughter-in-law, in every home which we [14]now have and also which we may acquire in the future, as (one) ought to treat a daughter-in-law, [15]as (is) the established custom in the land.

And a declaration was made (by us), Bab and Piduk. We have received [16]Ralik—I, Bab, and (I), Piduk—(as) a wife (**to be treated**) like a wife in every home which we now have [17]and also which we may acquire in the future, (as) a lady possessing authority, as (is) the established custom in the land. [18]And I, Bab, and I, Piduk, **shall not have the right** to make another (woman our) wife, **nor** [19]to keep a free (woman as a) concubine, to whom Ralik should not agree; and **if** I, Bab, or I, Piduk, [20]should make another (woman our) wife **or** keep a free (woman as a) concubine, to whom Ralik should not agree, **then** [21](we) shall give a fine to the royal treasury of twenty dinars of struck gold and the same to the opposite party.

[22]And a declaration was made (by us), Ninduk son of Muzda-wanind, and Yamsh-bandag and Pap and Yat, [23]the sons of Ninduk, whose house is called Okhshbadugan, that we ourselves have requested Ralik, [24]and in this matter we are in agreement, and (we) **shall not have the right**—I, Ninduk, and I, Yamsh-[25]bandag, and I, Pap, and I, Yat—to assign duties and tasks to Ralik, **nor** (to) whatever (child) [26]may be born from Ralik. And whatever son may be born from Ralik, then she may have (him) for her own, [27]and may put (him) for hire in free service as (his) grandfather and father did; and whatever [28]daughter may be born, then in as much as the father and the mother and the family may agree to give (her) away, then [29](we) shall grant (their) wish. And **if** I, Ninduk, or I, Yamsh-bandag, or I, Pap, or I, Yat, [30]should do otherwise, **or** (**if** we) should assign duties or tasks to Ralik, **or** (**if** we) should claim thus, [31]with reference to Ralik's progeny, that (the child) is our slave-girl or slave rather than in free service [32]like (its) grandfather and father, **then** (we) shall give a fine to the royal treasury of twenty dinars of struck gold [33]and the same to the opposite party, and so too our claim and argument shall be [inva]lid.

[34]The dowry (consists of) one blanket, one pillow, one..., ... cloak(s), ... one ..., [35]four bracelets, two..., three pairs of shoes, two sheep, ... three measures of [wheat].

[36]...

(Sims-Williams, 2000: 32–35).

In this document, several witnesses signed the declaration that (1) Ralik is a wife of two brothers, Bab and Piduk; (2) accordingly, she became a daughter-in-law of Bag-farn, the father of these brothers; (3) she has a dowry with an appropriate description. The two additional declarations formulated as conditional statements: (1) **if** (καλδανο) Bab **or** (αλαζο) Piduk make another woman their wife **or** (αλαζο) they keep a free woman as their concubine, to whom Ralik should not agree, **then** they shall give a penalty (of twenty dinars of struck gold) to Ralik **and** (οδο) to the royal treasury; (2) **if** (καλδανο) Ninduk **or** (αλαζο) Yamsh-bandag **or** (αλαζο) Pap **or** (αλαζο) Yat assign duties or tasks to Ralik **or** (αλαζο) to the child who may be born from Ralik, **then** they shall give a penalty (of twenty dinars of struck gold) to Ralik **and** (οδο) to the royal treasury.

Thus, we see that marriage contracts in Akkadian, Aramaic, Greek, and Bactrian were composed within one and the same legal hermeneutics, existing for more than two thousand years at the time of the Bactrian contracts and well traced retrospectively. In marriage contracts there were one main declarative and a series of additional declarations formulated as implications and all these declaratives were confirmed by several seals of witnesses. In the trial records (transcripts) of Sumerian-Akkadian legal tradition continuing in Hellenistic and partly Hellenized states of Central and South Asia, we find already explicit and logically correct applications of different logical inference rules including *modus ponens*, *modus tollens*, and many others (Schumann, 2019b, 2020a). Meanwhile, these trial records are free from rhetoric tropes and logical fallacies and contain only a purely logical argumentation based on the usage of propositional logic.

1.3.7 Conclusions

Logical competence consists of the methods of implementing logical knowledge that are correct from the standpoint of logical norms and it is not our natural (innate) ability. These methods have different historical forms and can be found in the texts of some early states. Moreover, the earliest examples of perfect logical competence are found only in the documents of Mesopotamia, starting with legal documents and business correspondence of the Akkadian period (since the 24th century B.C.). By the time of the appearance of Aristotle's treatises on logic in the 4th century B.C. there was already a long history of the perfect logical competence within the Mesopotamian legal hermeneutics expanded later through Hellenistic states. The analysis of various forms of manifestation of this competence allows us to find out the nature of logical knowledge in its practical applications.

References

Abraham, M., Gabbay, D.M. and Schild, U. 2009. Analysis of the talmudic argumentum: a fortiori inference rule (kal vachomer) using matrix abduction. *Studia Logica* 92(3): 281–364.

Abraham, M., Gabbay, D., Hazut, G., Maruvka, Y.E. and Schild, U. 2010a. *The Textual Inference Rules Klal uPrat How the Talmud Defines Sets*. London: College Publications.

Abraham, M., Gabbay, D. and Schild, U. 2010b. *Non-deductive Inferences in the Talmud*. London: College Publications.

Abraham, M., Gabbay, D. and Schild, U. 2013. *Principles of Talmudic Logic*. London: College Publications.

Adamatzky, A. 2007. Physarum machine: Implementation of a Kolmogorov-Uspensky machine on a biological substrate. *Parallel Processing Letters* 17(4): 455–467.

Adamatzky, A. and Schubert, T. 2014. Slime mold microfluidic logical gates. *Materials Today* 17(2): 86–91.

Adamatzky, A. and Mayne, R. 2015. Actin automata: Phenomenology and localizations. *International Journal of Bifurcation and Chaos* 25(2): 2015, doi: 10.1142/S0218127415500303.

Adamatzky, A. (ed.). 2017. *Advances in Unconventional Computing*. Springer International Publishing.
Adams, Ch.D. (ed.). 1868. *The Genuine Works of Hippocrates*. New York: William Wood and Company.
Al'churron, K.E. and Bulygin, Ye.V. 2013. "Normativnyye sistemy" i drugiye raboty po filosofii prava i logike norm [«Нормативные системы» и другие работы по философии права и логике норм]. Saint Petersburg.
Allen, J. 2001. *Inference from Signs. Ancient Debates about the Nature of Evidence*. Oxford: Clarendon Press.
Alonso-Sanz, R. and Adamatzky, A. 2016. Actin automata with memory. *International Journal of Bifurcation and Chaos* 26(1).
Al-Shamari, M.Kh.J. and Al-Jalili, M. 2020. Two old Babylonian marriage contracts from Isin. *Iraq* 82: 125–137.
Arnauld, A. and Nicole, P. 1662/1992. *La logique ou l'Art de penser*. Gallimard.
Asmis, E. 1996. Epicurean Semiotics. pp. 155–185. *In*: Manetti, G. (ed.). *Knowledge through Signs*. Brussels: Brepols.
Baker, G.S. 2020. *Critical Thinking. A Profound Handbook to Improve Your Memory and Logical Skills, Level Up Your Decision Making and Problem Solving*. Sannainvest Ltd.
Barbieri-Low, A.J. and Yates, R.D.S. 2015. *Law, State, and Society in Early Imperial China. A Study with Critical Edition and Translation of the Legal Texts from Zhangjiashan Tomb No*. 247. Vols. 1–2. Leiden: Brill.
Bechert, H. 1989. The problem of the determination of the date of the historical Buddha. *Wiener Zeitschrift für die Kunde Südasiens* 33: 93–120.
Bechert, H. (ed.). 1991. *The Dating of the Historical Buddha = Die Datierung des historischen Buddha*. Abhandlungen der Akademie der Wissenschaften in Göttingen, philologischhistoriche Klasse, 3rd series, Nr. 189, 194. Göttingen: Vandenhoeck & Ruprecht.
Bolzano, B. 1837. *Wissenschaftslehre*. Seidel: Sulzbach.
Bonabeau, E., Dorigo, M. and Theraulaz, G. 1999. *Swarm Intelligence: From Natural to Artificial Systems*. Oxford University Press.
Boole, G. 1847. *The Mathematical Analysis of Logic*. Cambridge: MacMillan.
Boole, G. 1854. *An Investigation of the Laws of Thought*. London: Walton & Maberly.
Brandom, R. 2000. *Articulating Reasons: An Introduction to Inferentialism*. Harvard University Press.
Burnyeat, M.F. 1982. The origins of non-deductive inference. pp. 193–238. *In*: Barnes, J. et al. (eds.). *Science and Speculation*. Cambridge: Cambridge University Press.
Burrow, T. 1940. *A Translation of the Kharosthi Documents from Chinese Turkestan*. James G. Forlong Fund, Vol. XX. London: The Royal Asiatic Society.
Bury, R.G. (tr.). 1976. *The Complete Works of Sextus Empiricus*. 4 vols. Loeb Classical Library. Cambridge: Harvard University Press.
Byeong-uk, Y. 2014. Numeral classifiers and the white horse paradox. *Frontiers of Philosophy in China* 9(4): 498–522.
Byosiere, S.-E., Chouinard, P.A., Howell, T.J. and Bennett, P.C. 2020. Illusion susceptibility in domestic dogs. *Ethology* 126(10): 949–965.
Caldwell, E. 2018. *Writing Chinese Laws. The Form and Function of Legal Statutes Found in the Qin Shuihudi Corpus*. Routledge.
Chandler, C. 2006. *Philodemus. On Rhetoric*. Books 1 and 2. New York & London: Routledge.
Chelpanov, G. 1917. *Uchebnik logiki* [Учебник логики]. 9th edn. Moscow.
Chen, J., Hong, H. and Stein, J.C. 2001. Forecasting crashes: Trading volume, past returns, and conditional skewness in stock prices. *Journal of Financial Economics* 61(3): 345–381.
Chén, Z. 1937. *Gōngsūn Lóngzi jíjiě* 公孫龍子集解. Shanghai: Shangwu yinshu guan.
Cheng, Ch. 1997. Philosophical significance of gongsun long: a new interpretation of theory of zhi as meaning and reference. *Journal of Chinese Philosophy* 24(2): 139–178.
Chmielewski, J. 2009. *Language and Logic in Ancient China: Collected Papers on the Chinese Language and Logic*. Warsaw: PAN.
Church, A. 1996. *Introduction to Mathematical Logic*. Princeton, New Jersey: Princeton University Press.
Clegg, I.L., Rödel, H.G. and Delfour, F. 2017. Bottlenose dolphins engaging in more social affiliative behaviour judge ambiguous cues more optimistically. *Behavioural Brain Research* 322: 115–122.

Clegg, I.L.K. 2018. Cognitive bias in zoo animals: An optimistic outlook for welfare assessment. *Animals* 8(7), doi: 10.3390/ani8070104.

Cohen, J. 1992. *An Essay on Belief and Acceptance*. Oxford: Clarendon Press.

Cooper, R., Mukai, K. and Perry, J. (eds.). 1990. *Situation Theory and Its Applications*, Vol. 1. Stanford: CSLI Publications.

Copi, I.M., Cohen, C. and McMahon, K. 2014. *Introduction to Logic*. 14th edn. Pearson Education Limited.

Correia, V. 2011. Biases and fallacies: The role of motivated irrationality in fallacious reasoning. *Cogency: Journal of Reasoning and Argumentation* 3(1): 107–126.

Crews-Anderson, T.A. 2007. *Critical Thinking and Informal Logic*. Tirril: Humanities-Ebooks.

Culbertson, L.E. 2009. *Dispute Resolution in the Provincial Courts of the Third Dynasty of Ur*. A dissertation submitted in partial fulfillment of the requirements for the degree of Doctor of Philosophy (Near Eastern Studies) in the University of Michigan.

Culbertson, L. 2015. Local courts in centralizing states: the case of Ur III Mesopotamia. pp. 185–202. *In*: Emberling, G. (ed.). *Social Theory in Archaeology and Ancient History*. Cambridge University Press.

Daoudi, S., Badihi, G. and Buchanan-Smith, H.M. 2017. Is mixed-species living cognitively enriching? Enclosure use and welfare in two captive groups of tufted capuchins (*Sapajus apella*) and squirrel monkeys (*Saimiri sciureus*). *Animal Behavior and Cognition* 4(1): 72–90.

De Lacy, Ph.H. and De Lacy, E.A. (eds.). 1941. *Philodemus: On Methods of Inference. A Study in Ancient Empiricism*. Philadelphia: American Philological Association.

Dedekind, R. 1933. *Was sind und was sollen die Zahlen?* Siebente Auflage. Braunschweig: Friedr. Vieweg & Sons.

Dorigo, M. and Gambardella, L.M. 1997. Ant colonies for the travelling salesman problem. *Biosystems* 43(2): 73–81.

Dutt, N. 1984. *Gilgit Manuscript*, Vols. 1–4, Calcutta, Srinagar 1939–1959. Vol. III: Mūlasarvāstivādavinayavastu, part 1–4. Part 1: Bhaiṣajyavastu; 2nd edn., Delhi.

Eemeren, F. van, Grootendorst, R., Blair, A. and Willard, C.A. (eds.). 1986. *Argumentation*. Dordrecht: Foris Publications.

Eysenck, M., Mogg, K., May, J., Richards, A. and Mathews, A. 1991. Bias in interpretation of ambiguous sentences related to threat in anxiety. *Journal of Abnormal Psychology* 100(2): 144–150.

Fackler, O.T. and Grosse, R. 2008. Cell motility through plasma membrane blebbing. *Journal of Cell Biology* 181: 879–884.

Foucault, M. 2005. The *Order of Things. An Archaeology of the Human Sciences*. Taylor and Francis e-Library.

Franks, C. 2012. The Realm of the Sacred, wherein We may not draw an inference from something which itself has been inferred: A reading of Talmud Bavli *Zevachim* folio 50. *History and Philosophy of Logic* 33(1): 69–86.

Fraser, Ch. 2020. Paradoxes in the School of Names. *In*: Fung, Y. (ed.). *Dao Companion to Chinese Philosophy of Logic*. Springer.

Frauwallne, E. 1973. *History of Indian Philosophy*. Vol. I. *The Philosophy of the Veda and of the Epic – the Buddha and the Jina – the Sāmkhya and the Classical Yoga-system*. Introduction by Univ. Prof. Dr. Leo Gabriel. Translated from original German into English by V.M. Bedekar. Delhi: Motilal Banarsidass Publishers.

Frege, G. 1918/1919. Der Gedanke. Eine logische Untersuchung. pp. 58–77. *In*: *Beiträge zur Philosophie des deutschen Idealismus*. Band 1. Erfurt: Stenger, S.

Frege, G. 1934. *Grundlagen der Arithmetik*. Breslau: Marcus.

Freimann, A. 1917. Die Hebräischen Kommentare zu den 13 Middot des Rabbi Ismael. *In*: Krauss, S. (ed.). *Festschrift Adolf Schwarz*. Berlin and Wien.

Fung, Y. (ed.). 2020. *Dao Companion to Chinese Philosophy of Logic*. Springer.

Fung, Y.-L. 1948. *A Short History of Chinese Philosophy*, edited by D. Bodde. New York and London: The Free Press.

Gangopadhyaya, M. (tr.). 1971. *Vinītadeva's Nyāyabindu-ṭīkā Reconstructed from Tibetan into Sanskrit*. Calcutta: Indian Studies: Past & Present.

Getmanova, A.D. 2011. *Uchebnik logiki* [Учебник логики]. 8th edn. Moscow: КНОРУС.

Gilovich, T., Griffin, D. and Kahneman, D. (eds.). 2002. *Heuristics and Biases: The Psychology of Intuitive Judgment*. Cambridge University Press.
Gödel, K. 1931. Über formal unentscheidbare Sätze der Principia Mathematica und verwandter Systeme, I. *Monatshefte für Mathematik und Physik* 38(1): 173–198.
Gokhale, P.P. (ed.). 1993. *Vādanyāya of Dharmakīrti. The Logic of Debate*. Shakti Nagar, Delhi: Sri Satguru Publications, A Division of Indian Books Centre.
Goldberg, J. and von Nitzch, R. 2001. *Behavioural Finance*. John Wiley & Sons.
Graham, A.C. 1986. *Studies in Chinese Philosophy and Philosophical Literature*. The Institute of East Asian Philosophies, National University of Singapore.
Greengus, S. 1969. The old babylonian marriage contract. *Journal of the American Oriental Society* 89(3): 505–532.
Hansen, C.D. 1983. *Language and Logic in Ancient China*. Michigan: University of Michigan Press.
Hattori, M. (tr.). 1968. *Dignāga, On Perception, being the Pratyakṣapariccheda of Dignāga's Pramāṇasamuccaya from the Sanskrit fragments and the Tibetan Versions*. Cambridge, Massachusetts: Harvard University Press.
Heaton, J.B. 2002. Managerial optimism and corporate finance. *Financial Management* 31(2): 33–45.
Hegel, G.W.F. 1975. *Lectures on the Philosophy of World History,* Introduction (Cambridge Studies in the History and Theory of Politics.) Cambridge University Press.
Hegel, G.W.F. 1984. *Lectures on the Philosophy of Religion*. The Regents of the University of California.
Herbart, J.F. 1813. *Lehrbuch zur Einleitung in die Philosophie. Zweiter Abschnitt. Die Logik*.
Herbart, J.F. 1816. *Lehrbuch der Psychologie*. Königsberg: A.W. Unzer.
Herbart, J.F. 1824–1825. *Psychologie als Wissenschaft, neu gegrundet auf Erfahrung, Metaphysik und Mathematik*. Königsberg: A.W. Unzer.
Heymans, G. 1905. *Die Gesetze und Elemente des wissenschaftlichen Denkens. Ein Lehrbuch der Erkenntnistheorie in Grundzügen*. Leipzig: Barth.
Hilbert, D. and Ackermann, W. 1949. *Grundzüge der theoretischen Logik*. Dritte, verbesserte Auflage. Berlin, Heidelberg: Springer-Verlag.
Hilbert, D. and Bernays, P. 1968. *Grundlagen der Mathematik*. Band 1. Zweite Auflage. Berlin, Heidelberg, New York: Springer-Verlag.
Hilbert, D. 1987. *Foundations of Geometry*, 10th edn., translated by Leo Unger. La Salle, Ill.: Open Court, 2nd edn.
Hirschfeld, H.S. 1840. *Der Geist der talmudischen Auslegung der Bibel. Der erste Teil. Halachische Exegese. Ein Beitrag zur Geschichte der Exegese und zur Methodologie des Talmud's*. Berlin: M. Simion in Berlin.
Hirschfeld, H.S. 1847. *Der Geist der ersten Schriftauslegungen. Oder: Die Hagadische Exegese. Ein Beitrag zur Geschichte der Exegese und zur Methodologie des Midrasch*. Berlin: M. Simion in Berlin.
Hoffbauer, J.C. 1792. *Analytik der Urtheile und Schlüße*.
Holtz, Sh.E. 2014. *Neo-Babylonian Trial Records*. Society of Biblical Literature Atlanta.
Howard, S.R., Avarguès-Weber, A., Garcia, J.E., Stuart-Fox, D. and Dyer, A.G. 2017. Perception of contextual size illusions by honeybees in restricted and unrestricted viewing conditions. *Proceedings of the Royal Society B: Biological Sciences* 284(1867): 20172278, doi: 10.1098/rspb.2017.2278.
Hu, Sh. 2003. Mòzi· xiǎoqǔ piān xīngǔ 墨子 · 小取篇新诂 (New Research of Xiaoqu·Mozi). *In*: *Hú Shì quánjí* 胡适全集 (Collected Works of Hu Shi), Vol. 1. Hefei: Anhui Education Press.
Huang, X., MacEvoy, S.P. and Paradiso, M.A. 2002. Perception of brightness and brightness illusions in the macaque monkey. *Journal of Neuroscience* 22(21): 9618–9625.
Hunt, A.S. and Edgar, C.C. 1932. *Select Papyri*. Vol. I: Private Documents. Loeb Classical Library 266. Cambridge, MA: Harvard University Press.
Ilyenkov, E. 1974. *Dialectical Logic*. Moscow: Politizdat (in Russian).
Indraccolo, L. 2017. The "white horse is not horse" debate. *Philosophy Compass* 12(10): e12434.
Jacobs, L. 1961. *Studies in Talmudic Logic and Methodology*. London.
Jakob, L.H. 1791. *Grundriss der allgemeinen Logik*. 2nd edn.
Johnson, R. and Blair, J. 1983. *Logical Self-defense*. Toronto: McGraw-Hill.
Johnston, I. (tr.). 2010. *Book of Master Mo*. The Chinese University Press.
Jones, J.D. and Adamatzky, A. 2010. Towards Physarum binary adders. *Biosystems* 101(1): 51–58.

Jones, J.D. 2015. Towards lateral inhibition and collective perception in unorganised non-neural systems. pp. 103–122. *In*: Pancerz, K. and Zaitseva, E. (eds.). *Computational Intelligence, Medicine and Biology: Selected Links*. Springer.

Kelley, D. 2014. *The Art of Reasoning: An Introduction to Logic and Critical Thinking*. 4th edn. New York: W.W. Norton & Company, Inc.

Kennedy, G.A. 2007. *Aristotle. On Rhetoric: A Theory of Civic Discourse*. 2nd edn. Oxford University Press.

Kern, L.H., Mirels, H.L. and Hinshaw, V.G. 1983. Scientists' understanding of propositional logic: An experimental investigation. *Social Studies of Science* 13(1): 131–146.

Kiesewetter, J.G.C. 1791. *Grundriss einer allgemeinen Logik nach Kantischen Grundsätzen*.

King, P. and Shapiro, S. 1995. History of logic. *In*: *The Oxford Companion to Philosophy*. Oxford: Oxford University Press.

King, R. 1995. Early Advaita Vedānta: The date and authorship of the Gauḍapādīyakārikā. *Indo-Iranian Journal* 38(4): 317–355.

Kneale, W. and Kneale, M. 1962. *The Development of Logic*. Oxford: Clarendon.

Korolkov, M.V. 2013. *Zòuyànshū* ("Sbornik sudebnykh zaprosov"): *paleograficheskiye dokumenty drevnego Kitaya*. Moscow: Nauka (in Russian).

Krug, W.T. 1806. *Denklehre oder Logik*.

Kunda, Z. 1990. The case for motivated reasoning. *Psychological Bulletin* 108(3): 480–498.

Kurtz, J. 2011. *The Discovery of Chinese Logic*. Leiden: Brill.

Lapointe, S. and Pincock, Ch. 2017. Introduction. *In*: *Innovations in the History of Analytical Philosophy*. Houndmills: Palgrave.

Lapointe, S. (ed.). 2019. *Logic from Kant to Russell. Laying the Foundations for Analytic Philosophy*. London, New York: Routledge, Taylor & Francis.

Lemanski, J. (ed.). 2020. *Language, Logic, and Mathematics in Schopenhauer*. Basel, Schweiz: Birkhäuser.

Levi, S. 1896. Notes on the Indo-Scythians. *Indian Antiquary* 32: 381–389.

Liang, Q. 1999. *Liángqǐchāo quánjí* 梁启超全集 (Collection of Liang Qichao), Vol. 6. Beijing: Beijing Publishing House.

Maass, J.G.E. 1793. *Grundriss der Logik*.

Malvania, D. (ed.). 1955. *Paṇḍita Durveka Miśra's Dharmottarapradīpa: Being a Sub-commentary on Dharmottara's Nyāyabinduṭīkā, a Commentary on Dharmakīrti's Nyāyabindu*. Patna: K.P. Jayaswal Research Institute.

Mansel, H.L. 1851. *Prolegomena Logica: An Inquiry into the Psychological Character of Logical Processes*.

Mayne, R. and Adamatzky, A. 2015. Slime mould foraging behaviour as optically coupled logical operations. *International Journal of General Systems* 44(3): 305–313.

Mo, Sh. 1981. Mòzi· xiǎoqǔ piān luójí de tǐxì 墨子 · 小取篇逻辑的体系 (Logic System in Mozi·Xiaoqu). *In*: *Zhōngguó luójíshǐ lùnwénxuǎn* 中国逻辑史论文选 (Collection of Theses on History of Chinese Logic) (1949–1979). Beijing: Sanlian Publishing House.

Mortari, C.A. 2001. *Introdução à Lógica*. São Paulo: Editora UNESP: Imprensa Oficial do Estado.

Müller, F.M. 1878. *Lectures on the Origin and Growth of Religion as Illustrated by the Religions of India.* Delivered in the chapter house, Westminster Abbey, in April, May, and June 1878. London: Longmans, Green, and co. Williams and Norgate.

Müller, F.M. 1883. *India: What can it teach us?* A course of lectures delivered before the University of Cambridge by F. Max Müller, K.M. text and foot-notes complete. With an introduction and notes by Prof. Alexander Wilder, M.D. London: Longmans, Green, and Co. Williams and Norgate.

Müller, M. 2001. On Indian logic. pp. 59–74. *In*: Ganeri, J. (ed.). *Indian Logic. A Reader.* Richmond: Curzon Press.

Neelis, J. 2011. *Early Buddhist Transmission and Trade Networks: Mobility and Exchange Within and Beyond the Northwestern Borderlands of South Asia*. Brill.

Neusner, J. 2002. *Rabbinic Judaism's Generative Logic* (Academic Studies in the History of Judaism). Global Academic Pub.

Oldenberg, H. (ed.). 1880. *The Vinaya Piṭakam: One of the Principal Buddhist Holy Scriptures in the Pâli Language: The Cullavagga*. Vol. II. London: Pali Text Society.

Ozbakir, L., Baykasoglu, A. and Tapkan, P. 2010. Bees algorithm for generalized assignment problem. *Applied Mathematics and Computation* 215(11): 3782–3795.

Parrish, A., Brosnan, S. and Beran, M. 2015. Do you see what I see? A comparative investigation of the Delboeuf illusion in humans (*Homo sapiens*), rhesus monkeys (*Macaca mulatta*), and capuchin monkeys (*Cebus apella*). *Journal of Experimental Psychology. Animal Learning and Cognition* 41, doi: 10.1037/xan0000078.

Pepperberg, I.M., Vicinay, J. and Cavanagh, P. 2008. Processing of the Müller-Lyer illusion by a grey parrot (Psittacus erithacus). *Perception* 37(5): 765–781.

Pollard, T.D. and Borisy, G.G. 2003. Cellular motility driven by assembly and disassembly of actin filaments. *Cell* 112: 453–465.

Porton, G.G. 1977. *The Traditions of Rabbi Ishmael*. 2. Exegetical comments in Tannaitic collections. Leiden: Brill.

Povarnin, S.I. 1916. *Logika* [Логика]. Petrograd.

Rainey, A.F. 1996. *Canaanite in the Amarna Tablets a Unguistic Analysis of the Mixed Dialect Used by the Scribes from Canaan*. Vol. I Orthography, Phonology Morphosyntactic Analysis of the Pronouns, Nouns, Numerals. BRILL Leiden: Brill.

Ravitsky, A. 2004. Halakhic arguments as dialectical arguments and exegetical principles as aristotelian τόποι in maimonides' philosophy. *Tarbits* 73: 219.

Ravitsky, A. 2009. Talmudic Methodology and Aristotelian Logic: David ibn Bilia's commentary on the thirteen hermeneutic principles. *The Jewish Quarterly Review* 99: 184–199.

Ravitsky, A. 2011. Saadya gaon and maimonides on the logic and limits of legal inference in context of the karaite-rabbanite controversy. *History and Philosophy of Logic* 32: 29–36.

Ravitsky, A. 2018. Yaʿqūb al-Qirqisānī on human intellect, legal inference, and the meaning of the aristotelian syllogism. *Journal of Jewish Thought & Philosophy* 26: 149–173.

Rhys Davids, T.W. and Carpenter, J.E. (eds.). 1890. *Digha-Nikaya of the Sutta-Pitaka*, Vol. I: Silakkhandavagga, Suttantas 1–13. London: Pali Text Society.

Rhys Davids, T.W. and Rhys Davids, C.A. (trs.). 1899–1921. *Dialogues of the Buddha*, 3 vols.

Rickert, H.J. 1896. *Die Grenzen der naturwissenschaftlichen Begriffsbildung*. Vol. 1. Freiburg, Leipzig: J.C.B. Mohr.

Rickert, H.J. 1899. *Kulturwissenschaft und Naturwissenschaft*. Tübingen, Freiburg: J.C.B. Mohr.

Robinson, E. and Roiser, J. 2016. Affective biases in humans and animals. pp. 263–286. *In*: Robbins, T.W. and Sahakian, B.J. (eds.). *Translational Neuropsychopharmacology*. Springer International Publishing.

Roelofs, S., Boleij, H., Nordquist, R.E. and van der Staay, F.J. 2016. Making decisions under ambiguity: Judgment bias tasks for assessing emotional state in animals, *Frontiers in Behavioral Neuroscience* 10: 119, doi: 10.3389/fnbeh.2016.00119.

Ross, W.D. (ed.). 1959. *Ars Rhetorica. Aristotle*. Oxford: Clarendon Press.

Roth, M.T. 1995. *Law Collections from Mesopotamia and Asia Minor*. With a contribution by Harry A. Hoffner, Ir.; edited by Piotr Michalowski. Scholars Press Atlanta, Georgia.

Russell, B. 1993. *Our Knowledge of the External World as a Field for Scientific Method in Philosophy*. New York: Routledge.

Sachau, E.C. (ed.). 1910. *Al-Beruin's India: An Account of the Religion, Philosophy, Literature, Geography, Chronology, Astronomy, Customs, Laws, and Astrology of India about 1030 AD*. 2 vols. London: Kegan Paul, Trench, Trubner & Co. Ltd.

Sakiyama, T. and Gunji, Y.-P. 2013. The Müller-Lyer illusion in ant foraging. *PLOS ONE* 12(8): 1–12.

Salomon, R. 1998. *Indian Epigraphy. A Guide to the Study of Inscriptions in Sanskrit, Prakrit, and the Other Indo-Aryan Languages*. New York, Oxford: Oxford University Press.

Santacà, M., Miletto Petrazzini, M.E., Agrillo, C. and Wilkinson, A. 2019. Can reptiles perceive visual illusions? Delboeuf illusion in red-footed tortoise (*Chelonoidis carbonaria*) and bearded dragon (*Pogona vitticeps*). *Journal of Comparative Psychology* 133, doi: 10.1037/com0000176.

Schayer, S. 2001. On the method of research into Nyaya. pp. 102–109. *In*: Ganeri, J. (ed.). *Indian Logic. A Reader*. Richmond: Curzon Press.

Schelling von, F.W.J. 1907. *Werke*. Band 3. Leipzig.

Schultz, J. 1789. *Prüfung der Kantischen Critik der reinen Vernunft*. Vol. 1.

Schumann, A. 2013a. On two squares of opposition: The Leśniewski's style formalization of synthetic propositions. *Acta Analytica* 28: 71–93.

Schumann, A. 2013b. Logical cornestones of judaic argumentation theory. *Argumentation* 27: 305–326.

Schumann, A. 2015. *p*-adic valued logical calculi in simulations of the slime mould behaviour. *Journal of Applied Non-Classical Logics* 25(2): 125–139.

Schumann, A. 2016. Toward a computational model of actin filament networks. pp. 290–297. *In*: *Proceedings of the 9th International Joint Conference on Biomedical Engineering Systems and Technologies (BIOSTEC 2016)—Volume 4: BIOSIGNALS, Rome, Italy, February 21–23, 2016.* doi: 10.5220/0005828902900297.

Schumann, A. 2017. On arithmetic functions in actin filament networks. *In*: *10th EAI International Conference on Bio-inspired Information and Communications Technologies (formerly BIONETICS).* ACM, doi: 10.4108/eai.22-3-2017.152402.

Schumann, A. 2018. Decidable and undecidable arithmetic functions in actin filament networks. *Journal of Physics D – Applied Physics* 51(3): 034005.

Schumann, A. and Kuznetsov, A.V. 2018. Foundations of mathematics under neuroscience conditions of lateral inhibition and lateral activation, *International Journal of Parallel, Emergent and Distributed Systems* 33(3): 237–256.

Schumann, A. and Pancerz, K. 2019. *High-level Models of Unconventional Computations. A Case of Plasmodium*. New York: Springer International Publishing.

Schumann, A. 2019a. On the origin of Indian Logic from the viewpoint of the Pāli Canon. *Logica Universalis* 13: 347–393.

Schumann, A. 2019b. Did the neo-Babylonians construct a symbolic logic for legal proceedings? *Journal of Applied Logics – IfCoLog Journal of Logics and Their Applications* 6(1): 31–82.

Schumann, A. 2019c. *Behaviourism in Studying Swarms: Logical Models of Sensing and Motoring*. New York: Springer International Publishing.

Schumann, A. 2020a. Legal argumentation in Mesopotamia since Ur III. *Journal of Argumentation in Context* 9(2): 126–154.

Schumann, A. (ed.). 2020b. *Swarm Intelligence: From Social Bacteria to Humans*. CRC Press, Taylor & Francis Group.

Schumann, A. and Arif, S. 2021. Proto-Hindu Buddhist Deities Maheśvara and Hārītī in Gandhāra. *Journal Asiatique* 309(2): 231–248.

Schwarz, A. 1897. *Die Hermeneutische Analogie*. Vienna.

Schwarz, A. 1901. *Der Hermeneutische Syllogismus*. Vienna.

Schwarz, A. 1909. *Die Hermeneutische Induktion*. Vienna.

Schwarz, A. 1913. *Die Hermeneutische Antinomie*. Vienna.

Schwarz, A. 1916. *Die Hermeneutische Quantitätsrelation*. Vienna.

Schwarz, A. 1921. *Der Hermeneutische Kontext*. Vienna.

Sharma, P. (tr.). 2014. *Caraka-saṃhitā: Agniveśa's treatise refined and annotated by Caraka and redacted by Dṛḍhabala: Text with English translation*. Vol. 1. Varanasi: Chaukhambha Orientalia.

Shastri, Dh.N. 1976. *The Philosophy of Nyāya-Vaiśeṣika and Its Conflict with the Buddhist Dignāga School: Critique of Indian Realism*. Delhi: Bharatiya Vidya Prakashan.

Shchedrovitsky, G. 1964. *Methodological Problems of System Research*. Moscow: 'Znanie' Publishing House (in Russian).

Shirakawa, T., Yokoyama, K., Yamachiyo, M., Gunji, Y.P. and Miyake, Y. 2012. Multi-scaled adaptability in motility and pattern formation of the Physarum plasmodium. *International Journal of Bio-Inspired Computation* 4: 131–138.

Shirakawa, T., Sato, H. and Ishiguro, S. 2015. Constrcution of living cellular automata using the Physarum polycephalum. *International Journal of General Systems* 44: 292–304.

Shirakawa, T., Gunji, Y.-P., Sato, H. and Tsubakino, H. 2020. Diversity in the chemotactic behavior of Physarum Plasmodium induced by bi-modal stimuli. *International Journal of Unconventional Computing* 15(4): 275–285.

Siccardi, S. and Adamatzky, A. 2015. Actin quantum automata: Communication and computation in molecular networks. *Nano Communication Networks* 6(1): 15–27.

Siccardi, S. and Adamatzky, A. 2016. Quantum actin automata and three-valued logics. *IEEE Journal on Emerging and Selected Topics in Circuits and Systems* 6(1): 53–61.

Siccardi, S., Tuszynski, J.A. and Adamatzky, A. 2016. Boolean gates on actin filaments. *Physics Letters A* 380(1): 88–97.

Sigwart von, Ch. 1895. *Logic*. Vol. I: The Judgment, Concept, and Inference. 2nd edn., translated by Helen Bendy. London: Swan Sonnenschein & Co; New York: Macmillan & Co.

Sims-Williams, N. 2000. *Bactrian Documents from Northern Afghanistan I: Legal and Economic Documents*. Oxford: Oxford University Press.

Sims-Williams, N. 2007. *Bactrian Documents from Northern Afghanistan II: Letters and Buddhist Texts*. London: Nour Foundation.

Sjoberg, E. 2017. Logical fallacies in animal model research. *Behavioral and Brain Functions* 13(1): 3.

Smith, R. (ed.). 1997. *Topics Books I and VIII: With Excerpts from Related Texts*. Oxford, England: Clarendon Press.

Spruijt, B.M., van den Bos, R. and Pijlman, F.T. 2001. A concept of welfare based on reward evaluating mechanisms in the brain: Anticipatory behaviour as an indicator for the state of reward systems. *Applied Animal Behaviour Science* 72(2): 145–171.

Stenning, K. and van Lambalgen, M. 2008. *Human Reasoning and Cognitive Science*. Cambridge, MA: MIT Press.

Stenning, K. and van Lambalgen. M. 2011. Reasoning, logic, and psychology. *Wiley Interdisciplinary Reviews. Cognitive Science* 2(5): 555–567.

Swetz, F.J. 2003. Leibniz, the *Yijing*, and the religious conversion of the Chinese. *Mathematics Magazine* 76(4): 276–291.

Tachikawa, M. 1981. *The Structure of the World in Udayana's Realism. A Study of the Lakṣaṇāvalī and the Kiraṇāvalī*. Dordrecht, Boston, London: D. Reidel Publishing Company.

Talisse, R. and Aikin, S. 2004. Two forms of the straw man. *Argumentation* 20(3): 345–352.

Tarkatirtha, A.M., Taranatha Nyayatarkatirtha and Tarkatirtha, H.K. (eds.). 1982. *Nyāyadarśanam* (*Nyāya-sūtra with four commentaries, the Nyāya-sūtra-bhāṣya of Vātsyāyana, the Nyāya-vārttika of Uddyotakara, the Nyāya-vārttika-tātpārya-ṭīkā of Vācaspati Miśra, and the Vṛtti of Viśvanātha*). Calcutta Sanskrit Series 18. Calcutta, 1936–1944. Reprint Kyoto.

Thapar, R. 1997. *Aśoka and the Decline of the Mauryas*. Delhi: Oxford University Press.

Tredennick, H. and Forster, E.S. (trs.). 1960. *Aristotle. Posterior Analytics. Topica*. Loeb Classical Library 391. Cambridge, MA: Harvard University Press.

Tverky, A. and Kahneman, D. 1974. Judgment under uncertainty: Heuristics and biases. *Science* 185: 1124–1131.

Tversky, A. and Kahneman, D. 2008. Extensional versus intuitive reasoning: the conjunction fallacy in probability judgment. pp. 114–135. *In*: Adler, J.E. and Rips, L.J. (eds.). *Reasoning: Studies of Human Inference and Its Foundations*. Cambridge University Press.

Van Haastert, P.J. and Devreotes, P.N. 2004. Chemotaxis: Signalling the way forward. *Nature Reviews Molecular Cell Biology* 5: 626–634.

Veenhof, K.R. 1995. In accordance with the words of the Stele: Evidence for Old Assyrian legislation. *Chicago-Kent Law Review* 70(4): 1717–1744.

Vidyabhusana, S.Ch. 1921. *A History of Indian Logic* (*Ancient, Mediaeval and Modern Schools.*). Delhi, Patna, Varanasi: Motilal Babarsidass.

Vinogradov, S.N. and Kuz'min, A.F. 1954. *Logika* [Логика]. 8th edn. Moscow: Государственное учебно-педагогическое издательство Министерства просвещения РСФСР.

Walton, D. 1989. *Informal Logic*. Cambridge: Cambridge University Press.

Watson, B. (tr.). 2003. *Zhuangzi: Basic Writings*. Columbia University Press.

Weiss, I.H. (ed.). 1862. *Sifra*. Vienna.

Wèizǔ, S. and Gǔfù, G. (eds.). 1997. *Qín Hàn jīnwén huìbiān* 秦漢金文匯編. Shanghai: Shanghai Shudian.

Whiting, J.G., de Lacy Costello, B.P. and Adamatzky, A. 2014. Slime mould logic gates based on frequency changes of electrical potential oscillation. *Biosystems* 124: 21–25.

Willetts, R.F. (ed.). 1967. *The Law Code of Gortyn*. De Gruyter, pp. 37–50.

Windelband, W. 1912. *Die Prinzipien der Logik in Enzyklopadie der Philosophischen Wissenschaften*.

Wright, W.F. and Bower, G.H. 1992. Mood effects on subjective probability assessment. *Organizational Behavior and Human Decision Processes* 52(2): 276–291.

Yaron, R. 1958. Aramaic marriage contracts from elephantine. *Journal of Semitic Studies* 3(1): 1–39.

Zinoviev, A. 1957. On the working out of dialectics as Logic. *Voprosy Filosofii* 4 (in Russian).

Chapter 2

Some Ways of Discovering Logical Competence in Social Practices

Logical competence as a practice of thinking, devoid of logical errors and fallacies, is not an innate ability, since in its natural form it is not separable from cognitive biases but is a special social practice of making logically correct decisions, which once arose then subsequently developed and spread through direct training. The chapter demonstrates the effectiveness of the structuralist methodology for analyzing the dynamics of logical competence as a social practice. First, it is shown how this methodology may be applied for analyzing myths of different peoples (Section 2.1). Second, how it could be used for tracing ontological pictures belonging to different philosophical traditions (Section 2.2). Third, how the same methodology might be applied for tracing the history of logical competence (Section 2.3).

2.1 Methodology for Determining the Structural Similarity of Different Mythological Cycles of Different Peoples

2.1.1 Problem Setting

It is possible to find many similarities between some Indo-Aryan and Mesopotamian mythological cycles, e.g., see Hopkins (1905), Sinha (2014). Nevertheless, an appropriate comparative study might seem to be quite speculative, considering such different Indo-Aryan and Mesopotamian cultures and religious practices. But we can ignore a content of myths and cults and just focus on a similarity between structures in both mythological narratives.

The idea of analyzing similarities not between contents, but between structures of different myths came first to the mind of Vladimir I͡Akovlevich Propp (1968). He has introduced the notion of *narratemes* as simple units of mythological stories. A recombination of narratemes with different structures gives different myths. The idea of Propp to consider structures of myths and their similarities was developed later by Claude Lévi-Strauss (1955) who has introduced *mythemes* instead of narratemes to emphasize a function of elementary description in different mythological stories. But the meaning of narratemes and mythemes is quite close. According to Lévi-Strauss, there are syntagmatic as well as paradigmatic relations among the mythemes. Let M_{11}, M_{12}, M_{21}, … M_{ij} be mythemes. Then i is their paradigmatic order and j is their syntagmatic order. Hence, we obtain a matrix

$$M = \begin{bmatrix} M_{11} & \cdots & M_{1j} \\ \vdots & \ddots & \vdots \\ M_{i1} & \cdots & M_{ij} \end{bmatrix}$$

to explicate the structure of myth M by distinguishing its mythemes with various paradigmatic and syntagmatic relations. Let us notice that at some places of this matrix there can be zeros. It means that at an appropriate place there is no mytheme.

Let O and O' be two mythological cycles belonging to different cultures (e.g., belonging to the Mesopotamian and Indian religions, respectively):

$$O = \begin{bmatrix} O_{11} & \cdots & O_{1j} \\ \vdots & \ddots & \vdots \\ O_{i1} & \cdots & O_{ij} \end{bmatrix} \quad O' = \begin{bmatrix} O'_{11} & \cdots & O'_{1m} \\ \vdots & \ddots & \vdots \\ O'_{k1} & \cdots & O'_{km} \end{bmatrix}$$

At some cells of O and O' we can have zeros, too. We say that there is a (partial) *structural similarity* of O to O' if and only if we can construct a (partial) *bijective mapping* (or bijection) f from O to O', with preserving the syntagmatic and paradigmatic orders of O and O'. Let us recall that the bijective mapping f in mathematics means a function between the elements of O and O', where each element of O is paired with exactly one element of O', and each element of O' is paired with exactly one element of O. And our additional property is as follows: if the non-zero mytheme O_{xy} follows the non-zero mytheme $O_{x'y'}$ by the syntagmatic and/or paradigmatic order in O, then the mapped non-zero mytheme $f(O_{xy})$ follows the mapped non-zero mytheme $f(O_{x'y'})$ by the syntagmatic and/or paradigmatic order in O'.

In this section, I will show that a structural similarity is observed between the Hindu (*Purāṇic*) mythological cycle of dying-and-rising deity *Nārāyaṇa* and the West-Semitic (Ugaritic, Phoenician, Biblical) mythological cycle of dying-and-rising deity *Ba'al*.

Let us present the main mythemes from the *Ba'al* cycle in the following table:

The Ultimate Level	*Ba'al* is the highest **O11**	*'Anat* is the wife of *Ba'al* **O12**
The Level of Gods	*'El* is the creator of the world, but he is subordinated to *Ba'al* **O21**	*'El* is subordinated to *'Anat* **O22**
The Level of Monsters	*Leviathan*, the many-headed snake, is subordinated to *Ba'al* **O31**	*Leviathan*, the many-headed snake, is subordinated to *'Anat* **O32**

In the matrix form:

$$\mathbf{O} = \begin{bmatrix} \mathbf{O11} & \mathbf{O12} \\ \mathbf{O21} & \mathbf{O22} \\ \mathbf{O31} & \mathbf{O32} \end{bmatrix}$$

This set of mythemes is bijectively mapped into the set of mythemes from the *Nārāyaṇa* cycle presented as the following table:

The Ultimate Level	*Nārāyaṇa* is the highest **O'11**	*Lakṣmī* is the wife of *Nārāyaṇa* **O'12**
The Level of Gods	*Brahmā* is the creator of the world, but he is subordinated to *Nārāyaṇa* **O'21**	*Brahmā* is subordinated to *Lakṣmī* **O'22**
The Level of Monsters	*Śeṣa*, the many-headed snake, is subordinated to *Nārāyaṇa* **O'31**	*Śeṣa*, the many-headed snake, is subordinated to *Lakṣmī* **O'32**

Or as the following matrix:

$$\mathbf{O'} = \begin{bmatrix} \mathbf{O'11} & \mathbf{O'12} \\ \mathbf{O'21} & \mathbf{O'22} \\ \mathbf{O'31} & \mathbf{O'32} \end{bmatrix} = \begin{bmatrix} f(\mathbf{O11}) & f(\mathbf{O12}) \\ f(\mathbf{O21}) & f(\mathbf{O22}) \\ f(\mathbf{O31}) & f(\mathbf{O32}) \end{bmatrix},$$

where f is a bijection from **O** into **O′**.

Additionally, to the structuralist approach established by Propp and Lévi-Strauss, we can introduce a stronger requirement to structural similarities among myths. In the new approach proposed first in this section, to show a similarity between two different myths with the same motif, we will appeal to the notion of *homotopy equivalence between mythological spaces*. The notion of homotopy equivalence is taken from topology. Two spaces (figures) X and Y are considered homotopy equivalent if there is a structural similarity between them: one can be obtained from another through its continuous transformations. For example, a cup with a handle is continuously transformed into a ring, but a cup with a handle cannot be continuously transformed into the cup without a handle, see Fig. 2.1.1.

Figure 2.1.1. The two figures above are homotopy equivalent, but the two figures below are not.

The intuitive meaning of applying the idea of homotopy equivalence to comparing two different mythological cycles is as follows. We can observe a structural similarity between two myths, but their detalization can be so different—in one myth some mythemes can be given in greater details, in another myth the same mythemes can become less detailed. Hence, the detalization of the same mythemes can be different. Nevertheless, we can assume that there are continuous transformations from one detailed part to other details as in Fig. 2.1.1. The matter is that in different cultures we deal with different social and even climatic contexts. For example, we find a flood myth in Mesopotamia as well as in India, but the climatic context of the same myth is different. In India, there is a rainy season (summer monsoon), lasting from June to September. In Mesopotamia, the flood of the Tigris and Euphrates rivers begins in the spring (March–April), when the snow in the mountains starts to melt. So, the two floods (in India and Mesopotamia) are periodic, but in India it is connected to the summer monsoon and seems to be more natural, because the cause of the flood (rain) is more evident. As a result, the flood myth in India is related to the idea of *pralaya*, a periodic destruction of the world. In Mesopotamia, it is rather a random act of destruction because of the divine will.

In finding a homotopy equivalence between two myths X and Y, there are the following requirements: (i) the number of characters/actors of both mythological cycles must be the same (e.g., in the *Nārāyaṇa* cycle the main actors are presented by the four: *Nārāyaṇa, Lakṣmī, Brahmā, Śeṣa,* in the *Ba'al* cycle by the same number: *Ba'al, 'Anat, 'El, Leviathan*); (ii) in the myth X, for the main actors there are the following properties: F_X (function), L_X (location), A_X (attributes), R_X (relations among the actors); in the myth Y, for the main actors there are the following properties: F_Y (function), L_Y (location), A_Y (attributes), R_Y (relations among the actors); and there are the following continuous transformations from the properties of X into the properties of Y giving their homotopy equivalence: $F_X \equiv F_Y$, $L_X \equiv L_Y$, $A_X \equiv A_Y$, $R_X \equiv R_Y$. It means that there can be differences in details of properties, but structurally (topologically) they are the same.

In this section, I will show that between the *Nārāyaṇa* cycle and the *Ba'al* cycle, there is a homotopy equivalence. It is a much stronger structural similarity of both myths than a similarity proposed by Propp and Lévi-Strauss which can be applied there, too. What does the structural similarity between both myths mean? According to the research program *Phylogenetics of Myths and Folktales* proposed by Yuri Berezkin (Thuillard et al., 2018), any structural similarity between myths means that between the possessors of both myths there was a deep cultural contact in the past which caused borrowing the mythological cycle and adopting it to the own cultural context implying differences in detail. These differences are called *mutations* by Berezkin. The more mutations, the earlier the contact was. Consequently, it is possible to present the different myths on a phylogenetic tree, according to the following assumptions (Thuillard et al., 2018: 409): (i) mythological motifs are transmitted unchanged over time and space except for minor transformations that are called mutations; (ii) a new mythological motif appears only once, therefore its appearance in other cultures means borrowing.

Thus, by Berezkin's research program, the homotopy equivalence between the *Nārāyaṇa* myth and the *Ba'al* myth should be interpreted as a fact of deep intercultural contacts of Indo-Aryans and West Semites in the past. The mythological cycle around *Ba'al* is dated to the second millennium and it is well traced in the Ugaritic, Luwian, and Hittite texts—in the cuneiform texts of Syria and Anatolia of that time, e.g., see about mentioning *Ba'al* in the Hittite texts in (Laroche, 1968). Nevertheless, the *Nārāyaṇa* cycle can be textually dated only since the 2nd century A.D.—since the time of early *Purāṇa*s, i.e., much later. But in the *Nārāyaṇa* myth, we observe many mutations (many differences in details) which are avoided only by continuous transformations. So, the homotopy equivalence is a *direct proof* that the *Nārāyaṇa* cycle is much older than the 2nd century A.D. It is difficult to date this cycle precisely. There are two possibilities: (i) to date the contact of the Indo-Aryans with the West Semites to the Bronze Age (i.e., to ca. the second millennium B.C.) that can have taken place during the Mitanni state in northern Syria and southeast Anatolia, where the political elite had some 'Indo-Aryan' roots; (ii) to date it to the Achaemenid conquest of North India (ca. 518 B.C.–327 B.C.), when Aramaic (the West Semitic language) became one of the official languages of North India. In any case, the *Nārāyaṇa* myth must be dated strongly B.C.

2.1.2 Homotopy Equivalence between Two Myths

Let us remark that *Nārāyaṇa* is mentioned as the Creator of the Universe and the Supreme Lord in the *Mahābhārata*:

> *Nārāyaṇa* of immeasurable energy, the Infinite, the Immaterial, the Uncreate, the Old, the Eternal, and the Spirit of these universes without limits. *Nārāyaṇa* approved of everything (*Mahābhārata* I: 199; Ganguli, 1883–1896, I: 393).

> He that is called *Nārāyaṇa* is older than the oldest ones (*Mahābhārata* VII: 201; Ganguli, 1883–1896, VII: 482).

In addition, *Śeṣa* (*Vāsuki*) is mentioned in the *Mahābhārata*, too:

> It is ruled over by *Vāsuki*, the king of the *nāga*s. That *Śeṣa* dwelleth here, who, in consequence of his ascetic austerities of the foremost order, is able to support this earth with all her vastness. His body is like that of a white mountain. He is decked in celestial ornaments. He hath a thousand heads. His tongues are blazing like flames of fire, and he is endued with great strength (*Mahābhārata* V: 103; Ganguli, 1883–1896, V: 210).

Lakṣmī and *Brahmā* are mentioned in the same context with *Nārāyaṇa*: *naranārāyaṇau brahmā yamaḥ sthāṇuś ca pañcamaḥ* (*Mahābhārata* II: 3, 13); "*Lakṣmī* in the house of *Viṣṇu*", *viṣṇukule lakṣmīr* (*Mahābhārata* V: 102, 8); "*Nārāyaṇa* with *Lakṣmī*", *nārāyaṇo lakṣyāṃ* (*Mahābhārata* V: 115, 10).

It is difficult to date the *Mahābhārata* as a whole text, since its different motifs and parts were composed surely at different times, B.C. as well as A.D. So, sometimes there are mentioned some phenomena appearing only at the time of the Gupta Empire (i.e., from the 4th century A.D. to the 6th century A.D.). For instance, there are mentioned some dangerous enemies of Gupta, such as *hāra-hūṇa* (*Mahābhārata* II: 29, 11; II: 47, 19; III: 48, 21), meaning the White Huns or Hephthalites (Bactrian: ηβοδαλο), or *hūṇa* (*Mahābhārata* VI: 10, 64), meaning Huns. It is worth noting that some enemies of Gupta fought on the side of Kaurava:

> *maheṣvāso gautamaś citrayodhī* ***śakaiḥ kirātair yavanaiḥ pahlavaiś*** *ca* (*Mahābhārata* VI: 20, 13)

> The great bowman, Gautama, knowing all modes of warfare, accompanied by the Indo-Scythian [dynasty] (*śaka*), the Kidara [dynasty] (*kirāta*), the Hellenistic [dynasty] (*yavana*), and the Sasanian [dynasty] (*pahlava*).

In the *Mahābhārata* there are mentioned the cities of Antioch (*antākhi*), Roma (*roma*) and Alexandria in Egypt (*yavanapura*):

> ***antākhīṃ*** *caiva* ***romāṃ*** *ca* ***yavanānāṃ puraṃ*** (*Mahābhārata* II: 28, 49).

In any case, the first images of *Nārāyaṇa* are dated to the time of the Gupta Empire—the time of the early *Purāṇa*s, e.g., see one of the earliest images of *Nārāyaṇa* resting on *Śeṣanāga*, the king of all snakes in Cave 3, Badami cave temples, the Bagalkot district in northern part of Karnataka, India.

Nevertheless, by applying the structuralist approach proposed above we are going to show that the *Nārāyaṇa* myth can be dated strongly B.C. even maybe to the Bronze Age. Let us consider some fundamental properties in the mythological space of *Nārāyaṇa* which remain unchangeable in the stories: F_N (function), L_N (location), A_N (attributes), R_N (relations to other actors). As concerns F_N, it should demonstrate a place of N (*Nārāyaṇa*) among all other deities and a place of the narration about him among all other narrations. The location L_N should fix the main features of the abode of N, such as the top of maintains or the ocean. The attributes A_N should express the main characteristics of N. The relations to other actors R_N should introduce the main actors with whom N has his contacts and all the subordinations among them.

Now, let us involve the mythological space of *Ba'al*, B (Ugaritic: *ba'lu*; Akkadian: *bēlu*) with its significant properties in West Semitic narrations: F_B (function of B), L_B (location of B), A_B (attributes of B), R_B (relations of B to other actors). And let us show that the image of *Nārāyaṇa*, N, and the image of *Ba'al*, B, are *homotopy equivalent*, i.e., there exist continuous transformations from F_N to F_B, from L_N to L_B, from A_N to A_B, from R_N to R_B, proving that they are homotopy equivalent: $F_N \equiv F_B$; $L_N \equiv L_B$; $A_N \equiv A_B$; $R_N \equiv R_B$.

2.1.2.1 Structural Similarity between Functions of Main Actors of Two Myths: $F_N \equiv F_B$

In the narration of *Nārāyaṇa* as well as the narration of *Ba'al* (another possible way of spelling in Ugaritic: *Ba'lu*) we deal with the act of creation and further destruction of the world in the context of the same water, i.e., the first and last possible narration. This homotopy equivalent property $F_N \equiv F_B$ describes the function of the West Semitic myth about *Ba'al* as well as the function of the Hindu myth about *Nārāyaṇa*.

Table 2.1.1. The homotopy equivalence between the Hindu and the West Semitic myth of creation.

Hindu Myth of Creation	***Semitic Myth of Creation***
Brahmā and Nārāyaṇa on the waters	*'Ĕlohīm ('El) and Ba'al on the waters*
"At the beginning, the **waters** were present everywhere, on the surface of the earth. When this (visible world) had been annihilated and when the winds had become quiet, nothing could be known. In that vast ocean-like expanse of water, when the whole world consisting of mobile and immobile beings had perished, the only entity present there was that the lord *Brahmā* with thousand eyes and thousand legs. At that time *Brahmā* named *Nārāyaṇa* lay asleep in the waters. He had a thousand heads" (*Brahmāndapurāna* I: 5, 1–3; Tagare, 1958: 36–37).	"In the beginning God [*'Ĕlohīm*] created the heaven and the earth. And the earth was without form, and void; and darkness was upon the face of the deep. And the Spirit of God moved upon the face of the **waters**" (*Genesis* 1: 1–2)
	Ba'al is a dying and rising fertility god giving water, his main epithet: *rkb 'rpt* "rider of the **clouds**" (Rahmouni, 2008: 288).
Nārāyaṇa having commanded the heaven and the earth	*YHWH (Ba'al) having commanded the heaven and the earth*
"And Nārada knew that the exalted and holy *Nārāyaṇa*, also called Śambhu the lord of the universe, having commanded all the celestials thus, had taken his birth in the race of *yadus* and that foremost of all perpetuator of races, having sprung from the line of the *andhakavṛṣṇis* on earth was graced with great good fortune and was shining like the moon herself among stars" (*Mahābhārata* II: 35; Ganguli, 1883–1896, II: 75).	"A land which the Lord thy God (*YHWH 'Ĕlōheḵā*) careth for: the eyes of the Lord thy God (*YHWH 'Ĕlōheḵā*) are always upon it, from the beginning of the year even unto the end of the year" (*Deuteronomy* 11: 12).
	Some important epithets of *Ba'al* showing that he is a supreme lord: *'al'iyn b'l* "*Ba'al* the mighty one" (Rahmouni, 2008: 53); *mlkn* "our king", *ṯpṭn* "our judge" (Rahmouni, 2008: 55); *'in d 'lnh* "without equal (lit. there is none above him)" (Rahmouni, 2008: 85); *'al'iyn qrdm* "the mightiest of the heroes" (Rahmouni, 2008: 49); *zbl b'l'arṣ* "the prince, lord of the earth" (Rahmouni, 2008: 162); *'ly* "the supreme one" (Rahmouni, 2008: 258).
At the end the earth becoming flooded with water and Nārāyaṇa resting on the waters	*At the end the earth becoming flooded with water and YHWH (Ba'al) moving upon the waters*
"When at the close of four thousand *yugas* the Earth thus became **flooded with water**, like one vast sea, and all mobile creatures were hushed in death, and the sun and the moon and the winds were all destroyed, and the Universe was devoid of planets and stars, the Supreme Being called *Nārāyaṇa*, unknowable by the senses, adorned with a thousand heads and as many eyes and legs, became desirous of rest" (*Mahābhārata* III: 270; Ganguli, 1883–1896, III: 531).	"And I will establish my covenant with you; neither shall all flesh be cut off any more by the **waters of a flood**; neither shall there any more be a flood to destroy the earth" (*Genesis* 9: 11).

Table 2.1.1 contd. ...

...Table 2.1.1 contd.

Hindu Myth of Creation	***Semitic Myth of Creation***
Nārāyaṇa is described as a sleeping-and-rising god giving waters.	*Ba'al* is described as a sleeping(dying)-and-rising god: *b'l mt my l'im* "*Ba'al* is dead, what (shall become of) the peoples" (Rahmouni, 2008). At this moment, "*Ba'al* has no house as (do) the gods, (no) courtyard as (do) the sons of *'Aṯirat*, (no) dwelling (as does) *'El*, (no) shelter (as do) his sons" (Rahmouni, 2008: 120). But, *'Anat* the sister (wife) of *Ba'al* threatens *'El* in order to win his permission for the construction of a palace for the storm-god (Smith, 1994: 31): "And [Virgin *'Anat*] said: 'Bull *'El* [my father] will give me an answer, he will give me an answer, or [I shall give him an answer]: [I] shall trample him like a lamb to the ground; [I shall make] his gray hair run with blood, the gray hair of his beard [with gore], if he does not give a house to *Ba'al* like the gods, [and a dwelling] like the sons of *'Aṯirat*!' [She stamped her f]eet and the earth [shook]" (Wyatt, 2002: 84). "And Virgin *'Anat* replied: 'Your word, *'El*, is wise: you are everlastingly wise; a life of good fortune is your word. Our king is Valiant, *Ba'al* is our ruler, there is none (who is) above him. We should all carry his chalice, we should all carry his cup'" (Wyatt, 2002: 87). After his rising, *Ba'al* gives water for everyone: "And now may *Ba'al* also luxuriate with his rain, may he luxuriate luxuriantly with abundant water in a downpour (?). And may he give his thunder in the clouds, may he flash to the earth lightning" (Smith, 1994: 6).

Let us notice that while *Ba'al* as the Creator is manifested as *'El*, *'Il*, *'Ilu* (Hebrew: *'Ĕlohīm*), *Nārāyaṇa* as the Creator is manifested as *Brahmā*. In both narratives (i) at the beginning, the Supreme Lord is resting on waters (He sleeps or is dead); (ii) He controls the celestial motions and takes care on his race; (iii) at the end, He destroys the earth by flood. See Table 2.1.1.

2.1.2.2 Structural Similarity between Locations of Main Actors of Two Myths: $L_N \equiv L_B$

Both supreme lords are associated to the highest top among the waters. See Table 2.1.2.

Table 2.1.2. The homotopy equivalence between the location of *Nārāyaṇa* and of *Ba'al.*

Location of Nārāyaṇa	***Location of Ba'al***
Nārāyaṇa is located at the highest top among the waters	*'El and Ba'al are located at the highest top among the waters*
"And the region of *Nārāyaṇa* lieth resplendent to the **east of the Meru**, where, O child, that lord of all creatures, the self-create primal Cause of the universe, having manifested all beings, looketh splendid of his excellent grace" (*Mahābhārata* III: 162; Ganguli, 1883–1896, III: 333).	"For *'El* at the **springs of the Rivers**, Amid the **streams of the Deeps** She [*A.Sch.*: *'Anat*] came to *'El*'s **mountain** and entered the tent of the King, the Father of Years" (Smith and Pitard, 2009: 79–80).
	On the one hand, *Ba'al* is "rider of the **clouds**" (Rahmouni, 2008: 288). On the other hand, "*Ba'al* ascends the **mountain**, *Ba'al* sits upon the **throne**" (Rahmouni, 2008: 95).
"At the close of the past (or *pādma*) *kalpa*, the divine *Brahmā*; endowed with the quality of goodness, awoke from his night of sleep, and beheld the universe void. He, the supreme *Nārāyaṇa*, the incomprehensible, the sovereign of all creatures, invested with the form of *Brahmā*, the god without beginning, the creator of all things; of whom, with respect to his name *Nārāyaṇa*, the god who has the form of *Brahmā*, the imperishable origin of the world, this verse is repeated, 'The **waters** are called *Nārā*, because they were the offspring of *Nārā* (the supreme spirit); and as in them his first (*ayaṇa*) progress (in the character of *Brahmā*) took place, he is thence named *Nārāyaṇa* (he whose place of moving was the waters)'" (*Viṣṇupurāṇa* I: 4; Wilson, 1840: 27).	*'El* is seated in his watery abode when *'Anat* and *'Aṯirat* approach him (Smith, 1994: 11). *'El*'s abode is described: "amidst the channels of the **Double Deeps**". It means that *'El*'s abode is at the meeting place of the two cosmic oceans, the upper waters, and the lower waters (Smith, 1994: 225).
"*Nārāyaṇa*, the primeval existent, and eternally enduring, seated in the heart of the **stellar sphere**, is the supporter of all beings" (*Viṣṇupurāṇa* II: 9; Wilson, 1840: 232).	The importance of *Ba'al*'s rain is described, when *'Aṯirat* reacts to *'El*'s granting his permission for *Ba'al* to build his palace by saying: "So now may *Ba'al* make his rain abundant, May he make the water greatly abundant in a downpour, And may he give his voice in the **clouds**, May he flash to the **earth lightning**" (Smith and Pitard, 2009: 81).

2.1.2.3 Structural Similarity between Attributes of Main Actors of Two Myths: $A_N \equiv A_B$

Both supreme lords are associated with the divine waters. See Table 2.1.3.

Table 2.1.3. The homotopy equivalence between the main attribute of *Nārāyaṇa* and of *Ba'al*.

Waters as the Main Attribute of Nārāyaṇa	***Waters as the Main Attribute of 'El or Ba'al***
Devine waters are investigated in yoga meditation	*Devine waters are investigated in prayer*
"That Being *Viṣṇu* became enlightened, following the path of meditation. He was delightfully surprised. "O what is that True entity?" From the body of *Viṣṇu* who thus exerted himself, **water-currents of various sorts** began to flow as a result of *Śiva*'s *Māyā*. O great sage, the Supreme *Brahman* **in the form of divine waters** pervaded the entire void. A mere contact with the same is destructive of sins. *Viṣṇu*, the weary person went to sleep amidst the **waters**. He was in that blissful state of delusion for a long time. As approved in the Vedas, his name came to be established as *Nārāyaṇa* (Having water as abode). Excepting for that Primordial Being there was nothing then" (*Śivapurāṇa* II: 6, 50–54; Shastri, 1970: 198).	"Then the **channels of waters** were seen, and the foundations of the world were discovered at thy rebuke, O Lord, at the blast of the breath of thy nostrils" (*Psalms* 18: 15). "The **voice of the Lord is upon the waters**: the God of glory thundereth: the **Lord is upon many waters**" (*Psalms* 29: 3). "Thou visitest the earth, and waterest it: thou greatly enrichest it with the **river of God, which is full of water**: thou preparest them corn, when thou hast so provided for it" (*Psalms* 65: 9). "Save me, O God; for the **waters are come in unto my soul**" (*Psalms* 69: 1).

2.1.2.4 *Structural Similarity between Relationships of Main Actors of Two Myths:* $R_N \equiv R_B$

The following main characters were introduced in the mythological space of *Nārāyaṇa* to accompany him: (i) *Śeṣa*, also known as *Śeṣanāga* or *Ādi Śeṣa*, the king of all snakes, the many-headed serpent usually pictured with five or seven heads; (ii) *Lakṣmī*, the wife and female form (*śakti*) of *Nārāyaṇa*; (iii) *Brahmā*, the creator of the world. In the Ugaritic *Ba'al* cycle, *Ba'al* (*Ba'lu*) is accompanied by the following three characters: (i) *Leviathan*, *Litan*, *Litanu* (Ugaritic: *ltn*), e.g., "I smote the twisting (/twisted) serpent (*ltn*), the dominant one who has seven heads" (Rahmouni, 2008: 144); (ii) *'Anat*, *'Anatu* (Ugaritic: *'nt*) a virgin (*btlt 'nt*) who is the sister and the "cow of *Ba'al*"; (iii) *'El*, *'Il*, *'Ilu* (Ugaritic: *'il*), the creator of the world. The Ugaritic myth was transmitted in the Bible, too. For instance, in the text there are several occurrences of *Leviathan* struck down by *YHWH* (the Jewish analogue of *Ba'al*) and we see the two aspects of the Lord originally appeared as *'El* and *Ba'al* in the West Semitic mythology. Notice that the character of *Leviathan* is known as *Hydra* in the Greek mythology. The Biblical *Leviathan* is many-headed, also: "the heads of dragon(s)" (Hebrew: *rš'y tnynym*) (*Psalms* 74: 13) and "the heads of *Leviathan*" (Hebrew: *rš'y lwytn*) (*Psalms* 74: 14). In the Ugaritic texts the epithet of *Leviathan* (*Litan*) is as follows: "who has seven heads" (Ugaritic: *d šb't r'ašm*). This epithet of the serpent occurs in the Akkadian omens, too: "a seven-headed snake" (Akkadian: *ṣēru 7 qaqqadāšu*; MUŠ 7 SAG.DU.MEŠ-*šú*) (Rahmouni, 2008: 303). Between the three characters of *Śeṣa*, *Lakṣmī*, and *Brahmā* and the three characters of *Leviathan*, *'Anat*, and *'El* there exists the homotopy equivalence, see Table 2.1.4.

Table 2.1.4. The homotopy equivalence between the three main additional characters of the Hindu myth on creation and of the West Semitic myth on creation.

The Three Main Additional Characters of the Hindu Myth on Creation: Śeṣa, Lakṣmī, Brahmā	***The Three Main Additional Characters of the Semitic Myth on Creation: Leviathan, 'Anat, 'El***
Śeṣa as many-headed serpent subordinated to Nārāyaṇa	*Leviathan as many-headed serpent subordinated to Ba'al (YHWY)*
"And the serpent *Śeṣa*, looking terrible with **his thousand hoods**, and shining with the splendour of ten thousand suns, and white as the Kunda flower or the moon or a string of pearls, or the white lotus, or milk, or the fibres of a lotus stalk, served for his conch" (*Mahābhārata* III: 270; Ganguli, 1883–1896, III: 531–532).	*YHWH* strikes *Leviathan*: "Thou didst divide the sea by thy strength: thou brakest the **heads** of the dragons in the waters" (*Psalms* 74: 13). *Ba'al* strikes *Leviathan*: "Though you smote *Leviathan* the wriggling serpent, finished off the writhing serpent, Encircler-with-**seven-heads**, the skies will be hot, they will shine" (Wyatt, 2002: 87).
"He who is the Source and Creator of the Universe, viz., the Eternal and unfading *Viṣṇu*, He who is called by *munis* crowned with ascetic success as the Supreme Lord of the Universe, that Being of great holiness, then lay in *yoga* sleep on the wide hood of the **Snake** *Śeṣa* of immeasurable energy, and the Creator of the Universe, that highly-blessed and holy *Hari*, knowing no deterioration, lay on the hood of that **Snake** encircling the whole Earth and as the Deity lay asleep on that bed, a lotus, endued with great beauty and effulgence equal unto that of the Sun, sprang from his navel" (*Mahābhārata* III: 202; Ganguli, 1883–1896, III: 419).	*YHWH* strikes *Leviathan*: "In that day the Lord with his sore and great and strong sword shall punish *leviathan* the piercing serpent, even *leviathan* that crooked serpent; and he shall slay the *dragon* that is in the sea. In that day sing ye unto her, A vineyard of red wine" (*Isaiah* 27: 1–2). "Thou brakest the **heads** of *leviathan* in pieces, and gavest him to be meat to the people inhabiting the wilderness" (*Psalms* 74: 14). *Ba'al* strikes *Leviathan*: "When you struck down *Leviathan*, the fleeing snake, annihilated the twisting snake, the powerful one with seven head" (Smith and Pitard, 2009: 252).

Table 2.1.4 contd. ...

...Table 2.1.4 contd.

The Three Main Additional Characters of the Hindu Myth on Creation: Śeṣa, Lakṣmī, Brahmā	***The Three Main Additional Characters of the Semitic Myth on Creation: Leviathan, 'Anat, 'El***
"The adorable god, the lord *Nārāyaṇa*, pervading all things, lives, O *brāhmaṇa*, in a quadruple form: he is possessed as well as devoid of attributes. His first form is inscrutable, the wise behold it bright, it is covered with garlands of flame; it is the acme of perfection to devotees; it is both far and it is near, it is to be understood as transcending attributes; when called Vasudeva, it is seen devoid of egoism; its shape, colour, etc., are not real but imaginary; it is indeed always pure, it is the sole form of pre-eminence. His second form, called *Śeṣa*, supports the earth below with its head; it is described as being characterized by the quality of darkness; it belongs to the brute creation. His third form is active, and devoted to the preservation of creatures, it is to be considered as consisting chiefly of the quality of goodness, it is the fashioner of virtue. His fourth form abides in water; it lies on a **serpent as its bed**; its attribute is passion; and it is always indeed active. The third form of *Viṣṇu*, which is assiduously intent on the preservation of creatures, always maintains righteousness on the earth. It destroys the haughty Asuras, the exterminators of righteousness; it protects the gods, and holy men, who are devoted to the preservation of righteousness" (*Mārkaṇḍeyapurāṇa* 4: 44–53; Pargiter, 1904: 21).	"Canst thou draw out *Leviathan* with an hook? Or his tongue with a cord which thou lettest down?" (*Job* 41: 1). "O Lord, how manifold are thy works! in wisdom hast thou made them all: the earth is full of thy riches. So is this great and wide sea, wherein are things creeping innumerable, both small and great beasts. There go the ships: there is that *leviathan*, whom thou hast made to play therein. These wait all upon thee; that thou mayest give them their meat in due season. That thou givest them they gather: thou openest thine hand, they are filled with good" (*Psalms* 104: 24–28).
Lakṣmī as the wife of Nārāyaṇa	*'Anat as the wife of Ba'al*
"The god is one with all male—the goddess one with all female, rivers. The lotus-eyed deity is the standard; the goddess seated on a lotus the banner. *Lakṣmī* is cupidity; *Nārāyaṇa*, the master of the world, is covetousness. Oh thou who knowest what righteousness is, *Govinda* is love; and *Lakṣmī*, his gentle spouse, is pleasure. But why thus diffusely enumerate their presence: it is enough to say, in a word, that of gods, animals, and men, *Hari* is all that is called male; *Lakṣmī* is all that is termed female: there is nothing else than they" (*Viṣṇupurāṇa* I: 8; Wilson, 1840: 61).	Some epithets of *'Anat*: *btlt*(*m*) "maiden"; *b'l yḥmdnh* "Ba'lu lusts for her" (Rahmouni, 2008: 132).

Table 2.1.4 contd. ...

...Table 2.1.4 contd.

The Three Main Additional Characters of the Hindu Myth on Creation: Śeṣa, Lakṣmī, Brahmā	***The Three Main Additional Characters of the Semitic Myth on Creation: Leviathan, 'Anat, 'El***
"This is the contemplation-sleep of the lord of the world, and the **Great Illusion that comes from** *Hari* [*A.Sch.*: *Lakṣmī*]; by it the world is completely deluded. Verily **she, the adorable goddess, Great Illusion**, forcibly drawing the minds even of those who know, presents them to delusion. By **her** is created this whole universal both moveable and immoveable; **she** it is who when propitious bestows boons on men with a view to their final emancipation, **she** is knowledge supreme; **she** is the eternal cause of final emancipation, and the cause of the bondage of worldly existence; **she** indeed is the queen over all lords. The king spoke: 'Adorable Sir! Who then is **that goddess** whom you style *Mahā-māyā*? How was **she** born, and what is **her** sphere of action, O *brāhmana*? And what is **her** disposition, and what is **her** nature, and whence did **she** originate, the **goddess-all** that I wish to hear from you, O you most learned in sacred knowledge!' The *ṛṣi* spoke: '**She** exists eternally, embodied as the world. By **her** this universe was stretched forth. Nevertheless, **her** origin is in many ways; hear it from me. When **she** reveals **herself** in order to accomplish the purposes of the gods, it is then said in the world that **she** is born; **she** is also named the Eternal One. While the adorable lord *Viṣṇu*, stretching *Śeṣa* out, wooed the sleep of contemplation at the end of the *kalpa*, when the universe was converted into absolute ocean, then two terrible Asuras named Madhu and Kaiṭabha, springing from the root of *Viṣṇu*'s ear, sought to slay *Brahmā*'" (*Mārkaṇḍeyapurāṇa* 81: 41–51; Pargiter, 1904: 469–470).	In this passage the goddess *'Anat*, the wife and sister of *Ba'al*, describes a number of *Ba'al*'s foes that she has defeated in the past: "Surely I struck down *Yamm*, the Beloved of *'El* (*mdd 'il*), Surely I finished off River, the Great God (*'il rbm*), Surely I bound *Tunnanu* and destroyed (?) him, I struck down the *Twisty Serpent* (*Leviathan*), **The Powerful One with Seven Heads**. I struck down Desi[re] (*'ar*[*š*]), the Beloved of *'El* (*mdd ilm*), I destroyed Rebel (*'tk*), the Calf of *'El* (*'gl 'il*). I struck down Fire (*'išt*), the Dog of *'El* (*klbt 'ilm*), I annihilated Flame (*ḏbb*), the Daughter of *'El* (*bt 'il*) That I might fight for silver, inherit gold" (Smith and Pitard, 2009: 53).
"*Brahmā* the *Prajāpati* stood on the lotus that grew from *Viṣṇu*'s navel; and seeing those two fierce Asuras and sleeping Janārdana, and standing with heart solely thereon intent, in order to awaken *Hari*, extolled that Sleep of contemplation which had made its dwelling in *Hari*'s eyes-the lord of splendour extolled *Viṣṇu*'s Sleep, which is **Queen of the universe, the supporter of the world, the cause of permanence and dissolution**, full of reverence, incomparable" (*Mārkaṇḍeyapurāṇa* 81: 51–53; Pargiter, 1904: 470).	
Brahmā is subordinated to Nārāyaṇa	*'El is subordinated to Ba'al and 'Anat, his wife*

Table 2.1.4 contd. ...

...Table 2.1.4 contd.

The Three Main Additional Characters of the Hindu Myth on Creation: Śeṣa, Lakṣmī, Brahmā	***The Three Main Additional Characters of the Semitic Myth on Creation: Leviathan, 'Anat, 'El***
"The essential properties of existent things are objects of observation, of which no foreknowledge is attainable; and creation, and hundreds of properties, belong to *Brahmā*, as inseparable parts of his essence, as heat, oh chief of sages, is inherent in fire. Hear then how the deity *Nārāyaṇa*, in the person of *Brahmā*, the great parent of the world, created all existent things" (*Viṣṇupurāṇa* I: 4; Wilson, 1840: 21–22).	*'Anat* enters the dwelling of *'El*: "*'A[na]t* the cow of *Ba'al*, The[n] [she set her f]ace [towards *'El*] at the source of the river[s], [amid]st the [sprin]gs of the two [deeps]. She rolled back the tent of *'El* and came to the [pavi]lion of the **King, the Father of [the Bright One]**. [She] came to the territory, the tent of the creator" (Wyatt, 2002: 84). Then she defeated him.
"*Brahmā*, the creator of the universe, who sprang from the lotus that grew from the navel of *Nārāyaṇa*" (*Viṣṇupurāṇa* IV: 6; Wilson, 1840: 292).	*'El* (*'Ilu*) is named *'ab šnm* "father of years". This epithet occurs eight times in the Ugaritic corpus. It corresponds to the Biblical Aramaic *'tyq ywmyn* "Ancient of Days" (*Daniel* 7: 9) and to the Arabic divine epithet *qadīmun* "Ancient One" (Rahmouni, 2008: 19).
"The parent and guide of me and of all spheres is the supreme *Nārāyaṇa*, who is *Brahmā*, the lord of the lord of patriarchs; the eldest of the eldest born; one with minutes and hours; one with time; having form, though indiscrete" (*Viṣṇupurāṇa* V: 1; Wilson, 1840: 493).	

2.1.3 Conclusions

Hence, I have shown the homotopy equivalence between the Hindu myth on creation and the West Semitic one by appealing to the Ugaritic *Ba'al* cycle with its later receptions in the Bible. Ugarit, the site of an ancient port city, located in northern Syria, was destroyed in ca. 1200 B.C. The Ugaritic myth is much more archaic than the *Purāṇ*ic myth on *Nārāyaṇa* and, therefore, much less philosophic. The latter was known in the *Mahābhārata*, the large epic book that played an important ideological role in the Gupta Empire that existed from approximately 320–550 C.E. Probably, this book was finished and published by the generous patronage from the Gupta dynasty, as well as the first recension of the *Shahnameh* was written by the patronage from the Samanid dynasty in 994 C.E. The matter is that the *Mahābhārata* has established the ideals of the Hindu monarchy incorporated first by the Guptas, and the *Shahnameh* has presented the ideals of the Persian monarchy with a long tradition.

So, *Ba'al* is accompanied by his sister *'Anat* depicted as a bird. His rising is symbolized by his victory over *Leviathan*, the seven-headed snake. The similar structure has seen in the myth on *Nārāyaṇa* who is accompanied by *Lakṣmī*, his wife and female form (*śakti*), and by *Śeṣa*, the many-headed king of all snakes usually pictured with five or seven heads, on whom *Nārāyaṇa* rests.

Describing the myth of *Ba'al* and the myth of *Nārāyaṇa* is different, but we can detect there a homotopy equivalence—functional similarities in characters and their subordinations. I claim that the homotopy equivalence tells us about a possible influence of the *Ba'al* cycle on the *Nārāyaṇa* cycle, most probably, at the time of the Mitanni state (from ca. 1600 to ca. 1350 B.C.).

2.2 Methodology for Determining the Structural Similarity of Different Ontological Pictures of Different Philosophical Traditions

2.2.1 Problem Setting

In this section, by analogy with the mytheme, introduced in Subsection 2.1.1, an ontologeme is now defined as an elementary unit in a philosophical ontological picture (Schumann, 2021c). In relation to ontologemes, a syntagmatic and paradigmatic order is also introduced. As a result, it is possible to analyze the structure of ontological pictures of various philosophical traditions to study their similarities and differences. As an example of this new methodology, this section shows that Plato's dialogue *Phaedrus* contains an ontological system that was repeated with the preservation of the syntagmatic and paradigmatic orders among the ontologemes of the early Pure Land Sūtras. This structural similarity in ontologies between the *Phaedrus* and the early Pure Land Buddhism takes place in the context of historical facts that some Hellenistic artifacts and the earliest Pure Land texts coexisted in the same region of Gandhāra (now several regions of Pakistan, such as the Swat Valley) during the reign of the Hellenistic or at least partly Hellenized states there.

Since the 19th century many analogies have been found[1] between the Platonic philosophy, on the one hand, and the philosophies of the Hindu school of *advaita vedānta* and the Buddhist school of *yogācāra*,[2] on the other hand. These analogies cause some philosophical speculations that (i) Plato might have been influenced by some Oriental and even, perhaps, Indian traditions and, indeed, he traveled a lot (he visited Italy, Egypt, Phoenicia and some other countries and was impressed by certain Oriental teachings)—this idea was believed by some philosophers (Bussanich, 2005; Ram, 2005; Deussen, 1930); (ii) the Indian monistic philosophy could have been grounded under the influence of Greek ideas. Really, the *yogācāra* appeared in Gandhāra—the country which was home to several Hellenistic or Hellenized states whose official language was Greek from 326 B.C. to 127 A.D., though this claim has not been generally accepted till date. However, both (i) and (ii) are mere speculations, because the Greek and Indian philosophical traditions had quite different religious contexts and different social doctrines.

There is possibly another extreme position in the history of ancient philosophy (apart from (i) and (ii)) according to which there were no interinfluences at all among different ancient traditions (such as Greek and Indian). According to this position, an influence may be proved only by way of detecting direct quotations and commentaries,[3] but Indian philosophers did not comment on the Greek ones

[1] The classical work about this analogy was written by Deussen (1930), the more recent researchers are Bussanich (2005) and Ram (2005).

[2] Both schools were established in India approximately since the 2nd century A.D.

[3] For example, Vidyabhusana (1921) found that there are direct quotations in the *Nyāyasūtra* from some early Madhyamaka and Yogācāra texts.

(at least by means of quotations), and Greek philosophers did not comment on (or at least did not quote from) the Indian ones. This opinion is responsible for that the comparative study between the Platonic and Indian philosophy is still considered marginal, although the similarities between two are quite evident. Therefore, there are just a few secondary references on this subject.

Nevertheless, hidden quotations with no references to their authors can sometimes be found in philosophical texts, e.g., in the *Nyāyasūtra* one can see many quotations from the doctrinal books of two Buddhist schools, *madhyamaka* and *yogācāra*, without references to them. Hence, we may still expect to detect some interinfluences through a textual analysis focusing on such hidden quotations, though the latter may be explored only in the cases where the texts in question are written in the same language.

In this section, I propose a new methodology to find interinfluences in ancient traditions of different languages. This methodology, called the "archaeology of philosophy", is an extension of the structural analysis of myth proposed by Claude Lévi-Strauss (1955). Let us remember that in his analysis there was introduced an elementary unit, called *mytheme*, which is defined as a narrative unit which generally relates character, event, and theme, and which can explain the essence of some elementary episode. Then Lévi-Strauss writes out mythemes on cards and places the cards in such a way that, while fixing the syntagmatic relations among the mythemes, these mythemes would also find some paradigmatic connections among themselves. Let M_{11}, M_{12}, M_{21}, … M_{ij} be mythemes. Then i is their paradigmatic order and j is their syntagmatic order. So, we obtain a matrix

$$M = \begin{bmatrix} M_{11} & \cdots & M_{1j} \\ \vdots & \ddots & \vdots \\ M_{i1} & \cdots & M_{ij} \end{bmatrix}$$

to describe the myth M by analyzing its mythemes with their paradigmatic and syntagmatic relations. Notice that at some places of this matrix we can have zeros. It means that at an appropriate place there is no mytheme.

In relation to ancient philosophies, if we are going to reconstruct their ontological pictures following the structural approach, we should focus on *ontologeme*s (instead of mythemes)—minimal units in ontological descriptions. Let O and O' be two ontological pictures belonging to different cultural traditions (for example, belonging to the Greek and Indian religious philosophies, respectively):

$$O = \begin{bmatrix} O_{11} & \cdots & O_{1j} \\ \vdots & \ddots & \vdots \\ O_{i1} & \cdots & O_{ij} \end{bmatrix} \quad O' = \begin{bmatrix} O'_{11} & \cdots & O'_{1m} \\ \vdots & \ddots & \vdots \\ O'_{k1} & \cdots & O'_{km} \end{bmatrix}$$

At some places of O and O' we can have zeros, too. We say that there is a (partial) *structural similarity* of O to O' if and only if we can construct a (partial) injective mapping f from O to O', with preserving the syntagmatic and paradigmatic orders of O and O'. Let us recall that the injective mapping f in mathematics means a function

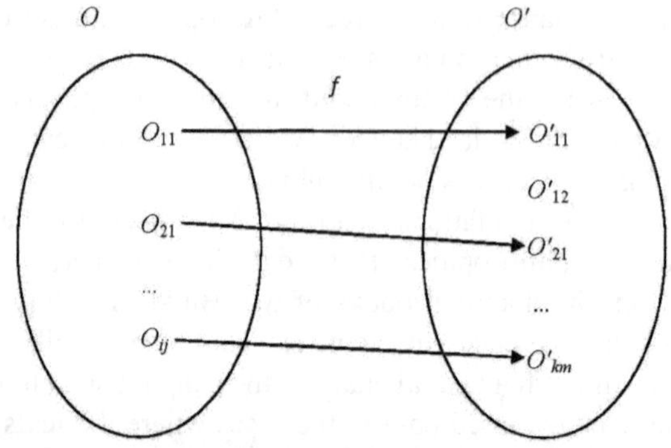

Figure 2.2.1. The mapping from O into O'.

that preserves distinctness: it never maps distinct elements of its domain O to the same element of its codomain O'. And our additional property is as follows: if the non-zero ontologeme O_{xy} follows the non-zero ontologeme $O_{x'y'}$ by the syntagmatic and/or paradigmatic order in O, then the mapped non-zero ontologeme $f(O_{xy})$ follows the mapped non-zero otologeme $f(O_{x'y'})$ by the syntagmatic and/or paradigmatic order in O'. Please see an example, Fig. 2.2.1.

In this example, $f(O_{11}) = O'_{11}$, $f(O_{21}) = O'_{21}$, and $f(O_{ij}) = O'_{km}$. Hence, the syntagmatic and paradigmatic orders are preserved. The mapping f is partial if some elements from O are not mapped to the elements of O'. In this case we say that there is a *partial structural similarity* of O to O'.

The main claim of this section (Subsection 2.2.2) is that there is at least partly a *structural similarity* of the Platonic dialogue *Phaedrus* as an appropriate set of ontologemes (O) to the early Pure Land Sūtras as the second set of ontologemes (O'). It means that we can construct at least partly an injective mapping f from O to O', with preserving their syntagmatic and paradigmatic orders of their ontologemes. It is the first step in our reasoning.

The second step (Subsections 2.2.3 and 2.2.4) is to show that for some mapped non-zero ontologemes $f(O_{xy})$ there is archaeological evidence confirming the existence of syncretic culture of O and O' at the place of $f(O_{xy})$. It is to show that there is direct archaeological evidence that cultural contexts of O and O' coexisted in the same region, i.e., that there was a joint syncretic culture uniting the contexts of O and O', and the cultural context of O was older than the context of O'.

From these two premises (a) there is a structural similarity of O to O'; (b) for some mapped ontologemes $f(O_{xy})$ from O there is archaeological evidence of cultural syncretism of $f(O_{xy})$, we are concluding that the text of O (i.e., the *Phaedrus*) could have influenced the text of O' (i.e., the early Pure Land Sūtras) in fact. This conclusion is not definitive, but it is plausible with a very high grade of probability, revealing a similar epistemological status to the conclusions yielded by Mill's methods which are quite close to our approach.

This methodology is new and is applied for the first time to demonstrate that the ideas of Plato's *Phaedrus* could have partly influenced the Pure Land Buddhism at its starting point. First, we can construct an injective mapping from the set of ontologemes of the *Phaedrus* to the set of ontologemes of the first Pure Land Sūtras preserving their syntagmatic and paradigmatic orders (Subsection 2.2.2). Second, the religious doctrines referring to celestial or astral realms, the contexts for both the *Phaedrus* and first Pure Land Sūtras, coexisted in the same region of Gandhāra, where the Pure Land Buddhism emerged (Sections 2.2.3 and 2.2.4).

According to this methodology, it is enough to have a structural similarity between ontological descriptions rather than the same descriptions by way of quotations. For instance, in the *Phaedrus* there is introduced an ontologeme: "the army of gods and spirits, arrayed in eleven squadrons" (*Phaedrus* 247a),[4] while in the Pure Land Sūtras there is introduced another ontologeme: "innumerable buddhas of the ten directions" (*Sukhāvatīvyūhasūtra* 268a; Takakusu 1988, T. 12, No. 360). So, two entourages are different, but we are considering not the descriptions themselves, but the availability of these descriptions, between which we can construct an injective mapping: there is an entourage for the highest divinity in the text of *O* and there is an entourage for the highest divinity in the text of *O'*. In the meanwhile, in the Pure Land Sūtras there exist many principal descriptions such as *mantra*s for which there are no counterparts in the *Phaedrus*, while almost all ontologemes of the *Phaedrus* can find appropriate counterparts in the Pure Land Sūtras. So, structurally the *Phaedrus* is simpler and may be (at least partly) injectively mapped into the set of ontologemes of early Pure Land Sūtras with preserving their syntagmatic and paradigmatic orders.

2.2.2 Comparison of Plato's Phaedrus with the Pure Land Sūtras

In Greek mythology, the pure lands are called the *fortunate islands* or *islands of the blessed* (μακάρων νῆσοι). The designations occurred in two senses, as (i) a terrestrial pure realm (a real 'island'); (ii) a celestial pure realm (a divine 'island'). Both are localized in the West, the first is in the Atlantic Ocean. For example, in the first meaning the *fortunate islands* are called *Atlantis* (Ἀτλαντὶς νῆσος, "island of Atlantis") by Plato in his works *Timaeus* and *Critias*: according to Plato, this island in former times embodied his ideal republic. In the second sense, they are called the *Elysian fields* (Ἠλύσιον πεδίον) by Homer in the *Odyssey* (4.563) for the afterlife for good men.

Hesiod (Ἡσίοδος; ca. 750–650 B.C.), the Greek poet, describes the fortunate isles as follows (Evelyn-White, 1995):

> But when earth had covered this generation also, *Zeus* the son of *Cronos* made yet another, the fourth, upon the fruitful earth, which was nobler and more righteous, a god-like race of hero-men who are called demi-gods, the race before our own, throughout the boundless earth. Grim war and dread battle destroyed a part of them, some in the land of *Cadmus* at seven-gated *Thebe* when they fought for the flocks of *Oedipus*, and some, when it had brought them in ships over the great sea gulf to *Troy* for rich-haired *Helen*'s sake: there death's end enshrouded a part of them. But to the others father *Zeus* the son of *Cronos* gave a living and an abode apart from men and made them dwell at the ends of earth. And they live untouched by sorrow in the

[4] Tr. by Harold N. Fowler (1925).

> islands of the blessed along the shore of deep swirling Ocean, happy heroes for whom the grain-giving earth bears honey-sweet fruit flourishing thrice a year, far from the deathless gods, and *Cronos* rules over them; for the father of men and gods released him from his bonds. And these last equally have honour and glory (*Hes. Op*. II: 156–169b).

As we find out, these islands (i) were not involved into the Trojan War, (ii) give "honey-sweet fruit flourishing thrice a year", (iii) are ruled by *Cronos*, the deity who ruled before *Zeus*, i.e., life at the islands is outside of time (outside of *rūpaloka*) and outside of terrestrial realms (outside of *kāmaloka*). The worlds of *rūpaloka* and *kāmaloka* are ruled by *Zeus*, the 'Sky-father'.

Hence, the Greek term for the pure lands had the meaning of 'isles' (νῆσοι) and 'fields' (πεδίον). The Buddhist term had the meaning 'level' (*bhūmi*) and 'field' (*kṣetra*). The Greek main pure land is localized in the West. The main pure land of Buddhism, *Sukhāvatī*, is localized in the same direction, i.e., in the West. Let us recall that there are the following five pure lands in Buddhism, called "Buddha fields" (*buddhakṣetra*, 佛土, *fótǔ*):

1. *Abhirati* ("manifest joy"), presided over by *Akṣobhya* (East).
2. *Śrīmat* ('magnificent'), presided over by *Ratnasambhava* (South).
3. *Sukhāvatī* ("blissful realm"), presided over by *Amitābha* (West).
4. *Karmaprasiddhi* ("accomplishing perfect action"), presided over by *Amoghasiddhi* (North).
5. *Akaniṣṭha* ('highest'), presided over by *Vairocana* (Centre).

Sukhāvatī is the most important pure land described in the Pure Land Sūtras. After that the Buddhist idea of this pure land influenced the idea of *Vaikuṇṭha*—a pure land in the *Vaiṣṇava* tradition. The pure land of the latter was imagined much later than the pure land of Buddhism. *Vaikuṇṭha* is defined as a celestial home of Viṣṇu, located in the Capricorn constellation that is visible in the night sky moving gradually westward from different positions. Thus, we see here the same direction of West for localizing the pure land.

In the *Śrī-brahma-saṁhitā* (written after the 16th century, but reflecting a much older tradition), the pure lands are described as *Gokula* in Zenith surrounded by four other pure lands (Thakura, 1985):

> The superexcellent station of *Kṛṣṇa*, which is known as *Gokula*, has thousands of petals and a corolla like that of a lotus sprouted from a part of His infinitary aspect, the whorl of the leaves being the actual abode of *Kṛṣṇa*. <...> There is a mysterious quadrangular place named *Śvetadvīpa* surrounding the outskirts of *Gokula*. *Śvetadvīpa* is divided into four parts on all sides. The abode of *Vāsudeva*, *Saṅkarṣaṇa*, *Pradyumna* and *Aniruddha* are separately located in each of these four parts. These four divided abodes are enveloped by the fourfold human requirements such as piety, wealth, passion and liberation, as also by the four Vedas, viz., *Ṛg*, *Sāma*, *Yajur* and *Atharva*, which deal with the *mantra* and which are the bases of achievements of the fourfold mundane requirements (*Śrī-brahma-saṁhitā* 5: 1–26).

So, we read here about the five pure lands in a similar way as they are in the Pure Land Buddhism: in Zenith, West, East, North, and South.

Let us show that we have several direct analogues for the Buddhist pure lands in Platonic texts which were obviously written much earlier (namely, by about four or five centuries) than the first Pure Land Sūtras and these Platonic texts expressed a Greek theological doctrine about heaven adopted by most Greeks and continued and supported up to the end of Hellenistic civilization as such. In this Platonic teaching about pure lands, we also face a celestial context, though the astrological turn is missing here. To show the similarity of both pictures, let us compare the most significant ontologemes (fragments) from Plato's *Phaedrus* cited from Fowler (1925) and appropriate fragments from the Pure Land Sūtras cited from Inagaki (2003).

The main ontologemes from the *Phaedrus* are summed up in Table 2.2.1.

Table 2.2.1. The main ontologemes from the *Phaedrus*.

The Ultimate Level	Zeus is the highest **O11**	The light of Zeus gives enlightenment **O12**	There is a pure land of Zeus **O13**	Zeus teaches the divine intelligence as a true path **O14**
The Level of Gods	Zeus is followed by other gods **O21**	Other Greek gods can teach **O22**	Zeus reveals the absolute knowledge in this land for all beings **O23**	The meal of Zeus for all higher beings **O24**
The Level of Human Beings			Zeus shows the ultimate way for the deliverance from endless reincarnations. In these reincarnations the soul can spend three successive periods of a thousand years **O33**	Only philosophers can follow the path of Zeus and attain the ultimate reality **O34**

In the matrix form:

$$\begin{bmatrix} \mathbf{O11} & \mathbf{O12} & \mathbf{O13} & \mathbf{O14} \\ \mathbf{O21} & \mathbf{O22} & \mathbf{O23} & \mathbf{O24} \\ \mathbf{0} & \mathbf{0} & \mathbf{O33} & \mathbf{O34} \end{bmatrix}$$

We can show that the set of these ontologemes is injectively mapped into the set of ontologemes taken from the early Pure Land Sūtras presented as Table 2.2.2.

In the matrix form, we obtain the following result of mapping preserving the syntagmatic and paradigmatic orders:

$$\begin{bmatrix} \mathbf{O'11} & \mathbf{O'12} & \mathbf{O'13} & \mathbf{O'14} \\ \mathbf{O'21} & \mathbf{O'22} & \mathbf{O'23} & \mathbf{O'24} \\ \mathbf{0} & \mathbf{0} & \mathbf{O'33} & \mathbf{O'34} \end{bmatrix}$$

Table 2.2.2. The main ontologemes from the early Pure Land Sūtras.

The Ultimate Level	Amitāyus is the highest *f*(**O11**) = **O'11**	The light of Amitāyus gives enlightenment *f*(**O12**) = **O'12**	There is a pure land of Amitāyus *f*(**O13**) = **O'13**	Amitāyus teaches *dharma* *f*(**O14**) = **O'14**
The Level of Gods	Amitāyus is followed by other buddhas *f*(**O21**) = **O'21**	Other buddhas can teach *f*(**O22**) = **O'22**.	Amitāyus reveals an absolute knowledge in this land for all beings *f*(**O23**) = **O'23**	The meal of Amitāyus *f*(**O24**) = **O'24**
The Level of Human Beings			Amitāyus shows the ultimate way for the deliverance from endless reincarnations. In these reincarnations it is possible to spend many thousands of *koṭi*s of *kalpa*s *f*(**O33**) = **O'33**	Only in the practice of Pure Land Buddhism it is possible to attain the ultimate reality *f*(**O34**) = **O'34**

Now let us substantiate these ontologemes with quotes:

O11. Zeus is the highest: "The great leader in heaven" (*Phaedrus* 247a).

f(**O11**) = **O'11.** Amitāyus is the highest: "The majestic light of Buddha Amitāyus is the most exalted" (*Sukhāvatīvyūhasūtra* 270a; Takakusu, 1988, T. 12, No. 360).

O21. Zeus is followed by other gods: "Followed by an army of gods and spirits, arrayed in eleven squadrons" (*Phaedrus* 247a). His main attendants are Ares (*Idem* 252c), associated with war, and Apollo (*Idem* 253b; 265b), associated with wisdom.

f(**O21**) = **O'21.** Amitāyus is followed by other buddhas: "When you see Amitāyus you will also see innumerable buddhas of the ten directions" (*Amitāyurdhyānasūtra* 344a; Takakusu, 1988, T. 12, No. 365). Mahāsthāmaprāpta, i.e., Vajrapāṇi (大勢至, *Dàshìzhì*),[5] associated with war and power, and Avalokiteśvara (觀世音, *Guānshìyīn*), associated with wisdom and mercy, are his main attendants (*Amitāyurdhyānasūtra* 344c), see Fig. 2.2.2.

O12. The light of Zeus gives enlightenment: "The sight of perfect and simple and calm and happy apparitions, which we saw in the pure light, being ourselves pure and not entombed in this which we carry about with us and call the body, in which we are imprisoned like an oyster in its shell" (*Phaedrus* 250c).

f(**O12**) = **O'12.** The light of Amitāyus gives enlightenment: "The light of Amitāyus shines brilliantly, illuminating all the buddha lands of the ten directions" (*Sukhāvatīvyūhasūtra* 270b; Takakusu, 1988, T. 12, No. 360).

[5] He was depicted in India conforming to the Hellenistic iconography of Heracles (a bearded athletic man holding a club and a lion skin) at the time the Pure Land Buddhism emerged.

O22. Other Greek gods can teach: "Those who are included among the twelve great gods and are accounted leaders, are assigned each to his place in the army" (*Phaedrus* 247a).

f(**O22**) = **O'22.** Other buddhas can teach: "Having visualized these innumerable buddhas you will receive from each the prediction of your future buddhahood" (*Amitāyurdhyānasūtra* 343c; Takakusu, 1988, T. 12, No. 365).

O13. There is a pure land of Zeus: "There are many blessed sights and many ways hither and thither within the heaven, along which the blessed gods go to and fro attending each to his own duties; and whoever wishes, and is able, follows, for jealousy is excluded from the celestial band" (*Phaedrus* 247a).

f(**O13**) = **O'13.** There is a pure land of Amitāyus: "In that buddha land, the earth is composed of seven kinds of jewels—namely, gold, silver, beryl, coral, amber, agate, and ruby—that have spontaneously appeared. The land itself is so vast, spreading boundlessly to the farthest extent, that it is impossible to know its limit" (*Sukhāvatīvyūhasūtra* 270a; Takakusu, 1988, T. 12, No. 360).

O23. Zeus reveals an absolute knowledge in this land for all beings: "For the colorless, formless, and intangible truly existing essence, with which all true knowledge is concerned, holds this region and is visible only to the mind, the pilot of the soul" (*Phaedrus* 247c).

f(**O23**) = **O'23.** Amitāyus reveals an absolute knowledge in this land for all beings: "That buddha land, like the realm of unconditioned *nirvana*, is pure and serene, resplendent, and blissful. The *śrāvakas*, *bodhisattvas*, heavenly beings, and humans there have lofty and brilliant wisdom and are masters of the supernatural powers. They are all of one form, without any differences, but are called 'heavenly beings' and 'humans' simply by analogy with states of existence in other worlds" (*Sukhāvatīvyūhasūtra* 271c; Takakusu, 1988, T. 12, No. 360).

O33. Zeus shows the ultimate way for the deliverance from endless reincarnations. In these reincarnations the soul can spend three successive periods of a thousand years: "For each soul returns to the place whence it came in ten thousand years; for it does not regain its wings before that time has elapsed, except the soul of him who has been a guileless philosopher or a philosophical lover; these, when for three successive periods of a thousand years they have chosen such a life, after the third period of a thousand years become winged in the three thousandth year and go their way; but the rest, when they have finished their first life, receive judgment, and after the judgment some go to the places of correction under the earth and pay their penalty, while the others, made light and raised up into a heavenly place by justice, live in a manner worthy of the life they led in human form. But in the thousandth year both come to draw lots and choose their second life, each choosing whatever it wishes. Then a human soul may pass into the life of a beast, and a soul, which was once human, may pass again from a beast into a man. For the soul which has never seen the truth can never pass into human form" (*Phaedrus* 249a–249b).

f(**O33**) = **O'33.** Amitāyus shows the ultimate way for the deliverance from endless reincarnations. In these reincarnations it is possible to spend many thousands of *koṭis* of *kalpas*: "Injustice and vice inevitably follow and are allowed to run their course unchecked until evil *karma* accumulates to the limit. Before they expect their lives to end people meet sudden death and fall into evil realms, where they will suffer excruciating torments for many lives. They will not be able to escape for many thousands of *koṭi*s of *kalpa*s" (*Sukhāvatīvyūhasūtra* 275a–275b; Takakusu, 1988, T. 12, No. 360).

O14. Zeus teaches the divine intelligence as a true path: "Now the divine intelligence, since it is nurtured on mind and pure knowledge, and the intelligence of every soul which is capable of receiving that which befits it, rejoices in seeing reality for a space of time and by gazing upon truth is nourished and made happy until the revolution brings it again to the same place" (*Phaedrus* 247d).

f(**O14**) = **O'14.** Amitāyus teaches *dharma*: "When Amitāyus preaches the *dharma* to *śrāvaka*s and *bodhisattva*s, they all assemble in the seven-jeweled lecture hall" (*Sukhāvatīvyūhasūtra* 273c; Takakusu, 1988, T. 12, No. 360).

O24. The meal of Zeus for all beings: "But when they go to a feast and a banquet, they proceed steeply upward to the top of the vault of heaven, where the chariots of the gods, whose well matched horses obey the rein, advance easily, but the others with difficulty" (*Phaedrus* 247a); "in the revolution it beholds absolute justice, temperance, and knowledge, not such knowledge as has a beginning and varies as it is associated with one or another of the things we call realities, but that which abides in the real eternal absolute; and in the same way it beholds and feeds upon the other eternal verities, after which, passing down again within the heaven, it goes home, and there the charioteer puts up the horses at the manger and feeds them with ambrosia and then gives them nectar to drink" (*Phaedrus* 247e).

f(**O24**) = **O'24.** The meal of Amitāyus: "If, when I attain buddhahood, *bodhisattva*s in my land who would make offerings to buddhas through my divine power should not be able to reach immeasurable and innumerable *koṭi*s of *nayuta*s of buddha lands in as short a time as it takes to eat a meal, may I not attain perfect enlightenment" (*Sukhāvatīvyūhasūtra* 268b; Takakusu, 1988, T. 12, No. 360); "by the Buddha's power, *bodhisattva*s of that land go to innumerable worlds of the ten directions, in as short a time as it takes to eat a meal, in order to pay homage and make offerings to the buddhas and World-honored Ones" (*Sukhāvatīvyūhasūtra* 273c).

O34. Only philosophers can follow the path of Zeus and attain the ultimate reality: "And therefore it is just that the mind of the philosopher only has wings, for he is always, so far as he is able, in communion through memory with those things the communion with which causes God to be divine" (*Phaedrus* 249c).

f(**O34**) = **O'34.** Only in the practice of Pure Land Buddhism it is possible to attain the ultimate reality: "Anyone who sincerely desires birth in the Land of Peace and Bliss is able to attain purity of wisdom and supremacy in virtue. You should not follow the urges of the passions, break the precepts, or fall behind others in the practice of the Way" (*Sukhāvatīvyūhasūtra* 275b; Takakusu, 1988, T. 12, No. 360).

Thus, I have shown that the ontological descriptions from the *Phaedrus* and from the Pure Land Sūtras are *structurally similar* if we ignore different cultural contexts (e.g., in the *Phaedrus* there are no *mantra*s).

Figure 2.2.2. A Bodhisattva (perhaps Avalokiteśvara) with "Buddha ears" holds the trident in his right hand and stands on a lotus. We see the halo surrounding his entire figure. His crown is typical for *bodhisattvas*. This stone is located near the *stūpa* called Kanika Chorten (in honor of the name of Kaniṣka) in the Sani Monastery. *Period*: late Kuṣāṇas or the early Hūṇas. *Date*: ca. from the 4th century to the 6th century A.D. *Geography*: Zanskar in Jammu and Kashmir. *Medium*: Stone. **By courtesy:** Sergii Mazurkevich.

Plato is rather succinct, concise, and laconic in his narrative. Instead of describing the pure lands in detail, he just declares that in the pure lands "there are many blessed sights and many ways hither and thither within the heaven" (*Phaedrus* 247a). Nevertheless, almost for each Platonic ontological image we can explore an appropriate ontological image in the Pure Land Sūtras. For example, for the image "philosophers, they see the pure land of Zeus" there is an analogical image "Buddhists who follow the way of Amitāyus and visualize his pure lands". Plato and the authors of the Pure Land Sūtras proposed similar ontologies with a similar kernel of structures, at the same time the Buddhist ontology is more extensive and complex. Both ontologies introduce a kind of teaching about heaven within the framework of philosophical monism. Moreover, in both ontologies we find an *eternalism* claiming that each may attain actuality and then become eternal. The point is that in the celestial cycles of both Zeus and Amitāyus there is no place for repetitions, hence there is no place for astrology as for a science studying these repetitions. Each human being can become the follower of Zeus or Amitāyus and preserve his or her individuality forever.

2.2.3 Fragments from Lost Greco-Buddhist Texts

We can show that some ontologemes mentioned above and formulated first in the 5th–4th centuries B.C. are well confirmed later archaeologically at several sites in Bactria and Gandhāra (today's Afghanistan and Pakistan) dated from the 3rd century B.C. to the 5th century A.D. It means that they have been disseminated through Hellenism in later cultures of North India. Let us start with ontologeme **O34**: "Only philosophers can follow the path of Zeus and attain the ultimate reality".

There is a philosophical fragment (Rapin, 1992) written in Greek, excavated in Aï Khanum (Afghanistan), the first capital of Greeks in North India before its relocation to Taxila (Punjab), and dated to ca. 140–120 B.C., which contains a lot of Platonic terms such as 'ideas' (ἰδέαι), 'reason' (αἰτία), "each individually" (καθ' ἑκάστην), 'equalities' (ἰσότης), "the first reason of reasons" (πρῶτον τῶν αἰτίων), etc. Meanwhile, we find some quite Buddhist descriptions, too: (i) "generating and destructing all the intelligent entities are forever" (γένεσιν εἶναι καὶ τὴν φθορὰν ἀΐδιον τὴν τῶν αἰσθητῶν ἀναγκαῖον), which hints at the wheel of reincarnation (*saṃsāra*) mentioned as a significant ontologeme in the *Phaedrus*, too; (ii) "nothing of nothing" (οὐθέν οὐθενός) which hints at the concept of *śūnyatā*. Thus, knowledge of Platonic discourse in the territories of today's Afghanistan and Pakistan at the time of Hellenistic dynasties in India is confirmed, since we have this document, the direct evidence containing the pure Platonic term ἰδέαι.

In addition to **O34**, there is another confirmed ontologeme, namely *f*(**O34**) = **O'34**: "Only in the practice of Pure Land Buddhism it is possible to attain the ultimate reality". The point is that the earliest Pure Land Sūtras were written in the region of Bactria, Gandhāra, and Kashmir, where Greek and Gāndhārī were the two official languages from 326 B.C to 127 A.D. While Greek was used for edicts, business, and tax documents, Gāndhārī was reserved for philosophy and religion. The fall of the last Greek dynasty that ruled in Eastern Punjab occurred during ca. 10–30 A.D. But Hellenism as a cultural mix was carried on by various Śaka dynasties, first of all by the Kuṣāṇa Kingdom and the Western Kṣatrapa Kingdom. They continued to use the Greek language as an official language. Only in the 2nd century A.D. did they decide to replace Greek by Bactrian in the Kuṣāṇa Empire and by a Prakrit in the Western Kṣatrapa Kingdom. But both continued to use the Greek letters in their texts. So, in 127 A.D., according to the edict of Kaniṣka (the Rabatak inscription), the Greek language (Bactrian: ιωναγγο, 'Ionian') was replaced by the Bactrian proper (Bactrian: αριαο, 'Aryan'), and the latter existed as a new official language of Gandhāra up to the 8th century A.D., still using the Greek alphabet. In the 5th century A.D., Gāndhārī was replaced by Sanskrit and Sanskrit remained an official language up to the fall of the Hindu Shahi dynasty in 1026.

Hellenism was highly preserved in the world of the Bactrian language. For instance, here are some deities of Kuṣāṇas according to their inscriptions in Bactrian and/or Greek: Ανημοσ (Greek: Ἄνεμοι), Ζαοοσ (Greek: Ζεύς), Ηρακιλο (Greek: Ἡρακλῆς), Ηλιοσ (Greek: Ἥλιος), Ηφαιστοσ (Greek: Ἥφαιστος), Νανα (Greek: Νάνα; Sumerian: dNA.NA.A), Οανινδο ("Victory"), Ορλαγνο ("Mars"); Ριþτι, Ριþτο (Athena), Σαληνη (Greek: Σελήνη), Σαραπο (Greek: Σάραπις), along with many others. These deities are undoubtedly Hellenistic. Furthermore, in the world of Bactrian language, the Hellenistic standards of business documentation and many forms of Hellenistic legality continued to be applied.

For these reasons, the world of Bactrian language remained the main island of Hellenistic legacy in India after the fall of the last Greek dynasty in the 1st century A.D. This legacy was continued even after 400 A.D. (the fall of the Kuṣāṇa Empire). It may be readily shown thanks to the extant artifacts and text fragments.

Almost immediately after its emergence (i.e., since ca. the 2nd century A.D.), the Pure Land Sūtras started to be translated from Gāndhārī into Chinese by some Kuṣāṇa scholars such as Lokakṣema. Many fragments of Buddhist scriptures dated from the 1st century A.D. to the 5th century A.D. were excavated in Gandhāra. These fragments show that it was exactly in this region, that the Mahāyāna Buddism has arisen. For instance, the small birch-bark fragment in the Schøyen collection (MS 2381-40), copied in the script Gilgit/Båmiyan type I, contains the text that corresponds to Chapter 3 of the *Lotus Sūtra*, the next two fragments (MS 2381-2, MS 2382-271) correspond to Chapter 3, and the four fragments (MS 2381-1a, -1b, -20, 82) to Chapter 22 of this *sūtra* (Matsuda, 1999).

Among the texts in Gāndhārī or Hybrid Sanskrit excavated in Afghanistan and Pakistan, there are some Buddhist texts written in the Bactrian language and in the Greek script see (Braarvig and Liland, 2010). They contain homages to a series of Buddhas and Bodhisattvas, including Candrabhānu and Lokeśvararāja, who appear in the Pure Land Sūtras. One of these Bactrian documents, which are preserved, is a manuscript dated to the 5th century A.D. At that time the Greeks were completely assimilated and represented just one of the higher Indian casts (*Yavana*). This manuscript is made from leather which was not typical for Indian texts at all (the Indians preferred palm leafs or birch barks for writing) and is a mark of the Hellenistic style of copying books. Thus, it is one of the direct pieces of evidence that some Hellenized forms of the Mahāyāna Buddism could have been transmitted up to the 5th century A.D. It is a good archaeological confirmation for the ontologeme *f*(**O34**) = **O'34**: "Only in the practice of Pure Land Buddhism it is possible to attain the ultimate reality". This finding denoted by BBM (Bactrian Buddhist Manuscript) describes homage to different Buddhas and *bodhisattva*s, consisting of the names which occur in the Pure Land Sūtras. Let us quote this BBM (Braarvig and Liland, 2010: 72):

> *Recto*
>
> ρ'α' (=101)
>
> οτο σιδασο μαρ̣[ο πονιαδο ταδο]β̣ αν[δ]α̣γ̣ο •••σαγο̣ δο̣ χτο χοαοανδο̣
> And irrespective of [merit (*puṇya*) could be] here, [so] may [become …]-*bandag* (and) […]*sag-dukht* […]
>
> αλο μαδο αλο π[ιδο αλο λογδο αλο πορ]α̣νο αλ̣ο̣ χ̣ο̣[α]υο αλο βραδρανο αλο̣
> with (their) mothers, f[athers, daughters, son]s, sisters (and) brothers, (and) with
>
> οισπο παρολ̣ ω[γοσιγο αβι]ρ̣α̣δ̣ο οτο σ̣ιδ̣[ασ]ο μα̣ρ̣ο μισ̣ο πονιαδο
> all (their) decease[ed relatives, obtai]n (it); and irrespective of further merit (*puṇya*) could be here,
>
> ταδο πιριþτο [ ]δ̣ο̣ ολο αλο̣ μ̣α̣δ̣ο αλο πιδο [α]λ̣ο χοαυ̣[ο]
> so [may] (our) heirs [with their husbands] (and) wives, (and) the mothers, fathers, sisters,

αλο βρα̣δ̣[α]ρανο̣[α]βι̣ ρ[αδο] ο̣το̣ σ̣ ι̣ δ̣α̣σο̣ μ̣αρο̣ μ̣ ι̣ σ̣ο πονοαδο
(and) brothers (of the latter) […] obtain (it); and irrespective of further merit (*puṇya*) could be here.

τ̣ αδο σ̣ιδο̣ <μανο> μ̣ ω̣[σ]ο̣ μ̣[ ]αρ[ ]ο οαυ̣α̣μ̣ ο̣ αλο α̣σ̣το οτα̣γ̣ ο οισ-
so whatever [servants] now [belong] to my […] dwelling or home – may they all

π̣[ο] βοδ̣[••]γο α̣β̣[ιραδο οτο σιδασο μαρο μισο]π̣ογ̣α̣δο̣ τ̣α̣δ̣ο •[••] βανο̣βανδα̣<γο>
ob[tain it; and irrespective of further] merit (*puṇya*) may be here, so may […]*ban-bandag*,

κ̣[ι]δο λιβο τ̣ ο̣[ ]•[ ]δ̣ο αβιρ̣αδ̣[ο] να̣μω̣ο βο̣-
who [copied] the text, […] obtain (it). Homage to the bu-

Verso

δ̣δ̣ο σιδασ̣ο μαρ[ο μισο πονοαδο ταδο]...........σ̣ομαρ̣γο̣........ μο̣ρδο
ddha. Irrespective of [further *puṇya* may be] here, [so may …] (both) dead

οδο ζοοα̣γ̣δαγο̣[ ]..............ο γ̣α̣μ̣ωο **σ̣α̣ο̣κ̣ομα̣νο**
and living [… obtain it]. Homage to **Śākyamuni**

βοδδο ναμωο̣[ν]α̣μωο̣ **λ̣ω̣γ̣ο̣ασφαροραζο βοδδ̣ο** γ̣α̣-
buddha, homage [to … buddha, h]omage to **Lokeśvararāja buddha**, ho-

μωο σιρ̣ογρ•[ ]ρ̣β̣......... β̣οδδο ναμωο **ραδαγ̣ο̣κ-**
mage to […][… buddha, homage to …] [...] buddha, homage to **Ratno-**

ωταμο βοδδ̣[**ο** ναμωο]ο β[ο]δ̣δ̣ο̣ ναμωο **σανδ̣α̣ροβανο β̣ο̣-**
ttama/Ratnaketu buddh[**a**, homage to …] buddha, homage to **Candrabhānu bu-**

δδο ναμωο ρ[να]μ̣ω̣ο̣ **ρ̣αδανοζαυο βοδδο̣** ναμ-
ddha, homage to R[atna(?)-… buddha, ho]mage to **Ratna** […] **buddha**, hom-

ω̣ο **ραδανο•**[**σο**]**μηρο̣βηραμο βοδδο** ναμω̣-
age to **Ratna**[… buddha, homage to **Su**]**meru** […] **buddha**, homage

ο̣ **σ̣ομηρ̣οκ̣•**[ ]βοζο̣ β̣ο̣δδο γ̣α̣μ̣ωο̣ ζι•ο̣ βο̣δδο [ ]
to **Sumeruk**[**alpa buddha**, homage to …][…] buddha, homage to […] buddha […].

As we see, this text tells us about a pure land (*buddhakṣetra*, 佛土, *fótŭ*), where we can settle together with our wives, mothers, fathers, daughters, sons, sisters, brothers, and with all their relatives irrespective to our merits. It is an idea which is very close to the Pure Land Buddhism, or it is one of its earliest forms.

Many names from this short document are significantly corrupted. For instance, we can read just a name fragment ραδανο*[…] βοδδο (*Ratna*-[…] Buddha). In the Mahāyāna Sūtras, there are several names corresponded to this fragment: *Ratna*-candra, *Ratna*-abhibhāsa, Sapta-*ratna*-bhivṛṣḥta, *Ratna*-ketu, *Ratna*-śrī, and so on. Also, there is *Ratnā*-kara ("My Jewel"), the Tathāgata of the Eastern quarter of the pure lands (*dpang-skong-phyag-brgya-pa*, Kangyur volume 68, folio 1b).

Another corrupted name that is possibly read as [...σο]μηρρ βηραμο βοδδο ([...*Su*] *meru-Brahma* Buddha) can be a name that exposes a superiority over Brahma as well as Sumeru, the Lord of the mountains, e.g., *Abhibhūya-sumerubrahmā* Buddha (from Sanskrit: "the Buddha of Superiority over Brahma and Sumeru (the Lord of Montains)") probably to denote a Tathāgata of the Zenith pure lands. Perhaps, [...*Su*] *merubrahma* Buddha corresponds to the name 須彌天冠, *Xūmí-tiānguān*, "Heaven Crowned over Mount Sumeru" of the *Sukhāvatīvyūhasūtra* (266c23; Takakusu, 1988, T. 12, No. 360). Another Tathāgata of the Zenith pure lands is presented by *Sumerukalpa* Buddha (須彌等曜, *Xūmí-děngyào*, "Glorious like Mount Sumeru", 266c23; Takakusu, 1988, T. 12, No. 360). And it is a possible reading of the corrupted name σομηροκ*[...] βοδδο (*Sumeruk*[*alpa*] Buddha).

Some names from BBM can be well read and all of them have appropriate correspondences in the *Sukhāvatīvyūhasūtra* reconstructed by us in Subsection 2.2.2. They occur among the names of former buddhas. Thus, there is a complete name σαοκομανο βοδδο (*Śākyamuni* Buddha), the Tathāgata of the North pure lands. Also, there is the following complete name: λωγρασφαροραζο βοδδο (*Lokeśvararāja* Buddha, 世自在王, *Shìzìzài-wáng*, "World-Sovereign King"). It is one of the most important names in the *Sukhāvatīvyūhasūtra*. This buddha is mentioned as a teacher of the bodhisattva Dharmākara:

> ... a buddha named *Lokeśvararāja*, the Tathāgata, Arhat, Fully Enlightened One, Possessed of Wisdom and Practice, Well-gone One, Knower of the World, Unsurpassed One, Tamer of Beings, Teacher of Gods and Humans, and Enlightened and World-honored One (*Sukhāvatīvyūhasūtra* 267a; Takakusu, 1988, T. 12, No. 360).

The *Lokeśvararāja* Buddha possesses sacral knowledge about the pure lands (*Sukhāvatīvyūhasūtra* 267c; Takakusu, 1988, T. 12, No. 360).

The next complete names are as follows: ραδαγρκωταμο βοδδο (*Ratnottama* Buddha or *Ratnaketu* Buddha), σανδαροβανο βροδδο (*Candrabhānu* Buddha), ραδανοζαυο βοδδρ (*Ratnajaya* Buddha). All three are mentioned in the *Sukhāvatīvyūhasūtra*, too (266c23; Takakusu, 1988, T. 12, No. 360):

> *Ratnottama* or *Ratnaketu* (寶炎, *Bǎoyán*, "Best Jewel" or "Flame of Jewels"),
>
> *Candrabhānu* (月光, *Yuèguāng*, "Moon Light"), and
>
> *Ratnajaya* (威神, *Wēishén*, "Majestic Victory").

The complete name ραδανοζαυο βοδδρ (*Ratnajaya* Buddha) is mentioned as "Precious Conqueror" in the Japan Pure Land Sūtra *Ninnōkyō* (the *Sūtra of Benevolent Kings*) and this Bodhisattva is associated with the North-West pure lands there, see Visser (1935: 127).

So, the text BBM is devoted to a pure land, and it is written within the syncretic Greco-Buddhist culture. It is one of the earliest excavated documents of Pure Land Buddhism (maybe the earliest one) and it is demonstrating some Hellenistic (more precisely Greco-Buddhist) roots of this teaching. It is good archaeological evidence for the mapped ontologeme *f*(**O34**) = **O'34**: "Only in the practice of Pure

Land Buddhism it is possible to attain the ultimate reality" that it existed within the syncretic Hellenistic-Buddhist culture of Bactria and Gandhāra.

2.2.4 Some Direct Evidences of Influences of the Greek Astrology on the Indian One at the Time of the Early Sūtras of Mahāyāna

The Pure Land Buddhism has no astrological aspects, but it contains a teaching on the heaven universe with its division into Zenith, East, West, North, and South. Consequently, there is a celestial dimension of this doctrine. Therefore, we can reconstruct an astral context for the Pure Land Buddhism and check whether we can trace it back to some Hellenistic influences as well. This context shall be an archaeological and textual confirmation for the ontologemes **O11** "Zeus is the highest" and **O21** "Zeus is followed by other gods" (represented by constellations) as well as for the mapped ontologemes *f*(**O11**) = **O'11** "Amitāyus is the highest" and *f*(**O21**) = **O'21** "Amitāyus is followed by other buddhas" (represented by constellations).

Let us recall that in the Greek-Macedonian astrology the following names for the 12 months were used: Δίος (October), Ἀπελλαῖος (November), Αὐδυναῖος or Αὐδναῖος (December), Περίτιος (January), Δύστρος (February), Ξανδικός or Ξανθικός (March), Ξανδικός Ἐμβόλιμος (intercalated 6 times over a 19-year cycle), Ἀρτεμίσιος or Ἀρταμίτιος (April), Δαίσιος (May), Πάνημος or Πάναμος (June), Λώιος (July), Γορπιαῖος (August), Ὑπερβερεταῖος (September), Ὑπερβερεταῖος Ἐμβόλιμος (intercalated once over a 19-year cycle).

In the Buddhist inscriptions written in the Kharoṣṭhī script and in the Gāndhārī language excavated in the region conventionally called Gandhāra and dated from the 1st century A.D. to the 3rd century A.D., the following eight Greco-Macedonian months have been recently identified after decoding: Ἀρτεμίσιος, Δαίσιος, Πάνημος, Λώιος, Γορπιαῖος, Ἀπελλαῖος, Αὐδυναῖος, and Ξανδικός (Salomon, 2011).

Let us illustrate a context of such inscriptions. For instance, in one of the Senior pots of the *Kaniṣka* era (the 1st–2nd centuries A.D.) we find two inscriptions. The first one reads:

> 1. *[sa]ba[tsa]ra [ba](*ḍa)[śa](*mi) ma[se] a[vadu]nake sa[ste]hi (*paca) hi i[śa] (*kṣuṇa)mi [prati]tha[vi]? [matrapi]tra-p[uya]e sarva[satva]na [p] u(*ya)[e]*
>
> 2. *r(*o)haṇasa masumatraputrasa*

> In the year [twelve], in the month *Avadunaka*, after (*five) days; at this time [this] was established in honor of [his] father and mother, in honor of all beings; [donation] of *Rohaṇa*, son of *Masumatra* (Salomon, 2003: 76).

It is worth noting that "the first *Kuṣāṇa* century" (or "the first *Kaniṣka* era") is a date 127 A.D. So, the year 12 means 127 + 12 = 139 A.D.

The second inscription, on the lid, is as follows:

> *[sa]batsara* 10 [2] *mas[u] a[vadu] saste* 4 1 ? ? ? ? ? *rohaṇeṇa masumatraputreṇa thu[ba]m[i] sava[satvaṇa pu](*ya)*

> Year 12, month *Avadu[naka]*, after 5 days, (*established?) by *Rohaṇa*, son of *Masumatra*, in the *stūpa*, in honor of all beings (Salomon, 2003).

Hence, *Avadunaka* is understood by Richard Salomon as the month of Αὐδυναῖος. In the next fragment from Gandhāra, *Khsaṃdika* is examined by him as Ξανδικός:

> *saṃ* 20 4 *khsaṃdikas*[*a*] *di* [2] *thubaṃmi baütaaṇami acaryaṇa dharmagutakaṇa parigrahami*

> (Given in) the year 24, day 2 (?) of *Khsaṃdika* [Ξανδικός] to the *stūpa* at *Baütaaṇa*, in the possession of the *Dharmaguptaka* masters (Salomon, 2011: 168).

The year 24 means 127 + 24 = 151 A.D.

In the Wardak inscription, dated to the *Kaniṣka* year 51 (178 A.D.) during the reign of Huviṣka, the month name is also loaned from Greek: *masye arthamisiya sastehi* 10 4 1 "in the month *Arthamisiya* [Ἀρτεμίσιος], after fifteen days" (Konow, 1929: 170). A similar dating formula with a Greek month name appears in the next findspot of the Senior pot: *saṃbatsarae aṭhaviśatihi* 20 4 4 *mase apelae sastehi daśahi* 10 *iśa kṣunaṃmi* "in the year twenty-eight, 28 [*A.Sch*.: i.e., 155 A.D.], in the month *Apelaa* [Ἀπελλαῖος], after ten days, 10; at this time..." (Salomon, 2003: 76–77). Another example of the Greek month name, decoded by Sten Konow: *saṃ* 10 4 4 *masye arthamisiya sastehi* 10 *iś*[*e*] *kṣunaṃm*(*r*)*i* "Year 18 [*A.Sch*.: i.e., 145 A.D.], in the month *Arthamisiya* [Ἀρτεμίσιος], after 10 days; at this time..." (Konow, 1929: 151–152).

It is worth noting that in the *Kaniṣka* era (after 127 A.D.) the Greco-Macedonian month names were preferred in opposition to the Indian ones. In the meanwhile, they also occur much earlier, e.g., in the Taxila copper plate of Patika (Konow, 1929: 28), dated in the month *Panema* [Πάνημος].

One may conclude that the Greek-Macedonian calendar used in Gandhāra is also direct evidence that a kind of Greek astrology and astronomy was adopted by the Buddhist community at that time, since any calendar takes a significant part in compatible astrological and astronomical lore.

Furthermore, the Medieval Hindu astrology used a lot of loanwords from Greek: (i) an angular sign: *kendra* (κέντρον); (ii) Sun: *heli* (ἥλιος); (iii) diameter: *jāmitra* (διάμετρον); (iv) the planet Jupiter: *jyau* (Ζεύς); (v) the planet Mars: *āra* (Ἄρης), and many others.

Let us note that one of the first treatises of Indian astrology is represented by the *Yavanajātaka* ("*Jātaka* of the Greek") recently reconstructed by David Pingree (1981). This treatise was a translation from a book written in Greek and devoted to astrology and this translation was made first in the 2nd century A.D. by Yavaneśvara and later versified by Sphujidhvaja in the 3rd century A.D. It is from this book that the Greek astrological terminology was adopted in the Medieval Hindu astrology together with some ways of calculations.

Starting from the Sumerian period (ca. 3000–2500 B.C.), we have many ancient texts devoted to the matters of astronomy, astrology, and calendar systems, and all these texts do not separate these disciplines like it occurs nowadays (in the modern culture). The point is that in all known ancient cultures astronomy, astrology, and calendar systems belong together as one system of knowledge. The celestial (astronomical/astrological) aspect is also prominent in the Platonic works, especially in his descriptions of the pure lands. The same aspect is readily traced among the artifacts of Gandhāra (e.g., in the Greek names of months in the Buddhist inscriptions, in many occurrences of the following two names in Greco-Bactrian: Ζαοου in genitive (Greek: Ζεύς), Ηρακιλο (Greek: Ἡρακλῆς) who have a strong celestial/astronomical/astrological meaning, etc.). As a result, by applying the methods of the "archaeology of philosophy" we may conclude that there was a form of religious syncretism in a lost general joint system of calendar, astronomy, and astrology as a cultural context of **O11**, **O21**, *f*(**O11**) = **O'11**, *f*(**O21**) = **O'21**.

The Greek astrology was not only a tool for divination, but also an ontological picture of the world, where Zeus is a supreme lord (the ontologemes **O11** "Zeus is the highest" and **O21** "Zeus is followed by other gods"). Thus, the heaven was regarded as the territory of Zeus: ἐν Διὶ πατρί ("in the heaven of Zeus"). He established the signs in heaven and marked out the constellations to set the seasons and time and to "give to men their right signs" allowing them to work "for the planting of trees and for casting all manner of seeds:"

> From *Zeus* [Διὸς] let us begin; him do we mortals never leave unnamed; full of *Zeus* are all the streets and all the market-places of men; full is the sea and the havens thereof; always we all have need of *Zeus*. For we are also his offspring; and he in his kindness unto men giveth favourable signs and wakeneth the people to work, reminding them of livelihood. He tells what time the soil is best for the labour of the ox and for the mattock, and what time the seasons are favourable both for the planting of trees and for casting all manner of seeds. For himself it was who set the signs in heaven, and marked out the constellations, and for the year devised what stars chiefly should give to men right signs of the seasons, to the end that all things might grow unfailingly. Wherefore him do men ever worship first and last (Aratus, *Phaenomena* 1–15; Mair, 1921: 381).

This text belongs to Aratus (Ἄρατος ὁ Σολεύς; ca. 315 B.C./310 B.C.–240 B.C.), a Greek didactic poet, who wrote a prominent hexameter poem on astrology, the *Phaenomena* (Φαινόμενα καὶ Διοσημεῖα). It is the oldest extant Greek in-depth description of all the constellations and their meanings. He was probably a Phoenician, being a native of Soli in Cilicia or Tarsus, and a friend of the Stoic philosopher Zeno who was certainly a Phoenician. Aratus was a disciple of the Peripatetic philosopher Praxiphanes (Πραξιφάνης) in Athens. He was invited to the court of Antiochus I Soter (Ἀντίοχος Α΄ ὁ Σωτήρ; ca. 324/3–261 B.C.), the king of the Hellenistic Seleucid Empire that at the height of its power included central Anatolia, Persia, the Levant, Mesopotamia, and today's Kuwait, Afghanistan, and parts of today's Pakistan and Turkmenistan. So, Aratus visited in the 3rd century B.C. the state that included countries with the Indo-Aryan culture.

Thus, in the Greek astrology Zeus is the Supreme Lord (**O11** and **O21**). In the *Bṛhat-parāśara-horā-śāstra* (a foundational compilation of Indian astrology written not earlier than in the 7th century), the heaven is treated as the territory of Nārāyaṇa

(Śrī Viṣṇu) and, therefore, Nārāyaṇa in the Hindu astrology plays the role of Zeus (the *Vaiṣṇava* mapping for **O11** and **O21**):

> *Śrī Viṣṇu* who is the Lord (of all matters), who has undefiled spirit, who is endowed with the three *guṇa*s, although he transcends the grip of *guṇa*s (*gunātīta*), who is the Author of this Universe, who is glorious, who is the Cause and who is endowed with valour, has no beginning. He authored the Universe and administers it with a quarter of his power (*Bṛhat-parāśara-horā-śāstra* 1: 9–11, translated by R. Santhanam, 1984).

In the Pure Land Sūtras, there is no astrology at all, but these texts are an essential addition to *lokanirdeśa* (the Buddhist teaching on the structure of the universe) and, in the same way as Zeus for the Greeks and Nārāyaṇa for the *Vaiṣṇava*s, Amitāyus is regarded as a supreme lord of the heaven (*f*(**O11**) = **O'11** and *f*(**O21**) = **O'21**).

In the Greek (Latin) astrology, Zeus (Jupiter) has a celestial companion—Heracles (Hercules) who was traditionally considered a constellation named *The Kneeler*:

> *Ingeniculus* [*A.Sch.*: the Kneeler, i.e., Heracles] rises in the last degrees of Pisces. It is called in Greek *Engonasis*. Those who are born with this star rising will be wise, clever, trained in various tricks, liars who deceive people with different kinds of plots; they are always aggressive and display unbridled hostility (Julius Firmicus Maternus VIII: 17, 4; Bram, 1975: 280).

> Right there in its orbit wheels a Phantom form, like to a man that strives at a task. That sign no man knows how to read clearly, nor what task he is bent, but men simply call him On His Knees [*A.Sch.*: Ἐν γόνασιν or Ἐγγόνασιν, "the Kneeler"]. Now that Phantom, that toils on his knees, seems to sit on bended knee, and from both his shoulders his hands are upraised and stretch, one this way, one that, a fathom's length. Over the middle of the head of the crooked Dragon [Δράκοντος], he has the tip of his right foot (Aratus, *Phaenomena* 63–69; Mair, 1921: 385–387).

> But the Phantom On His Knees sinks all save knee and left foot beneath the stormy ocean (Aratus, *Phaenomena* 592–593; Mair, 1921: 429).

The same constellations including the Kneeler are enumerated by Geminus of Rhodes (Γεμῖνος ὁ Ῥόδιος; the 1st century B.C.), the Greek astronomer and mathematician, see (3.8). Eratosthenes of Cyrene (Ἐρατοσθένης; ca. 276 B.C.–ca. 195/194 B.C.), the Greek poet, mathematician, geographer, astronomer, the chief librarian at the Library of Alexandria, associated the Kneeler with Heracles (Οὗτος, φασίν, Ἡρακλῆς ἐστιν ὁ ἐπὶ τοῦ Ὄφεως βεβηκώς, *Cat.* 1.4). Dionysius of Halicarnassus (Διονύσιος Ἀλεξάνδρου Ἁλικαρνᾱσσεύς; ca. 60 B.C.–after 7 B.C.), the Greek historian and teacher of rhetoric, noted that Heracles must have become immortal because of his virtue (διαμείψαντα τὴν θνητὴν φύσιν ἀθάνατον γενέσθαι δι᾽ ἀρετήν, *Antiquitates Romanae* I: 40, 2).

In the Hindu astrology, the celestial location quite like the Greek (Latin) constellation of Heracles (Hercules) is called Abhijit, the 28th *nakṣatra* (constellation). In the *Sārāvalī*, the Hindu astrological fundamental book, written by Kalyāṇavarman in ca. 800 A.D., Abhijit is a sign to become a king:

> *jāyate'****bhijiti*** *yaḥ śubhakarmā bhūpatirbhavati so'tulavīryaḥ |*
> *nīcaveśmakulajo'pi naro'smin rājayoga iti na vyapadeśaḥ ||*

> One born in *Abhijit nakṣatra* will become a lord of the earth though with auspicious action, unequaled strength, and born of a family of an inferior dwelling. There is a *Rāja yoga* in this, even for a man with no name (*Sārāvalī* 86; Subrahmanya Sastri, 1907, tr. by Michael Douglas Neely).

The point is that it is believed that Kṛṣṇa was born under this constellation. Abhijit is a *bhakti* and *avatāra* of Viṣṇu. Thus, this sign allows a righteous person to attain heaven:

> *dugdhaṃ tv abhijite yoge dattvā madhu ghṛtāplutam |*
> *dharmanityo manīṣibhyaḥ svargaloke mahīyate ||*

> By making a gift under the conjunction called *Abhijit*, of milk with honey and ghee unto men of wisdom, a righteous person attains to heaven and becomes an object of attention and honour there (*Mahābhārata* XIII: 64, 27; Ganguli, 1883–1896).

This sign allows the human beings to attain high knowledge, too:

> *śrāddhaṃ tv abhijitā kurvan vidyāṃ śreṣṭām avāpnuyāt*

> By doing it under the constellation *Abhijit* one attains to high knowledge (*Mahābhārata* XIII: 89, 11; Ganguli, 1883–1896).

Hence, the constellation of Abhijit is a Hindu analogue for the Greek constellation of Heracles even according to the celestial location, prognoses, and connotations approximating those of Heracles, the Kneeler. In the *Bhāgavata-purāṇa* (written after the 8th century A.D.), we read:

> *saṃvatsaro 'smy animiṣām ṛtūnāṃ madhu-mādhavau |*
> *māsānāṃ mārgaśīrṣo 'haṃ nakṣatrāṇāṃ tathābhijit ||*

> Among all the divisions of Time, I am a year (*saṃvatsara*) and among all seasons, I am the spring. Among the months [of a year], I am *Mārgaśīrṣa*, and among all constellations of stars, I am *Abhijit* (*Bhāgavata-purāṇa* XI: 16, 27).

In the region of today's Afghanistan, Tajikistan, and Pakistan, and even in some lands of North India such as Mathura, the excavations produced a great number of images of Hellenistic deities dated from the 3rd century B.C. to the 4th or even 5th century A.D. For instance, the images of Zeus and Heracles were quite often minted with accompanying Greek inscriptions or without them on the coins of various dynasties of North India: Indo-Greeks, Indo-Parthians, Indo-Scythians, Kuṣāṇas, and their assorted satraps. Usually, Zeus was depicted as standing and holding Nike (the Hellenistic goddess of victory) or an enthroned deity, see, Fig. 2.2.3. Heracles was depicted conventionally as a deity holding a club and a lion skin, see Figs. 2.2.4 and 2.2.5. In the 2nd century A.D., the Kuṣāṇa Empire was quite extensive, including the

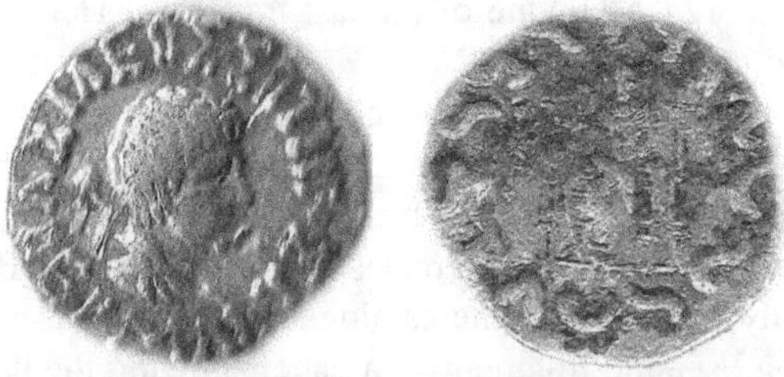

Figure 2.2.3. Posthumous imitation of Hermaios (Ἑρμαῖος, ca. 90–70 B.C.), the Indo-Greek king of the Eucratid Dynasty. This posthumous issue is attributed to the Indo-Scythian or Kuṣāṇa rulers of Hermaios' former territories. *Material*: Silver drachm (16 × 18 mm, 1.94 g). *Obverse*: The diademed and draped bust of Hermaios right; the Greek legend around: ΒΑΣΙΛΕΩΣ ΣΩΤΗΡΟΣ/ΕΡΜΑΙΟΥ. *Reverse*: Zeus enthroned slightly left, making the gesture of benediction, and holding the scepter; the Gāndhārī legend around in the Kharoṣṭhī script: *maharajasa tratarasa/heramayasa*.

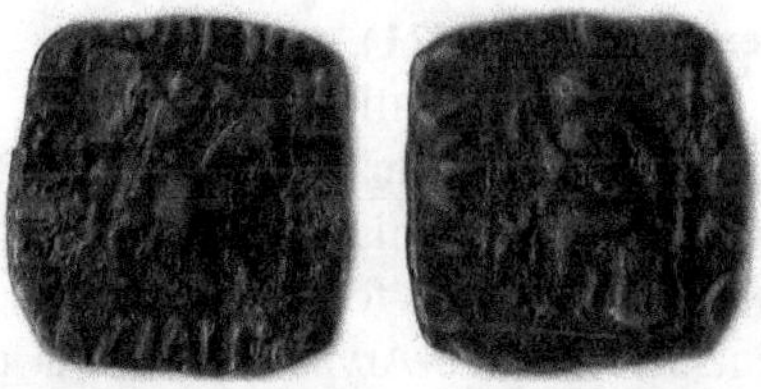

Figure 2.2.4. Azilises (Greek/Bactrian: ΑΖΙΛΙCΟΣ, ca. 60–35 B.C.), the Indo-Scythian king of Gandhāra. *Material*: Copper 1/4 unit (17 × 18 mm, 2.58 g). *Obverse*: The diademed king mounted on a horse turned right, within the square border; the corrupted Greek legend around: ΒΑΣΙΛΕΩΣ/ΒΑΣΙΛΕΩΝ/ΜΕΓΑΛΟΥ/ΑΖΙΛΙCΟΥ. *Reverse*: Heracles seated on a rock, holding a club in his right hand, a monogram below left all within the square border; the corrupted Gāndhārī legend around in the Kharoṣṭhī script: *maharajasa/mahatasa/ayiliṣasa*.

Figure 2.2.5. Huviṣka (Bactrian: ΟΟΗþΚΙ, ca. 140–180 A.D.), the king of Kuṣāṇa dynasty who was the ruler of the Kuṣāṇa Empire at the peak of its power covering today's Afghanistan, Pakistan, Tajikistan, North of India, and South of Turkmenistan. *Material*: Copper tetradrachm (24 mm, 9.99 g). *Obverse*: The king riding an elephant facing right, holding an elephant goad; the corrupted Bactrian legend around: þΑΟΝΑΝΟ þΑΟ ΟΟΗþΚΙ ΚΟþΑΝΟ. *Reverse*: Hercules facing left with a halo, holding a club and a lion skin in the left hand, and holding the diadem for the king in his right hand; a tamgha on the left; the corrupted Bactrian legend on the right: ΗΡΚΙΛΟ.

territories of North India up to Varanasi as well as the whole lands of Afghanistan, Pakistan, Tajikistan, along with some portions of Turkmenistan. And at that time, Zeus and Heracles remained quite popular. Zeus was spelled Ζαοοσ ('Zeus') in Greco-Bactrian and Heracles was called Ηρακιλο ('Herakilo') or Ηρκιλο ('Herkilo'). It is worth remembering that Bactrian with Gāndhārī were two main languages of the

Kuṣāṇa Empire since 127 A.D. One of the last images of Heracles on the Kuṣāṇa coins is exemplified in Fig. 2.2.5.

In Gandhāra, Heracles came to be considered a protector of Buddhism—in the *Sukhāvatīvyūhasūtra* (Takakusu, 1988, T. 12, No. 360), he is called *Mahāsthāmaprāpta*, 大勢至, *Dàshìzhì*—by one of the names of Vajrapāṇi. Now there is a consensus among historians that Heracles is Vajrapāṇi for the Buddhists, and this opinion is commonly expressed in the catalogues of most museums, including the British Museum, e.g., see the fragment of a panel showing the haloed Buddha and Vajrapāṇi (Heracles) behind him with other figures attending the Buddha in British Museum, number: 1947,1016.1. In Greek mythology, Heracles, born as a human being, ascended to heaven after all his labors and, finally, became a constellation. This means that now Zeus is followed by him on the sky (**O21**), in the pure lands in the Platonic meaning. So, even the character of Heracles who accompanies Zeus in heaven is now the same as the character of Vajrapāṇi who accompanies Amitāyus in the pure lands of the latter (*f*(**O21**) = **O'21**).

To sum up, we can trace back the influences of Greek astrology on its Hindu counterpart, and within this cultural context Zeus was treated as Nārāyaṇa for the *Vaiṣṇava*s (and Amitāyus for the Buddhists) as well as Heracles was treated as Abhijit for the *Vaiṣṇava*s (and Vajrapāṇi for the Buddhists). This shows a cultural diffusion between Hellenistic and Indo-Aryan religions at the time of writing the first Pure Land Sūtras which is an archaeological confirmation for **O11**, **O21**, *f*(**O11**) = **O'11**, *f*(**O21**) = **O'21**. Within this diffusion, some Hellenistic philosophical ideas such as the Platonic teaching about heaven from the *Phaedrus* could have put down roots in Gandhāra indeed.

2.2.5 Conclusions

In this section, I have drawn the conclusion that the teaching about heaven from the *Phaedrus* could have influenced the teaching about heaven of the Pure Land Sūtras. In drawing this conclusion, I have applied the new methodology called "archaeology of philosophy". This conclusion can be represented by the following total according to the Mill's methods:

1. The Platonic dialogue *Phaedrus* (*Ph*) is a teaching about heaven within the framework of monistic philosophy (*GTH*).
2. The *Phaedrus* (*Ph*) has structural similarity to the Pure Land Sūtras (*PLS*).
3. The Pure Land Sūtras (*PLS*) are an example of Buddhist teaching about heaven within the framework of monistic philosophy (*BTH*).
4. The *Phaedrus* (*Ph*) is a Greek teaching about pure lands (*GH*).
5. The Pure Land Sūtras (*PLS*) are a Buddhist teaching about pure lands (*BH*).
6. The earliest fragments of Pure Land Buddhism (*PLS*) appeared in Gandhāra at the time of Hellenistic influence and Greco-Buddhist syncretic culture (*G*).

7. The Greek teaching about heaven or pure lands (*GH*) has a Greek astrological context (*A*).
8. The same Greek astrological context (*A*) is reconstructed in Gandhāra at the time of Hellenistic influence and Greco-Buddhist syncretic culture (*G*).

Therefore, quite probably, the Platonic teaching about heaven within the framework of monistic philosophy (*GTH*) with the Greek teaching about pure lands (*GH*), could have influenced the Buddhist teaching about heaven within the framework of monistic philosophy (*BTH*), along with the Buddhist teaching about pure lands (*BH*) in Gandhāra (*G*) in the Greek astrological context (*A*).

2.3 Methodology for Identifying the Origins and Genesis of Logical Competence through the Detection of Structural Similarity in Reasoning

2.3.1 Problem Setting

As we have just seen, the structuralist methodology may be used not only to identify the dynamics of the spread of myths among different peoples, but also to determine the dynamics of the spread of philosophical ontological pictures belonging to different philosophical traditions and emerging in different languages. This section discusses the possibility of tracking the dynamics of the development of logically correct forms of reasoning—from the simplest ones presented in the use of elementary inference rules of propositional logic, such as *modus ponens* and *modus tollens*, to complex combinations of inference rules and up to the reasoning of modal logic.

Historical examples of correct reasoning lend themselves more easily to the structuralist analysis than myths or philosophical ontologies. The matter is that as an elementary unit of reasoning (that is, as an analogue to mytheme from a mythological cycle or to ontologeme from a philosophical picture of the world), one logical step in inference is presented by using, first, some statements as logical axioms or assumptions and, second, some inference rules. And it is precisely the same understanding of the elementary unit of reasoning that is present now in symbolic (mathematical) logic. Let us call this simple element of reasoning a *logeme*. So, it is an axiom or hypothesis (for example, the implication "If *A*, then *B*", which is considered true), as well as an expression obtained from an axiom or hypothesis by applying some logical rules (for example, *modus ponens*). In this case, the analogue of the syntagmatic orders of mythemes (or ontologemes) in one reasoning is given as a syntagmatic order of logemes, namely a chain of statements, each of which is derived from some previous ones by using inference rules. On the other hand, the analogue of the paradigmatic orders of mythemes (or ontologemes) is given as a paradigmatic order of logemes as a choice of specific inference rules—as a frequency of using different inference rules in the same reasoning (i.e., a depth of one inference within a given reasoning using different rules).

The structure of logemes sets a logically correct reasoning only if this structure does not have contradictions within the framework of modern (classical) propositional logic. In other words, this structure must be consistently formalized by the means of modern propositional logic. For example, there should be a correct understanding of the implication, according to which the conditional proposition "If *A*, then *B*" is logically identical to the conditional proposition "If not-*B*, then not-*A*", which means that there should be an understanding that *modus ponens* and *modus tollens* can perform only in pairs. The logically correct understanding of implication is first encountered in texts in the Sumerian and Akkadian languages (from the 24th to the 4th centuries B.C.). But, for example, in the logical treatise *Kathāvatthu* from the Pāli Canon (dated between the 2nd century B.C. and the 2nd century A.D.), this understanding is absent, which means that the structure of logemes does not meet the requirements of modern propositional logic (Schumann, 2019a).

The section shows that logically correct logeme structures first appeared no later than during the III dynasty of Ur (from about 2112 B.C. to 2004 B.C.) in legal documents in the Sumerian and Akkadian languages (Schumann, 2019b, 2020, 2021a). Then this structure has been made obviously complicated until the time of the New Babylonian Kingdom (626–539 B.C.) in texts of the same orientation—this includes various legal documents, such as court records and business correspondence. Moreover, there was a clear continuity even in the composition of the texts themselves, including the methods of logical reasoning.

Then the structures of logemes, already in highly developed forms of New Babylonian reasoning, were inherited in texts in Aramaic and Ancient Greek, which can be traced thanks to the judicial archive found on the island of Elephantine, located on the Nile River (Schumann, 2019b). In these texts (dating from the 5th century B.C.), the complex logeme structure was inherited along with the composition of the legal acts. Subsequently, a similar logeme structure and a similar composition of legal documents can be found in the archives (dated to the 2nd–8th centuries A.D.) in Bactrian and Sogdian languages (Eastern Middle Iranian languages), which, together with Ancient Greek and Aramaic, served as a *lingua franca* of international trade on the Silk Road.

Thus, the logeme structures of different legal traditions (such as New Babylonian, Achaemenid, Hellenistic, Indo-Scythian) have strong similarities and their presence suggests that these traditions were linked by a relationship of succession. This section substantiates that logically correct logeme structures appeared for the first time in the Mesopotamian legal culture no later than during Ur III and were subsequently developed up to the New Babylonian culture in the New Babylonian dialect of Akkadian. Then they developed within the framework of the Achaemenid culture in the Aramaic language, although they continued to use Akkadian as well. After that they were supported in the Hellenistic culture in Ancient Greek and then in the Eastern Iranian (Indo-Scythian) culture in the Bactrian and Sogdian languages.

The same similarity between the logeme structures within the framework of genealogical analysis can be traced in philosophical texts of the ancient world of the historical period. For example, logical determinism in the philosophy of the Early Stoa (the 3rd century B.C.) is repeated in similar logeme structures of the philosophical discourse belonging to the Madhyamaka, Yogācāra, and Nyāya schools (from the 2nd to the 6th century A.D.)—see Chapter 4.

Nevertheless, logeme structures are never given abstractly in everyday life. In their abstract form, they can be studied only in logical treatises. But in life they are incorporated into complex speech acts of argumentation in accordance with language-game rules of appropriate communicative communities (*Kommunikationsgemeinschaften*), see (Apel, 1973). Even mathematical logic and set theory were formed in specific communities with a possibility of extra-logical norms of communication. For instance, Stanisław Zaremba (1863–1942), a Polish mathematician, became the object of bullying among the Polish logicians after he declared that logic and set theory are just auxiliary disciplines of mathematics, not the main ones, as was the case according to the program of Zygmunt Janiszewski (1888–1920), Wacław Sierpiński (1882–1969), and Stefan Banach (1892–1945). However, much later it turned out that Zaremba was right. As such, mathematics cannot be reduced to logic and set theory. This discussion around Zaremba initiated by Jan Łukasiewicz (1878–1956) and published mainly in the Polish journal *Przegląd Filosophiczny*, from 1916 to 1918, shows that mathematical logic is implemented into appropriate complex argumentative speech acts of mathematicians, where bullying was also possible.

Hence, in human practice, even mathematical logic is a part of specific language-games among mathematicians and some forms of extra-logical rhetorical persuasiveness are always possible, as the case of Zaremba shows. It means, it is a part of argumentation theory.

Modern argumentation theory, focusing on reasoning in speech acts, has become an important addition to formal or symbolic logic developed by mathematicians such as Gottlob Frege, Bertrand Russell, David Hilbert, and others. One of the first studies in modern argumentation theory was presented by Howard Kahane's monograph *Logic and Contemporary Rhetoric. The Use Of Reason in Everyday Life* (1971) which touches on the arguments that students face directly in their private life—it was "a course relevant to everyday reasoning, a course relevant to the arguments they hear and read about race, pollution, poverty, sex, atomic warfare, the population explosion, and all the other problems faced by the human race in the second half of the twentieth century" (Kahane, 1971: vii). Consequently, informal logic became another possible name for the theory of argumentation to show its practical aspects in everyday life without mathematical formulas. According to Kahane, the subject of informal logic consisted of studying the ways of putting forward arguments in newspapers, in which some current political and social problems of that time were discussed.

In the years 1936–1939 Chaïm Perelman tried to use Frege's metaphysics to analyze logical paradoxes and value propositions: moral, ethical, and legal ones. As a result of these explorations of various values of morality, ethics, and law, he extended logical positivism, allowing practical reasoning to be applied to decision making through dialogue and argumentation. During his research conducted together with Lucie Olbrechts-Tyteca since 1948, Perelman has distinguished a new rhetoric (*la nouvelle rhétorique*), dating back to Aristotle's *Topics* (Τοπικά/*Topica*) and *Rhetoric* (Ῥητορική/*Ars Rhetorica*), from symbolic logic, dating back to Aristotle's *Posterior Analytics* (Ἀναλυτικὰ Ὕστερα/*Analytica Posteriora*). In their joint monograph *Traité de l'argumentation : La nouvelle rhétorique* (1958), the research goal was to create an argumentation theory that complements the already developed mathematical (formal) logic, beginning with Frege and Russell. Such a theory must deal with discourses that employ value judgments, i.e., with discourses in which neither empirical verification nor formal evidence can be made in a standard way. The new rhetoric of Perelman and Olbrechts-Tyteca became one of the first versions of this theory along with Kahane's approach (1971) and some other attempts of logicians who started to deal with argumentation in real speeches. According to the authors of the new rhetoric, the subject of informal logic is focused on specific examples of arguments that have taken place in scientific and legal discourses. Based on this definition of the research subject, the scientific objective is to describe those patterns of reasoning that can be effectively applied in practice. As a result, Perelman and Olbrechts-Tyteca's argumentation theory is not normative—it does not establish principles on which to reason. The basic idea was that symbolic logic is normative, while argumentation theory is not.

As we can see, informal logic in the version of Kahane, Perelman and Olbrechts-Tyteca is not a normative, but a descriptive theory. Meanwhile, informal logic arose after mathematical (or symbolic) logic in the second half of the 20th century and it was a form of emancipation of logical thinking from a logical formalism.

The early Wittgenstein, in his *Tractatus Logico-Philosophicus*, presented symbolic logic in the version of classical propositional logic as a complex of different norms in the sense of Immanuel Kant's transcendental conditions of knowledge. It means that, according to Wittgenstein, we can cogitate only within the structure of propositional logic. This logic is a family of norms for thinking. But the later Wittgenstein relativized this view and started to claim that there may exist different speech situations with a plenty of various norms, including those outside the norms of propositional logic. These speech situations with different norms were called by him language-games, believing that various norms are like different game rules. His examples of language-games are as follows: "Giving orders, and obeying them—Describing the appearance of an object, or giving its measurements—Constructing an object from a description (a drawing)—Reporting an event—Speculating about an event—Forming and testing a hypothesis—Presenting the results of an experiment in tables and diagrams—Making up a story; and reading it—Play-acting—Singing catches—Guessing riddles—Making a joke; telling it—Solving a problem in practical arithmetic—Translating from one language into another—Asking, thinking, cursing, greeting, praying" (Wittgenstein, 1958: 11–12). Each speech situation from this list presupposes its own norms for the use of the language. Consequently, "the meaning of a word is its use in the language" (Wittgenstein, 1958: 20), as this meaning can be different in different speech situations.

Karl-Otto Apel (1973) tried to define the norms (rules) of language-games as especial transcendental or *a priori* conditions of communication. These conditions are being summarized within the framework of linguistic competence or "intersubjective agreement (*Verständigung*) on the use of language." Linguistic competence is required for participating in appropriate communication. It is a sum of language rules for the corresponding speech situation. But with such a limited understanding of competence only as linguistic, metacommunication is impossible, that is, a communication about the very use of language, i.e., communication about the language structure as such. Accordingly, it is necessary to assume the existence of the higher form of competence. Apel calls it communicative (1973)—it is needed for understanding the purpose of various speech acts. The linguistic competence is syntactic and semantic, and the communicative competence is pragmatic. The grammatical (linguistic) one is responsible only for the formation of correct statements, in each specific case it should manifest itself as an openly controlled competence in the application of the rules of syntax and semantics, while the communicative competence appears as an openly controlled one in the application of mainly pragmatic rules. The communicative competence always presupposes the linguistic competence, but the reverse is not always true. This is due to the fact that pragmatic relations, overbuilding on syntactic and semantic ones, include them. The communicative competence is actualized in the process of socialization, when a person begins to feel in himself the corresponding normative consciousness (Apel, 1973).

Thus, the later Wittgenstein and then Apel were one of the first philosophers who paid attention that non-logical language-games (which are outside mathematical logic) also have some strict norms. These norms are studied in transcendental pragmatics, according to Apel (1973), and in universal pragmatics, according to Jürgen Habermas (1988).

Among different complex speech acts, there are utterances aimed at achieving a rationally motivated consensus. These acts are argumentative, and there is a specific communicative competence for participating in them. In argumentative speech situations, there can be various types of claims to significance (*Geltungsansprüche*): claims to truth, effectiveness, representativeness, etc. (Apel, 1973). In these claims, we find a complex of logical and behavioral (social) conditions for the possibility of dialogue and reaching an appropriate agreement. Without obeying some logical and behavioral norms, it is impossible to achieve communicative competence in argumentative speech acts. The complex of such norms is studied in pragma-dialectics developed by Frans H. van Eemeren and Rob Grootendorst (1984, 1988, 2004). So, according to pragma-dialectics, argumentation theory is normative to the same extent as symbolic logic, but it is a much more general approach to reasoning than symbolic logic, since it studies the communicative competence in any argumentation.

Thus, according to Kahane, Perelman and Olbrechts-Tyteca, the modern argumentation theory is not normative, but, according to van Eemeren and Grootendorst, it is. This section will offer an answer to the following question. What is the theory of argumentation, is it normative or descriptive? For this answer, the genesis of the most ancient forms of argumentative speech acts will be analyzed. The analysis will be carried out by means of the method "archaeology of knowledge" proposed by Michel Foucault (1969) with focusing on tracking the dynamics of the development of logically correct forms of reasoning—from the simplest ones presented in the use of elementary inference rules of propositional logic, such as *modus ponens* and *modus tollens*, to complex combinations of inference rules.

So, logeme structures will be considered within the framework of different complex speech acts in the pragma-dialectical approach to argumentation founded by

Frans H. van Eemeren and Rob Grootendorst (2004). They have justified that logic is applicable in speeches if, and only if, there are some communicative rules/norms accepted within a suitable language-game and these norms are involved to constitute a critical discussion or argumentation in an appropriate community. According to the norms of critical discussion, there should be illocutionary (cognitive-emotional) and perlocutionary (behavioral) conditions, which are accepted by members of the communicative community, for justifying or refuting a proposition based on inference and to accept only standpoints presented as a conclusion, drawn without logical fallacies (cf. van Eemeren and Grootendorst, 2004: 10). Thus, the members of the community should be taught or trained how to evaluate the argumentation as a reasonable critic (van Eemeren and Grootendorst, 2004: 2).

To sum up, logical competence as an ability to draw conclusions without logical fallacies can exist only in argumentation processes supported by appropriate communicative communities. The earliest forms of these communities were represented by networks of traders that first appeared in Mesopotamia since the early dynastic period (ca. 2900–2350 B.C.), within which a legality institute arose, including business contracts, business correspondence, and legal proceedings concerning economic issues, where the earliest logical competence developed. This institute became well organized for the period of the third dynasty of Ur (from the 22nd to the 21st century B.C.). At that time, the argumentative structure of legal proceedings consisted of the following necessary parts: recording standpoints of protagonist and antagonist, comparisons of testimonies given by the protagonist and the antagonist, depositions given by the protagonist, and rebuttals performed by the antagonist. In these disputes, there were used only logical inference rules (such as *modus ponens*, *modus tollens*, introducing conjunction, deleting conjunction, etc.) applied to facts supported by at least two witnesses or to legal documents (e.g., business contracts). Consequently, Babylonian merchants preferred logical tools in their business correspondences, too. Therefore, the business contracts and business correspondence satisfied the main argumentative stages of legal proceedings (Schumann, 2020).

Since Ur III, the argumentative structure of legal proceedings, business contracts, and business correspondence has become canonical in Mesopotamia. These documents were mainly composed in Akkadian. In the Achaemenid Empire (from 550 to 330 B.C.) the Aramaic language tended to replace Akkadian as an international language of legality for traders, but maintaining the same logical and argumentative standards in dispute resolutions. All the legal documents in Aramaic continued the tradition of Akkadian legality. The same has taken place with Hellenistic states since Alexander the Great (from 356 to 323 B.C.)—the Ancient Greek language tended to replace Aramaic as an international language of legality for traders, but maintaining the same logical and argumentative standards and preserving the fundamentals of composing legal documents. Since the 2nd century A.D., in the territories of today's North India, Afghanistan, Pakistan, and Tajikistan, Bactrian (one of East Iranian languages) replaced Ancient Greek as a language of legality for traders.

Hence, in the late antiquity from the 1st to the early 5th century A.D., the trade networks propagated strong logical and argumentative standards in the language-games of merchants' communities. Therefore, we can find out not only the (at least partly) isomorphic logeme structures in their texts, but also the isomorphic illocutionary and perlocutionary structures of suitable speech acts used in the communication among the traders where the logical competence was first widely involved. This is a good sample of argumentation theoretically reconstructed in accordance with the pragma-dialectical standards, and then archaeological evidence justifying the pragma-dialectics as such. The point is that in legal documents of Mesopotamian tradition implemented in trading networks of the Silk Road, we observe the same logeme structures in the same illocutionary and perlocutionary structures of argumentative speech acts. And this is traced only in the communities of merchants participating in the Silk Road. This was only possible because of the strong norms of argumentation they shared together. This provides an archaeological proof that pragma-dialectics is correct, argumentation theory is normative, and its earliest norms can be found in legal documents from late antiquity.

In Subsection 2.3.2, some features of the world-economy of late antiquity are examined. In Subsection 2.3.3, main characteristics of the legal documents in Sogdian and Bactrian are defined. In Subsection 2.3.4, it is proved that there is an injective mapping from the illocutionary and perlocutionary structures of Akkadian legal texts into the same structures of Aramaic documents. Then the latter structures are injectively mapped into the illocutionary and perlocutionary structures of Ancient Greek legal texts. And the latter are injectively mapped into the structures of Bactrian documents. So, we face the following continuous tradition: (1) argumentative speech acts in Akkadian; → (2) argumentative speech acts in Aramaic; → (3) argumentative speech acts in Ancient Greek; → (4) argumentative speech acts in Bactrian (and Sogdian). This demonstrates that the same language-game with supporting the same high logical competence and the same norms of argumentation was transmitted via the trade networks of the world-economy. These norms we can reconstruct are consonant with the norms developed in pragma-dialectics, which proofs that this theory of argumentation is even archaeologically confirmed.

2.3.2 The Silk Road in the Classical Antiquity

A unified world-economy with a wide labor division among different remote regions and among various peoples has been confirmed at least since the Bronze Age. For example, the Akkadian word *šušānu* (Assyrian: *susānu*; 'horse-trainer') is of the same root with the Old Indo-Aryan word *aśvasani*, which means that the horse-trainers in the Near East were rather of proto-Aryan origins. The *Kikkuli Text* composed in Hittite and dated to the 14th century B.C. is an instruction how to train chariot horses and its author is named ^{I}*Ki-ik-ku-li li* ^{LÚ}*a-aš-šu-uš-ša-an-ni ŠA* KUR ^{URU}*Mi-it-ta-an-ni* ("Kikkuli, the horse trainer from the land Mittani [the state in present-day Syria, whose political elite was of proto-Aryan origin at that time]") who was, very probably, a proto-Aryan, therefore he used some words like numericals close to Old Indo-Aryan in his treatise.

For the first time, Fernand Braudel (1992) has defined the term 'world-economies' to denote systems which exist beyond political, religious, and national boundaries. Then Immanuel Wallerstein (1976) has introduced the term 'world-systems' to consider world-economies along with the cultural life of their nations. In turn, André Gunder Frank convincingly proved that the various remote regions have been connected to each other by close economic relations since the Bronze Age (Frank and Gills, 1992, 1993). In this section, I examine the period of late antiquity from the 1st to the late 4th century A.D., when there was a world-system, which linked the Indo-Scythian kingdoms with Egypt by the Silk Road. I am going to show that at this time logical competence, as a part of legal hermeneutics, was widely expanded within this system.

The matter is that from the early 1st to the late 4th century A.D., different Indo-Scythian (Indo-Iranian) clans (mainly, Sogdians, Bactrians, and Western Kṣatrapas) controlled the Asian part of the Silk Road. The Sogdians settled in the region between the Amu Darya and the Syr Darya (today's Uzbekistan and Tadjikistan). The Bactrians inhabited present-day Afghanistan, and the Western Kṣatrapas controlled the country round the Indus (today's Pakistani province of Sindh and the Indian province of Gujarat). All these clans were mainly Buddhists. Furthermore, the Sogdians and Bactrians became the primary propagators of Buddhism in China in the 2nd–3rd centuries A.D. This Silk Road connected China with Egypt. Dio Chrysostom (Δίων Χρυσόστομος; ca. 40–ca. 115 A.D.) mentioned Bactrians, Sogdians, and Indians (the latter were mostly from Sindh and Gujarat) among various ethnic groups well represented in Alexandria, the largest city in Egypt:

> Not merely Greeks and Italians and people from neighbouring Syria, Libya, Cilicia, nor yet Ethiopians and Arabs from more distant regions, but even Bactrians and Scythians [*A.Sch.*: Sogdians] and Persians and a few Indians [*A.Sch.*: from Sindh and Gujarat] (*Discourses* III: 32, 40; Cohoon and Crosby, 1940: 211).

In the *Periplus Maris Erythraei* (Περίπλους τῆς Ἐρυθρᾶς Θαλάσσης), dated to the 1st century A.D., the Silk Road is described as follows:

> At the northernmost point, where the sea ends somewhere on the outer fringe, there is a very great inland city called Thina [China] from which silk floss, yarn, and cloth are shipped by land via Bactria [*A.Sch.*: present-day Afghanistan] to Barygaza [*A.Sch.*: the main port of today's Bharuch District in Gujarat] and via the Ganges River back to Limyrikê [*A.Sch.*: the present-day Malabar Coast of Kerala] (Casson, 1989, §64: 91).

In Chinese sources, such as the *Sānguózhì* (三國志) or *Memoir of the Three Kingdoms*, published in the 3rd century A.D., the Sogdians are mentioned as having good contacts with the Romans of the Bosporan Kingdom on the Black Sea through the mediation of the Sarmatians:

> The kingdom of Yǎncài [奄蔡; *A.Sch.*: the Sarmatians—an Iranian-speaking nomadic people around the Aral Sea] which is also called Ālán [阿蘭]. They have all the same customs as the people of Kāngjū [康居; *A.Sch.*: the Sogdians]. To the west it adjoins Dàqín [大秦; *A.Sch.*: the Roman Empire, represented by the Bosporan Kingdom in this description]. To the southeast, Kāngjū [康居; *A.Sch.*: Sogdiana] (Chavannes, 1905: 558–559).

Hence, in different sources, such as Chinese and Roman chronics, the two routes by which silk was traded from the 1st to the 5th century A.D. are described: (i) the Central Asian caravan route from the Tarim Basin and via Sogdiana and Bactria to Persia and to the Bosporan Kingdom, and (ii) the maritime route via the Western Kṣatrapa Kingdom (Sindh and Gujarat) and Ceylon to Egypt (De La Vaissière, 2005: 86).

The merchant societies of the Indo-Scythians (in the wide meaning) or Indo-Iranians had a complex horizontal and hierarchical social structure covering the Asian trade network. So, in Ancient Letter II, written in Sogdian, the following three hierarchical levels are mentioned: (1) Varzakk son of Nanai-thvār, having a residence at Samarkand, (2) Nanai-vandak, the manager of the network located in Gansu in China, and (3) some local agents: Armat-sāch at Jiuquan, Arsāch at Wuwei, etc. (De La Vaissière, 2005: 48).

Different clans (dynasties) of the Indo-Scythians controlling the Asian part of the Silk Road are mentioned in the *Mahābhārata* among enemies of the Pāṇḍavas. In fact, they were enemies of the Gupta dynasty and were conquered by the Guptas in ca. 400 A.D. This battle on 'Kurukshetra' between the Indo-Scythians and the Guptas is described as follows:

> *trīṇi sādisahasrāṇi duryodhanapurogamāḥ |*
> *śakāḥ kāmbojabāhlīkā yavanāḥ pāradās tathā ||*
> *kuṇindās taṅgaṇāmbaṣṭhāḥ paiśācāś ca samandarāḥ |*
> *abhyadravanta śaineyaṁ śalabhāḥ pāvakaṁ yathā ||*
> (*Mahābhārata* VII: 97, 13–14).

> Three thousand horsemen led by Duryodhana, as well as Śakas [the Indo-Scythians controlling Sindh, Gujarat, and South Punjab at that time], Kāmbojas [the Iranian-speaking tribe from Badakhshan], Bactrians [the Iranian-speaking people from Afghanistan], Greeks [the Greek clan of East Punjab, assimilated at that time by the Indo-Scythians], Indo-Parthians [the Iranian-speaking people from South Khorasan and South Baluchistan], Kuṇindas [the Indo-Scythian vassals from Himachal Pradesh], Taṅgaṇas [the Indo-Scythian vassals from Uttarakhand], Ambaṣṭhas [the Indo-Scythian vassals from central districts of Bihar] and Paiśācas [a kind of ghosts-demons] with Mandaras [the inhabitants of the mountain Mandara, a hill in Banka district in Bihar], ran towards Śaineya like moths towards a fire.[6]

The same peoples are mentioned in another fragment:

> *sindhusauvīragāndhārakāmbojayavana*
> (*Mahābhārata* XI: 22, 11).

> These Sindhusauvīras [the people from the country round the Indus, controlled by the Indo-Scythian dynasty of Western Kṣatrapas at that time], Gāndhāra [the people from Peshawar in Pakistan who were founders of Greco-Buddhism], Kāmboja [the people from Badakhshan] and Greeks [the people from East Punjab].

[6] These are the modern names of the regions.

In the meanwhile, the Kuṣāṇas were mentioned as 'Tocharians': *tuṣāra/tukhāra* (*Mahābhārata* III: 48, 20); in the Chinese chronics: 月氏; *yuèzhī*. This dynasty came from the Tarim Basin, they existed from the 1st to the 4th century A.D. and ruled today's Afghanistan, Pakistan, Tajikistan, and North India at the peak of their power. They belonged to the same Indo-Scythian political and cultural space, e.g., they spoke Bactrian (Eastern Middle Iranian) and were mainly Buddhists in the Gāndhāran version.

Around 400 A.D., all the Indo-Scythian dynasties of North and West India, including the Western Kṣatrapas and the Kuṣāṇas fell, except for the Sogdians. (Only the Sogdians remained highly influential in Central Asia until the 7th–8th century A.D.). And this tragic event of the decline of the Indo-Scythians in India is mentioned in the *Mahābhārata* as an epic fall of all the enemies of the Pāṇḍavas. This period of the late 4th and early 5th century A.D. was also the time of the decline of the Western Roman Empire. The almost parallel fall of both: on the one hand, the kingdoms of the Kuṣāṇas and the Western Kṣatrapas in Asia, and on the other hand, the Western Roman Empire in Europe, significantly weakened the trade network of the Silk Road. The maritime route from the Western Kṣatrapas (Sindh and Gujarat) and Gāndhāra to Egypt was suspended. For this fact there is good archaeological evidence. For example, one of the most important intermediate ports connecting Egypt and Western India was located on the island of Socotra. On this island, there were excavated many graffiti, written by the merchants participating in the maritime Silk Road: 192 in a Prakrit (often close to Gāndhārī) and in the Brāhmī script of Western Kṣatrapas, 1 in Gāndhārī and in the Kharoṣṭhī script, 1 in Bactrian in the Greco-Bactrian script and in parallel in Brāhmī, 3 in Greek, 1 in Palmyrene Aramaic, 8 in Aksūmite (Ancient Ethiopian), 11 in South Arabic, 2 in South Arabic or Aksūmite (Strauch, 2012: 30). These graffiti show which ethnic groups participated in the maritime trading networks, but all of them are strongly dated by the 5th century A.D. and not later. There is also textual evidence, e.g., according to Cosmas Indikopleustus (Κοσμᾶς Ἰνδικοπλεύστης, 6th century C.E.), see (McCrindle, 1964), in the 6th century C.E. there was no trade with Egypt from Western India.

It is worth noting that this collapse of international trade by the 5th century A.D. led to the end of the classical antiquity and consequently, to the beginning of the Middle Ages in Eurasian history, when the Islamic culture already became a universal mediator of Eurasian trade routes with its rise in the 11th–12th centuries A.D.

The Silk Road of late antiquity (i.e., from the 1st to the late 4th century A.D.) is being well studied from the point of view of world-systems analysis. Initially, this phenomenon was regarded within the framework of Marxist-oriented theories focusing on mere economic exchanges. But recently it has been detected that the intercultural flows were even more significant (Neelis, 2011: 15).

For example, the cross-cultural trade of the Silk Road started to be considered as a way of "Buddhist transmission along long-distance routes in South and Central Asia" as a sample of "religious markets" (Neelis, 2011: 16). The fall of the Indo-Scythian dynasties in North and West India in ca. 400 A.D. was the beginning of the fast decline of Buddhism there. The Kuṣāṇa Empire was highly urbanized and its cities and centers such as Taxila, Swāt, Balkh, Sirkap, Sirsukh, Mathurā and many others also were important centres of Buddhist education (Lam, 2013), see Fig. 2.3.1.

Figure 2.3.1. A Bodhisattva (perhaps Maitreya) with four arms holds (i) prayer beads, (ii) a flower, (iii) a water pot. This statue is located in the Mulbekh village. *Period*: the late Kuṣāṇas or more probably Hūṇas. *Date*: ca. from the late 4th century to the late 7th century A.D. *Geography*: Ladakh in Jammu and Kashmir. *Medium*: Stone. **By courtesy:** Sergii Mazurkevich.

At that time, Buddhism was an essential part of religious economies (markets) with a special way of commodity distribution among devotees, patrons, and Buddhist monks (Stark and Finke, 2002: 39–45). So, according to Buddhism, any "material gift" (*āmiṣadāna*), given to Buddhist monks, is an exchange in fact that is returned as a "gift of the law" (*dharmadāna*) and bears a "religious merit" (*puṇya*) back. Therefore, we could find Buddhist shrines and monasteries on all the routes for commercial exchange of that time showing a kind of "symbiotic relationships between religious institutions and economic networks" (Neelis, 2011: 19). This symbiosis is readily demonstrated by a huge number of inscriptions at all Buddhist sites made by donors including many merchants at this period from the 1st to the early 5th century A.D. In these inscriptions, the Buddhist merchants were designated by different titles such as *śresṭhin*/*seṭhin*, *sārthavāha*/*sathavāha* (caravan leaders participating in long-distance networks), *vaṇij*/*vāniya* (traders of basic goods), *negama* (members of guilds of traders or artisans), *hairaṇyaka*/*heraṇika* (treasurers), *gandhika* (perfumers or shopkeepers), *vyāvahārika* (businessmen) (Neelis, 2011: 24–25). The Buddha himself is called by the merchant title of "caravan leader" (*sārthavāha*/*satthavāha*), e.g., see in the *Saṃyutta-nikāya* 1: 192.

Thus, the rise of the Eurasian transnational trade routes, connecting China with Egypt and the Roman Empire from the 1st to the late 4th century A.D., was provided by the Indo-Scythian dynasties in Asia and the Roman Empire in Europe. Simultaneously, there was a rise of Buddhism in India and Central Asia. And the fall of the Indo-Scythian kingdoms and their vassals (who were the main royal donors and protectors of Buddhism of that time) implied the decline of the Silk Road of classical antiquity, and the decline of Indian Buddhism. Nevertheless, there is one more cultural trend which significantly changed after the Indo-Scythian fall. It was a support and continuation of Hellenistic traditions in Central and South Asia, which were then completely interrupted.

Figure 2.3.2. Two bronze coins minted by Kaniṣka I, the king of the Kuṣāṇas in the 2nd century A.D., to show the change of official languages during his reign. On the obverse of the first coin, we see the diademed king standing facing, holding a spear, and sacrificing at a Mazdean altar at left, the Greek legend around: ΒΑCΙΛΕΥC ΒΑCΙΛΕΩΝ ΚΑΝΗÞΚΟΥ, "King of Kings Kaniṣka", tamgha at right. On the obverse of the second coin, we find the same picture of the king but with the Bactrian legend around: ÞΑΟ ΚΑΝΗÞΚΙ, "Shah Kaniṣka". On the reverse of the first coin, the sun god stands to the left with a sun-like shining halo and a gesture of blessing, the Greek legend to the left: ΗΛΙΟC ('Helios'). On the reverse of the second coin, the same picture of the sun god but with the Bactrian legend: ΜΙΟΡΟ ('Mithra').

Figure 2.3.3. Two silver drachmas, minted by Nahapāna, the king of the Western Kṣatrapas in the first to early second century A.D., to show the change of official languages during his reign. On the obverse of the first coin, we find the Greek legend around the diademed bust of the king: ΒΑCΙΛΕΟC ΝΑ[ΗΑΠΑΝΑC ΒΑΓ]ΟΥ, "King Nahapana Baga [Baga means 'God' in Old Iranian]". On the obverse of the second coin, we see the same diademed bust but with the Prakrit legend around in the Greek script: ΡΑΝΝΙΩ ΞΑΗΑΡΑ[ΤΑC ΝΑΗΑΠΑΝΑC], "King Kṣaharata Nahapana". On the reverse of the two coins, the stylized thunderbolt of Athena Alkidemos (with the Buddhist meaning of *vajra*) at left, and an arrow at right, the Brāhmī legend clockwise on the right margin: *rajño kṣaharatasa nahapānasa*, and the Kharoṣṭhī legend anti-clockwise on the left margin: *raño caharatasa nahapanasa.*

One of the important Hellenistic traditions that took root in Central and South Asia was represented by Hellenistic forms of legality with advanced legal hermeneutics, including ways of applying the logical competence to resolve interpersonal conflicts via legal proceedings in the form of critical discussion. Ancient Greek was one of the main official languages in most Indo-Scythian kingdoms until the early 2nd century A.D. Then it was replaced by Bactrian, but in the Greek script in the

Kuṣāṇa Empire, and by a Prakrit, but in the Greco-Roman script in the Western Kṣatrapa Kingdom (Figs. 2.3.2, 2.3.3). Therefore, it is expected that in legal documents written in Bactrian and Sogdian we might find much evidence of direct continuation of Hellenistic forms of legality with high logical competence.

Hence, there is much evidence of merchant societies of very different ethnic origins participating in the Silk Road of late antiquity. These societies had some features of free associations (*freie Associationen*) in the sense of Jürgen Habermas, i.e., the features of groups of interests which have not yet been institutionalized. According to Habermas, the networks of free associations forms a public sphere (*Öffentlichkeit*) in which individuals can freely discuss and identify societal problems (Habermas, 1962). The merchant societies of late antiquity identified societal problems only of economic nature, but discussed them in accordance with the norms of strict argumentation of the Mesopotamian legal culture. These societies were communicative communities in the sense of Apel, supporting the communicative competence required for argumentation containing logeme structures.

2.3.3 Legal Documents in Bactrian and Sogdian

In Sogdian, we have only a few business letters (Ancient Letters I–VIII), dated to the 4th century A.D., and a few letters and legal documents from Mount Mug (marriage contract Nov. 3, Nov. 4, agreement for the sale of half a cemetery building B-8, agreement for the lease of three mills B-4, receipts B-3, A-10, A-7), dated to the 8th century A.D. (Livshits, 2015), but much more legal documents were found in Bactrian including business correspondence and contracts dated until the 8th century A.D. (Sims-Williams, 2000, 2007). The Sogdian and Bactrian documents belong to the same legal tradition dating back to the Old Babylonian time. Therefore, they have a lot in common in their composition with each other and with Akkadian, Aramaic, and Ancient Greek legal texts. For instance, the formulae of greetings, addresed to a person of the higher social status, are similar in Sogdian and Bactrian: (i) "greetings (and) homage", λροδο ναμωσο in Bactrian, where λροδο is *drwd* in Parthian and *δrwth*, *δrwt'twh* in Sogdian, ναμωσο (ναμασο/ναμοσο) comes from the Old Indo-Aryan *namas* ('bowing', 'salutation')—the Buddhist term used in addressing to upper beings; (ii) "blessing (and) homage", *"pryw nm'cyw* or *"prywn nm'cyw* in Sogdian, where *"prywn* comes from the Avestan *āfriuuana* ('blessing') and *nm'cyw* from the same Old Indo-Aryan *namas* ('bowing'). These formulae illustrate that the Sogdians as well as the Bactrians were Buddhists who at the same time preserved some Mazdean (Iranian religious) traditions.

Among various peoples speaking different Iranian languages, Aramaic was the first *lingua franca* until the 3rd century B.C. and then Greek until the 1st–2nd century A.D. Therefore, we find in the Bactrian legal documents that some month names became loanwords from Aramaic (Sims-Williams and de Blois, 1996): (i) νεισανο (cf. Sogdian *nysnyc*, the name of the third month) from the Babylonian name of the first month *nisānu*, Aramaic and Hebrew *nīsān*; (ii) σιοανο (the name of the fifth month) from the Babylonian name of the third month *siwānu*, Aramaic and Hebrew *sīwān* (Sims-Williams, 2000: 223); (iii) αββο/αβανο (cf. Sogdian *"p'nc*, Middle Persian *ābān*, the name of the eighth month) from αββο 'water'

(Sims-Williams, 2000: 172) or from the Babylonian name of the fifth month *abu*, Hebrew *ʾāḇ*; (iv) þαβατο (the name of the first month?) from the Babylonian name of the eleventh month *šabāṭu*, Aramaic *šbāṭ*, Hebrew *šeḇāṭ* (Sims-Williams, 2007: 283); (v) αδορο (cf. Middle Persian *ādur*; the name of the second month?) from the Babylonian name of the twelfth month *adaru*, Hebrew *ʾădār* (Sims-Williams, 2007: 188). And one month name came from Greek (Sims-Williams and de Blois, 1996: 152–153): (vi) δρηματιγανο/δδρημιταγανο (cf. Sogdian *jymtyc*; the eleventh month) from the name of the Greek goddess Δημήτηρ, Δήμητρ- + suffix -γανο (Sims-Williams, 2000: 190). Meanwhile, some day names in Bactrian have a connection to Greek, too. For instance, ριþτο as a name of the twenty-sixth day of the month (Avestan: *arštāt*, 'justice' or 'honesty') belonged to a Bactrian (Kuṣāṇa) goddess who had the standard Hellenistic iconography of Athena on the Kuṣāṇa coins and associated with her.

The Bactrian legal documents began with dating to the same extent as Aramaic and Greek ones: χþονο (*khšono*) "[calendar] year, [regnal] year" (Aramaic: *šənā*; Khotanese: *kṣuṇa-*, "section, period of time").

Many stylistic devices and tropes used in the Bactrian legal documents were directly borrowed from Aramaic and Greek texts. For instance, the use of μισιδο ('moreover') in connection to conditional statements was borrowed from an appropriate Aramaic trope, which in turn goes back to the Akkadian *appūna* (*ap-pu-na*) and *appūnama* (*ap-pu-na-ma*), 'moreover', from the Canaanite (Palestinian) texts from the Amarna library:

> a thousand (and) ten thousand [2]greetings (and) homage (βηρ[2]αρο λροδο ναμωσο) from Wararan Springan [3]his servant. **Moreover** (μισιδο), here they report [4]that your lordship's son is alive. **If** [5]it be fate(d) that he be alive, [6]inform me quickly (Sims-Williams, 2007: 112).

Unfortunately, trial records in Bactrian have not been preserved, there are only business correspondence and legal declarations of various types including marriage and sales contracts. However, in the preserved texts we see a contextual usage of the term 'court', 'trial', which allows us to reconstruct the phenomenon of legal proceedings in Bactrian-speaking regions. For example,

> his [24]claim (and) argument before the judges (and) court (ταδδη[24]ιο χοανδο οισαρο αβο λαδοβαρανο αβο λαδο) (Sims-Williams, 2000: 112–113).

So, the Bactrian word λαδο meant 'law', 'lawsuit', 'trial', 'court', 'judgment', "legal statement". The Sogdian word *xṯy'k* [*xəṯyāk*]/*xty'k* [*xətyāk*]/*'xty'k* [əxtyāk] had the same meanings, which correspond to the same extent to the basic connotations of the Sumerian logogram DI (e.g., DI TIL.LA—"legal decision" or "settled law case"), the Akkadian *dīnum* (e.g., *dīnum gamrum*—"settled law case"), the Assyrian *dēnum*/*dēnu*, the Aramaic *dīnā'*, and the Hebrew *dîn*.

This Bactrian λαδο and the Sogdian *xṯy'k* simultaneously differ, first, from the Old Indo-Aryan *dharma* with the meaning "natural order" without legal proceedings and, second, from the Chinese *fǎ* (法) with the meaning 'standard' for a punishment code. Neither *dharma* nor *fǎ* meant a legal decision on the ground of critical discussion. But for both λαδο and *xṯy'k* this association with a critical discussion is basic. It is readily seen from the following main legal terminology found in Bactrian (Sims-Williams, 2000, 2007): (i) λαστανο ('dispute' or 'litigation'; cf. Khotanese: *lastana-*; Middle Persian: *dadestan*); (ii) αοισαρο/οισαρο/οοισαρο ("argument in the dispute"); (iii) ζινακο ('deduction'); (iv) λαδοβαρο (λαδο + suffix -βαρο; "judge as a mediator in the dispute"; cf. Parthian: *d'dbr*; Sogdian: *xtw* [*xətu*], *δ'tβr* [*δātβar*]); (v) χοανδο/σαχοανο ('claim'); (vi) αλβαρο/αλοβαρο ('court'; cf. Khotanese: *vara-*); (vii) ραζογολο ("royal tribunal"; from the Old Indo-Aryan *raja-kula-*, "royal family").

In the legal documents in Bactrian, the following word means making a legal declaration (Sims-Williams, 2000, 2007): πιδοοασ-/πιδοοησ-/πιδοοιησ- ("to declare", "to make a contract"), while πιδοναμγο/πιδοναμαγο/πηναμγο means 'signature' or "signed document". Any declaration was understood as a voluntary waiver of further claims and disputes. Therefore, the legal declaration was often accompanied by the implicit waiver of statements and discussions:

> without claim and without argument (αβησαχοανο οδο αβηαο[ι]σαρο) (Sims-Williams, 2000: 44–45);

> then his claim and argument [17]shall not be (considered) good nor valid in court (χο{ι}ανδο οδο οισαρδο αβο λαδο μα[17]χοζο οδο μαο{ο}αοαρο αβοδηιο) (Sims-Williams, 2000: 46–47);

> [20]from now to eternity, without rent, without tax, without exaction, without deduction, so that no-one [21]may be able to cause injury to you (χο[20]ζο αβοδδηιο ασο μωσο αβο ιαοιηδανο ζοριγο αβηιουαργο αβηιοτωγο αβηιουαγγο αβηωαβφαγγο κοαδδηνο υαρσο κι[21]σο οιζανδογαρο μαβοδδηιο οτο ναπαδοαχþανιμο αζο οιημο οδο ναζο) (Sims-Williams, 2000: 58–59);

> then our claim [26](and) argument shall not be valid in court in any way (αγγιτο ταδομηνο χοα[26]νδο οισαρο υαρσο αβο λαδο οαοαρο) (Sims-Williams, 2000: 60–61);

Hence, legal proceedings (courts) in Bactria and Sogdiana and in the regions of their vassals were understood as a critical discussion (where complex logeme structures are embodied) between two opposing parties mediated by judges. This understanding first arose in Mesopotamia at least in the Old Akkadian period (i.e., about 4300 years ago) and well traced in Ur III (i.e., about 4000 years ago). It is worth noting that legal proceedings were not understood in this way in other Bronze Age cultures such as the Chinese one. Moreover, in the Iron Age cultures that did not continue the Mesopotamian legality, we cannot find the same understanding. For example, in India, *dharma* as a law means a "natural way of things revealed in meditation" which has nothing to do with legal proceedings, but in the period of *śramaṇas* (*samaṇas*), non-orthodox religious monastic schools in India, beginning to be established after ca. 400 B.C., this way may have been discussed in doctrinal debates between different schools of thought. In China, *fǎ* (法) as a law means a "code of punishment", the Chinese legal proceeding which traced well from the Qin Empire (221–206 B.C.) was not a critical discussion between the opposing parties, but an inquiry into all circumstances of the case, organized by the administration.

Consequently, we observe that the world-system of the late antiquity, with an appropriate world-economy, has provided different regions not only with distributions of commodities, but also with distributions of forms of social communication. And one of the most significant forms of social communication in late antiquity was presented by the logical competence as a legal way of conflict resolution that was necessary for long-term international trades and intercultural relations. It shows that any sociality is not innate and is being extended through contacts. Meanwhile, logical competence and logical argumentation are not innate since they are forms of social contacts. About cognitive-scientific experiments proving this claim, please see Kern et al. (1983).

Logical competence was needed to correctly use some complex logeme structures for an automatic logical inference in the courts. The bearers of this competence from the 1st to the 5th century A.D. were merchants of various ethnic origins: Egyptians of the Roman Empire, Bactrians, Sogdians, Southern Arabs, Indians from Gāndhāra and the Western Kṣatrapa Kingdom, etc. They should have known how to draw up various legal documents, such as sales contracts and other legal declarations, that could be used in the courts as valid and strong arguments within the same argumentation standards.

In everyday life, logic is mainly presented as a critical discussion. It means that we can take some logeme structures from the argumentation process to study them from the point of view of symbolic logic, but initially these logeme structures are implemented within the framework of an appropriate complex speech act that has been explicated in the pragma-dialectical approach to argumentation as a critical discussion. This explication has been theoretically carried out in (van Eemeren and Grootendorst, 2004), but it is also well confirmed archaeologically. So, the very first sample of critical discussion is represented by the trial records of Ur III (Culbertson, 2009; Schumann, 2020). The legal decision, after an argumentation process between two disputants mediated by a judge, was called DI TIL.LA in Sumerian and *dīnum gamrum* in Akkadian.

In the trial records composed in Sumerian, Akkadian, Aramaic, Greek, and some other languages, we observe a continuation of the same legal tradition of critical discussion: (1) the Mesopotamian legality with complex logeme structures incorporated into argumentative speech acts, mainly, in Sumerian and Akkadian from the 21st century B.C. to the 5th century B.C. and even later in some local forms; (2) the Iranian legality (Achaemenid as well as other Iranian dynasties) with the same logical and argumentative structures in Aramaic from the 5th century B.C. to the 3rd century B.C. and even later in some local traditions; (3) the Hellenistic and then Hellenized legality with the same logical and argumentative structures in Ancient Greek from the early 4th century B.C. to the 5th century A.D. and later in the Byzantium; (4) the post-Hellenistic legality with the same logeme, illocutionary and perlocutionary structures in Bactrian, Sogdian, and Prakrit (first of all in Gāndhārī) from the 2nd to the 5th century A.D.—i.e., until the time of starting the decline of Indian Buddhism. Meanwhile, each next legal tradition is a continuation of the previous one: (1) → (2) → (3) → (4).

The trial records in all these languages reflect the following stages of the legal proceedings as critical discussions (Culbertson, 2009; Schumann, 2020):

(i) *confrontation stage*, when the scope and content of the dispute should be defined and the judge is responsible for specifying this scope within the framework of eventual applications of particular legal articles; (ii) *opening stage*, when the two opposing parties should define their starting points (which are claimed as facts or legal cases) and strategies of their proofs and the judge is responsible for finding appropriate legal rules of substantive and procedural laws and their interpretations for considering the starting points of the disputants; (iii) *argumentation stage*, when certain logeme structures are involved by the disputants to support the starting points and the evaluation methods and the judge is responsible for the logical correctness of the logeme structures and does not accept inferences containing logical fallacies; (iv) *concluding stage*, when the judge defines the result of the discussion and makes a legal decision as a logical inference based solely on facts and legal documents. The same stages are considered in the modern pragma-dialectical theory of legal argumentation (cf. Feteris, 2017: 222). Hence, we see a good archaeological confirmation of pragma-dialectics on the legal materials of the Silk Road of late antiquity.

2.3.4 Pragma-Dialectical Analysis of Legal Documents in Akkadian, Aramaic, Greek, and Bactrian

The main idea of pragma-dialectics is that any argumentation should be examined as a complex speech act consisting of some atomic illocutions and perlocutions (van Eemeren and Grootendorst, 1984; van Eemeren et al., 2007). (Let us remember that an illocution means an act performed by a speech and a perlocution means a result of this act.) Therefore, argumentation is a kind of social practice that may be normalized by means of some rules (van Eemeren and Grootendorst, 1988; Witek and Witczak-Plisiecka, 2019). Thus, the pragmatic dimension of discussion plays a crucial role (Becker, 2012; Bermejo, 2011; Oswald et al., 2020; Rocci, 2006), and various logical tools of argumentation are incorporated in appropriate speech acts of critical discussion and are too abstract outside this context. In a critical discussion, two opposing parties are involved to resolve a difference of opinion by achieving agreement on the acceptability or unacceptability of the standpoints formulated before the discussion and logic only is one of the means for obtaining this achievement.

The pragma-dialectical approach to argumentation sets an ideal model of critical discussion in which a difference of opinion may be resolved (van Eemeren and Grootendorst, 2004: 38). This model helps us to turn on "reflection on the procedures that are used in different argumentative practices" and to develop "the skills that are required for an adequate production, analysis, and evaluation of argumentative discourse" (van Eemeren and Grootendorst, 2004: 39).

In late antiquity, we find a legal tradition rooted in Babylonia and supporting very high standards of logical argumentation. We can prove that it is in fact a common tradition by showing that it is the same language-game with isomorphic structures of speech acts. For instance, the illocutionary and perlocutionary complex structures of contracts composed in Akkadian, Aramaic, Greek, and Bactrian are almost the same—indeed, their atomic illocutionary and perlocutionary acts of the same type are ordered by the same sequence: (i) date, place, and the names of judges; → (ii) parties; → (iii) judgment; → (iv) oath; → (v) withdrawal; → (vi) waiver of suit;

→ (vii) penalty; → (viii) reaffirmation; → (ix) signs, see Table 2.3.1. Let us consider this sequence more in detail:

i) *Date, place, and the names of judges*. Some assertives fixing the main attributes for the speech act: its date, place, and the names of the judges who transform this speech act into a complex legal declaration for the ultimate resolution without a critical discussion at the trial.
ii) *Parties*. Some assertives naming the two parties who agree to sign this legal declaration at the end. Hence, they agree to perform appropriate commissives that they accept the declaration.
iii) *Judgment*. An assertive expressing the joint standpoint (e.g., that something is sold or transferred in another way) which is assumed to be accepted by both parties without any critical discussion.
iv) *Oath*. Expressing a legal commissive (oath) by both parties that they share the joint standpoint indeed.
v) *Withdrawal*. One party utters a legal commissive of withdrawal that the owner (or another legal status) has just changed.
vi) *Waiver of suit*. Uttering a legal commissive that the joint standpoint is accepted by both parties without an eventual suit in the future.
vii) *Penalty*. Formulating some implications as legal declaratives that can be used at the trial: "If in the court one of both parties claims something contradicting the joint declaration (not-*A*), then the penalty for that is as follows (*B*)". This implication "If not-*A*, then *B*" is considered a legal statement from which, and from the fact of 'not-*A*', any court can infer the legal decision '*B*' directly by *modus ponens*. Then the absent of punishment '*B*' means that we can entail by *modus tollens* that we deal with '*A*'—with the joint declaration that indeed holds true.
viii) *Reaffirmation*. Repeating the joint declaration again through an appropriate commissive.
ix) *Signs*. Transforming the complex speech act into a suitable legal declaration.

We see that the complex speech act expressing a contract between the two parties is performed in such a way as to represent it as a legal declaration in the form of implications that can be used in critical discussions during the next possible legal proceedings with applying *modus ponens*, *modus tollens*, and other inference rules of propositional logic. Thus, in contracts in Akkadian, Aramaic, Greek, and Bactrian we observe the same agreement of traders on premises and discussion rules with a possibility of applying *modus ponens* and *modus tollens* to stage (vii) of the contracts. As a result, we have not only the same illocutionary-perlocutionary structure of contracts with the same sequence of atomic illocutions and perlocutions from (i) to (ix), but also a support of the same argumentation rules.

Could such a complex structure of contracts be identical by pure chance? Definitely no. This is the result of a common tradition that goes back to Mesopotamia. But the same structure of contracts means something else, namely, that the authors of these contracts shared common norms of communicative competence—they

Table 2.3.1. The chain of injective mappings of corresponding speech acts from Akkadian contracts (leftmost column) into Aramaic ones, then into Greek and Bactrian contracts (rightmost column).

Ca. 596 B.C. The contract between Iddin-Nergal, son of Šamaš-ēreš, and Nādin-aḫi, son of Ubār. *The Akkadian document from Nippur,* MMA 86.11.174 (Spar and von Dassow, 2000: 205–207).	**26 August, 440 B.C. The contract between Peu, son of Paḥe/Pakhoi, and Miptahiah, daughter of Mahseiah**. *The Aramaic document from Syene in Egypt*, TAD B2.8, Cowley 14 (Porten, 1996: 188–189).	**4 January, 136 B.C. The contract between Isias, daughter of Patepnebteus, and her husband Neoptolemos, son of Neoptolemos to Esaroeris son of Petoseiris**. *The Greek document from Syene in Egypt*, BGU 6 1249 (Porten, 1996: 424–425).	**11 March, 574 A.D. The contract between Aur. Victor, Paeion, and Tsia, children of Dios, and Tlou, his widow, and Aur. Iakobos son of Dios**. *The Greek document from Syene in Egypt*, P. Münch. 1 1 (Porten, 1996: 470–473).	**757 A.D. The contract between Hilitber, the sons of Bah, and Zard son of Wurol**. *The Bactrian document from Gandhāra*, Khalili collection, DOC 116 (Sims-Williams, 2000: 126–135).
		Date, Place, and the Names of Judges		
1–2Tablet of the uncultivated field, up to..., (in) the Marduk Gate district which is within Nippur.	RECTO 1On the 14th of Ab, that is day 19 of Pachons, year 25 of Artaxerxes the king.	RECTO 1[In the yea]r 34, Choiak [1]1, in Syene of the Thebaid, before Aristodemos (the) Notary Public.	RECTO 1[†]χμγ† 3In the [n]inth [y]e[ar] 2of the reign and consulship of our most divine master Flavius Justinus Minor, the 3eternal Augus[tus, Emperor], on Ph[a]menoth 15, in the seventh ind(iction).	1(It was) the year 525 [it corresponds to 757 A.D.], the month Pusig, when (this) sealed document, (this) purchase contract, was written here in Gandara [Gandhāra], in the presence of 2the god Wakhsh [*Oaḫšo*, the Indo-Scythian god of Amu-daria/Oxus], the granter of favours (and) fulfiller of wishes, great (and) wonderful, who has worship in Gandara, and in the presence of 3Wisbur-ziniy Samsitan, the steward of the Shaburan (family), and also in the presence of the assembled freemen of the district 4who were

Table 2.3.1 contd. ...

...Table 2.3.1 contd.

				present there amongst (them) and (who) bear witness concerning this matter, when a contract [5]to give and to sell was freely (and) willingly made
Parties				
[3–11]30 cubits, the upper long side on the north, adjacent to (the property of) Šamaš-iddin, son of Nabû-gāmil; 30 cubits, the lower long side on the south, adjacent to (the property of) Nuska-iddin, son of Illaya; 30 cubits, the upper short side on the east, ad[jacent] to (the property of) Šulaya, son of Nabû-nāṣir; 30 cubits, the lower short side on the west, adjacent to (the property of) Ninurta-īpuš, son of Ḫašdaya: total field of Nādin-aḫi, son of Ubār, in its entirety.	said Peu [2]son of Paḥe/Pakhoi, a builder of Syene the fortress, to Mibtahiah daughter of Mahseiah son of Jedania, [3]an Aramean of Syene of the detachment of Varyazata.	[2]Isias, also (called) Sennesis, daughter of Patepnebteus, (woman of) Afonti with, as her guardian, her [3][hus] band Neop[to] lemos son of Neoptolemos, a Cyrenean, [4]*taktomisthos* [3]of the (regiment) of Ptolemy and his sons.	[4]we, the Aurelians Victor and Paeion and Tsi[a], the wife of Hatres, who is in agreement with her, the [5]three being siblings of the same family, whose father (is) Dios, also (called) Pasaraei, (and) whose mother (is) [6]Tlou, who is convin[ced and joins in consenting and affirms everything [7]agreed to in this settlement, (all) originating from Syene, [8]providing [7]both a signatory below [8]signing on their behalf and witnesses bearing witness below to this [9]settlement acknowledgment, † to Aurelius Iakobos, our own brother born of the same father, [10]of the aforesaid Dios, also (called) Pasaraei, and of Tlou, coming from the same Syene, [11]greetings.	by me, Wurol, and by me, Hilitber, the sons of Bah, and by me, Zard [6]son of Wurol, inhabitant of Gandara..., me whose house they call Wurolan, the servant of the *khar* of Rob, on [7]the following terms:

Judgment				
[12–15]Iddin-Nergal, son of Šamaš-ēreš, named the price (of the field) as 52 shekels silver, and purchased (it) from Nādin-aḫi, son of Ubār, for its (full) price, including 1 shekel given as an extra payment for his fingernail (mark). [15–18]Total 53 shekels piece-silver (which) Nādin-aḫi, son of Ubār, has received from [Id]din-Nergal, son of Šamaš-ēreš, as the price of his property, [pa]id in full.	about the suit which we made in Syene, a LITIGATION about silver [4]and grain and clothing and bronze and iron – all goods and property – and (the) wifehood document.	[4]to Esaroeris son of Petosiris, (man of) Afonti without pay, [2][acknowl]edges [4](that she) has settled with him and Tagos, [5][his w]ife, about the petition (she) submitted against them in Y(ear) 34 to the [6]judges [5]in [6][Pto]lemais of the Thebaid, of whom Philinos was bailiff, through which (she) accused them [7][of] assault a[nd] bat[t]ery, which (offenses) she evaluated (at) [b]ronze, two tale[n]ts, and silv[e]r, a [8][hund]red [7]drachmas.	Since our mentioned father Dios, while he was alive, [12]had made you, the aforesaid Iakobos, [13]partner in his shipping business [12]as if you had reached manhood and had attained the [13]required age; and when [14]our [13]said [14]father Dios had come to the end of his life, [15]we received [14] (the advance) left by our blessed [15]father through me myself, the aforesaid Tlou, our mother; [16]and (since) [...] we instituted a suit against you, saying that there was left with you [17]something of the advance of the same blessed father of ours, and you, the said Iakobos, [18]assured us verbally that nothing of our father's property had been appropriated by you; [19]and since the contention was becoming lengthy, we chose the Board of the Devoted [20]Leaders of the Regiment of Syene, who, after they had heard the matters under dispute between us, decided [21]that you were in possession of nothing at all of our father's property. And as a warranty [22]and complete reconciliation of the contention initiated by us against you, and to forestall [23]any quarrel that might be initiated against you by us, they determined that we should receive a [24]sworn word that nothing of our father's property was left with you.	A large poll-tax and harvest-tax have been assessed to be given by me, and I had no other means in (my) house [8]which could have been given (as) poll-tax and harvest-tax. But we have a certain property (consisting of) water and land, [9]here in the plain of Askin, this (same) property which they call Nogfaman, has thus now been sold by me [10]to you, Wahran, and to you, Mir, the sons of Bek, the inhabitants of Asp, servants of the *ser*, for [11]sixty good, locally current Arab dirhams of silver; and (the price) has been received by us all complete.

Table 2.3.1 contd. ...

...Table 2.3.1 contd.

Oath				
	Then, the oath [5]came upon you and you swore to me about them by Sati the goddess. And my heart was satisfied [6]with that oath which you made to me about those goods.		These [25]matters, then, being decided by us and pleasing to us, after we had gotten from you the sworn and dreadful [26]oath upon the divine and immaculate and august relics, that [27]you have appropriated [26]absolutely nothing [27]of this same paternal property, we have come to this [28]acknowledgment of the settlement, by which we acknowledge, we the aforesaid Victor and [29]Paeion and Tsia, of our (own) free will and with self-chosen volition and guileless resolve, apart from [30]any guile or any fear or force or deception or compulsion or deceit, that first, we have received [31]and accepted from you the awesome oath upon the holy and blameless relics that nothing [32]of our father's is being retained by you, not even the least thing.	(and) we have become satisfied (and) content, (and) the price agreed we acknowledge (to be) profitable, not damaging. Now the boundary to the property whose name (is) Nogfaman is (as follows): to the east a property whose name (is) Piroz-bam; and to the west what is left of the Warawang vineyard (is) the boundary; [14]and to the south the property of Sinz Sawan (is) the boundary; and to the north the water of the ditch. There, which (is) in [15]Askin, (one) uses and employs fifty quarts of grain (as) seed, either a little more or a little less.
Withdrawal				
[18–20]He is paid, he is quit; [there shall be no] contesting (the sale); they shall not turn back [nor contest (the sale)] with each other.	and I withdrew from you from [7]this day and forever.		Secondly, we acknowledge that [33]we have and shall have [32]no [33]claim against you, the said Iakobos our brother, [34]concerning the advance left by our mentioned father, Dios, [35]or concerning any sort of property pertaining to our father's legacy, whether it was small [36]or large, thought of or not thought of, said or not said, whether it came into the open [37]or not, posted in this settlement or not posted.	[16]So the property has now been given and sold by me, just as it is described within with reference to the four boundaries, (both) water and land, (with) neighbourhood [17]and environs, from side to side and from boundary to boundary, without tax, without rent, without debt, without obligation on the part of anyone [18]in respect of this property which is described herein. You (sg.) have authority, from tomorrow (and) for (all) future time, and from [19]now until eternity – you, Wahran, and you, Mir, and your (sg.) brothers, sons (and) descendants – [20]to have and to hold (the property) yourselves, (and) afterwards to sell (it), to give (it) away, to pawn (it), to put (it for) hire, to exchange (it) for another (piece of) land, [21]to give (it) for a son's

				bride-price (or) with a daughter (as) a dowry, to make a monastery or temple, to make a place of burial or crematorium, whatever [22]may suit yourselves, just as owners (customarily) have authority over what is purchased with money (and) ancestral estates, so that no-one (else) [23]has authority to withhold (it), or to seize (it), or to commit violence, nor to produce a contract – neither an old nor a new (one).
Waiver of Suit				
	I shall not be able to institute against you suit or process – (against) you or son [8]or daughter of yours – in the name of those goods about which you swore to me.	[8]and (that she) will not bring suit against them [f]or the afore[s]aid nor for [9]al[l] (matters) set forth [in t]he petition in any way.	We shall absolutely not bring suit against [38]you, not we ourselves or our heirs or our successors or assigns nor [39]shall we bring action against you or your heirs or your successors or assigns, nor shall we sue [40]nor shall we violate or undermine this settlement now or on any occasion, [41]in any way or (on any) pretext or stratagem or sophistry, in court or out [42]of court because, as we said, we have been reconciled with you concerning all issues and [43]every seed of litigation and action relating to this case or which is able to relate (in the future) has been eliminated. [44]Furthermore, in addition to that, we have sworn the divine and awesome oath by almighty God [45]and by the victory and the piety of the most powerful masters of the world, the Fl(avians) Justinus Minor and [46]Aelia Sophia, the eternal Augusti, emperors, not to challenge you nor bring action against you nor [47]sue you any longer, nor to violate or undermine this settlement.	[24]If thus there should be anyone who might dispute (or) withhold (the property), fight, argue, (or) not allow (you) to have (and) hold (it), whether [25]men of Rob, or men of Bamyan, or Turks, or Arabs, or locals, or anyone else should be shown (to be) disputatious, then [26]I, Wurol, and I, Hilitber, and I, Zard, and our brothers, sons (and) descendants, guarantee thus: and [27]by us it has been guaranteed that we shall go to each (and) every lawsuit beside (you) and we shall cause (the property to be) released and detached from every dispute and from every retention; [28]if we do not go to each (and) every lawsuit beside (you) and do not cause (it to be) released and detached from every dispute, [29]or (if) we ourselves, the declarants, should dispute (or) withhold (the property), not allow (you) to have (and) hold (it), or [30]produce another contract (or) sealed document, (whether) an old or a new (one), then our claim and argument shall not be (considered) valid (and) authoritative in court.

Table 2.3.1 contd. ...

...*Table 2.3.1 contd.*

Penalty				
[20–28][If ever in future days any of the brothers, sons, family, relatives, or kin by marriage of the house of Nādin-aḫi, son of Ubār, should come up and bring suit regarding this field, cause (anyone else) to bring suit, alter (the contract), raise a claim, or produce a claimant, say] ing ["That field was not sold and the silver was] not received," the claimant of the field, as much as it is worth, shall repay up to 12 times the silver that (the seller) received.	If I institute against you [9]suit or process, or a son of mine or a daughter of mine institute against you (suit) in the name of that oath, I, Peu – or my children – [10]shall give to Mi(b)ṭaḥiah silver, 5 karsh by the stone(-weight)s of the king, without suit or without process.	And if she [10][or] some[one el]se on her behalf [9]contests (it), [10]let the suit be invalid for them and additionally let Isias compensate [11][Esaroeris and Tag]os (by) an ass[ess]ment of [bronze, f]iv[e t]alent[s and [12]pu]re silver, [x] dr[achmas, [11]h]oly to t[h]e kings.	But, if we should wish on any occasion [48]to bring action against you or proceed against you concerning this case, first, we shall be patently liable for (violation of) the [49]divine oath and (subject) to the risk and complaint of perjury; secondly, we shall recognize (that we must provide), because [50]of the attempt and [the violation], go[ld], two [*unciae*], i.e., go(ld), 2 *unc*(*iae*)	[31]and also we shall pay a fine to the treasury of the *bredag* of a hundred and twenty good, locally current Arab dirhams of silver, [32]and also we shall pay the same fine to the opponents.

Reaffirmation				
[29–32]In order that there be no change and no litigation, they (the contracting parties) have taken an oath by Enlil and Ninlil, their gods and creators; they have taken an oath by Ninurta(?) and DN, guardians of their health; they have taken an oath by King Nebuchadnezzar, their lord.	[11]and I am withdrawn from every suit or process.	[12][apart from the a]fore[said [13]being] [12]valid.	Also, nothing (shall) prevail against the [51]arrangements in this settlement, which, for your security, we have made valid and warranted and lawful, [52][wit]h the [sig]nature of the one signing on our behalf and of the witnesses below. In answer to the (formal) question, [53]we have expressed acknowledgment †. We have clearly received from you one, *solidus* in the weight of Syene, which [54]our mentioned father, Dios [53]gave [54]to you for the military (fee) of you his son, and we have no claim against you [55]concerning this matter because, as we said, we have accepted it from you and have so acknowledged †. (2nd hand) † We, the Aurelians Victor and Paeion [56]and Tsia, whose father (is) Dios, (also called) Pasaroei, (and) whose mother (is) Tlou, who is convinced [57]and joins in consenting, the aforem(entioned), have had this settlement drawn up and it is [58]satisfactory to us as aforem(entioned).	And after the fine has been paid, may (this) purchase contract be (considered) good (and) potent, [33]valid and authoritative as is written herein.
Signs				
[33]At the sealing of this tablet (the following witnesses were present): [34–40]Before Uzbu-Erra, son of Nabû-ušabši Before Iqīšaya, son of Ninurta-aḫa-iddin Before	Wrote Peṭeese son of Nabunathan this document [12]in Syene the fortress at the instruction of Peu son of Paḥe/Pakhoi. Witnesses hereto:	[13]Tran[sacted by Aristodemos the Notary P]ublic.	Fl. Iakob son of Basileides, former *actuar*(*ius*) of the regi(ment) of Syene, [59]having been asked, wrote on his behalf since they are not literate. (3rd hand) † I, Fl. Menas son of [60]Didymus, sol(dier) of the regiment of Syene, bear witness †. (4th hand) I, Fl. Apa Dios son of Martios, former [61]*actuarius* of the regiment of Syene, bear witness to the acknowledgment, having heard it from [62]the contracting parties †. (5th hand) † I, Fl. Isak son of Iakobos, sol(dier) of the regiment of Syene, bear witness † [63](6th hand) † I, Fl. Komes son of Paamios, sol(dier) of the regiment of Syene, bear witness †. [64](6th hand) Finished by me, Victor son of Petros, on Phamenoth 15, the [65]seventh ind(iction), in Syene † VERSO [66](1st hand) † Settlement in the na(mes) of Victor and Paeion and Tsia, sib[lings of the same family], to Iakobos.	[Originally sealed with three sealings: (A) thong and sealing with fingernail impression intact; (B) string and sealing with seal impression intact; (C) string intact, sealing missing].

Table 2.3.1 contd. ...

...Table 2.3.1 contd.

Šamaš-erība, son of Bēl-aḫḫē-erība Before Tabnêa, son of Nergal-ēṭir Before Nergal-aḫḫē-iddin, son of Enlil-iddin Before Ninurta-ašarēdu, son of Šamaš-ēreš Before Balāssu, son of Šamaš-iddin [41]And the scribe: Ninurta-zākir-šumi, son of... [42-43]Kār-Bēl-mātāti, Du'ūzu day 18 year 9, Nebuchadnezzar king of Babylon [44-45][Fingern]ail of Nādin-aḫi, son of Ubār, impressed as his seal.	Nabureʿi son of Nabunathan. [13]Luḥi son of Mannuki. ʿOdnahar son of Duma. Nabureʿi son of Vashtan. [Endorsement.] [14]Deed of quittance which Peu wrote for Mibṭaḥiah.			

Original Texts				
Obv. 1. ˹*ṭup*˺-*pi* A.ŠÀ *ki-šub-bu-ú* *a-di* x x *u* x [x] **2**. KI^{ti} KÁ.GAL- ^{d}AMAR.UTU *šá qé-reb* ^{d}EN. LIL^{ki} **3**. 30 *ina* 1 KÙŠ UŠ AN.TA IM.SI.SÁ ÚS.SA. DU **4**. ^{Id}UTU- MU DUMU ^{Id}AG-*ga-mil* **5**. 30 *ina* 1 KÙŠ UŠ KI.TA IM.U$_{18}$. LU ÚS.SA.DU **6**. ^{Id}*Nuska*-SUM.NA DUMU ^{I}*Il-la-˹a˺* **7**. 30 *ina* 1 KÙŠ SAG.KI AN.TA IM.KUR.RA ÚS.[SA.DU] **8**. ^{I}*Šu-la-a* DUMU ^{Id}AG-*na-ṣir* **9**. 30 *ina* 1 KÙŠ SAG.KI KI.TA IM.MAR.TU ÚS.SA.DU **10**. ^{Id}MAŠ-DÙ^{uš} DUMU ^{I}*Haš-da-a* **11**. PAB A.ŠÀ šá ^{I}SUM	**1**. *b* ///\ *l'ḇ hw* *ywm* /// /// /// *lp̱ḥns šnt* /// // *ʾrṯḥššš mlkʾ* *ʾmr p̱yʾ* **2**. *br* *p̱ḥy ʾrrḵyl lswn* *byrtʾ lmbṭḥyh* *brt mḥsyh br* *ydynʾ* **3**. *ʾrmyʾ* *zy swn ldgl* *wryzt ʿl dynʾ* *zy ʿḇdn ḇswn* *np̱rt ʿl ksp̱* **4**. *w ʿḇwr wlbwš* *wnḥš wp̱rzl kl* *nksn wqnyn* *wsp̱r ʾdyn* *mwmʾh* **5**. *mṭʾh* *ʿlyky wymʾty* *ly ʿlyhm bsty* *ʾlhth wṭḇ lbby* **6**. *bmwmʾh* *dkʾ zy ʿḇdty ly* *ʿl nksyʾ ʾlḵy* *wrḥqt mnky* *mn* 7. *ywmʾ* *znh w ʿd ʿlm lʾ* *ʾkhl ʾgrnky dyn* *wdbb ʾnty wbr* **8**. *wbrh lḵy bšm* *nksyʾ ʾ*	**1**. [ἔτου]ς λδ Χοιὰχ [ ]α ἐν Συήνηι τῆς Θηβαίδος ἐπ᾽ Ἀριστοδήμου ἀγορανόμου **2**. [ὁμολ]ογεῖ Ἰσιὰς ἡ καὶ Σεννῆσις Πατεπνεβτεῦτος ἀφώντισσα μετὰ κυρίου τοῦ αὐτῆς **3**. [ἀν] δ̣ρὸς Νεοπ[το] λέμου τοῦ Νεοπτολέμου Κυρηναίου τῶν Πτολεμαίου καὶ τῶν υἱῶν **4**. [τακ]τομίσθου Ἐ̣σαροήρει Πετοσίριος ἀφωντεῖ ἀμίσθωι συλλελύσθαι αὐτῶι τε καὶ Ταγῶτι **5**. [τῆι γ] υναικὶ αὐ[το]ῦ ὑπὲρ ἧς ἐνέβαλεν κατ᾽ αὐτῶν ἐντεύξεως ἐν τῶι λδ (ἔτει) τοῖς ἐν **6**. [Πτο]λεμαίδι	RECTO **1**. † χ̣ μγ † **2**. † βασιλείας καὶ ὑ(∗)πατείας τοῦ θειοτάτου ἡμῶν δεσπότου Φλαυί(∗)ου Ἰ(∗)ουστίνου Νέου τοῦ **3**. αἰωνίου Αὐγούστ[ου Αὐτοκράτορος ἔ]τ̣[ους ἐ]νάτου Φ[α]μενὼθ ιε τῆς ἑβδόμης ἰ(∗)νδ(ικτίονος) **4**. Αὐρήλιοι Βίκτωρ καὶ Παείων καὶ Τσί[α] γαμετὴ Ἀτρῆτος τοῦ καὶ συναινοῦντος αὐτῇ οἱ **5**. τρεῖς ὁμογνήσιοι ἀδελφοὶ ἐκ πατρὸς Δίου τοῦ καὶ Πασαραεὶ ἐκ μητρὸς **6**. Τλοῦ τῆς καὶ πειθομ[ένης καὶ συνευδοκούσ] ης καὶ διαβεβαιωσαμένης πάντα **7**. τὰ ὁμολογηθέντα ταύτῃ τῇ διαλύσει ὁρμώμενοι ἀπὸ Συήνης οἱ καὶ ἑξῆς ὑπογραφέα **8**. παρέχοντες τὸν ὑπὲρ αὐτῶν ὑπογράφοντα καὶ μάρτυρας τοὺς ἑξῆς μαρτυροῦντας τῇδε τῇ **9**. διαλυτικῇ ὁμολογίᾳ. † Αὐρηλίῳ Ἰ(∗)ακώβῳ ἰ(∗)δίῳ ἡμῶν ὁμοπατρίῳ ἀδελφῷ **10**. ἐκ τῶν προγεγραμμένων Δίου τοῦ καὶ Πασαραεὶ καὶ Τλοῦ ὁρμωμένῳ ἀπὸ τῆς αὐτῆς Συήνης **11**. χαίρειν· ἐπειδὴ ἡνίκα ἐν τοῖς ζῶσι τυχάνει(∗) ὁ μνημονευθεὶς(∗) ἡμῶν πατὴρ Δῖος **12**. εἶχεν σὲ τὸν προγεγραμμένον Ἰ(∗)άκωβον ὡσανεὶ ἀνδρωθέντα καὶ ἥκοντα εἰς τὴν **13**. δέουσαν ἡλικίαν συνεργαζόμενον αὐτῷ(∗) τῇ ναυτικῇ τέχνῃ, καὶ τοῦ προειρημενουο(∗) **14**. πατρὸς ἡμῶν Δίου τέλει τοῦ βίου χρησαμένου τὴν ὑπολειφθεῖσαν τῷ μακαρίῳ ἡμῶν **15**. πατρὶ παρειλήφαμεν δι᾽ ἐμοῦ αὐτῆς τῆς προειρημένης Τλοῦ μητρὸς ἡμῶν **16**. καὶ ε[......].. τὴν ἐξαγωγήν σοι ποιησάμενοι φάσκοντες ὑπολείπεσθαί σοί **17**. τι τῆς προχρείας τοῦ αὐτοῦ μακαρίου ἡμῶν πατρός, καὶ σὺ ὁ προλεχθεὶς Ἰ(∗)άκωβος διεβε- **18**. βαιώσω(∗) λέγων· μηδέν τι ἰ(∗)διοποιεῖσθαί(∗) σοι τῶν πατρῴων ἡμῶν πραγμάτων, **19**. καὶ τῆς φιλονικείας(∗) μακρᾶς γενομένης ᾑρησάμεθα τὸ κοινὸν τῶν καθοσιωμένων(∗) **20**. πρώτων ἀριθμοῦ Συήνης, οἵτινες ἀκούσαντες τὰ μεταξὺ ἡμῶν ἀμφιβαλλόμενα συνῖδαν(∗) **21**. σὲ μηδὲν τὸ σύνολον κατέχοντα τῶν πατρῴων ἡμῶν πραγμάτων καὶ εἰς βεβαίωσιν **22**. δὲ κ(αὶ) πᾶσαν ἀπαλλαγὴν τῆς(∗) κινηθείσης σοι παρ᾽ ἡμῶν φιλονικεις(∗) καὶ ἀποσόβησιν **23**. πάσης ἐρεσχειλίας(∗) κινηθησομένης(∗) κατὰ σοῦ παρ᾽ ἡμῶν ὥρισαν(∗) ὥστε ἡμᾶς δέξασθαι **24**. ἐνώμοτον λόγον ὡς μηδὲν ὑπολείπεσθαί σοι τῶν πατρῴων ἡμῶν πραγμάτων. Τούτων **25**. τοίνυν	**1**. αχϸονο ϕ′ κ′ ε′ μαυο ποσιγο καλδο ναβιχτο μολραγο χιρσοβωστιγο μαλαβο γανδαρο πισο **2**. βαγο οαχϸο λαδοιανο βορδογαλαγο στοργο οαρσοχονδδιγο ακιδαβο γανδαρο σαπισινδο οδο πισο οι\σβοροζινιιο **3**. σαμοσιτανο ϸαβορανο φρομαλαρο οδο πισο μισο υαμαχονδο ωδαγοσιγο ζαδοκαρ\ο **4**. ακιδο οαλο πιδο μιλαγγο αρμαστινδο ατασιμο∗ σαχοανο ζανινδο καλδο πιδοοασαδο χογαμο **5**. χοσινδο λαδο οδο παραλαδο μανο οορωλο οδο μανο υιλιτοβηρο βαυο πορανο οδο μανο ζαρδο **6**. οορωλο πορο γανδαριγο χοιαχο ακιδομο χανο οορωλανο γιρλινδο ρωβοχαρο μαρηγο πιδο **7**. δογγο ιαοαρο ακιδομο στοργο γαζιτο οδο βαριτο λαδο σαχσαδο ταδομο αβο χανο υαρσο ανδαρο ιαο\οανο **8**. νιστο σταδο ακιδο γαζιτο οδο βαριτο λαδο ϸιαδδηιο ασιδο μαχο αστο αββο οδο ζαμιγο αγγαρα\γο **9**. ιωγο μαλαβο ιασκισαγγο ραγο ιειο αγγαραγο ασιδο νωγοφαρνανο γιρλινδο ταδομο ωσο παραλαδο **10**. αβφαγο οαυρανο ρ̣δαβφαγο μιρο βηκο πορανο ιασπιιο σηρο μαρηγο πιδο δδραχμο ταζαγο σιμι\γγο **11**. ϸιζγο ωδαγο ρ̣ι ζινδδιγο χοατο ξ′ οδο μαχο αγγιτινδο οισο σποριγο ϸαδο υονοσινδο κιρδαμο πιδρ̣\σαχτο **12**. οαυαγο φρογαοο ζαναμο ναγο μηγαρο ωσο αστο αβο αγγαραγο ασιδο νωγοφαρνανο ναμο παδουα̣-μο\νδο **13**. ασο μιροσανο αγγαραγο ασιδο πιρωζοβαρνο ναμο οδασο μιρονφαρανο οαραο̅α̅γ̅γο ροζγο ρηγο παδρ̣\ υαμονδο **14**. οδασο νιμορωσανο∗

Table 2.3.1 contd. ...

...Table 2.3.1 contd.

NA-ŠEŠ A ^{I}*Ú-bar ana gi-mir-˹ti˺* 12. ^{I}MU-^{d}U+GUR A ^{Id}UTU-KÁM *it-ti* ^{I}SUM. NA-ŠEŠ 13. A ^{I}*Ú-bar ki-i* 52 GÍN KÙ.BABBAR KI.LAM *im-bé-e-ma* 14. *i-šam ši-mi-šú a-di* 1 GÍN KÙ.BABBAR *šá* UMBIN- *šú* 15. *a-ki-i* KA DIRI *na-ad-nu* PAB 53 GÍN KÙ.BABBAR 16. KÙ.PAD.DU TIL^{ti} ^{I}MU-ŠEŠ A ^{I}*Ú-bar ina* ŠU^{II} 17. [^{I}M] U-^{d}U+GUR A ^{Id}UTU-APIN^{eš} ŠÀM É-*šú* 18. [SUM.N]A *ma-ḫir a-pil za-ka ru-gúm-ma-a* 19. [*ul i-ši u*]*l* ˹*i*˺*-tur-ru-ma a-na a-ha-me*[*š*] 20. [*ul i-rag-gu-mu ma-*	*lby zy ymʾty ly ʿlyhm hn grytby* 9. *dyn wdbb wgrky brly wbrʾly bšm mwmʾh dky ʾnh pyʾ wbny* 10. *ʾntn lmṭḥyh ksp kršn /// // bʾbny mlkʾ wlʾ dyn wlʾdbb* 11. *wʾnh rḥyq mn kldyn wdbb ktb pṭʾsy br nbwntn sprʾ znh* 12. *bswn byrtʾ kpm pyʾ br pḥy šhdyʾ bgw nbwrʿy br nbwntn* 13. *lwḥy br mnky ʿwdnhr br dwmʾ nbwrʿy br wštn* **Endorsement**. 14. *spr mrḥq zy ktb pyʾ lmbṭ[ḥyh]*	τῆς Θηβαίδος δικασταῖς ὧν εἰσαγωγεὺς Φιλῖνος δι᾽ ἧς ἐνεκάλει αὐτοῖς 7. [περὶ] ὕβρεως κ[αὶ] πλ[η]γῶν ἃ ἐτιμήσατο [χ] αλκοῦ ταλά[ν] των δύο καὶ ἀργυρί[ο] υ δραχμῶν 8. [ἑκατ]ὸν καὶ μὴ ἐπελεύσεσθαι ἐπ᾽ αὐτοὺς [ὑ]πὲρ τῶν προγεγρα[μ] μένων μήδ᾽ ὑπὲρ τῶν 9. [διὰ τ] ῆς ἐντεύξεως δηλουμένων πάν̣τ̣[ω]ν κατὰ μηθένα τρόπον, ἐὰν δ᾽ ἐπέλθῃ 10. [ἢ ἄλλ]ος τι[ς] ὑπ[ὲ]ρ αὐτῆς [ἢ] τ᾽ ἔφοδ[ος] αὐτοῖς ἄκυρ[ο] ς ἔστω καὶ προσαποτεισάτω Ἰσιὰς 11. [Ἑσαροήρει καὶ	δοξάντων κ(αὶ) ἀρεσθέντων ἡμῖν κομισαμένοις παρὰ σοῦ τὸν ἐνώμοτον καὶ φρικτὸν **26**. ὅρκον ἐπὶ τῶν θείων καὶ ἀχράντων καὶ σεπτῶν κειμηλίων, ὡς μηδὲν παντελῶς **27**. συμφεροποιησαμένου τῶν αὐτῶν πατρικῶν ἡμῶν πραγμάτων, εἰς ταύτην ἐληλύθαμεν **28**. τὴν ὁμολογίαν τῆς διαλύσεως, δι᾽ ἧς ὁμολογοῦμεν ἡμεῖς οἱ προγεγραμμένοι Βίκτωρ καὶ **29**. Παείων καὶ Τσία ἑκουσίω(∗) γνώμῃ καὶ αὐθαιρέτῳ βουλήσει καὶ ἀδόλῳ προαιρέσει δίχα **30**. παντὸς δόλου καὶ φόβου τινὸς καὶ βίας καὶ ἀπάτης καὶ ἀνάγκης καὶ συναρπαγῆς πρῶτον μὲν δεδέχθαι **31**. καὶ εἰληφέναι παρὰ σοῦ τὸν σεβάσμιον ὅρκον ἐπὶ τῶν ἁγίων καὶ ἀμώμων κειμηλίων ὡς μηδὲν **32**. τοῦ πατρὸς ἡμῶν διακατέχοντος μέχρι εὐτελοῦς τινος, ἔπειτα δὲ ὁμολογοῦμεν μηδένα **33**. λόγον ἔχειν πρὸς σὲ τὸν προειρημένον Ἰ(∗)άκωβον τὸν ἀδελφὸν ἡμῶν μηδὲ ἕξειν **34**. περὶ τῆς προχρείας τῆς ὑ(∗)πολειφθείσης παρὰ τοῦ μνημονευθέντος Δίου πατρὸς ἡμῶν **35**. μηδὲ περὶ οἱουδήποτε πράγματος προσήκοντος τῇ πατρῴᾳ ἡμῶν κληρονομίᾳ, μικροῦ **36**. ἢ μεγάλου νοηθέντος ἢ μὴ νοηθέντος λεχθέντος ἢ μὴ λεχθέντος ἐλθόντος εἰς μέσον **37**. ἢ μὴ ἐλθόντος ταγέντος ταύτῃ τῇ διαλύσει ἢ μὴ ταγέντος, ἁπλῶς οὐκ ἐπελευσόμεθά **38**. σοι οὐχ ἡμεῖς αὐτοὶ οὐ κληρονόμοι ἡμῶν οὐ διάδοχοι οὐ διακάτοχοι μηδὲ 39. ἐνάξωμέν σοι μηδὲ κληρονόμοις σοῖς ἢ διαδόχοις ἢ διακατόχοις μηδὲ ἐγκαλέσωμεν **40**. μηδὲ παραβαίνωμεν ἢ παρασαλεύσωμεν ταύτην τὴν διάλυσιν ἐν τῷ νῦν καὶ ἐν μηδενὶ καιρῷ **41**. κατὰ μηδένα τρόπον ἢ ἀφορμὴν ἢ μέθοδον ἢ εὑρεσιλογίαν ἐν δικαστηρίῳ μηδὲ ἐκτὸς **42**. δικαστηρίου διὰ τὸ ὡς προείπαμεν ἀπηλλάχθαι ἡμᾶς πρὸς σὲ περὶ πάντων καὶ **43**. ἐσβέσθαι πᾶν σπέρμα δίκης καὶ ἐναγωγῆς ἁρμοζούσης ταύτῃ τῇ ὑποθέσει ἢ καὶ ἁρμόζαι(∗) δυναμένης. **44**. καὶ πρὸς ἐπὶ τούτοις ἐπωμοσάμεθα τὸν θεῖον καὶ σεβάσμιον ὅρκον τοῦ τε παντοκράτορος θεοῦ **45**. καὶ νίκης καὶ εὐσεβείας τῶν κρατίστων δεσποτῶν τῆς οἰκουμένης Φλ(αυίων) Ἰ(∗)ουστίνου Νέου καὶ **46**. Αἰλίας Σοφᾶς τῶν αἰωνίων Αὐγούστων Αὐτοκρατόρων, μὴ ἐπιφυῆναί σοι μηδὲ ἐνάγειν μηδὲ **47**. ἐγκαλεῖν ἔτι μηδὲ παραβῆναι μηδὲ παρασαλεῦσαι ταύτην τὴν διάλυσιν. vac.? εἰ δὲ βουληθείημεν <ἐν> καιρῷ πώποτε **48**. ἐνάγειν σοι ἢ ἐπεξιέναι σοι ταύτης ἕνεκεν τῆς ὑποθέσεως, π̣ρ̣ῶ̣τ̣ο̣ν μὲν ἔνοχοι φανησόμεθα τῷ	σινζο σαοανο αγγαραγο παδουαμονδο οδασο αχρανο κινο αββο οαλο ασιδαβο **15**. ιασκινο βαοιδο οδο οαζιδο τοχμανο ιαοο πανζαςο ν′ καβσο αλο καςοκο φριιαδαρο αλο κασοκο καμβαταρ\ο **16**. ταδομο ωσο λαδο οδο παραλαδο αγγαραγο σαγονδαβο σοφαρο παδουαμονδο βανδαρο ναβισιδο αββο οδο ζαμιγο αυαγγο **17**. οδο ναυαγγο ασο κωσο αβο κωσο οδαςο υαμονδο αβο υαμονδο βηγοτωγο βηυαργο βηπαρο βηπιδδιστο ασο οισο **18**. κισο πιδδιγο αγγαραγο ασιδαβο μασκο ναβισιδο παδαχφανηιο ασο ανδορονιγο ρωσο παρσο μαβαρο ζαμανο οδαςο **19**. μωσο αβο ιαοηδανο ζοραγο τοο οαυρανο οδο τοο μιρο οδο ταοαγγο βραδο πορο φροζανδανανο χοαδο **20**. χοαρδο οδο δδριγδο παρσο παραλαδο αμβαγδο ναβαγο ωσταδο οιχαρηγανο κιρδο αλο ανδαρο ζαμιγο μιγδο αβο **21**. πορο οολοβαρο λαδο ναβανδο λογδο ναμαγγο βαυαρο οδο βαγολαγγο κιρδο λαχμιγο οδο λαχβατανιγο κιρδο ασιδο **22**. υαρσο χοαδο ριμαδο σαγονδο χοβαοανδο αβο δδραχμο χιρσαγο βονο πιδαργανο παδαχφανινδο ακιδο υαρσο κισο **23**. πιδοκιφτο οδο ζιδο οδο γαμανο κιρδο ναγο παδαχφανινδο οδο ναγο πωστιγο αζγαδο ναγο καβογγο ναγο νωγο Line: (22’) πωστιγο αζγαδο ναγο καβογγο ναγο νωγο ναγο χιρσοβωστιγο ναγο **24**. καλδο δογγο ασταδο υαρσο κισο ακιδο αχασαδο πιδοκαρλαδο ιωλαδο οισηραδο

ti-ma ina ar-kát *u*$_{4}$*-mu*] **21**. [*ina* ŠEŠ^{meš} DUMU^{meš} *kim-tu*$_{4}$ *ne-su-tu*$_{4}$ *u sa-lat*] **22**. [*šá* É ^{I}SUM.NA-ŠEŠ A ^{I}Ú*-bar šá* E$_{11}$] **23**. [*a-na* UGU A.ŠÀ *šu-a-ti i-dab-bu-* *bu ú-šad-ba-bu*] **24**. [BAL^{ú} *ú-pa-* *qa-ru u pa-qir-* *a-nu ú-šar-šu-ú* *um-ma*] **Rev**. **25**. [A.ŠÀ MU^{meš} *ul* *na-din-ma kàs-pi*] ˹*rul ma-ḫir*˺ **26**. [*i-qa*]*b-*˹*bu*˺*-ú* *kàs-pi im-ḫu-ru* ^{lú}*pa-qir-a-nu* **27**. A.ŠÀ *ki-i šà ub-* *ba-lu a-di* 12-TA. ÀM **28**. *i-ta-* *nap-pal* **29**. *a-na* *la-né-e u la si-ḫu* *šà* <DIŠ> *niš* ^{d}50 *u* ^{d}*Nin-líl*		Ταγ]ῶτι ἐπ[ίτιμ] ον [χαλκοῦ τ] ά̣λα̣ν̣[τα π]έν[τε καὶ ἱ]ερὰς τ[οῖ] ς βασιλεῦσιν **12**. [ἀργυρίου ἐπισήμ]ου δρ[αχμὰς -ca.?- χωρὶς τοῦ τὰ π]ρ[ογ] εγραμμέ[να] κ̣ύρια **13**. [εἶναι. -ca.?-] κεχ[ρημάτισται δι' Ἀριστοδήμου ἀγοραν]όμου	**49**. θείῳ ὅρκῳ καὶ τῷ τῆς ἐπιορκείας(∗) κινδύνῳ καὶ ἐγκλήματι, ἔπειτα δὲ ἐπιγνωσόμεθα λόγῳ ἐπιχει- **50**. ρήσεως καὶ [παραβάσεως] χρ[υσοῦ ὀγκίαι](∗) δύο γί̣ (νονται) χρ(υσοῦ) ὀγκ(ίαι)(∗) β μ̣ε̣τ̣ὰ̣ κ̣α̣ὶ̣ τοῦ μ η θ ὲν ἰ(∗)σχύειν κατὰ τῶν **51**. συντεταγμένων ταύτῃ τῇ διαλύσει, ἥνπερ εἰς σὴν ἀσφ(άλειαν) πεποιήμεθα κυρίαν καὶ βεβαίαν καὶ ἔννομον **52**. [με]θ' [ὑπ] ο̣γραφη(∗) τοῦ ὑπὲρ ἡμῶν ὑπογράφοντος καὶ τῶν ἑξῆς μαρτυρούντων καὶ ἐπερωτηθέντες **53**. ὡμολογήσαμεν(∗) †. δῆλον δὲ ὅτι ἐδεξάμεθα παρὰ σοῦ νομισμάτιον ἓν ζυγῷ Συήνης, ὅπερ δέδωκεν **54**. σοὶ ὁ̣ μνημονευθεὶς ἡμῶν πατὴρ Δῖος ὑπὲρ στρατευσίμου τοῦ υἱ(∗)οῦ σοῦ καὶ οὐδένα λόγον ἔχομεν πρὸς σὲ **55**. περὶ τούτου διὰ τὸ ὡς προείπαμεν εἰληφέναι παρὰ σοῦ καὶ οὕτ[ω]ς ὡμολογήσ(αμεν) †. (hand 2) † Αὐρηλι(∗)(∗) Βίκτωρ καὶ Παείω[ν] **56**. καὶ Τσία ἐκ πατρὸς Δίου Πασαροεὶ ἐκ μητρὸς Τλοὺ τῆς πιθομένης(∗) **57**. καὶ σηνευδοκουης(∗) οἱ προκ(είμενοι) ἐθέμεθα ταύτην τὴν δ̣ιάλυσ̣η̣ν(∗) καὶ στοι- **58**. χεῖ ἡμεῖν(∗) ὁς(∗) πρόκ(ειται). Φλ(αύιος) Ἰ(∗)ακὼβ Βασιλείδου ἀπὸ ἀκτουαρ(ίων) ἀριθ(μοῦ) Συήνης **59**. ἀκξιοθὶς(∗) ἐκγραυε(∗) ὑπὲρ αὐτου(∗) γρμματα(∗) μὴ εἰτότον(∗). (hand 3) † Φλ(αύιος) Μ̣ηγ̣ᾶ̣ς **60**. Διδύμου στρ(ατιώτης) ἀριθμοῦ Σήνης(∗) μαρτυρῶ †. (hand 4) Φλ(αύιος) Ἄπα Δῖος Μαρτίου ἀπὸ **61**. ἀκτουαρίων ἀριθμο(ῦ) Συήνης μαρτυρῶ τῇ ὁμολογίᾳ ἀκούσας παρὰ **62**. τῶν θεμένων †. (hand 5) † Φλ(αύιος) Ἰ(∗)σὰκ Ἰ(∗)ακώβου στρα(τιώτης) ἀριθμ(οῦ) Συήνης μαρτυρῶ †. **63**. (hand 6) † Φλ(αύιος) Κόμες Πααμίου στρ(ατιώτης) ἀριθμοῦ Σήνης(∗) μαρτυρῶ †. **64**. (hand 7) † δι' ἐμοῦ Βίκτορος Πέτρου ἐτελειώθη Φαμενὼθ ιε τῆς ἑβδόμης ἰνδ(ικτίονος) ἐν Συήνῃ †. VERSO (hand 1) † διάλυσις ἐπ' ὀ̣ν(όματος) Βίκτορος καὶ Παείωνος καὶ Τσίας ἀδελ[φῶν ὁμογνησίων] π̣ρὸς Ἰ(∗)άκωβον.	χο̣αρδο οδο δδριγδο ναγο υιρζαδο αλο **25**. ρωβιγο αλο βαμοιανιγο αλο τορκο αλο ταζιγο αλο ωδαγοσιγο αλο ανδαρο υαρσο κισο αχασογαργο πηδαγο **25**. βοο͞α δο ταδο **26**. οαλαζο οορωλο οδαζο υιλιτοβηρο οδαζο ζαρ̣ δο οδο μαχαγγο βραδο πορο φροζ̣ανδανανο δογγο πιτιαμο οδο **27**. μαχο πιταδδηιο ακιδαβο οισο λαδο λαδο ναβανδο ϸαοαμο οδασο οισο αχασο οδασο οισο πιδοκαρδο χοηδαγο οδαβ{ο}\υηβινδο **28**. κιραμο καλδαβο ο̣ισο λαδο̣ λαδο ναβανδο ναγο ϸαοαμο οδασο οισο αχασο χοηδαγο οδαβυηβινδο ναγο **29**. κιραμο οδαλο μαχο χοαδο πιδοοασινδδιγανο αχασαμο πιδοκαρλαμο χοαρδο δδριγδο ναγο υιρζαμο αλο ανδαρ\ο **30**. Πωστιγο μολραγο αζβαραμο καβογγο αλο νωγο ταδο μαχο χοανδο οδο οισαρο αβο λαδο οαυοαρο παρμανο μαγο **31**. βοο͞αδο οδο μισο τωζαμο ταοανο αβο γαζνο βρηγανο̣ δ̣δραχ̣μο ταζαγο σιμιγγο ϸιζγο ωδαγο οιζινδδιγο σαδο ρʹ οι\στο **32**. κʹ οδο μισο δαρφηλο ταοανο αβο παδοιαρδδιγανο τωζαμο οδο παρσο τογδο ταοανο χιρσοβωστιγο χοζο οηζο **33**. οαυοαρο οδο παρμανο βοοαδο σαγονδαβο μ̣α̣σκο ναβισιδο

played the same language-game in which contracts should be compiled only within the framework of this structure. The records of court sessions in Bactrian and Sogdian have not been preserved, but we see that stage (vii) ('penalty') can be said to adumbrate any rules for an envisaged future legal dispute and, hence, bears a reference to something like a critical discussion, in which implications are involved and, therefore, the inference rules of *modus ponens* and *modus tollens* are applied, as we know from the trial records in Akkadian, Aramaic, and Ancient Greek (Culbertson, 2009; Schumann, 2019b, 2020, 2021a).

We can reconstruct the argumentative rules of trials on the basis of legal documents. These rules are adopted from van Eemeren and Grootendorst (2004: 136–157), but with some simplifications. In particular, only 7 out of 15 rules can immediately be adopted.

Rule 1. The propositional content of the assertives by which a standpoint is expressed (or its negation calling the standpoint into question) is assumed to be checkable by the court (cf. van Eemeren and Grootendorst, 2004: 136). There should be some appropriate verified facts or past legal documents that directly concern the matter of either the propositional content of the standpoint or its negation.

Rule 2. At the trial the discussant has a challenge to defend his or her standpoint (van Eemeren and Grootendorst, 2004: 137, 143). In pragma-dialectics we are not limited to classical propositional logic there, but in the court records in Akkadian, Aramaic and Ancient Greek we find only the application of logical inference rules of propositional logic to verified facts and to valid legal documents, without rhetoric figures or reasoning by analogy. Therefore, we can assume that the court should check there the correctness of reasoning from the point of view of deductibility.

Rule 3. The discussant is obliged to defend the standpoint at the trial so long as he or she does not retract it, and so long as he or she has not successfully defended it against the other discussant based on agreed premises and discussion rules (cf. van Eemeren and Grootendorst, 2004: 139), but in trial records in Akkadian, Aramaic, and Ancient Greek, the agreed premises were understood only as facts and valid legal documents verified by the court and the discussion rules were understood as a requirement to apply only inference rules of propositional logic without rhetoric tropes or reasoning by analogy.

Rule 4. The plaintiff fulfills the role of protagonist in the argumentation stage of the legal proceeding, and the defendant fulfills the role of antagonist (cf. van Eemeren and Grootendorst, 2004: 142).

Rule 5. The protagonist may always defend the standpoint by performing a complex speech act of argumentation, where only logical inference rules are applied and only verified facts and valid legal documents are involved as premises. The antagonist may always attack the propositional content or the justificatory or refutatory force of the argumentation by applying only logical inference rules and through references to contradicting verified facts and valid legal documents. This rule is actually rule 6 in (van Eemeren and Grootendorst, 2004: 144), but again we remember that pragma-dialectics is not limited to classical propositional logic as a method of reasoning. Nevertheless, in court protocols in Akkadian, Aramaic, and Ancient Greek, reasoning strongly follows the standards of classical propositional logic.

Rule 6. The protagonist (antagonist) has successfully defended (rejected) the propositional content of a complex speech act of argumentation if the intersubjective identification procedure made by the court yields a positive (negative) result based on verified facts and valid documents. This rule is actually rule 7 in (van Eemeren and Grootendorst, 2004: 147–148). But it is worth noting that in pragma-dialectics, the intersubjective identification procedure is not carried out by a third party (the court), but by the protagonist and the antagonist themselves in joint scrutiny (see van Eemeren and Grootendorst, 2004: 146).

Rule 7. The protagonist (antagonist) has successfully defended (rejected) a complex speech act of argumentation if the intersubjective inference procedure made by the court yields a positive (negative) result based on correct inference rules for the justification (refutation) of the standpoint as a logical conclusion. This rule is actually rule 8 in (van Eemeren and Grootendorst, 2004: 150).

Thus, the argumentation rules from 1 to 7 represent one of the simple cases of critical discussion defined in (van Eemeren and Grootendorst, 2004: 136–157). So, the contracts are composed in the way as to be applied as new valid documents for inferring a punishment by *modus ponens* if another party of the contract refutes its propositional content in the future. Evidently, in pragma-dialectics the logic can be any.

We have many fragments of trial records in Akkadian, Aramaic, and Ancient Greek to observe that they have similar logeme structures as well as similar order of illocutionary acts of the same type. It allows us to reconstruct the communicative competence of their authors and to recreate rules 1–7. Nevertheless, we do not have trial records in Bactrian and Sogdian. But we have many fragments of legal documents of other types in these languages, and they have a similar structure to documents in Akkadian, Aramaic, and Greek. From this it follows that the court records in Bactrian and Sogdian must be equally similar to Akkadian, Aramaic and Greek ones.

Furthermore, the contracts (see Table 2.3.1) are composed in the way as to be applied as new valid documents for inferring a punishment by *modus ponens* if another party of the contract refutes its propositional content in the future (stage (vii)). We observe, therefore, a complex language-game of merchants in which the critical discussion is understood as a culmination of different intersubjective means of conflict resolutions, including the means of propositional logic and legal hermeneutics participating in drawing up legal documents such as contracts.

The structural similarity to the contracts, having the sequence from (i) to (ix), is also detected in texts written, for example, in Demotic (Egyptian), such as matrimonial property arrangements (Porten, 1996: 346–347), exchanges of stipends between two priests (Porten, 1996: 351–355), withdrawals after a lawsuit (Porten, 1996: 360–362), etc. These documents also start with the date, but the Egyptian calendar was used exclusively prior to 483 B.C. and after this year we find the synchronous Babylonian and Egyptian month dating (Porten, 1996: 81). In these

Table 2.3.2. The isomorphic speech acts of letters in Akkadian, Aramaic, Greek, and Bactrian.

The mid-14th century B.C. The letter from Rib-Hadda, the king of Byblos, to the Egyptian official Amanappa. *The Akkadian document from Amarna*, BM 29797 (Moran, 1992: 153–154).	**The late 5th century B.C. The letter from PN to Jedaniah, Mauziah, Uriah, and the Troop**. *The Aramaic document from Elephantine in Egypt*, Sachau Plate 11 (Porten, 1996: 127–129).	**29 December, 137 B.C. The letter of Asklepiades to Philinos**. *The Greek document from Syene in Egypt*, BGU 6 1248 (Porten, 1996, pp. 422–423).	**From ca. 330 to ca. 365 A.D. The letter from Warahrān the great Kušānšāh to Khwadew-wanind**. *The Bactrian document from Gandhāra*, Khalili collection, DOC 79 (Sims-Williams, 2007: 52–54).
Internal Address			
[R]ib-[Hadda sa]ys to his [lord], king of all countries, Great King:	RECTO [1]To my lords Jedaniah, Mauziah, Uriah, and the Troop, [yo]ur servant [PN	RECTO [1]Asklepiades to Philinos, the bailiff for the judges from the Panopolite (nome) and for allo[tted] [2]places.	From Dukht-anosh the princess, the noble …, [and from] Warahran the great Kushan-shah, to [2]Khwadew-wanind,
Salutations			
May the [L]ady of Gubla grant p[owe]r to the king, my lord. I fall at the feet of my lord, my Sun, 7 times and 7 times.	The welfare of my lords may the gods, all (of them)], [2]seek after at all times. It is well for us here.	greetings.	a thousand (and) ten thousand (times), many greetings. [And] (since) I have heard that you are healthy, therefore [3]I am happy; but then I would be more happy when I might hear from you (of your) even better health.
Complaint			
[7–14]Why do you not send back word to me that I may know what I should d[o]? I sent a man of mine to my lord, and both his horses were taken. A second man – a man of his – was taken, [and] a tablet of the king was not put [i]n my man's hand. Listen t[o m]e! [15–20]Wh<y> are you negligent so that your land is being taken? Let it not be said in the days of the commissioners, "The ᶜApiru have taken the entire country!" Not so shall it be said in the days <of the commissioners>, or you will not be able to take it back.	**Now**, every day that [...] [3]he complained to the investigators. One Zivaka, he complained to an investigator...[...] [4]we have since the Egyptians a bribe to them give. And from (the time) that [...] [5]of the Egyptians before Arsames, but thievishly act.	You have written to us that we should take a copy of the petition s[ubmit]ted [3]by Isias to give to the accused Esoroeris and Tagos and [3]sum[mon] them [4]to be present at the court in five days with, in addition, the [5]documents [4]relating to t[he] [5]proceedings.	[4]**Moreover**, you should know that Dathsh-mareg the eunuch has appealed to me (saying) that my corn, [5](for) which I was responsible [in respect of] the locusts – because of your shepherds, sheep [6]were let into the corn and they ate and spoiled the corn, and your shepherds [7]struck Dathsh-mareg's brother and nephew. And they came here without... and they claim thus, [8]that (it was) at your command. I do not believe this statement, because [9]I have heard that you are a just man.

Reports and Instructions			
[*Report I and Instructions*:] [21–29]**Moreover**, I have written for a garrison and horses, but they are not given. Send back word to me, or like Yapaḫ-Hadda and Zimredda I will make an alliance with ᶜAbdi-Asirta and stay alive. [*Report II and Instructions*:] **Moreover**, now that over and above everything else Ṣumur and Bit-Arḫa have defected, [30–37]may [yo]u put me in Yanḫamu's charge so he will give me grain to eat th<at> I may guard for him the king's city. May the king also give the order and release my man. His family are very upset with me, (saying) day and night, "You gave our son to the king." So, release him, especially him. [38–42](The *other* is a citizen of Ibirta.) He is, I assure you, in Yanḫamu's house. [*Report III and Instructions*:] **Moreover**, tell Ya(n)ḫamu, "I declare Rib-Hadda to be in your charge and whatever <ha>ppens to him to be yo[ur] responsibility." [43–51]May the troops on campaign not fall upon me. And so, I write, "**If** you do not tell him this, I will abandon the city and go off. **Moreover**, **if** you do not send word back to me, I will abandon the city and go off, together with the men who are loyal to me."	[*Report I and Instructions*:] **Moreover**, [...] [6]the province of Thebes and thus say: Mazdayasna/A Mazdean is an official of the province [...] [7]we are afraid because we are fewer by 2. [*Report II and Instructions*:] **Now**, behold, they favored [...]. [8]Had we revealed our presence to Arsames prior to this, this(!) wou[ld] not [*have been done to us*...] [9]he will report our affairs before Arsames. Pisina pacifies us [... *So whatever*] [10]you will find – honey, castor oil, string, rope, leather skins, BOARDS [... — *send us since*] VERSO [11]they are full of anger at you. [*Report III and Instructions*:] Pasu, son of Mannuki came to Memphis and...[...] [12]and the investigator. And he gave me silver, 12 staters and happy with it [*am I*...] [13]Hori gave me when they detained him because of the pitcher. Tiri... said: "[...] [14]at the order of the king and they detain them. And the damage of Arsames and the compensation of Djeḥ[o...] [15]and Ḥori whom they detained."	And how we managed the summonses [6]to Tagos in person, but to Esoroeris, (who was) away from home, through Eponychos the police official, [it would be] (too) lo[ng] [7]to report.	[*Report I and Instructions*:] **Now** I have sent Dathsh-mareg's [10]brother and nephew thither into your presence. So, you should make judgement for Dathsh-mareg's [11]brother and for (his) nephew in respect of the eating and spoiling of the corn and in respect of [12]the injury (which they received) from the shepherds. [*Report II and Instructions*:] And **if** you know that [13]I ought not to take notice of (their) statement in this matter, then send the shepherds hither into my presence, [14]and then I myself shall order (that) whoever may take notice of this statement [15]should be disbelieving (of) anything (said) by my servants in this matter, and [16]then I shall punish them; and **if** evil should have been committed by your shepherds, [17]then I shall order (a letter) to be written to you, and then you should impose a judgment against the shepherds (and) for (my) servants; [18]and **if** anything [should be] otherwise, please write to me now, so that I may know [19]that I should demand compensation for the damage from (my) servants. [*Report III and Instructions*:] And I have heard that you [20]claim thus, that I myself am,... by my servants. Therein you just [21]put... to the ruler..., you claimed and did concerning this.

Table 2.3.2 contd. ...

...Table 2.3.2 contd.

Final Greetings			
[51–57]Also for your information: Ummaḫnu (along with Milkuru, her husband), the maidservant of the Lady [of] Gu[bl]a, … *p*[*ow*]*erful* [*pray*]*s* [*t*]*o the L*[*ady of Gubla for the king, my lord*].	[17]To (*sealing*) my lords Jaadaniah, Mauziah, y[our] se[rvant PN].	[8]Farewell. [Date:] L 34, Choiak 5. VERSO [9]To Philinos.	[22]To Khwadew-wanind, a thousand (and) ten thousand (times), many greetings, (as is) fitting.
Original Texts			
Obv. 1. [ᴵ*Ri*-]*ib*[-ᵈIŠKUR *i*]*q*-⸢*bi*⸣ ⸢*a*⸣[-*na*] **2**. [EN-*š*]*u* LUGAL KUR.KUR.KI.MEŠ LUGAL GAL **3**. [ᵈN]IN *ša* URU *Gub-la ti-di-*in_4 **4**. [KAL.G]A *a-na šàr-ri* EN-ia **5**. *a-na* GÌR.MEŠ EN-*ia* ᵈUTU-*ia* **6**. *7-šu 7-ta-an am-qú-ut* **7**. *a-na mi-ni la-a tu-te-*⸢*ru*⸣-[*n*]*a* **8**. *a-wa-*ta_5 *a-na ia-a-ši ù* **9**. *i-de ip-ša ša i-pu-š*[*u*] **10**. LÚ-*ia ut-ta-ši-ir a-na ma-*[*ḫa*]*r* **11**. [E]N-*ia ù la-qú* 2 ANŠE.KUR.RA-*šu* **12**. ⸢*ù*⸣ LÚ *ša-nu la-qí* LÚ-*šu* **13**. ⸢*ù*⸣ *ṭup-pí* LUGAL *la-a na-di-*⸢in_4⸣ **14**. [*i-n*]*a qa-at* LÚ-*ia ši-*⸢*mé*⸣ ⸢*ia*⸣-*ši* **15**. [*a-n*]*a mi-‹ni› qa-la-ta ù tu-*⸢*ul*⸣-⸢*qú*⸣ **16**. ⸢KUR⸣-*ka ú-ul yu-uq-ba i-na* **17**. UD.KAMV.MEŠ LÚ.MEŠ MÁŠKIM *la-qú* LÚ.MEŠ GAZ.MEŠ **18**. *ka-li* KUR.KUR.ḪI.A *ú-ul ka-a-ma* **19**. ⸢*yu*⸣-*uq-bu i-na* UD. KAMV.MEŠ‹-*ka*› **20**. *ù la-a ti-le-ú la-qa-ši* **21**. ⸢*ša*⸣-*ni-tam aš-ta-pár a-na* LÚ.MEŠ *ma-ṣa-ar-ti* **22**. *ù a-na* NŠE.KUR.RA.MEŠ *ù la-a* **23**. ⸢*tu*⸣-*da-nu-na šu-te-ra a-wa-*ta_5 **24**. *a-na ia-ši ù i-pu-ša a-na-ku* **25**. *ki-ta it-ti* ᴵ ÌR-A-*ši-ir-ta* **26**. ⸢*ki*⸣-*ma* ᴵ*Ia-pa*ᵈIŠKUR *ù* ᴵ*Zi-*⸢*im*⸣-*re-*⸢*da*⸣ **27**. *ù* ⸢*bal*⸣-*ṭá-ti ša-ni-tam šum-ma* **Lo.ed. 28**. *ap-pu-na-ma a-nu-ma pa-aṭ-ra* **29**. ⸢URU⸣ *Ṣu-mu-ra ù* URU É-*Ar-*[*ḫ*]*a*(?) **Rev. 30**. [*t*]*a-din-ni i-na qa-at* **31**. ᴵ*Ia-an-ḫa-mi ù ia-dì-na* **32**. ŠE.*IM*.	**1**. *ʾl mrʾ ydnh mʿwzyh ʾwryh whyl ʾbdk*[*m šlm mrʾy ʾlhyʾ klʾ*] **2**. *yšʾlw bkl ʿdn šlm ln tnh kʿnt bkl ywm zy*[…] **3**. *hw qbl ptyprsn ḥd zywn hw qbl ptprsʾ*[…] **4**. *ʾyty ln bzy mṣryʾ šḥd lhn yhbn wmn zy* […] **5**. *zy mṣryʾ qdm ʾršm lhn gnbyt ʿbdn ʾp*[…] **6**. *mdnt nʾwkn ʾmrn mzdyzn hw pqd lmdynʾ*[…] **7**. *ndḥl bzy zʿyrn ʾnḥnh b// qdmn kʿn hʾ sbrw* […] **8**. *hnlw glyn ʾnpyn ʿl ʾršm lkn lʾ kznh hw*[*h ʿbyd ln*] **9**. *yʾmr mlyn qdm ʾršm pysn mhšdk ʾnpyn*[…]	**1**. Ἀσκληπιάδης Φιλίνωι τῶι εἰσάγοντι τοῖς ἀπο(∗) του(∗) Πανοπολίτου καὶ τοις(∗) μεμερισ[μένων] **2**. τοποις(∗) δικασταῖς χαίρειν. ἔγραψας ἡμῖν λάβοντα ἀντίγραφον εὐτεύξεως(∗) τῆς ἐ[μβληθει-] **3**. σης(∗) ὑπὸ Ἰσιάδος δοῦναι τοῖς ἐγκαλουμένοις Ἐσοροήρει καὶ Ταγῶτι καὶ παρ[αγγεῖλαι] **4**. παραγίνεσθαι ἐπὶ τὸ κριτήριον ἐν ἡμέραις πέντε ἔχοντας καὶ τὰ πρὸς τ[ὴν] **5**. κατάστασιν δικαιώματα. καὶ ὃ[ν] ἂν τρόπον οἰκονομήσαμεν(∗) παρηγγελκότε[ς] **6**. Ταγῶτι μὲν ἐνωπίωι, Ἐσοροήρει [δ] ἐ ἀπ' οἰκίας δι' Ἐπωνύχου φυλακίτου μακ[ρὸν ἦν] **7**. διασαφῆσαι. **8**. ἔρρωσθε (ἔτους) λδ Χοιὰκ ε *Verso* **9**. Φιλίνωι	**1**. Ασο δοχτ [οα]γ ωþο δοχþολ ο γδο μαζα δε þ [••* οδασο] ο α ρα ρα γ ο στο ργο κοþαγ οþα ο αβο χο-**2**. αδηοοανι[ν]δο υαζαρο βηοαρο λροδο υαβι σο [οτο]μο ναγατο σιδο λρογηιο ταδο þαδ-**3**. ημο ασιδανο οαλο þαταρανο καλ δανο ασαþα [γο*]μισο þαρδαρο λροστανο ναγαυαγ ο μ ι-**4**. σιδο τ ο ο* ζαγ ο κοαδο δ αϑþομαρηγο ι þαβαστα γ ο α βα μαγο πιδοοαυαδο κοαδομο ιαοα ρδ-**5**. αοι σιδομο[* α]σ ο* μαλαχανο ωζαδο ταδανο ι ταοι χο βανανο πιδοραλο ι ποσο αβο ιαοαρδα ο [ι] **6**. υιρτηιο οτοιηνο ιαοαρδαοι χοαρδηι ο οδο [να]þτ η ι ο* οτα γ ο* ταοι χοβανανο αβο δαϑ-**7**. þομαρηγο βραδο οδαβο βραυριγο ζιδο οτο αβη [•] αριγο* μαλο αγαδινδο οτο ιϑαχοηνιγ -**8**. δο κοαδο ταοι φρομανιγο ταδο αζο ειο σαχοανο οαροαροι νακιρημο πιδο ι σαγο **9**. κοαδο μανο ναγατο κοαδο τοο μαρδο* λαδιγηιο ωσιδο μανο α βο δαϑþομα ρη-**10**. γο βραδο οδαβο βραυριγο ταρο πισοαφαγο φοþτ αδ ο ταδο τοο φαρο δαϑþομαρηγο **11**. βραδο οδο φαρο ι βραυριγο πιδο ιαοαρ[δ] α ο ι χο α ραοι οδο ναþταοι οδο πιδο ζ-**12**. αχμο ασο χοβανανο λαδο κιρο οδο αλδο τοο ιϑαζανηιο κοαδο ταλο ι σαχ-**13**. οανο παρογατο ναþιιημο ταδο αβο χοβανανο μαρο πισοαμαγο φοþτιιο **14**. ταδο αζο χοαδο φρομιιανο κιδο ι σαχοανο παρογασαδο ταλιδο ασο ι μανο **15**. μαρηγανο σιγισο αν{ν}αοαραδο ***Verso, continuation*** **16**. ταδο ιηνο παδοφρασο κιρανο οδο αλδο ασο ι ταοι χοβανανο βιζδανο* κιρ-**17**. δ δο ταδο αζο αβαφαγο ναβιχτο φρομιιανο ταδο ασο ι χοβανανο φαρο ι μαρηγο λα δ ο []* **18**. ωστιιο οδο αλδο σιγισ ο ι ερο οιζα γγο δ α ρ[αδ] ο ταδ ο μο ωσ ο γ α βιχτ ο φ(ρ)ομιι ο ατο ζα γ α -**19**. γ ο σαγωνδο ασο ι μαρηγο ι μιυγαρο οιαναργιφο χωζανο οτο μανο ναγατο κοαδο το-**20**.

ḪI.A *a-na a-ka-li-ia* **33**. *a-na-ṣa-ra* URU LUGAL *a-na ša-a-šu*	**10**. *tškḥwn dbš tqm ḥtlyn ḥbln mšky ṣl šp̱n*[…] ***Verso*** **11**. *mlyn lbtk̠m ʾth p̱sw br mnky lmnp̱y whn* […] **12**. *wpṯp̱rsʾ wyhb̠ ly ksp̱ sttrn // wḥdh mnh*[*ʾnh*] **13**. *ḥwry yhb̠ ly kzy klwhy ʿl kdʾ ʾmr tyry* *** […] **14**. *bṣwt mlkʾ wklyn lhn wnzq ʾršm wkp̱r ṣḥ*[ʾ] **15**. *wḥwry zy klw bywm* /// /// *lp̱ʾp̱y mṯw ʾg̠rtʾ*[…] **16**. *nʿbd mlh* **17**. *ʾl mrʾy yʾdnyh mʿwzyhʿ*[*b̠dk̠m*]		ο ι̣ θαχοηνηιο̣ κοαδο ασ̣ο ι μανο μα̣ρηγ̣ανο χο̣α̣δο̣ ••γ̣ο* βο̣οη̣μο ṭ α̣λ̣ ι̣ δο υαρσο ṭ α̣οι αβο̣ **21**. ι χαρο μορη̣ [ο*]α̣βα̣ṭ α̣γ̣ ο* ωστα̣δο τατο χοα̣γ̣ [δο ο]ṭ ο̣ * π̣ ι̣ δ̣ ο ι κ̣ιρδο ***At the bottom of the page,*** **address** **22**. Α̣βο χοα̣ δ̣ η [ο]οα̣νινδο υαζαρο βηο̣ α̣ ρο λ̣ ρο[δο]υ̣ αβ[ι]σο νια̣ β̣γ̣ο̣ *

documents we may see various ethnicons: Jew, Egyptian, Caspian, Khwarezmian, and so on. Hence, the influence of Babylonian jurisprudence took place much wider than it is discussed in this section—we focus only on the languages of international trade and diplomacy of Eurasia in antiquity: Akkadian, Aramaic, Ancient Greek, Bactrian, Sogdian, etc.

But not only do legal documents demonstrate structural similarities between texts written in different languages, but also the same similarities are observed in some business, administrative, and diplomatic correspondence in different languages, belonging to peoples participating in international trading networks. One of the typical sequences of illocutions and perlocutions embodied in these letters is as follows: (i) internal address; → (ii) salutations; → (iii) complaint; → (iv) reports and instructions; → (v) final greetings, see Table 2.3.2. In the third stage (complaint) there is formulated a problem which should be solved and in the fourth stage (reports and instructions) some facts are enumerated, based on which an argumentation is proposed by means of some inference rules, such as *modus ponens*, applied to mentioned facts to solve the problem. We face a sample of critical discussion there without appealing to courts but drawn up within the framework of the strong argumentation standards of legal proceedings: mainly facts and deductive inference rules should be taken in reasoning.

2.3.5 Conclusions

Summing up, different legal documents in Akkadian, Aramaic, Ancient Greek, Bactrian, Sogdian have a similar illocutionary and perlocutionary structure and a similar logeme structure—mainly, implications are used, and *modus ponens* and *modus tollens* are applied. The application of these rules can be formalized by means of symbolic logic, while in argumentation theory we can recreate the following arguments based on implications and used in legal issues, including contracts (see Table 2.3.1) and business correspondence (see Table 2.3.2):

1) *Argument from commitment* (cf. Walton, 1995: 55), formulated as a promissory oath: (1) if the protagonist/antagonist takes the oath A in the contract or in the business letter (but it is), then he or she is committed to proposition A; (2) if he or she is committed to proposition A (but it is), then he or she must support A. This scheme is follows: If the oath A, then A. But the oath A is contained in the legal document. Hence, by *modus ponens* we have that A.
2) *Argument from position to know* (cf. Walton, 1995: 61), pronounced as an evidentiary oath: if the protagonist/antagonist asserts the proposition A (but it is), then he or she knows that A is true. The scheme is follows: If the assertion A, then A. But we see the assertion A in the legal document. Then we obtain A by *modus ponens*.
3) *Argument from established rule* (cf. Walton, 1995: 91): if doing A is the established rule according to business commitments or other legal obligations and responsibilities of the protagonist/antagonist (but it is), then he or she must do A. The scheme: If the rule A, then A. But the rule A occurs in the legal document. Then A by *modus ponens*.

4) *Argument from causation* (cf. Walton, 1995: 73): if there is a causal connection "If the event *A*, then the event *B*" and the protagonist/antagonist confirms that *A* takes place, then he or she should also confirm *B*. Its scheme is *modus ponens*: If *A*, then *B*. But *A*. Therefore, *B*.
5) *Argument from observing a sign* (cf. Walton, 1995: 47): if there is a connection "If the sign *A*, then the event *B*" and the protagonist/antagonist confirms that *A* takes place, then he or she should also confirm *B*. Again *modus ponens*: If *A*, then *B*. But *A*. Therefore, *B*.

All arguments from this list are also found in court records in Akkadian, Aramaic and Ancient Greek. This shows that the same arguments were used in legal texts of various types and belonged to the same tradition of argumentation, first rooted in Mesopotamia but flourishing in late antiquity among the merchant societies along the Silk Road.

The normative approach to argumentation proposed within the framework of pragma-dialectics is well confirmed by the material of legal documents in Akkadian, Aramaic, Ancient Greek, Bactrian, and Sogdian. Here we may reconstruct a complex language-game maintained by the traders of late antiquity, in which we see similar (sometimes even isomorphic) complex speech acts associated with critical discussions to resolve interpersonal conflicts mediated by logic and legal hermeneutics. Hence, the earliest samples of correct logical reasoning are detected in the international communities of merchants of the antique period. These ways of resolving conflicts were a very efficient social practice, therefore this practice was widely disseminated through the Silk Road networks. Consequently, the earliest logical and argumentative techniques were propagated due to the world-system of late antiquity.

According to Karl Marx, there is an economic basis of society, consisting of production forces and relations. Meanwhile, all cultural phenomena, including political institutions and laws, constitute the ideological superstructure of economic foundations. Briefly, it means that economic (production) relations are primary, and culture is secondary. According to Jürgen Habermas (1962), the public sphere (*Öffentlichkeit*) consisting of networks of free associations at least must be primary. But what do we see in the trading communities along the Silk Road in late antiquity? What was primary, their trading networks or the production of various goods?

The large-scale crisis of the world-economy occurred from the 5th century A.D. due to the rupture of many trade routes and the disappearance of many communities of merchants. This was caused, first of all, by the fall of the Indo-Scythian dynasties in Asia in ca. 400 A.D. The world-economy was able to fully recover only by the 9th century A.D. already within the framework of the emergence of networks of Muslim merchants. As we can see, the presence of communicative communities that form a stable trading network is a necessary condition for the development of the world-economy. The interaction within the network is built according to the norms of strict argumentation and unified standards for drafting legal documents. Consequently, here the primary ones were not production relations but cultural forms of interconnection, including, first of all, the logical competence and legal hermeneutics, which allowed merchants to establish broad business contacts that stimulated the production forces.

We can say that members of merchant societies in late antiquity performed rational or communicative actions in the sense of Habermas (1988), i.e., actions that can be justified or criticized within the framework of generally accepted argumentative norms. Rational action is not argumentative, but it can become one at any moment, in other words, communicative action may not lead to this or that argumentative discussion. But if the latter is the case, then the members of the communicative community must thematize the disputed claims to significance in a certain way and try to resolve them with some arguments.

Based on the legal documents of late antiquity, we can recreate the communicative competence of the merchants of that time and the corresponding rules of argumentation. These rules can be seen as a simplification of the rules developed in pragma-dialectics, but their main difference is that the arguments had some limited schemes of propositional logic, mostly based on *modus ponens* and *modus tollens*. In pragma-dialectics, the logic can be any.

References

Apel, K.-O. 1973. *Transformation der Philosophie*. 2 Bände. Frankfurt am Main: Suhrkamp.

Becker, T. 2012. The pragmatics of argumentation. pp. 257–272. *In*: Schalley, A.C. (ed.). *Practical Theories and Empirical Practice: A Linguistic Perspective*. John Benjamins.

Bermejo, L. 2011. *Giving Reasons: A Linguistic-pragmatic Approach to Argumentation Theory*. Vol. 20. Dordrecht: Springer.

Bostock, J. and Riley, H.T. 1855. *The Natural History. Pliny the Elder*. London: H.G. Bohn.

Braarvig, J. and Liland, F. 2010. *Traces of Gandhāran Buddhism. An Exhibition of Ancient Buddhist Manuscripts in the Schøyen Collection*. Oslo: Hermes Publishing.

Bram, J.Rh. 1975. *Ancient Astrology. Theory and Practice*. New Jersey: Noyes Press Park Ridge.

Braudel, F. 1992. *Civilization and Capitalism, 15th–18th Century*. Vol. 3: *The Perspective of the World*. Oakland.

Bussanich, J. 2005. The roots of Platonism and Vedānta: Comments on Thomas McEvilley. *Hindu Studies* 9: 1–20.

Casson, L. (ed.). 1989. *The Periplus Maris Erythraei*. Princeton: Princeton University Press.

Chavannes, É. 1905. Les pays d'Occident d'après le Wei lio. *T'oung pao* 6: 519–571.

Cohoon, J.W. and Crosby, H.L. (trs.). 1940. *Dio Chrysostom. Discourses 31–36*. Loeb Classical Library 358. Cambridge, MA: Harvard University Press.

Culbertson, L.E. 2009. *Dispute Resolution in the Provincial Courts of the Third Dynasty of Ur.* A dissertation submitted in partial fulfillment of the requirements for the degree of Doctor of Philosophy (Near Eastern Studies) in the University of Michigan.

De La Vaissière, É. 2005. *Sogdian Traders: A History*. Translated by J. Ward. Boston: Brill.

Deussen, P. 1930. *The Philosophy of the Vedanta in Its Relations to the Occidental Metaphysics*. Adyar, Chennai: Theosophical Publishing House.

Eemeren, van F.H. and Grootendorst, R. 1984. *Speech Acts in Argumentative Discussions: A Theoretical Model for the Analysis of Discussions Directed towards Solving Conflicts of Opinion*. Dordrecht, Netherlands: Foris Publications.

Eemeren, van F.H. and Grootendorst, R. 1988. Rules for argumentation in dialogues. *Argumentation* 2(4): 499–510.

Eemeren, van F.H. and Grootendorst, R. 2004. *A Systematic Theory of Argumentation. The Pragma-dialectical Approach*. Cambridge: Cambridge University Press.

Eemeren, van F.H., Houtlosser, P. and Henkemans, A.F.S. 2007. *Argumentative Indicators in Discourse: A Pragma-dialectical Study*. Vol. 12. Springer Science & Business Media.

Evelyn-White, H.G. (tr.). 1995. *Hesiod. Works and Days*. Harvard University Press.

Feteris, E.T. 2017. *Fundamentals of Legal Argumentation. A Survey of Theories on the Justification of Judicial Decisions*. 2nd edn. Springer.

Foucault, M. 1969. *L'archéologie du savoir.* Paris: Éditions Gallimard.

Fowler, H.N. (tr.). 1925. *Plato in Twelve Volumes*, Vol. 9. Cambridge: Harvard University Press.

Frank, A.G. and Gills, B. 1992. The five-thousand-year world system: An interdisciplinary introduction. *Humboldt Journal of Social Relations* 18: 1–79.

Frank, A.G. and Gills, B. 1993. *The World System: Five Hundred Years or Five Thousand*? London.

Ganguli, K.M. (tr.). 1883–1896. *Mahabharata of Krishna Dwaipayana Vyasa*. Books 1–18.

Gorea, M. 2012. Palmyra and Socotra. pp. 447–485. *In*: Strauch, I. (ed.). *Foreign Sailors on Socotra. The Inscriptions and Drawings from the Cave of Hoq*. Bremen: Hempen Verlag.

Habermas, J. 1962. *Strukturwandel der Öffentlichkeit.* Neuwied.

Habermas, J. 1988. *Theorie des kommunikativen Handelns.* Bde 1–2. Baden-Baden.

Hopkins, E. 1905. The fountain of youth. *Journal of the American Oriental Society* 26: 1–67.

Inagaki, H. (tr.). 2003. *The Three Pure Land Sutras*. Numata Center for Buddhist Translation and Research.

Kahane, H. 1971. *Logic and Contemporary Rhetoric. The Use Of Reason in Everyday Life*. Belmont, SA: Wadsworth.

Kern, L.H., Mirels, H.L. and Hinshaw, V.G. 1983. Scientists' understanding of propositional logic: An experimental investigation. *Social Studies of Science* 13(1): 131–146.

Konow, S. 1929. *Kharoshthi Inscriptions with the Exception of Those of Asoka. Corpus Inscriptionum Indicarum*, Vol. 21. Calcutta: Government of India.

Lam, R. 2013. Kuṣāṇa Emperors and Indian Buddhism: Political, economic and cultural factors responsible for the spread of Buddhism through Eurasia. *South Asia: Journal of South Asian Studies* 36(3): 434–448.

Laroche, E. 1968. *Textes mythologiques hittites en transcription, 2e partie: Mythologie d'origine etrangere. Revue Hittite et Asianique*. Tome XXVI.

Lévi-Strauss, C. 1955. The structural study of myth. *The Journal of American Folklore* 68(270): 428–444.

Livshits, V.A. 2015. *Sogdian Epigraphy of Central Asia and Semirech'e*. Translated from the Russian by T. Stableford. Edited by N. Sims-Williams. London: School of Oriental and African Studies.

Mair, A.W. and Mair, G.R. (trs.). 1921. *Callimachus: Hymns and Epigrams. Lycophron, Aratus*. London: William Heinemann.

Matsuda, K. 1999. *New Sanskrit Fragments of the Saddharmapuṇḍarīkasūtra in the Schøyen Collection*. Norway.

McCrindle, J.W. (tr.). 1964. *The Christian Topography of Cosmas, an Egyptian Monk* (Works issued by the Hakluyt Society 1, 98), 1897, repr. New York.

Moran, W.L. (ed.). 1992. *The Amarna Letters*. Baltimore: The Johns Hopkins University Press.

Neelis, J.E. 2011. *Early Buddhist Transmission and Trade Networks: Mobility and Exchange within and beyond the Northwestern Borderlands of South Asia*. Leiden: Brill.

Oswald, S., Greco, S., Pollaroli, C., Miecznikowski, J. and Rocci, A. (eds.). 2020. *Argumentation and Meaning: Semantic and Pragmatic Reflexions*. John Benjamins.

Pargiter, F.E. (tr.). 1904. *The Mārkaṇḍeya-Purāṇam*. Sanskrit text, English translation with notes and index of verses. Delhi: Parimal Publications.

Perelman, Ch. and Olbrechts-Tyteca, L. 1958. *Traité de l'argumentation : La nouvelle rhétorique*. Paris: Presses Universitaires de France.

Pingree, D. 1981. Jyotiḥśāstra. *In*: Jan Gonda (ed.). *A History of Indian Literature*, Vol. VI Fasc. 4. Wiesbaden.

Porten, B. 1996. *The Elephantine Papyri in English: Three Millennia of Cross Cultural Continuity and Change*. Leiden, New York, Koln: Brill.

Propp, V.I͡A. 1968. *Morphology of the Folktale*. 2nd edn. University of Texas Press.

Rahmouni, A. 2008. *Divine Epithets in the Ugaritic Alphabetic Texts*. Translated by J.N. Ford. Leiden-Boston: Brill.

Ram, I. 2005. Plato and Vedanta. *Phronimon: Journal of the SA Society for the Greek Philosophy and the Humanities* 6(1): 59–69.

Rapin, C. 1992. *La Trésorerie du palais hellénistique d'Aï Khanoum. L'apogée et la chute du royaume grec de Bactriane*. Paris: De Boccard.

Rocci, A. 2006. Pragmatic inference and argumentation in intercultural communication. *Intercultural Pragmatics* 3(4): 409–442.

Salomon, R. 2011. The Macedonian Month Xandikos in Gandhāran Inscriptions. *Studia Orientalia* 110: 165–170.

Salomon, R. 2003. The senior manuscripts: another collection of gandhāran buddhist scrolls. *Journal of the American Oriental Society* 123(1): 73–92.

Santhanam, R. (tr.). 1984. *Brihat Parasara Hora Sastra*. Vol. 1. Ranjan Publications.

Schumann, A. 2019a. On the origin of Indian Logic from the viewpoint of the Pāli Canon. *Logica Universalis* 13: 347–393.

Schumann, A. 2019b. Did the Neo-Babylonians construct a symbolic logic for legal proceedings? *Journal of Applied Logics – IfCoLog Journal of Logics and Their Applications* 6(1): 31–82.

Schumann, A. 2020. Legal argumentation in Mesopotamia since Ur III. *Journal of Argumentation in Context* 9(2): 126–154.

Schumann, A. 2021a. Towards the definition of logical competence. *Cogency* 13(2): 7–40.

Schumann, A. 2021b. Nāgārjunian-Yogācārian modal logic versus aristotelian modal logic. *Journal of Indian Philosophy* 49(3): 467–498.

Schumann, A. 2021c. Plato and pure land buddhism. *Platonic Investigations* 15(2): 110–141.

Shastri, J.L. (ed.). 1970. *The Śiva-Purāṇa*. Vol. I. Delhi-Varanasi-Patna: Motilal Banarsidass.

Sims-Williams, N. 2000. *Bactrian Documents from Northern Afghanistan I: Legal and Economic Documents*. Oxford: Oxford University Press.

Sims-Williams, N. 2007. *Bactrian Documents from Northern Afghanistan II: Letters and Buddhist Texts*. London: Nour Foundation.

Sims-Williams, N. and de Blois, F. 1996. The bactrian calendar. *Bulletin of the Asia Institute* 10: 149–165.

Sinha, K. 2014. The Devas, the Asuras and Gilgamesh: Exploring the cross-cultural journey of the myths. *Vedic Studies* 6: 360–401.

Smith, M.S. (tr.). 1994. *The Ugaritic Baal Cycle*. Vol. I. Introduction with Text, Translation and Commentary of KTU 1.1–1.2.

Smith, M.S. and Pitard, W.T. (tr.). 2009. *The Ugaritic Baal Cycle*. Vol. II. Introduction with Text, Translation and Commentary of KTU/CAT 1.3–1.4. Leiden: Brill.

Spar, I. and von Dassow, E. 2000. *Cuneiform Texts in the Metropolitan Museum of Art*. Vol. III: *Private Archive Texts from the First Millennium B.C.* New York: The Metropolitan Museum of Art.

Stark, R. and Finke, R. 2002. Beyond Church and Sect: Dynamics and stability in religious economics. pp. 31–62. *In*: Jelen, T.G. (ed.). *Sacred Markets, Sacred Canopies: Essays on Religious Markets and Religious Pluralism*. Lanham, Md.: Rowman & Littlefield.

Strauch, I. (ed.). 2012. *Foreign Sailors on Socotra. The Inscriptions and Drawings from the Cave of Hoq*. Bremen.

Subrahmanya Sastri, V. (tr.). 1907. *Sārāvalī. Bombay*.

Tagare, G.V. (tr.). 1958. *Brahmānda Purāna*. Part I. Delhi, Varanasi, Patna: Motilal Banarsidass.

Takakusu, J. 1988. *Taishō shinshū Daizōkyō* 大正新脩大藏經. Vol. 12. Tokyo.

Thakura, Bh.S. (tr.). 1985. *Śrī Brahma Saṁhitā*. Bhaktivedanta Book Trust.

Thuillard, M., d'Huy, J., Berezkin, Y. and Le Quellec, J.-L. 2018. A large-scale study of world myths. *Trames Journal of the Humanities and Social Sciences* 22(4): 407–424.

Vidyabhusana, S.Ch. 1921. *A History of Indian Logic* (*Ancient, Mediaeval and Modern Schools.*). Delhi, Patna, Varanasi: Motilal Babarsidass.

Visser, M.W.de. 1935. *Ancient Buddhism in Japan*. E.J. Brill.

Wallerstein, I. 1976. A world-system perspective on the Social Sciences. *The British Journal of Sociology* 27: 343–352.

Walton, D. 1995. *Argumentation Schemes for Presumptive Reasoning. Series: Studies in Argumentation Theory*. Routledge.

Wilson, H.H. (tr.). 1840. *Viṣṇu Purāṇa, A System of Hindu Mythology and Tradition*. London.

Witek, M. and Witczak-Plisiecka, I. 2019. *Normativity and Variety of Speech Actions*. Brill.

Wittgenstein, L. 1958. *Philosophical Investigations,* trans. by G.E.M. Anscombe. New York: Macmillan.

Wyatt, N. 2002. *Religious Texts from Ugarit*. 2nd edn. Sheffield Academic Press.

Chapter 3

Conditions and Reasons for the Emergence of Logical Competence in the Sumerian-Akkadian Culture

The chapter defines the main parameters of the earliest manifestations of perfect logical competence in documents written in the Sumerian and Akkadian languages, starting with the III dynasty of Ur, i.e., from about 2112 B.C. to 2004 B.C. This competence is found primarily in two types of texts: (1) in legal texts, which include legal codes, judicial acts, powers of attorney, wills as well as business correspondence; (2) texts created for divination and commentaries on them.

3.1 Formation of Strict Norms of Logical Argumentation in the Sumerian and Akkadian Legal Culture

3.1.1 Problem Setting

In this section, it is shown that we can find some foundations of logic and legal argumentation in the tablets of Mesopotamia at least since the dynasty of Ur III. In these texts, we see the oldest correct application of logical inference rules (e.g., *modus ponens*). As concerns the legal argumentation established in Mesopotamia, we can reconstruct based on the tablets the following rules of dispute resolutions during trials: (1) There are two parties of disputants: (i) a protagonist who formulates a standpoint, and (ii) an antagonist who disagrees with the protagonist's standpoint and formulates an alternative statement. (2) There is a rational judge represented by high-ranking citizens who should follow only logical conclusions from facts and law articles as premises.

Traditionally, the first attempt to formulate argumentation theory is attributed to Aristotle (384–322 B.C.), the founder of the Peripatetic school of philosophy. He formulated the following three sciences about argumentation: (i) logic (represented in his books: *Categories*, *On Interpretation*, *Prior Analytics*, *Posterior Analytics*, where Aristotle analyzed "analytical arguments in which the truth of the premises is evident"); (ii) dialectic (represented in his *Topics* and *On Sophistical Refutations*, where "the art of regulated debate" was introduced); (iii) rhetoric (represented in his book *On Rhetoric*, where "the art of persuading an audience" is proposed) (van Eemeren and Grootendorst, 2004: 42).

It is assumed that since Aristotle we can deal with some real argumentative samples of various Greek and Latin authors. According to Aristotle, arguments are called *loci* (Greek: τόπος, Latin: *locus*, 'place'). Hence, the following example of "argumentation from the genus" (*locus a genere*) provided by Cicero in his *Topica* can be explored from the standpoint of today's argumentation theory: "Since all the silver was bequeathed to the wife, the coin which was left in the house must also have been bequeathed. For the species is never separated from its genus, as long as it keeps its proper name; coin keeps the name of silver; therefore, it seems to have been included in the legacy" (Cicero, *Topica* 3: 13; Hubbell, 2006). One of the arguments (*locus*) applied by Cicero is formulated as follows: "The species is never separated from its genus, as long as it keeps its proper name." The argumentative steps reconstructed in (van Eemeren et al., 2014: 90):

1. The standpoint: "The coin (*nummus*) which was left in the house must have been bequeathed to the wife."
2. The first premise based on the statement from the legal document (testament): "All the silver (*argentum*) was bequeathed to the wife."
3. The second premise: "The coin (*nummus*) belongs to all the silver (*argentum*)."
4. The arguments supporting (3): "Coin (*nummus*) keeps the name of silver (*argentum*)" and "The species is never dissociated from the genus as long as it keeps its proper name."
5. From (2) and (3) we infer (1) by the *Barbara* syllogism.

Hence, there are a lot of justifications for the claim that since Aristotle we have a kind of argumentation theory indeed.

In this section, I examine the Mesopotamian tradition of legal argumentation that is much older that the Rome tradition represented in the *Topica* of Cicero—it is at least about 1600 years older than Aristotle and Cicero's works. The point is that in Mesopotamia there were strong rules for dispute resolutions between a protagonist and antagonist by involving logical inference rules (first, *modus ponens*) and some rhetorical tropes (from the Greek τρόπος to denote different rhetorical tools of figurative speeches to convince an interlocutor emotionally—for the Babylonians tropes are represented, first, by rhetorical questions and *argumenta ad hominem*). In fact, it is unknown whether the Babylonians realized that they applied *loci* (arguments) in their resolutions. The problem is that there was not excavated yet a logical handbook in Sumerian or Akkadian such as the *Topica* of Cicero. Nevertheless, the rules for argumentation which are reconstructed based on the tablets are common for all legal documents excavated at least since the dynasty of Ur III. Therefore, we can safely affirm that there was a practice of legal argumentation since Ur III.

In the legal argumentation of Mesopotamia that can be reconstructed by us, there were used speech acts of the following forms:

1. Assertives by which the speaker or writer swears the truth of a proposition. These speech acts were performed as evidentiary oaths by disputants and their witnesses to put forward something as facts.
2. Directives by which the speaker or writer orders to do something. These speech acts occur in business correspondences between superiors and subordinates as well as in court protocols (trial records) between judges and defendants/claimants.
3. Commissives by which the speaker or writer promises to do something. They were carried out as promissory oaths. This was a necessary final step in many dispute resolutions. So, they can be called legal commissives.
4. Expressives by which the speaker or writer expresses a feel about something. They occur in business correspondences as rhetorical questions, but they never occur in court protocols.
5. Declaratives by which the speaker or writer calls a particular situation into being. These speech acts are legal decisions of courts.

In the way of van Eemeren et al. (2014: 533), we can formulate the main stages of legal argumentation in Mesopotamia (see Table 3.1.1).

For a testimony about something, disputants and their witnesses must swear an appropriate oral statement or demonstrate a written document. The possible oral statements are as follows (Culbertson, 2009): (i) the promissory oath (swearing to fulfill a promise) denoted by the Sumerian term *mu lugal* (*pàd*); (ii) the evidentiary oath (swearing to the veracity of their own statements) denoted by the Sumerian term *nam-erim*$_2$(*am*$_3$) (Lagaš) or *nam-erim*$_2$ *tar* (Umma). So, there was a requirement in the courts, first, to be held responsible for all the commitments during the court that they will be fulfilled indeed and, second, to be held responsible for all testimonies before the judges that they are true in fact—that they really take place. From the promissory and evidentiary oaths and by applying some logical inference rules, the judges inferred their verdicts.

Hence, the main claim of the section is that a logically perfect practice of legal argumentation with high and strong standards was first established in Mesopotamia and it is well researchable at least since the dynasty of Ur III. In this argumentation, we observe the oldest correct application of logical inference rules (e.g., *modus ponens*). The organization of the section is as follows. In Subsection 3.1.2, the methodology is discussed. In Subsection 3.1.3, the argumentative structure of business contracts, business correspondences, and trial records is examined to show their logical and rhetorical nature. Also, it is shown that after the judicial reform in Ur III the legal argumentation was unified for the whole state of Ur and the role of logic in argumentation is increased significantly. In Subsection 3.1.4, there are considered some typical examples of logical and rhetorical arguments used in business correspondence since the Old-Babylonian period. In Subsection 3.1.5, there are proposed logical reconstructions of typical legal proceedings since Ur III.

Table 3.1.1. Argumentative stages in the legal proceeding since Ur III.

I	Confrontation Stage
Assertive	Expressing a standpoint by a protagonist and antagonist in the form of evidentiary oaths.
Commissive	(i) Acceptance of a standpoint in the form of promissory oaths (legal commissive) or (ii) non-acceptance of a standpoint and then, starting a dispute according to the standards of legal argumentation.
II	**Opening Stage**
Directive	Challenging to defend a standpoint by facts of the following two kinds: (i) testimonies from at least two witnesses; (ii) legal documents signed by judges and witnesses.
Commissive	Acceptance of the challenge to defend a standpoint by providing testimonies and documents. Agreement on premises based on facts and agreement on discussion rules of legal argumentation. Decision to start a discussion.
III	**Argumentation Stage**
Directive	Requesting argumentation. During courts, there are possible only logical inference rules (such as *modus ponens*) from legal documents and oral testimonies. During business correspondence, in addition there are also possible rhetorical tropes such as rhetorical questions and *argumenta ad hominem*.
Assertive	Advancing argumentation. Judges and other officials and court witnesses serve as a rational judge in the dispute resolution. They can accept only logical ways of argumentation according to the standards of court proceeding.
Commissive	Acceptance or non-acceptance of argumentation by the rational judge. Fixing in the court records only the accepted argumentation.
IV	**Concluding Stage**
Commissive	Acceptance or non-acceptance of a standpoint by the rational judge. Uttering promissory oaths by the disputants that they will follow the decision of the rational judge.
Declarative	Establishing the result of the discussion by the rational judge as a new legal fact.

The fact that in Mesopotamia there existed some high standards of dispute resolutions is well known for the Assyriology community, see (Culbertson, 2009, 2015), but it is not yet for the Logic and Argumentation Theory community that in business correspondence and legal proceedings we observe a perfect logical competence of their authors and some sophisticated applications of rhetorical tropes and logical inference rules.

3.1.2 Archaeology of Logic for Mesopotamian Texts

In the history of philosophy, we study subjects within a living philosophical tradition by sharing a certain set of (philosophical, cultural, societal, etc.) values. For example, in the history of logic and argumentation theory there is such a set of values which grounds any historical research in that area. This set of values includes Eurocentrism—so, it is assumed that logic appeared in Greece since Aristotle, and its greatest achievements happened within the limits of the European culture. Accordingly, only some samples of the European thinking are understood by us as the doctrine of traditional logic.

However, there existed and till date exist philosophical traditions of logic and argumentation out of the European culture. It is possible to define the Indo-Buddhist logic/argumentation and Judaic logical hermeneutics as some alternative traditions. Hence, we can refuse the standard approach when a philosophical doctrine is substantiated through the fact of its living tradition (e.g., the history of logic is constituted through the fact of the living European logic since Aristotle).

As a new way of substantiation of the history of logic and philosophy, I have proposed a methodology of analyzing textual fragments composed for various purposes and in various cultures whether a kind of logical competence of their authors can be detected. Let us remember that this methodology is called "archaeology of logic". According to this methodology, we are not interested in the living philosophical traditions of logic, but we aim to reconstruct some latent traditions of logical competences—when and where they first appeared, in what forms they were established and inherited later.

Logic is understood as a complex of algebraic rules allowing us to obtain some statements from other statements, purely mechanically. It was learnt that the most ancient fragments whose authors possessed a perfect logical competence are represented by the texts in Sumerian and Akkadian (from the 22nd to the 5th century B.C.) (Schumann, 2017b, 2019a, 2020, 2021a, 2021b; Schumann and Sazonov, 2021). In Mesopotamia, there were the following three kinds of texts consisting of implications "If *A*, then *B*" in the logical meaning: (1) hermeneutical texts ("If in the text there is a sign/word *A*, then we should understand it as *B*"); (2) predictions or divinations ("If in the nature there is a sign/omen *A*, then we foresee an event *B*"); and (3) legal codes and trial judgments ("If someone has committed an act *A*, then he or she should be punished in the way of *B*"). In all these three types of texts, a difficult algebraic system of drawing logical conclusions purely mechanically could be easily reconstructed.

On the other hand, if to analyze logical fragments of the Pāli Canon (the earliest Buddhist texts composed from the 3rd century B.C. to the 3rd century A.D.), we find out a lack of well-developed logical competence, then some correct conclusions are regarded in parallel with sophisms and fallacies. Therefore, we see that the authors did not distinguish them (Schumann, 2018, 2019c). The only text of the Pāli Canon with a perfect logical competence is presented by the *Milinda-pañha* ("Questions of King Milinda"), created from the 1st century B.C. to the 1st century A.D. in Gandhāra at the time of Hellenistic domination in the North of India (Schumann, 2019c).

So, the "archaeology of logic" allows us to reveal centers of logical cultures and waves of disseminations of logical competence: (1) the most ancient logical culture is found out in the Sumerian and Akkadian texts (Schumann, 2017b, 2019a, 2020, 2021a, 2021b; Schumann and Sazonov, 2021) (from the 22nd to the 5th century B.C.) grounding the oldest sets of logemes; (2) this culture, even purely terminologically,[1]

[1] For example, the logical term 'particular' (*pərāṭ*) of the Talmud came from the Akkadian word *parāsu* with the same meaning and the logical term 'general' (*kəlāl*) came from the Akkadian words *kalû* and *kullatu* with the same meaning. Meanwhile, both terms ('particular' and 'general') played a significant role in logical reasoning of the Talmud.

is completely inherited in logical fragments of the Talmud (from the 4th century B.C. to the 5th century A.D.) continuing and developing the Babylonian logemes; (3) this culture also is inherited in Stoic logical fragments, e.g., in the *Topica* of Cicero (from the 3rd century B.C. to the 2nd century A.D.) continuing the Babylonian logemes in another way; (4) the Stoic fragments well correspond to the propositional-logical rules of Madhyamaka and Yogācāra schools of Indian philosophy (from the 2nd to the 8th century A.D.)—the Stoic logemes were similar to the Madhyamaka and Yogācāra logemes, therefore it is possible to admit either their common source of origin or an influence of logical competences of Stoicism on logical competences of Indian rational-logical philosophy since Nāgārjuna (i.e., since the 2nd century A.D.). Thus, a certain linear development of propositional logic with the common Sumerian-Akkadian root is traced back. But this tradition is reconstructed exclusively through analyzing the logical competences of authors without accepting or rejecting a philosophical doctrine and an appropriate set of values.

To sum up, we face the earliest samples of perfect logical competence among the Sumerian and Akkadian tablets. How can we reconstruct these logical fragments? According to Ludwig Wittgenstein, "meaning is use". So, to reconstruct the logical texts, we should understand appropriate speech situations and recreate three dimensions of speech: locutive (propositional content), illocutive (expressing an intended action, such as ordering, warning, or promising through emotions in uttering a propositional content), and perlocutive (reactions to a propositional content and its emotional context).

Hermeneutical texts (consisting of lists of "If in the text there is a sign/word *A*, then we should understand it as *B*") were short notes of lessons during which a teacher was explaining to pupils how to interpret signs and sometimes how to draw logical conclusions from several signs. In Akkadian, this oral lore was denoted *ša pî* ("of the mouth"), *ša pī ummâni* ("according to a scholar"), and *šūt pî* ("those of the mouth"). These expressions are very similar (even linguistically) to the Hebrew term [*tôrâh*] *še-be'al peh* ("[Scripture] that is upon the mouth"). According to the hermeneutical fragments found at different sites, we can guess that the teaching in the Akkadian culture was like the Talmudic way of teaching through sets of dialogues. So, this context for the hermeneutical tablets allows us to reconstruct locutive, illocutive, and perlocutive dimensions of suitable speeches.

Divinations (consisting of lists of "If in the nature there is a sign/omen *A*, then we foresee an event *B*") were devoted to a religious doctrine in Mesopotamia. This doctrine was based on obeying lists of signs/omens (Akkadian: *ittātu*; Sumerian: GISKIM.MEŠ) given by gods for us. The Akkadian term *alaktu* meant "to follow omens of gods". Later this term was transformed into the Hebrew *hălāḵāh* in the meaning "to follow commandments of the Lord". As we see, both terms *alaktu* and *hălāḵāh* are not only the words of the same root 'walking', but also words with similar technical meaning. This religious context should be considered in reconstructing locutive, illocutive, and perlocutive dimensions of predictions in Akkadian.

Legal codes and trial judgments (consisting of lists of "If someone has committed an act *A*, then he or she should be punished in the way of *B*") were used in legal argumentation and this section is concentrated on these documents. So, argumentation and dispute resolution are the main speech context for these tablets.

Now the question is which model of argumentation could be applied in analyzing legal fragments from Mesopotamia.

The substantial difference between the documents in Sumerian/Akkadian and the texts of subsequent cultures is that the emotionality and modality (i.e., a system of linguistic options to express illocutionary points) in speeches of Sumerian and Akkadian authors was completely poor and weak, as compared with the rhetorical texts in Greek, starting with Hellenism (i.e., since the 4th century B.C.), and especially as compared with the rhetorical texts in Latin. The point is that thinking in Babylonia was much more concrete and focused rather on reality than on feelings and opinions. Consequently, even the ancient rhetorical theory since Aristotle is inappropriate for the legal argumentation in Sumerian and Akkadian. If we take modern theories such as the new rhetoric developed by Chaïm Perelman and Lucie Olbrechts-Tyteca, the situation will be worse—these theories suppose much more forms of emotional expressions in our argumentation than it was in the classical rhetoric and most of these forms have no analogues in the documents from Mesopotamia at all.

As an example, let us consider argumentation schemes proposed by Douglas Walton in his book (Walton, 1995). We observe only a few analogues for them among Mesopotamian documents, for instance, we can detect the following schemes:

a) Ethotic argument, expressed as *argumentum ad hominem*: (1) if the antagonist trusts his own son/brother (but it is), then he can trust the protagonist, also; (2) if the protagonist as the 'son'/'brother' of the antagonist contends the standpoint A (but it is), this standpoint should be accepted by the antagonist as more plausible.
b) Argument from commitment, formulated as a promissory oath: (1) if the protagonist/antagonist takes the oath A (but it is), then he or she is committed to proposition A; (2) if he or she is committed to proposition A (but it is), then he or she must support A.
c) Argument from position to know, pronounced as an evidentiary oath: if the protagonist/antagonist asserts the proposition A (but it is), then he or she knows that A is true.
d) Argument from established rule: if doing A is the established rule according to business commitments or other legal obligations and responsibilities of the protagonist/antagonist (but it is), then he or she must do A.

The low level of emotionality and modality in Babylonian texts, including private correspondence, is explained by the high level of legal liability for all utterances expressing our subjectivity. Each disputant must have been responsible for all statements uttered by him or her during conversations even in his or her private correspondence. Each promise could have been evaluated as a promissory oath for which there exists a legal responsibility, and each assertion could have been evaluated as an evidentiary oath for which there exists a legal responsibility, too. So, the disputants preferred to discuss only facts and appealed in their reasoning rather to inference rules of propositional logic (first, to *modus ponens*) than to rhetorical tropes (emotional arguments). Therefore, the classical rhetoric since Aristotle was more emotional than the legal argumentation in Mesopotamia to the same extent as the new rhetoric of Perelman and Olbrechts-Tyteca became more emotional than the classical one.

Hence, almost all modern theories of argumentation cannot be applied in reconstructing the practice of legal argumentation in Babylonia, because of the essential cultural difference between us and the Babylonian authors. In my opinion, the pragma-dialectical approach (van Eemeren et al., 2014; van Eemeren and Grootendorst, 2004) is the best suitable theory that can be well used indeed. The matter is that, according to this approach, we can focus on social roles of interlocutors and their ways of expressing illocutionary points. Then we can define locutive, illocutive, and perlocutive levels of Mesopotamian speeches (see Table 3.1.1) and reconstruct some ways of their argumentation by distinguishing logical inference rules from rhetorical tropes. So, the main advantage of pragma-dialectical approach can be stressed as follows: (a) it is a process-oriented theory that can be used in explicating a purpose of logic in debates and a social role of logical inference rules, and we aim to explicate the purpose and social role of logic in the Mesopotamian legal discourse; (b) this theory has a social orientation allowing a reconstruction of social functions in speeches, e.g., legal settings being studied in the section; (c) in this theory it is emphasized that argumentation should be deductively valid, but the Babylonians used only deductive logical schemata.

3.1.3 Court Proceedings in Dispute Resolutions since Ur III

Legal documents expressing a kind of legal argumentation in Mesopotamia include the following two sets of texts: (i) tablets including a possible way of dispute resolution between a protagonist and antagonist without involving the community as a rational judge: business correspondences, records of contracts, sales or other types of transactions; (ii) court records, i.e., records of legal proceeding, assuming a conflict resolution between a protagonist and antagonist through involving the community as a rational judge in the resolution process. Some fragments of legal texts of both types are being excavated for different periods of Mesopotamia starting with the early dynastic period (2900–2350 B.C.). We know that both types have developed in a complementary fashion. On the one hand, business correspondences and records of transactions were composed considering an opportunity to file a lawsuit if the disputants regard something in their resolution as a violation of the law. On the other hand, each court in trial decisions was based on the documents of the first type: contracts and other records of transaction.

Legal proceedings assumed that the community is involved in the dispute resolution. Usually, the community was represented in this resolution by the governor and some additional judges. This way of settlement of the dispute was denoted by the expression in Sumerian *inim* (*al-til*) in the Old Akkadian period (2334–2154 B.C. according to the middle chronology or 2270–2083 B.C. according to the short chronology).[2] In Ur III, the Sumerian term *di* (Akkadian *dīnum*) began

[2] The short chronology dates the reign of Hammurabi to 1728–1686 B.C. according to a synchronization of chronologies of Aššur and Babylon. The middle chronology dates the reign of Hammurabi to 1792–1750 B.C. according to the 56/64 year Venus cycle based on tablet 63 of the *Enuma Anu Enlil* (U_4 AN.na ^{d}EN.LÍL.lá).

to denote a lawsuit, legal case, judgment, verdict, legal decision. In the meanwhile, the Sumerian term *di ti-la* started to denote a "case closed" or "verdict passed", i.e., it meant an outcome of a legal proceeding (*di*). The most important feature of *di ti-la* is that the judgment of the council participating in the court cannot be turned back: *di-til-la inim pu-uḫ-ru-*[*um-ma-ka šu* gi_4-gi_4 *nu-g̃al*$_2$] (Michalowski and Walker, 1989: 59). During Ur III, the term *di* referred to trial disputes rather than to final verdicts, while since the Old Babylonian period an appropriate Akkadian term *dīnum* referred to verdicts rather than to disputes (Culbertson, 2009).

The dynasty of Ur III was established by Ur-Nammu (Sumerian: *ur-*^{d}*nammu*), who ruled from the city of Ur from 2112 to 2095 B.C. (according to the middle chronology) or from 2047 to 2030 B.C. (according to the short chronology) and extended his control over many cities-states in southern Mesopotamia. His successors were: Shulgi (Akkadian: ^{d}*Šulgi*, ruled 2094–2047 B.C. or 2029–1982 B.C.); Amar-Sin (Akkadian: ^{d}*Amar-*^{d}*Sîn*, ruled 2046–2038 B.C. or 1981–1973 B.C.); Shu-Sin (Akkadian: ^{d}*Šu-*^{d}*Sîn*, ruled 2037–2029 B.C. or 1972–1964 B.C.); Ibbi-Sin (Sumerian: ^{d}*i-bi*$_2$-^{d}*suen*, ruled 2028–2004 B.C. or 1963–1940 B.C.)—notice that the dates of their rules are mentioned according to the middle chronology and the short chronology, respectively. In the state of Ur III, a unified authoritative (or even totalitarian) policy in respect to all aspects of both public and private life was established. Based on the fragments dated to this period, we see that all the local administrations of different cities started to be unified. Ur-Nammu even published the law code that became the first among all the codes in the world which are preserved till now, at least in some fragments. Some later codes which are known recently: (i) the Code ascribed to Lipit-Ištar, the 5th king of the dynasty of Isin who ruled 1934–1924 B.C. by the middle chronology or 1870–1860 B.C. by the short chronology; (ii) the Code written in the 19th century B.C. in Ešnunna, the city located north of Ur on the Tigris River; (iii) the Code belonging to Hammurabi (Akkadian: *Ḫammu-rāpi*), the 6th king of the first Babylonian dynasty who ruled from 1792 to 1750 B.C. by the middle chronology or from 1728 to 1686 B.C. by the short chronology.

So, in Ur III was created the first law code since the governance of Ur-Nammu and, consequently, all independent judicial systems belonging to the different cities, which were controlled by the new dynasty, began to be reformed and standardized. The goal of this social reform was to make a common transparent judicial system for all cities to ground a centralized power for all communities in the state. This judicial reform first affected the traditional procedures of legal argumentation in Mesopotamia to make them common.

It is worth noting that all legal proceedings in different cities of Mesopotamia had a similar structure of main actors participating in courts: a protagonist (a claimant applying to a court), an antagonist (a defendant in this court), and a rational judge (a council authorized to issue a court decision after evaluating whether the controversial standpoint stated by the claimant was successfully justified). The main difference lays in authorizing rational judges: each city had its own way of authorization on behalf of its rulers of cities. Now, in Ur III, the mechanism of authorization to be a rational judge in courts was harmonized: all the court councils could be convened only on behalf of the king of Ur III. Therefore, all evidentiary oaths and legal commissives (promissory oaths) performed before witnesses or a court could be taken only in the name of the king.

In the paper by Culbertson (2009), are analyzed the court records of two archives from governors' palaces in the cities of Umma (Sumerian: *umma*[ki]) and Girsu (Sumerian: *ĝir*$_2$*-su*[ki]), which were capitals of city-states conquered by Ur-Nammu and Shulgi. To incorporate both cities into the new judicial system, a significant change in the pattern of dispute resolution and court composition was installed. First, in the earliest court reports of both archives we see the following participants in each court proceeding: (i) the disputants (both protagonist and antagonist) and their witnesses; (ii) a governor; (iii) from zero to four representatives of the city elite who are members of the court council (i.e., who are judges); (iv) a scribe. Hence, at this stage a court decision could be issued only by one governor and, therefore, a rational judge for dispute resolutions was represented, first, by a governor and perhaps some additional judges from the urban elite (but in sum there were not more than the four additional judges).

Then at the time of Shu-Sin, the fourth king of Ur III, we find the following new distribution of roles in legal proceedings: (i) the disputants (both protagonist and antagonist) and their witnesses; (ii) a governor; (iii) between one and six representatives of the city elite who are additional judges; (iv) between one and two case reporters ("case witnesses"); (v) from zero to nine high-ranking men, perhaps even non-urban actors such as village headmen and estate owners who sign the final court record as witnesses like jurors ("men who stood at the case"); (vi) a scribe. Thus, many new actors were introduced into the court to make a legal decision more transparent and logical for all citizens. Now, no court decision could be taken only by the governor—there should also be an acceptance of at least one additional judge and one additional case reporter.

Since Ibbi-Sin, the fifth king of Ur III, the court process consisted of the following participants: (i) the disputants (both protagonist and antagonist) and their witnesses; (ii) a governor or his representatives; (iii) groups of judges; (iv) groups of "men who stood at the case"; (v) a military leader (general; Sumerian: *šagina*, Akkadian: *šakkanakku*), the non-urban actor also who was a member of the royal family or his local official (Akkadian: *ḫazannu*); (vi) a scribe. As we see, at this stage representatives of the royal family are newly appointed as members of court councils.

Hence, we observe the three stages in the development of the Sumerian trial system of Ur III: (i) a rational judge in court proceedings is represented by the governor and perhaps additional judges; (ii) he is represented not only by the governor and judges, but also by additional members—different high-ranking men who participate in analyzing arguments put forward by the disputants in the trial; (iii) he is represented by the governor, judges, high-ranking men, and by a member of the royal family or his local official. At the second stage, the role of city community has increased tremendously. At the third stage, the role of central control has grown.

This unification of legal proceedings and the increase of the community role allowed the court councils to focus on logical aspects of argumentation to develop the logical way of conclusion by applying inference rules—first, *modus ponens*. In the records of legal proceedings in Sumerian and Akkadian, we cannot find rhetorical tropes, such as rhetorical questions or *argumenta ad hominem*. This fact proves that since Ur III the difference of logical and rhetorical means of argumentation was readily distinguished. In trials, only logic was used, but in business correspondences there could be some rhetorical tropes in addition.

We cannot affirm that logical inference rules applied in trials were becoming more sophisticated from the first to the third stage of unification of legal proceedings in Ur III. The point is that the logical structure of business correspondence and legal records was almost the same. And logical inference rules remained almost identical, too—they were quite simple such as *modus ponens*. But the social role of logic indeed increased highly in trials. Before this reform any court decision, even based on logical inference rules, was considered as a subjective will of the governor. Later (at the second stage of the reform), it was examined as a decision of the whole community, and after that (at the third stage of the reform) as a decision of the whole human universe controlled by the king.

The institute of legal proceedings based on logic has disciplined eventual disputants. Consequently, they preferred logical tools in their correspondences and contracting procedures as well. Therefore, business contracts and business correspondence satisfied the main argumentative stages of legal proceedings, see Table 3.1.1. Indeed, the structure of business contracts in Sumerian in the 21st–20th century B.C. was as follows (Mercer, 1913):

1. The introduction of disputants, i.e., both parties involved to award an appropriate contract (writing their names).
2. Formulating a standpoint that will be considered acceptable for both disputants—the object and nature of the transaction.
3. Mentioning any "extra incidents connected with the transactions".
4. Performing a legal commissive by the disputants—expressing appropriate promissory oaths by them that the transaction is now valid for the disputants.
5. Writing the names of witnesses and judges who validate the contract by their declarative.
6. Date.

The business correspondence in Sumerian and Akkadian since the 22nd century B.C. had the following structure:

1. The introduction of disputants (writing the names of protagonist and antagonist).
2. Formulating a standpoint of the protagonist.
3. Quoting a standpoint of the antagonist.
4. Logical arguments (arguments based on inference rules such as *modus ponens*) for accepting the standpoint of the protagonist and for rejecting the standpoint of the antagonist.

5. Rhetorical arguments (arguments based on rhetorical tropes such as *argumenta ad hominem*) for accepting the standpoint of the protagonist and for rejecting the standpoint of the antagonist.
6. Performing a legal commissive by the protagonist—expressing an appropriate promissory oath.
7. Sign.
8. Date.

The structure of legal proceedings in Sumerian in the 21st–20th century B.C. reconstructed in (Lafont, 2000: 35–68):

1. The initiation of the legal process expressed by one of the three words: *inim-gar*, *di-dug*$_4$, *inim-dug*$_4$.
2. Recording both parties: first, the protagonist and antagonist (perhaps, including some additional entities cited as having been involved before the court).
3. The comparison of testimonies given by the protagonist and antagonist.
4. The deposition given by the protagonist.
5. The rebuttals performed by the antagonist.
6. The legal commissives, i.e., promissory oaths, performed by the protagonist or antagonist: *nam-erim*$_2$*-am*$_3$, *mu-lugal*.
7. Perhaps, the issue of return: *gi*$_4$.

In Molina (2000), another reconstruction of legal proceedings during Ur III was proposed:

1. The heading: *di-tila*.
2. Formulating a standpoint of the dispute—the object of litigation. If we have both parties: protagonist (claimant) and antagonist (defendant), then the standpoint is represented as a claim formulated by the protagonist. If we have only one party (antagonist or defendant), then the standpoint is expressed by the judge representing the injured party—first, the second option concerns the cases of legal persons, e.g., temples or king's palaces.
3. Noticing some relevant information to the prehistory of the standpoint (court case).
4. The nature of the standpoint.
5. The argumentation supporting the standpoint.
6. Respective testimonies performed by the witnesses and other proofs to support one of the two disputants. The protagonist aims to confirm the standpoint and the antagonist aims to deny it.
7. Yielding of one of the parties: either the protagonist's standpoint or the antagonist's rebuttals.
8. Taking a final legal decision—the resolution of the court case.

9. A legal commissive (promissory oath) to express an explicit acceptance of the judgment by the party who did not win the case.
10. Writing the name of the official (*maškim*) participating in the court.
11. Writing the names of the governor and judges participating in the process.
12. Writing the names of witnesses whose testimonies were applied.
13. Date.

As an instance, let us consider a court process typical for the legal archive in Sumerian from Lagaš in Ur III:

1. The heading: *di-tila*.
2. Formulating a standpoint of the dispute (denoted by A): PN_1 (denoting a personal name) bought a slave from PN_2 (denoting another personal name) for x shekels of silver. PN_1 is a protagonist here confirming this standpoint. PN_2 and/or the slave is an antagonist refuting the standpoint.

3.–4. The slave and/or PN_2 claims that the sale did not occur. Hence, according to the antagonist, A is false, i.e., we have $\neg A$. From this it follows that the slave was illegitimately enslaved by PN_1. Let us denote this statement by B. Then we see that the antagonist aims to apply *modus ponens*. Indeed:

If PN_1 did not buy the slave from PN_2 for x shekels of silver, then PN_1 does not possess the slave. PN_1 did not buy the slave from PN_2 for x shekels. Then, by *modus ponens*, we conclude that PN_1 does not possess the slave.

$$\frac{\neg A \Rightarrow B, \neg A}{B}$$

5.–7. The argumentation supporting the standpoint by witnessing:

There are two witnesses who swore that PN_1 paid for the slave in full. A

Two witnesses are enough as evidence for supporting the claim for all legal processes since Ur III.

8.–9. Taking the final legal decision: PN_1 bought a slave from PN_2 for x shekels of silver, in fact. From this statement we infer that the slave was legitimately enslaved by PN_1. Indeed:

If PN_1 bought the slave from PN_2 for x shekels of silver, then PN_1 possesses the slave. PN_1 bought the slave from PN_2 for x shekels. Then, by *modus ponens*, we conclude that PN_1 possesses the slave.

$$\frac{A \Rightarrow \neg B, A}{\neg B}$$

10. The name of the *maškim*.

11. The names of judges participating in the court proceeding.

12. The names of witnesses.

13. Date.

According to the opinion of Culbertson (2009), the text titled *di-tila* is distinguished from all other tablets of the Ur III period. So, *di-tila* or *di-bi til-am*$_3$ means "this legal case is closed". They contain legal protocols with an appropriate legal decision taken by judges after a trial discussion and this decision is complete, ultimate, and testified by *maškim* (the official). In the texts titled *a*$_2$*-dar*, we see only declarations of decisions. There are also the tablets where somebody "appeared before the judges" (PN *igi di-ku*$_5$ *bi*$_2$*-in-ḡar*, where PN denotes a personal name). In these texts, there is a first-person declaration by the expression "he/she declared" (*bi*$_2$*-in-du*$_{11}$, *inim-ḡar*, *di-du*$_{11}$, *inim-du*$_{11}$).

The term *di* was used in the texts where a kind of mediation is introduced for both parties. This mediation is performed by an arbitrator who should help in "finding fact" (Roth, 2001: 255). The goal of *di-tila* texts is in finding both law and facts.

To sum up, since Ur III, there are legal documents of two different types: business correspondences and court records, but we can reconstruct a common unified legal argumentation used in texts of both types.

3.1.4 Examples of Logical and Rhetorical Arguments from Business Correspondence

In the business correspondence in Old Akkadian, there were different expressions to denote that a standpoint begins in the letters: *na-e-a*; *en-ma*; *na-bé-a*.

Let us consider some examples from the business correspondence of the Old Akkadian period. So, in the letter from Abbāja to Dudu'a, excavated in Kiš and dated to the era of Sargon the Great, the first ruler of the Akkadian Empire in the 24th– 23rd centuries B.C., there are both rhetorical and logical arguments, see Example 3.1.1. Among the rhetorical arguments (tropes) there are rhetorical questions, *antiphrasis*, and *argumentum ad hominem*. Among the logical arguments there are implications assuming *modus ponens*.

The transliteration and German translation of this tablet are taken from Kienast (1995: 141–142). The German translation was translated then into English.

This fragment is one of the oldest samples of a business letter where the author composed the text obviously in the way that his claim can be used in court. So, the author (protagonist), Abbāja, formulated his standpoint: "Dudu'a should give 6,000 liters of barley to Abbāja for 2.76 grams of silver." It is expected that this assertive becomes later the subject-matter of the agreement between Abbāja and Dudu'a and, therefore, as a business commitment it can be used in court. This standpoint is examined as inferred by *modus ponens* form two conditional statements: "If Dudu'a needs money, Abbāja pays for 6,000 liters of barley" and "If Dudu'a agrees to sign the contract, Abbāja is ready to send the full amount of money." In the letter, Abbāja assumes that Dudu'a needs money, in fact, and he agrees to sign the agreement (transaction). Hence, it is inferred from both conditional statements that the subject-matter of the agreement should be represented by the standpoint of the protagonist. It is a logical part of the text. Its rhetorical part appeals, first, to an ethotic argument expressed as *argumentum ad hominem* that Abbāja is trustworthy like a son of Dudu'a: (1) if Dudu'a trusts his own son (but it is), then he can trust

Example 3.1.1. Business letter from Abbāja to Dudu'a dated to the 24th–23rd centuries B.C.

Transliteration	Translation	Argumentative Reconstruction
Obverse 1. *en-ma* 2. *Ab-ba-a-a* 3. *a-na* Du-du-a 4. *mi-núm* 5. *ù-la a-bí at-tá* 6. *ma a-na* 10 ŠE.GUR 7. *ù-la tá-qí-pá-an-ni* 8. É *e-rí* 9. *śum-ma* KÚ.BABBAR **Reverse** 10. *è-rí-šu-kà* 11. ⌈*a*?⌉-⌈*na*?⌉ 20 GUR 12. ⌈*šal*?⌉-*ša* GÍN ⌈*lu-śa*⌉-*bí-l*[*a*]-*kum* 13. *śum-ma*[a]-[*na*] SAG 14. [SA]G *lu-śa*-[*b*]*í*-<*la*>-*kum* 15. [*a*]*t-tá* ⌈*śu*?-*ṣí*⌉-*a-am* 16. ⌈*ù*⌉DUMU-⌈*kà*⌉ šu-up-[r]a-⌈am⌉-ma 17. [*śú-m*]a(?) *li-iš-me*	1. It is said by 2. Abbāja: 3. To Dudu'a: 4. Why 5. are you not [like] my father? 6.–7. What!? Do you not have confidence in me [to give me] 3,000 liters of barley? 8. Is [your] company bankrupt? 9. If silver 10.–12. is required from you by someone, I will send you [always] 2.76 grams [of silver] [even] for 6,000 liters [of barley]. 13.–14. If to come to the full amount, I will send you the full amount. 15. Give me the message. 16.–17. Furthermore: send it to your 'son' and [then] he shall hear [more details].	1.–3. Abbāja is a protagonist, Dudu'a is an antagonist. 4.–14. The standpoint of the protagonist in the form of assertive: Dudu'a should give 6,000 liters of barley to Abbāja for 2.76 grams of silver. Arguments used by the protagonist: (*a*) The rhetorical question by appealing to *argumentum ad hominem* that the antagonist can trust the protagonist: "Why are you not like my father?" (*b*) The rhetorical question as *antiphrasis* (one of the tropes) that the antagonist's company is not bankrupt and then, it can give at least 3,000 liters of barley: "Do you not have confidence in me to give me 3,000 liters of barley? Is your company bankrupt?" (*c*) The two legal commissives in the form of implications assuming *modus ponens*: (*i*) If the antagonist needs money, the protagonist pays for 6,000 liters of barley. Yes, the antagonist needs money. The protagonist is ready to pay. (*ii*) If the antagonist agrees to sign the contract, the protagonist is ready to send the full amount of money. The antagonist agrees. The protagonist sends. 15.–16. Expectation (directive) that the antagonist will accept the standpoint.

Abbāja, also; (2) if Abbāja as the "son of Dudu'a" contends the standpoint (but it is), this standpoint should be accepted by Dudu'a as more plausible.

The next letter from Iškundagān to Lugalra is excavated in Girsu and dated to the same Old Akkadian period, see Example 3.1.2. We see an argumentation, where first, the protagonist has rejected a logical argument proposed by the antagonist not to accept the standpoint and, second, the protagonist has put forward an own logical argument in the form of implication assuming *modus ponens*. Then the protagonist has formulated three legal commissives (oaths) as implications of the form: "If the event *A* occurs, then the protagonist makes the act *B*."

The transliteration and translation are taken from Kienast (1995: 89–91). The German translation was translated then into English.

In this document, there is no rhetorical part as in Example 3.1.1, but we see only logical arguments. Its author, Iškundagān, as a protagonist formulated the standpoint: "Lugalra should plant the field and herd the cattle." It is supposed that Lugalra cannot do it because of the Gutians: "If the Gutians are there, Lugalra cannot plant the field and herd the cattle, but they are there." Iškundagān rejected this reasoning by two conditional statements: "If the Gutians are there, then Lugalra leaves additional security at each ca. 5 kilometers and plant the field and herd the cattle" and "If the security notices the hostile troops, then the security can attack the Gutians and Lugalra brings the cattle temporarily to the city." At the end, Iškundagān articulated the three legal commissives (promissory oaths) in the form of implications which can be directly used in court: (*a*) "If again the Gutians continue to take away the cattle, then Lugalra provides the own compensation for the damage, but it is expected that the Gutians can continue"; (*b*) "If Iškundagān comes into the city, then he gives no silver to Lugalra for the cattle, but it is expected that Iškundagān can come into the city"; (*c*) "If Lugalra does not herd the cattle, Iškundagān requires the normal 'income' from Lugalra, but it is expected that Lugalra does not herd the cattle." So, if Lugalra does not follow the instructions from (*a*) to (*c*), Iškundagān will file a lawsuit. Then each court will take a decision just by *modus ponens* from the text of the letter. For example, the Gutians continue to take away the cattle indeed, and consequently, according to (*a*) Lugalra provides compensation for the damage.

Hence, it is readily seen that both letters as well as many others of the same era demonstrate the logical meaning of conditional sentences: each occurrence of implication assumes applying *modus ponens* for inferring. It is worth noting that implications are quite frequent in mathematical texts of Mesopotamia, too. Let us exemplify a mathematical context for using implications, see Example 3.1.3. This text is written at the time of the first dynasty of Babylonia (20th–16th century B.C.)—the transliteration and translation are taken from Friberg (2007: 61–62).

Thus, we face there a real mathematical proof by applying some inference rules, among which there occurs *modus ponens*. The full mathematical reconstruction of this reasoning is as follows:

1. Premise: 12 shekels is an interest from 60 shekels.
2. Mathematical fact (axiom): $12 = 60/5$.
3. From (1) and (2) we conclude that the interest is in 20%.

Example 3.1.2. Business letter from Iškundagān to Lugalra dated to the 24th–23rd centuries B.C.

Transliteration	Translation	Argumentative Reconstruction
Obverse 1. *en-ma* 2. *Iś-ku-un*^{d}*da-gan* 3. *a-na* Lugal-ra 4. GÁNA^{lam} *'à-ru-uš* 5. *ù* MÁŠ.ANŠE *ù-ṣú-ur* 6. *a-pu-na-ma* 7. Gu-ti-um-ma-mì 8. GÁNA^{lam} 9. *ù-la a-ru-uš* 10. *a táq-bí* 11. *a-na* ½ DA.NA.TA 12. *ma-aq-qá-ti* 13. *śu-ší-ib-ma* 14. *at-tá* 15. GÁNA^{lam} *'à-ru-uš* 16. *ki* GURUŠ.GURUŠ 17. *u-wa-kà-mu* 18. *ti-bu-tám* 19. *li-śé-ù-ni-kum-ma* 20. MAŠ.ANŠE *a-na* URU.KI^{lim} 21. [ś]*u-tá*-⌈*rí*⌉-*ib* **Reverse** 22. [*táq*$^{!}$-*b*]*i-ma* MÁŠ.ANŠE-*mì* 23. *Gu-ti-ù it-ru-ù* 24. *ù a-na-ku*$_{8}$ 25. *mí-ma ù-la a-qá-bi* 26. KÙ.BABBAR^{am} *a-na-da-kum* 27. [*a*]-*ni* 28. *na- 'à-aś Śar-kà-lí*-LUGAL^{ri} 29. *ù-má* 30. *śum-ma* MÁŠ.ANŠE 31. *Gu-ti-ù it-ru-ù* 32. *in ra-ma-ni-kà* 33. *lu tá-na-da-nu* 34. *a-na-lim-ma ki a-la-kam* 35. KÙ.BABBAR^{am} *<la> a-na-da-nu-kum* 36. *ù at-tá* MÁŠ.ANŠE 37. *ù-la tá-na-ṣa-ar* 38. *iš-pí-kí* 39. *kí-nu-tim* 40. *a-rí-iš-kà* 41. *mu-bi*$^{!}$ *lu ti-da*	1. It is said by 2. Iškundagān: 3. To Lugalra: 4. plant the field 5. and herd the cattle! 6.–10. Furthermore, you cannot say [yet again]: "The Gutians [are there]! I could not plant the field"! 11.–15. leave additional security at each (ca.) 5 kilometers and plant the field! 16.–21. Once them [i.e., your security] notices the [hostile] troops [then] they can attack [the Gutians] for you and [you] bring the cattle [temporarily] to the city. 22. You have said: "The cattle have been 23. led away by the Gutians" 24. and should I 25. say nothing at all [about this] 26. [and] give you (more) silver? 27. Now, however, 28. on the life of Šarkališarrī, 29. I swear to you hereby: 30.–33. If [again] the Gutians continue [to take away] the cattle, you will provide the own [compensation for the damage] indeed! 34. When I come into the city, 35. I will [certainly] give you no silver! 36.–37. In addition: [If] you do not herd the cattle, 38.–40. I will require the normal 'income' from you. 41. My instructions you should consider.	1.–3. Iškundagān is a protagonist, Lugalra is an antagonist. 4.–5. The standpoint of the protagonist in the form of assertive: Lugalra should plant the field and herd the cattle. 6.–10. Denying the following *modus ponens* for the non-acceptance of the standpoint: If the Gutians are there, Lugalra cannot plant the field and herd the cattle. The Gutians are there. Lugalra cannot plant the field and herd the cattle. 11.–21. Proposing the following two *modus ponens* for accepting the standpoint: (*i*) If the Gutians are there, then Lugalra leaves additional security at each ca. 5 kilometers and plant the field and herd the cattle. The Gutians are there. Lugalra leaves additional security at each ca. 5 kilometers and plant the field and herd the cattle. (*ii*) If the security notices the hostile troops, then the security can attack the Gutians and Lugalra brings the cattle temporarily to the city. The security notices the hostile troops. The security can attack the Gutians and Lugalra brings the cattle temporarily to the city. 22.–26. Denying the following assertive of the antagonist: "The cattle have been led away by the Gutians and Iškundagān should give more silver to Lugalra." 27.–40. The three legal commissives (promissory oaths) in the form of implications assuming *modus ponens*: (*a*) If again the Gutians continue to take away the cattle, then Lugalra provides the own compensation for the damage. The Gutians continue to take away the cattle. Lugalra provides the own compensation for the damage. (*b*) If Iškundagān comes into the city, then he gives no silver to Lugalra. Iškundagān comes into the city. He gives no silver to Lugalra. (*c*) If Lugalra does not herd the cattle, Iškundagān requires the normal 'income' from Lugalra. Lugalra does not herd the cattle. Iškundagān requires the normal 'income' from Lugalra. 41. The following directive from the protagonist: Lugalra should consider the instructions of Iškundagān.

Example 3.1.3. Mathematical fragment dated to the 20th–16th century B.C.

Transliteration	Translation	Argumentative Reconstruction
1. *a-na* 1 ma.na [1]2 gín *i-di-in-ma* / 2. máš ba.si.1.lá *li-id-di-kum* / 3. 1 ma.na [gar.r]a 12 *ṣ*[*í*]*-ip-tam* gar.ra / 4. ù 18 a.r[á *š*]*a* ba.si 1.lá *i-*[*na-di-nu-kum* gar.ra-ma] / 5. 1 ma.na *a-na* 12 máš íl 12 igi 12 *p*[*u-ṭ*]*ur-ma* 5 / 6. 5 *a-na* 18 a.rá *ša ša-ak-nu* íl 1 30 / 7. 1 30 sag kù.babbar / 8. *šum-ma* 1 30 sag kù.babbar 9. *a-na* 1 ma.na 12 *lu-ud-di-im-ma* / 10. 18 máš.bi 1 *li-id-di-nam* / 11. 1 ma.na *a-na* 12 *ṣí-ip-tim* íl 12 / 12. 12 *a-na* 1 30 sag kù.babbar íl 18 máš.bi / 13. ba.si 18.e[?] 1.lá en.nam 3	1. To 1 mina *12* shekels he gave, then 2. interest (as) "likeside-1-minus" may he give to you. 3. 1 mina *set*, 12, the interest, set, 4. and 18 (as) the fac*tor* that "likeside-1-minus" *will give to you set, then* 5. 1 mina to 12, the interest, raise, 12. The opposite of 12 *resolve*, then 5. 6. 5 to 18, the factor that was set, raise, 1 30. 7. 1 30 is the initial silver. 8. If 1 30 is the initial silver, (and) 9. for 1 mina 12 he may give, then 10. 18 its interest 1(?) may he give to me. 11. 1 mina to 12, the interest, raise, 12. 12. 12 to 1 30, the initial silver, raise, 18 its interest. 13. 18 (as) "likeside-1-minus" is what? 3.	1.–4. The first mathematical assumption (premise) in the form of assertive: let 12 shekels be an interest from 1 mina (60 shekels). 5. The first mathematical fact (axiom): to obtain 12 from 60, we should divide 60 by 5. It shows that 12 shekels give 20% of 1 mina (60 shekels). 6. The second mathematical fact (axiom): 18 gives 20% of 1 30. Let us notice that in Mesopotamia there was no today's decimal system. And instead of that, there was a sexagesimal (base 60) system. Let us convert 1 30 into a decimal number. It is 90. Hence, the second fact is as follows: 18 gives 20% of 90. 7. The second mathematical assumption (premise) in the form of assertive: we have 90 shekels (1 30 in the sexagesimal system). 8.–12. The logical consequence in the form of implication providing *modus ponens*: If we have 90 shekels and the interest for 60 shekels is equal to 12 shekels, then the interest for 90 shekels is equal to 18 shekels. We have 90 shekels and the interest for 60 shekels is equal to 12 shekels. The interest for 90 shekels is equal to 18 shekels. According to the second mathematical assumption, we have 90 shekels. According to the first mathematical assumption, the interest for 60 shekels is equal to 12 shekels. Now, we can introduce conjunction from both premises: we have 90 shekels and the interest for 60 shekels is equal to 12 shekels. Then from this conjunction and the logical consequence (lines 8.–12.) we can infer by *modus ponens*: The interest for 90 shekels is equal to 18 shekels. 13. For 18 we have $n \cdot n \cdot (n-1)$ with $n = 3$.

4. Mathematical fact (axiom): 18 = 90/5.
5. Premise: We have 90 shekels.
6. From (1) and (5) we introduce conjunction: "We have 90 shekels and the interest for 60 shekels is equal to 12 shekels."
7. From (3) and (6) we state: "We have 90 shekels, and the interest is in 20%."
8. From (4) and (7) we introduce implication: "If we have 90 shekels and the interest is in 20%, then the same interest for 90 shekels is equal to 18 shekels."
9. From (7) and (8) by *modus ponens*: "The interest for 90 shekels is equal to 18 shekels."

In the mathematical tablets from Mesopotamia, we do not find explicit formulations of logical inference rules, but we see their applications well. The same situation took place in Ancient Greece—in the *Elements* (Στοιχεῖα) attributed to Euclid (Εὐκλείδης), the Greek mathematician from Alexandria of Ptolemaic Egypt, and dated to ca. 300 B.C., we see a lot of correct applications of logical rules, but without their definitions. So, yet we did not find a logical handbook in Sumerian or Akkadian, but the fragments of tablets in these languages demonstrate a high logical competence of their authors with a correct application of implicit inference rules.

Mathematical texts from Mesopotamia such as Example 3.1.3 differ from business correspondence and legal proceedings by the lack of protagonists and antagonists, although *modus ponens* and other logical inference rules are applied widely to the same extent. For instance, in Example 3.1.2 we observe many implicit occurrences of *modus ponens*, but some of them are expressed as promissory oaths "If the event *A* occurs, then the act *B* is performed" that can be examined by court. In this case, there are needed witnesses or some documents for validating facts *A* for inferring court decisions *B* from promissory oaths ("If *A*, then *B*") by *modus ponens*. In the Babylonian mathematics, we deal only with impersonal facts for which we do not need witnesses or documents at all.

Since Ur III, the business correspondence became more sophisticated in its logical and rhetorical tools. In the Neo-Babylonian period (626–539 B.C.), we can read many tablets acquiring insidious new forms. Let us exemplify them by the private business letter from Nabû-nīrāru'a to Nabû-nādin-aḫi in relation to Šalim, the slave of Nabû-nīrāru'a, see Example 3.1.4. This text was excavated at Nippur—the transliteration and translation are taken from Cole (1996).

This tablet is written 18 centuries later than the documents of Example 3.1.1 and Example 3.1.2. Nevertheless, it is composed in the standards of Examples 3.1.1 and 3.1.2. For instance, an ethotic argument expressed as *argumentum ad hominem* is very similar to the ethotic argument of Example 3.1.1: (1) if Nabû-nādin-aḫi trusts his own brother (but it is), then he can trust Nabû-nīrāru'a, also; (2) if Nabû-nīrāru'a as the "brother of Nabû-nādin-aḫi" contends the standpoint (but it is), this standpoint should be accepted by Nabû-nādin-aḫi as more plausible. The difference between the ethotic argument of Example 3.1.4 and the ethotic argument of Example 3.1.1 is that now it is made explicitly by means of the following implication: "If Nabû-nādin-aḫi desires brotherhood and friendly relations, then he must hold the slave in confinement before Nabû-nīrāru'a can write and swear an oath concerning the slave."

Example 3.1.4. Business letter from Nabû-nīrāru'a to Nabû-nādin-aḫi dated to 626–539 B.C.

Transliteration	Translation	Argumentative Reconstruction
Obverse 1. *a-na* ^{md}PA-AŠ-ŠEŠ *qí-bi-ma* 2. *um-ma* ^{md}AG-ÉRIN.TAḪ-*ú-a* ŠEŠ-*kám*	1.–2. Say to Nabû-nādin-aḫi, thus says Nabû-nīrāru'a your brother.	1.–2. Nabû-nīrāru'a is a protagonist and Nabû-nādin-aḫi is an antagonist.
3. *um-ma-a a-na* ŠEŠ-*ia-a-ma*	3. Say to my brother:	3. The *argumentum ad hominem* before the argumentation: (*a*) Nabû-nādin-aḫi is called the brother of Nabû-nīrāru'a.
4. *ki-i áš-mu-ú um-ma* ^{m}*Šá-lim* 5. *ina pa-ni-ka la tu-maš-šar šú-ú-ma* 6. *a-na* 1+*en a-ḫi la i-ḫal-liq* 7. *a-di a-šap-pa-ram-ma* 8. MU DINGIR *ú-še-la-áš-šú*	4.–8. As soon as I heard — "You must not let Šalim out of your sight. *He* must not run away to someone else before I can write and swear an oath concerning him.	4.–11. The standpoint of Nabû-nīrāru'a is expressed as the conjunction consisting of the following four parts: (*i*) Šalim, the slave of Nabû-nīrāru'a, has run away. (*ii*) Nabû-nādin-aḫi has catched him. (*iii*) Nabû-nādin-aḫi must not let Šalim out of the sight before Nabû-nīrāru'a can write and swear an oath concerning the slave. (*iv*) After that, Nabû-nādin-aḫi must send the slave back to Nabû-nīrāru'a.
9. *ár-ki-šú šup-raš-šú ki-i* 10. ŠEŠ-*ú-tu ù* MUN.ḪI.A 11. *ṣe-ba-ta* LÚ *lu-ú ṣa-bit*	9.–11. After that, send him to me. If you desire brotherhood and friendly relations, let the man be held in confinement.	9.–11. The legal commissive in the form of implication assuming *modus ponens* to infer *iii* from *a*: (*b*) If Nabû-nādin-aḫi desires brotherhood and friendly relations, then he must hold the slave in confinement before Nabû-nīrāru'a can write and swear an oath concerning the slave. Nabû-nādin-aḫi desires brotherhood and friendly relations. He must hold the slave in confinement.
12. *a-di* IGI^{II}-*ia tam-mar* 13. *1+en ina lìb-bi* ŠEŠ.MEŠ-*e-šú* 14. *a-na lìb-*[*bi*] GEŠTU^{II}.MEŠ-*šú* 15. *la ir-ru-ub-ma la ú-šaḫ-laq-šú*	12.–15. Until you see me in person, not a single one of his brothers must come within earshot of him, that he may not help him to escape" —	12.–15. The directive in the form of request to infer *iii* from his directive by *modus ponens*: (*c*) Nabû-nādin-aḫi cannot allow a single one of the brothers of this slave to help the slave to escape. If *c* holds true, then *iii* (Nabû-nādin-aḫi does not let Šalim out of the sight before Nabû-nīrāru'a can write and swear an oath concerning the slave). This *c* is true. Nabû-nādin-aḫi must not let Šalim out of the sight.
16. *ki-i tal-tap-ra* 17. *um-ma a-bu-us-su aṣ-bat* 18. *man-nu ki-i* [*i*]-*kud-da-*[*áš-šum*]-*ma* 19. *ta-šap*[*par*]-*ra-ma lu a-šib* 20. *a-di at-ta tal-la-kám-ma* 21. *tab-ba-ku-uš* 1+*en-šú* 2-*šú*	16. you wrote (again) to me, 17. saying: "I have taken on the role of his father. 18.–19a. Whoever takes him into custody, you will send (him) to me; 19b.–21a. otherwise, let him stay put until you yourself come and take him away".	16.–21a. The antagonist comments on the standpoint: (*i'*) Nabû-nādin-aḫi has taken on the role of the father of the slave. (*ii'*) If someone takes the slave into custody, the slave will be sent to Nabû-nādin-aḫi, not to Nabû-nīrāru'a. (*iii'*) If no one takes the slave into custody, the slave stays put until Nabû-nīrāru'a comes and takes him away. Hence, from *i'* there is inferred *ii*. From *ii'* we infer *ii*. The statement of *iii'* contradicts to the conjunction of *iii* and *iv*. So, for the antagonist it is not enough that Nabû-nīrāru'a writes and swears an oath concerning the slave so that Nabû-nādin-aḫi himself sends the slave back to Nabû-nīrāru'a. To take the slave back, Nabû-nīrāru'a must come personally to take the slave away.
Lower Edge 22. *la kit-ta-a* 23. *aḫ-tar-ṣa-a* **Reverse** 24. *en-na a-šap-pa-rak-kám-ma* 25. *ul* [*ta*]-*qí-pan-ni*	21b.–23. Have I even once or twice unjustly made a withdrawal?	21b.–23. The rhetorical question as a claim that the protagonist always follows his obligation: (*d*) "Have I even once or twice unjustly made a withdrawal?" From *d* there is concluded that it is enough that Nabû-nīrāru'a writes and swears an oath concerning the slave so that Nabû-nādin-aḫi himself sends the slave back to Nabû-nīrāru'a.

<table>
<tr>
<td>26. ina maḫ-[ri]-i L[Ú] sar-ru-ti-[ia$_5$]
27. šá LÚ-tú-[ka] ki-i u-ṣab-bit
28. 1+en 5 KÙ.BABBAR ta-an-da-ḫar-šú-nu-tu
29. ina lìb-bi an-[ni]-i MUN.ḪI.A-a
30. ḫu-su-u[s] am-me-ni LÚ ḫal-qu
31. tu-tir-ram-ma a-na LÚ be-lí KÚR-ia
32. ta-nam-[din] ki-i mim-ma
33. te-ri-[šá]-an-ni šup-ram-ma
34. lu-še-bi-lak-ka ù ki-i
35. ku-tal-[lu*]-ta šú-ú
36. tu-kal da-šá-an-ni
37. šup-ru AD.MEŠ-ú-nu
38. a-ḫa-meš ku-ul-lu
39. ù 1+en [a]-na a-a-li
40. šá šá-ni-i ú-šu-uz-zu</td>
<td>24.–25. Now I’m sending a message to you because you didn’t believe me.
26.–27. Previously when I captured my kidnappers of your slaves,
28. you received them for (only) five (shekels) of silver each.
29.–30a. Remem[ber] this goodwill of mine,
30b.–32a. Why did you capture the runaway and are now giving him to my enemy?
32b.–34a. If you crave anything from me, write to me and I’ll send it to you.
34b.–37a. And if it is reserve-duty that you are holding him for—send rings.
37b.–38. Our fathers rule jointly,
39.–40. and they stand each as the confederate of the other.</td>
<td>24.–25. The declarative as a promise that the protagonist will follow his obligation: (e) “Now I’m sending a message to you because you didn’t believe me.”
26.–30a. The argument to infer by analogy that it is enough that Nabû-nīrāru’a writes and swears an oath concerning the slave so that Nabû-nādin-aḫi himself sends the slave back to Nabû-nīrāru’a: (f) Nabû-nīrāru’a captured the slaves of Nabû-nādin-aḫi previously and Nabû-nīrāru’a himself sends the slave back to Nabû-nādin-aḫi for only five shekels of silver each.
30b.–32a. The rhetorical question as a claim that the antagonist is not friendly (the contradiction to a): (g) “Why did you capture the runaway and are now giving him to my enemy?”
32b.–34a. The legal commissive in the form of implication to infer iii & iv by modus ponens: (h) If Nabû-nādin-aḫi craves anything from Nabû-nīrāru’a to send the slave back, Nabû-nīrāru’a will send this thing (maybe, silver) to Nabû-nādin-aḫi to receive the slave later. Nabû-nādin-aḫi writes that he needs something from Nabû-nīrāru’a. Nabû-nīrāru’a will send this thing (maybe, silver) to Nabû-nādin-aḫi to receive the slave later.
34b.–37a. The legal commissive in the form of implication to infer iii & iv by modus ponens: (j) If Nabû-nādin-aḫi holds the slave for reserve-duty (for a troop), Nabû-nādin-aḫi will send rings. Nabû-nādin-aḫi holds the slave for reserve-duty. Nabû-nādin-aḫi sends rings.
37b.–40. The new declarative to infer a: (k) The fathers of Nabû-nādin-aḫi and Nabû-nīrāru’a rule jointly and they stand each as the confederate of the other.</td>
</tr>
</table>

Another important difference between Example 3.1.4, on the one hand, and Examples 3.1.1 and 3.1.2, on the other hand, is that the standpoint of protagonist as well as the standpoint of antagonist is more composite in Example 3.1.4. So, the standpoint of protagonist consists of atomic statements from *i* to *iv*: (*i*) Šalim, the slave of Nabû-nīrāru'a, has run away; (*ii*) Nabû-nādin-aḫi has catched him; (*iii*) Nabû-nādin-aḫi must not let Šalim out of the sight before Nabû-nīrāru'a can write and swear an oath concerning the slave; (*iv*) after that, Nabû-nādin-aḫi must send the slave back to Nabû-nīrāru'a.

The standpoint of antagonist consists of atomic statements from *i'* to *iii'*: (*i'*) Nabû-nādin-aḫi has taken on the role of the father of the slave; (*ii'*) if someone takes the slave into custody, the slave will be sent to Nabû-nādin-aḫi, not to Nabû-nīrāru'a; (*iii'*) if no one takes the slave into custody, the slave stays put until Nabû-nīrāru'a comes and takes him away. In the meanwhile, the statement of *iii'* contradicts to the conjunction of *iii* and *iv*.

The protagonist (Nabû-nīrāru'a) formulates an additional ethotic argument in the form of rhetorical question: (*d*) "Have I even once or twice unjustly made a withdrawal?" As a result, *iii* and *iv* should follow from *d*. Then Nabû-nīrāru'a stresses the possibility to use this business letter as a business commitment in court: (*e*) "Now I'm sending a message to you because you didn't believe me." It is an argument from commitment: "If someone is committed to proposition *A*, then he or she should support *A*." Thus, *iii* and *iv* should follow from *e*.

There is an argument from analogy: (*f*) "Nabû-nīrāru'a captured the slaves of Nabû-nādin-aḫi previously and Nabû-nīrāru'a himself sends the slave back to Nabû-nādin-aḫi for only five shekels of silver each." It is assumed that Nabû-nādin-aḫi can make the same, i.e., *iii* and *iv* can be concluded from *f*.

After that Nabû-nīrāru'a formulates an additional ethotic argument in the form of rhetorical question: (*g*) "Why did you capture the runaway and are now giving him to my enemy?" In the explicit form, this argument is as follows: (1) if Nabû-nādin-aḫi is giving the slave to the enemy of Nabû-nīrāru'a (but it is), then Nabû-nīrāru'a cannot trust Nabû-nādin-aḫi; (2) if Nabû-nādin-aḫi under these circumstances contends the standpoint from *i'* to *iii'* (but it is), this standpoint should be accepted by Nabû-nīrāru'a as less plausible.

Then Nabû-nīrāru'a anounces the following two implications to infer *iii* & *iv* by *modus ponens* which can be used in court: (*h*) "If Nabû-nādin-aḫi craves anything from Nabû-nīrāru'a to send the slave back, Nabû-nīrāru'a will send this thing (maybe, silver) to Nabû-nādin-aḫi to receive the slave later" and (*j*) "If Nabû-nādin-aḫi holds the slave for reserve-duty for a troop, Nabû-nādin-aḫi will send rings."

To conclude, the business correspondence since the Old Akkadian period contains many different logical and rhetorical tools of legal argumentation. In trial proceedings, the rhetorical tools (such as *argumenta ad hominem*, i.e., ethotic arguments) were excluded. The verdicts were based only on logical arguments.

3.1.5 Examples of Logical Arguments from Legal Proceedings

Any legal decision is assumed to become the ultimate case for both parties and for the whole community. The judges must draw conclusions in their judgments only

Example 3.1.5. Record of proceedings against Ilamma dated to the 21st–20th century B.C.

Transliteration	Translation	Argumentative Reconstruction
Obverse 1. $^{[I]}$Nin-a-ni-sa$_{6}$ geme$_{2}$ I$_{3}$-lam-ma 2. I$_{3}$-lam-ma-še$_{3}$ 3. Ša$_{3}$-ge-bu$_{3}$-lu$_{5}$ 4. in-ši-sa$_{10}$ bi$_{2}$-in-du$_{11}$ 5. sa$_{10}$-a-aš nu-un-da-ge-en$_{6}$ 6. Ša$_{3}$-ge-bu$_{3}$-lu$_{5}$ 7. mu lugal in-pad$_{3}$ **Reverse** 1. tukum-bi u$_{4}$ 3-kam-ka 2. lu$_{2}$-inim-ma nu-mu-tum$_{2}$ 3. di [til-la] [he$_{2}$]-a bi$_{2}$-in-du$_{11}$ 4. l[u$_{2}$-inim-m]a nu-mu-da-tum$_{2}$ 5. Ni[n-a-ni-sa$_{6}$] ba-an-tum$_{2}$ 6. igi N[I$^{?}$-d]a$^{?}$-mu-še$_{3}$ 7. igi Arad$_{2}$-^{d}Nanna gu-za-la$_{2}$-še$_{3}$ 8. igi Ur-^{d}Nun-gal lu$_{2}$-kiĝ$_{2}$-gi$_{4}$-a lugal-še$_{3}$ 9. iti še-saĝ$_{11}$-ku$_{5}$ mu us$_{2}$-sa En-unu$_{6}$-gal ^{d}Inanna ba-huĝ	1.–4. "Šagebulu bought Ninanisa, slave woman of Ilamma, from Ilamma," he (Šagebulu) declared. 5. As for the purchase, he (Šagebulu) was not able to confirm it. 6.–r.3. Šagebulu has sworn by the name of the king: "If in three days I do not bring the witness(es), let it be a concluded case," he declared. 4.–5. He could not bring the witness(es), (so Ilamma) took Ninanisa with him. 6.–8. Before NIdamu? before Arad-Nanna, the 'throne-bearer'; before Ur-Nungal, envoy of the king. 9. Date.	1.–4. Šagebulu is a protagonist and Ilamma is an antagonist. The standpoint of the protagonist: (*i*) Šagebulu bought Ninanisa, slave woman of Ilamma, from Ilamma. 5. The protagonist has no argument (no purchase document to confirm *i*). 6.–r.3. The legal commissive (oath) of the protagonist in the form of implication assuming *modus ponens*: (*a*) If Šagebulu cannot bring witnesses in three days, the case will be closed. Šagebulu cannot bring the witness in three days. The case is closed. 4.–5. Šagebulu cannot bring witnesses in fact. From the oath of *a*, the judges conclude that the case is closed. The legal decision: the slave woman belongs to Ilamma. 6.–8. The judges who signed the decision: NIdamu? Arad-Nanna, the 'throne-bearer'; Ur-Nungal. 9. Date

from facts: (i) signed documents and (ii) oral testimonies of at least two witnesses. If a statement of disputant cannot be justified by facts, then this statement is rejected as false. Let us consider a court protocol of the time of Ur III, where a statement was denied by the court council, see Example 3.1.5—the transliteration and translation are taken from Molina (2008: 136).

This example shows an explicit way of applying *modus ponens* in court: (1) the protagonist promised that if in three days he does not bring the witnesses, let it be a concluded case ("If *A*, then *B*"); (2) "He could not bring the witnesses" (*A*); (3) "So, Ilamma took Ninanisa with him" (*B*). It is a standard of dispute resolution in Mesopotamia according to which each claimant applying to a court is obligatory to support his claim by witnesses or documents. If he does not have them properly, he must swear: "The case will be closed if I cannot bring witnesses." The point is that each court decision (*di-tila*) at the time of Ur III is ultimate and if the claimant cannot have witnesses in his claim and he cannot bring them quite later, he cannot file a lawsuit again.

Oral statements were ever preferable in courts—both the claimant, and the defendant should swear to claim about something as well as their witnesses should swear personally, also, but past legal documents can be accepted as evidence, too.

Example 3.1.6. Record of proceedings against ANsaga dated to the 21st–20th century B.C.

Transliteration	Translation	Argumentative Reconstruction
1. di til-[la] 2. ^{Id}nin-men-^{d}ba-ba$_6$ geme$_2$ 3. nam-egi$_2$-ni-du$_{10}$ [ama]-na 4. sam$_2^{am3}$ til-la-ni 5 giğ$_4$ ku$_3$-babbar-še$_3$ 5. AN-sa$_6$-ga dumu ^{d}utu-ki-du$_{10}$-ke$_4$ 6. in-ši-sa-a 7. ur-^{d}ba-ba$_6$ 8. lugal-amar-ku$_3$ 9. lu$_2$-inim-ma ša$_3$ dub-ba-ka 10. dub-bi ki di-ku$_5$-ne-še$_3$ 11. AN-sa$_6$-ga-a mu-de$_6$ 12. u$_3$ nam-egi$_2$-ni-du$_{10}$-e 13. geme$_2$ nu-u$_3$-gi$_4$-gi$_4$-da 14. mu lugal-bi in-pa$_3$ 15. geme$_2$ AN-sa-ga 16. ba-na-ge-en$_6$	1. Closed case: 2. Nin-menbaba, the slave: 3.–6. ANsaga, the son of Utukidu, bought from her mother Nameginidu for her full price of 5 shekels of silver. 7.–10. Ur-^{d}Baba and Lugal-Amarku are the witnesses in the tablet [that documents this sale]. 11.–14. ANsaga brought this tablet to the place of the judges and Nameginidu swore by the king that she will not take back the slave. 15.–16. The slave was confirmed to ANsaga.	1.–6. Nameginidu is a protagonist (clamant) and ANsaga is an antagonist (defendant). The standpoint of the protagonist: Nin-menbaba is a slave belonging to Nameginidu, not to ANsaga, the son of Utukidu. The legal decision: Nin-menbaba is a slave belonging not to Nameginidu, but to ANsaga. 3.–6. The argument supporting the legal case: (*a*) ANsaga bought Nin-menbaba from her mother Nameginidu for her full price of 5 shekels of silver. This argument assumes the following implication providing *modus ponens*: if *a*, then Nin-menbaba is the slave of ANsaga. ANsaga bought Nin-menbaba (*a* is true). Nin-menbaba is the slave of ANsaga. 7.–14. The validity of the argument is confirmed by the two witnesses: (*i*) Ur-^{d}Baba and Lugal-Amarku. Nevertheless, it is not an oral testimony, but it is a document signed by the witnesses. If *i*, then *a*. The validity of the argument is confirmed by the following legal commissive: (*ii*) Nameginidu (the former owner of the slave) swore by the king that she will not take back the slave. If *ii*, then *a*. 15.–16. The argumentation is considered valid.

So, there is a court decision from the city of Lagaš and dated to the time of Ur III with the direct reference to the past legal case, see Example 3.1.6—the transliteration and translation are taken from Culbertson (2009: 94).

In this tablet, there are two logical steps in inferring the court decision:

1) If there are two witnesses of transaction, the transaction is valid ("If *A*, then *B*"). Ur-^{d}Baba and Lugal-Amarku are two witnesses that ANsaga bought Nin-menbaba (*A*). Then this transaction is valid (*B*).
2) If ANsaga bought Nin-menbaba according to a validated document of transaction, then Nin-menbaba is the slave of ANsaga ("If *B*, then *C*"). According to the legal document of transaction checked at the court, ANsaga bought Nin-menbaba from her mother Nameginidu for her full price of 5 shekels of silver legally (*B*). Then Nin-menbaba is the slave of ANsaga (*C*).

Example 3.1.7. Record of proceedings against the daughter of E'urbi dated to the 21st–20th century B.C.

Tranliteration	Translation	Argumentative Reconstruction
1. dumu mumus e_2-ur_2-bi $arad_2$ lu_2-gu-la-ka-ra 2. i_3-bi_2-la lu_2-gu-la-ke_4-ne 3. inim in-ni-^{ga2}ĝar^{ar}-eš 4. lu_2-gu-la ti-la-a 5. igi-ni in-^{ga2}ĝar^{ar} 6. mu-lugal 7. dumu-munus e_2-ur_2-bi $arad_2$-[ga_2-ka] 8. ama-ar-gi_4-[bi i_3-ĝar] 9. [bi_2-in-du_{11}-ga] 10. dub lu_2-[gu-la-bi] 11. nin-du_{11}-ga [dumu] [e_2-ur_2]-bi $arad_2$-da-[ke_4] 12. igi sukkal-maḫ-$še_3$ mu-de_6 13. dumu e_2-ur_2-bi-ke_4-ne 14. ki i_3-bi_2-la lu_2-gu-la-ke_4-ne-ta 15. tug_2 ib_2-ur_3 16. ur-^{d}lamma dumu lu_2-17. mu maškim	1.–3. The heirs of Lu-Gula have raised a claim about the daughter of E'urbi, the slave of Lu-Gula. 4.–10. When Lu-Gula was alive, he appeared and swore by the king that, 'the daughter of E'urbi, my slave, I set free.' 11.–12. Nin-duga, the sister of E'urbi, brought the tablet of Lu-Gula to the Grand Vizier. 13.–15. The heirs of Lu-Gula abandoned their claim against the child(ren) of E'urbi [literally, the child(ren) of E'urbi are dropped by the heirs of Lu-Gula?]. 16. Ur-Lamma, the son of Lumu, was the *maškim*.	1.–3. The heirs of Lu-Gula are a protagonist (claimant) and the daughter of E'urbi is an antagonist (defendant). 4.–15. The standpoint of the protagonist: (*i*) The daughter of E'urbi is a slave of the heirs of Lu-Gula. The main argument of the protagonist: (*a*) The daughter of E'urbi was a slave of Lu-Gula. By *modus ponens* we infer *i* from *a*: If the daughter of E'urbi was a slave of Lu-Gula, then she is a slave of the heirs of Lu-Gula. The daughter of E'urbi was a slave of Lu-Gula. She is a slave of the heirs of Lu-Gula. 4.–12. The counterargument: (*b*) Nin-duga, the sister of E'urbi, brought the tablet of Lu-Gula to the Grand Vizier according to which when Lu-Gula was alive, he appeared and swore by the king that "the daughter of E'urbi, my slave, I set free." According to *b*, the statement of *a* is false. Hence, *i* cannot be inferred from *a*. 12.–15. The validity of the counterargument of *b* is confirmed by the following legal commissive: (*c*) the heirs of Lu-Gula abandoned their claim against the daughter of E'urbi. According to *c*, the standpoint expressed in *i* is false. 16. The sign by the *maškim*.

It is significant also that the trial decision should be ultimate. Therefore, Nameginidu (the former owner of the slave) swore by the king that she will not take back the slave. Hence, she cannot file a lawsuit again.

Another direct reference to the past legal case in a court judgment is taken from the city of Lagaš dated to the same Ur III, see Example 3.1.7—the transliteration and translation are taken from Culbertson (2009: 95).

In this document, the heirs of Lu-Gula are the protagonist (claimant) who state that the daughter of E'urbi is their slave. They propose to use *modus ponens* for the court decision: "If the daughter of E'urbi was a slave of Lu-Gula, then she is a slave of the heirs of Lu-Gula after the death of Lu-Gula according to the Law Code. The daughter of E'urbi was a slave of Lu-Gula. Now she is a slave of the heirs of Lu-Gula." So, the daughter of E'urbi is an antagonist (defendant) who objects not to the standpoint of the protagonist that she became the slave of the heirs of Lu-Gula, but she objects to the argument that she was a slave of Lu-Gula before his death. There is a difference between Example 3.1.7 and Example 3.1.6, and it demonstrates that in the practice of legal argumentation since Ur III, not only standpoints can be

Example 3.1.8. Record of proceedings against Nabû-iddin dated to 600 B.C.

Transliteration	Translation	Argumentative Reconstruction
Obverse 1. *ina* ITI.ŠU ^{Id}AG.SUM-*na* [*a*-]*na* 2. UNUG.KI *il-la-kám-ma* ^{Id}AMAR.UTU.SUR 3. LÚ *mu-kin-ni-šú ib-bak-kám-ma* 4. ^{Id}AG.MU *ú-ka-nu* 5. *ki-i ina* ITI.ŠU *la uk-tin-ú-šú* 6. ^{Id}AG.MU *za-ku-u ki-i* 7. ^{Id}AMAR.UTU.SUR LÚ *mu-kin-ni-šú* 8. *i-ta-ab-kám-ma* ^{Id}AG[.MU] **Lower Edge** 9. *uk-tin-ni mim-ma ma-*[*la*] 10. *šá* ^{Id}AMAR.UTU.SUR *ú-ka* [*n-nu-uš* X X X] **Reverse** 11. ^{Id}AG.MU *i-na* [*m-din*] 12. LÚ *mu-kin-ni* ^{I}ÌR ^{d}AG [A-*šú šá*] 13. ^{I}*mu-še-zib-*^{d}AMAR.UTU A ^{I}*mi-*[*ṣir-a-a*] 14. ^{I}*gi-mil-lu* A-*šú šá* ^{Id}AG.KAL 15. A LÚ ŠU.KU_6 *u* LÚ DUB.SAR 16. $^{I[d]}$in-nin-NUMUN.GÁL-*ši* A-*šú šá* 17. ^{Id}U.GUR.MU.DÙ A ^{Id}UTU-*ya* 18. UNUG.KI ITI.ŠE UD.19.KÁM 19. MU.4.KÁM ^{Id}AG.NÍG.DU.URÙ **Upper Edge** 20. LUGAL TIN.[TIR.KI]	1.–4. In the month of Dûzu, Nabû-iddinna shall come to Uruk and Marduk-ēṭir shall produce his witness, (who) shall give testimony against Nabû-iddin. 5.–6. If he has not confirmed it (the claim) in Dûzu, then Nabû-iddin is free (of any claim against him). 6.–11. If Marduk-ēṭir has produced his witness and has confirmed (a claim against) Nabu-[iddin], (then) whatever Marduk-ēṭir shall estab[lish against him], Nabû-iddin shall pa[y]. 12.–15. Witnesses: Arad-Nabû, son of Mušezib-Marduk, descendant of Miṣiraya, Gimillu, son of Nabû-udammiq, descendant of Bā'iru. 15.–20. And the scribe was Innin-zēr-ušabšî, son of Nergal-šum-ibni, descendant of Šamšīya. Done at Uruk in the month of Addaru, day 19 in the 4th year of Nebuchadnezzar, King of Babylon.	1.–4. The legal commissive (oath): "In the month of Dûzu, Nabû-iddin shall come to Uruk and Marduk-ēṭir shall produce his witness against Nabû-iddin." 5.–6. The case in the form of implication assuming *modus ponens* if Marduk-ēṭir will break this oath: If Marduk-ēṭir has not produced his witness against Nabu-iddin, then Nabû-iddin is free of any claim against him. Marduk-ēṭir has not produced his witness against Nabu-iddin. Nabû-iddin is free. 6.–11. The case in the form of implication assuming *modus ponens* if Marduk-ēṭir will hold this oath fulfilled: If Marduk-ēṭir has produced his witness against Nabu-iddin, then Nabû-iddin shall pay. Marduk-ēṭir has produced his witness against Nabu-iddin. Nabû-iddin shall pay. 12.–15. Witnesses of the promissory oath: (i) Arad-Nabû, son of Mušezib-Marduk, descendant of Miṣiraya, (ii) Gimillu, son of Nabû-udammiq, descendant of Bā'iru. 15.–20. The scribe's name and the date.

objected to, but arguments also. The defendant has a written testimony according to which when Lu-Gula was alive, he swore by the king that "the daughter of E'urbi, my slave, I set free." To make the court decision ultimate that the daughter of E'urbi is free indeed, the heirs of Lu-Gula abandoned their claim against her. Now, they cannot file a lawsuit again.

The argumentative structure of legal records remained almost the same until the time of Neo-Babylonian period (626–539 B.C.). In the same way, oral statements were regarded as preferable in courts. The next text was written at Uruk in the month of 19 Addaru (Hebrew: *Addar*, approximately March), year 4 of Nebuchadnezzar II (600 B.C.). It contains a court order to produce witness, see Example 3.1.8—the transliteration and translation are taken from Cole (1996: 59).

This example is very similar to Examples 3.1.6 and 3.1.7, although it was composed about 15 centuries later. The difference is that the trial judgment is not ultimate here—it is unknown whether there were witnesses or not. Therefore, the judgment is conditional: (1) "If Marduk-ēṭir has not confirmed his claim against Nabû-iddin in the month of Dûzu (Hebrew: *Tammûz*, it corresponds to June-July), then Nabû-iddin is free of any claim against him"; (2) "If Marduk-ēṭir has produced his witnesses and has confirmed a claim against Nabû-iddin, then whatever Marduk-ēṭir shall establish against him, Nabû-iddin shall pay." Hence, after the month of Dûzu, year 5 of Nebuchadnezzar II (601 B.C.) without witnesses produced by Marduk-ēṭir, Nabû-iddin will be considered free according to this legal document.

The next tablet was composed on 26 Šabāṭu (Hebrew: *Šəvat*, approximately January-February), year 2 of Nabonidus, the king of Babylon (554 B.C.), see Example 3.1.9. It contains a rejection of claim because it cannot be confirmed by facts (documents or oral testimonies) also—the transliteration and translation are taken from Holtz (2009: 23–24).

This text also is very similar to Examples 3.1.6 and 3.1.7 which are much older. The main difference is that now the arguments that the claimant has no witnesses are examined more explicitly. The difference from Example 3.1.8 is that the judgment is ultimate.

The claim is formulated as follows: "^{f}Bābunu and her children are members of the household of ^{m}Nabû-mukīn-apli, but ^{f}Bābunu is the sister of ^{m}Tērik-šarrūssu and she must come back to ^{m}Tērik-šarrūssu". This claim is rejected by the court based on the following logical conclusion consisting of seven logical steps:

1) The fact established by the court: "For the 40 years, ^{f}Bābunu served ^{m}Amurru-šumaiddinam, father of ^{m}Bēl-mukīn-apli, but he did not have any case or claimant against him."
2) Now we can introduce the following implication: "If (1) is true, then ^{f}Bābunu does not want to be a member of the household of ^{m}Tērik-šarrūssu."
3) From (1) and (2) we infer by *modus ponens*: "^{f}Bābunu does not want to be a member of the household of ^{m}Tērik-šarrūssu."
4) We introduce implication: "If (3) is true, then ^{f}Bābunu must remain in the household of ^{m}Amurru-šumaiddinam."
5) Then from (3) and (4) by *modus ponens*: "^{f}Bābunu must remain in the household of ^{m}Amurru-šumaiddinam."

Example 3.1.9. Record of proceedings against ^{m}Nabû-mukīnapli dated to 554 B.C.

Transliteration	**Translation**	**Argumentative Reconstruction**
1. *di-i-ni ša*$_{2}$ ^{m}*re-man-ni-*^{d}EN A-šu$_{2}$ *ša*$_{2}$ ^{m}*te-rik*-LUGAL-*ut-su* 2. *a-na muḫ-ḫi* ^{f}*ba-bu-nu u*$_{3}$ DUMU. MEŠ-*šu*$_{2}$ UN.MEŠ E$_{2}$ 3. *ša*$_{2}$ ^{md}NA$_{3}$-<*mu*>-*ki-in*-IBILA DUMU-*šu*$_{2}$ *ša2* ^{md}KUR. GALMU-*id-di-nam* 4. *it-ti* ^{md}NA$_{3}$-DU-IBILA *a-na ma-ḫar* ^{lu2}DI.KU$_{5}$.MEŠ 5. *ša*$_{2}$ ^{md}NA$_{3}$-*na-'-id* LUGAL TIN. TIR^{ki} *id-bu-bu um-ma* ^{f}*ba-bu-nu* 6. *ša*$_{2}$ *i-na* E$_{2}$-*ku-nu*^{2} *a-ḫa-ta-a ši-i* ^{lu2}DI.KU$_{5}$.MEŠ 7. ^{m}*re-man-ni-*^{d}EN *iš-ta-'-a-lu um-ma* ^{f}*ba-bu-nu* 8. NIN-*ka ul-tu im-ma-ti ki-i* E$_{2}$^{md}KUR.GAL-MU-MU 9. AD *ša*$_{2}$ ^{md}NA$_{3}$–DU-IBILA *ši-i* ^{m}*re-man-ni-*^{d}EN *iq-bi* 10. *um-ma* 40 MU.AN. NA.MEŠ *an-na-a-ti* ^{f}*ba-bu-nu* 11. NIN-*a* ^{md}KUR.GAL-MU-MU *ta-pal-laḫ di-i-ni a-na muḫ-ḫi-šu*$_{2}$ 12. *it-ti* ^{md}KUR.GAL-MU-MU AD *ša*$_{2}$ ^{md}NA$_{3}$-DU-IBILA *ad-di-bu-ub* 13. *u*$_{3}$ *a-di i-na-an-na iš-tu* E$_{2}$-*šu*$_{2}$ *la u*$_{2}$-*še-ṣi-iš* 14. ^{m}*re-ma-an-ni-*^{d}EN *mim-ma i-da-tu ša*$_{2}$ *di-i-ni a-na* UGU 15. ^{f}*ba-bu-nu it-ti* ^{md}KUR. GALMU-MU AD *ša*$_{2}$ ^{md}NA$_{3}$-DU-A 16. *id-bu-bu a-na* ^{lu2}DI.KU$_{5}$. MEŠ *la u*$_{2}$-*kal-li-im* 17. ^{lu2}DI.KU$_{5}$.MEŠ *dib-bi-šu*$_{2}$-*nu-ti iš-mu-ma mim-ma i-da-tu*$_{4}$ 18. *ša*$_{2}$ *di-i-ni la i-mu-ru-u iš-ta-lumu* 40 MU.AN.NA.MEŠ 19. *an-na-a-ti* ^{f}*ba-bu-nu* ^{md}KUR. GAL-MU-MU AD *ša*$_{2}$ ^{md}NA$_{3}$DU-IBILA 20. *tu*$_{3}$-*pal-laḫ man-ma di-i-ni u*$_{3}$ *paqa-ri* 21. *ina muḫ-ḫi-šu*$_{2}$ *la ir-ši* 22. ^{f}*ba-bu-nu u*$_{3}$ DUMU.MEŠ-*šu*$_{2}$ UN.MEŠ E$_{2}$ *ša*$_{2}$ ^{md}KUR. GALMU-MU 23. *a-na* ^{md}NA$_{3}$-DU-IBILA *id-di-nu*	1.–5. The case which ^{m}Rēmanni-Bēl son of ^{m}Tērik-šarrūssu argued against ^{m}Nabû-mukīnapli, regarding ^{f}Bābunu and her children, members of the household of ^{m}Nabû-mukīn-apli son of ^{m}Amurru-šuma-iddinam, before the judges of Nabonidus, king of Babylon, thus: 5.–6. "^{f}Bābunu, who is in your household, is my sister!" 6.–7. The judges questioned ^{m}Rēmanni-Bēl thus: 7.–9. "Since when has ^{f}Bābunu, your sister, been part of the household of ^{m}Amurru-šumaiddinam, father of ^{m}Nabû-mukīn-apli?" 9.–10. ^{m}Rēmanni-Bēl said thus: 10.–13. "For these past 40 years, ^{f}Bābunu, my sister has served ^{m}Amurru-šuma-iddinam. I argued a case regarding her against ^{m}Amurru-šuma-iddinam, father of ^{m}Nabû-mukīn-apli, but he has not let her go from his household until now!" 14.–16. ^{m}Rēmanni-Bēl did not show the judges any evidence of the case regarding ^{f}Bābunu which he argued against ^{m}Amurru-šuma-iddinam, father of ^{m}Nabû-mukīn-apli. 17.–18. The judges heard their arguments. They did not see any evidence of the case. They conferred. 18.–21. For these 40 years, ^{f}Bābunu served ^{m}Amurru-šumaiddinam, father of ^{m}Bēl-mukīn-apli. He did not have any case or claimant against him. 22.–23. They assigned ^{f}Bābunu and her children, the members of the household of ^{m}Amurru-šumaiddinam to ^{m}Nabû-mukīn-apli.	1.–5. ^{m}Rēmanni-Bēl son of ^{m}Tērik-šarrūssu is a protagonist (claimant) and ^{m}Nabû-mukīnapli son of ^{m}Amurru-šuma-iddinam is an antagonist (defendant). 1.–6. The standpoint of the protagonist: ^{f}Bābunu and her children are members of the household of ^{m}Nabû-mukīn-apli, but ^{f}Bābunu is the sister of ^{m}Tērik-šarrūssu and she must come back to ^{m}Tērik-šarrūssu. 6.–13. The argument of the protagonist: (*a*) for the past 40 years, ^{f}Bābunu has served ^{m}Amurru-šuma-iddinam, but before that she was a member of the household of ^{m}Tērik-šarrūssu. 14.–18. The judges state that the protagonist did not show the judges any evidence of the case regarding ^{f}Bābunu which he argued against ^{m}Amurru-šuma-iddinam, father of ^{m}Nabû-mukīn-apli. In other words, there is no validation of *a*. 18.–21. The counter-argument of the judges: (*b*) for the 40 years, ^{f}Bābunu served ^{m}Amurru-šumaiddinam, father of ^{m}Bēl-mukīn-apli, but he did not have any case or claimant against him. If *b* is true, then (*c*) she does not want to be a member of the household of ^{m}Tērik-šarrūssu. Then by *modus ponens*: If *c* is true, then she must remain in the household of ^{m}Amurru-šumaiddinam. The statement of *c* is true. She must remain. 22.–23. The legal decision: ^{f}Bābunu and her children, the members of the household of ^{m}Amurru-šumaiddinam, belong now to ^{m}Nabû-mukīn-apli. The point is that by *modus ponens* we conclude: If ^{f}Bābunu and her children are the members of the household of ^{m}Amurru-šumaiddinam (the counter-argument expressed by *b*), then they are members of the heirs of ^{m}Amurru-šumaiddinam. ^{m}Nabû-mukīn-apli is the sole heir of ^{m}Amurru-šumaiddinam. ^{f}Bābunu and her children are the members of the household of ^{m}Nabû-mukīn-apli.

6) We introduce implication again: "If [f]Bābunu and her children are the members of the household of [m]Amurru-šumaiddinam, then they are members of the heirs of [m]Amurru-šumaiddinam ([m]Nabû-mukīn-apli is the sole heir of [m]Amurru-šumaiddinam)."
7) From (5) and (6) by *modus ponens*: "[f]Bābunu and her children are the members of the household of [m]Nabû-mukīn-apli."

In Example 3.1.9, there are mentioned only steps (1) and (7). Other steps are assumed.

The next legal judgment was excavated in Uruk, and it is dated to 12 Addaru (Hebrew: *Addar*), year 3 of Cambyses II, the king of Babylon (22 March, 526 B.C.), see Example 3.1.10—the transliteration and translation are taken from Holtz (2014: 179–181).

This record is especially interesting, as it contains an implicit reference to a law article according to which for stealing one item belonging to the god or to the king, the thief should give the same item thirtyfold. In the *Hammurabi Code* this law article is formulated as follows:

> (vi 57–69) *šumma awīlum lu alpam lu immeram lu imēram lu šaḫâm ulu elippam išriq šumma ša ilim šumma ša ekallim adi* 30-*šu inaddin šumma ša muškēnim adi* 10-*šu iriab šumma šarrāqānum sa nadānim la išu iddâk.*

> **§8** If a man steals an ox, a sheep, a donkey, a pig, or a boat—if it belongs either to the god or to the palace, he shall give thirtyfold; if it belongs to a commoner, he shall replace it tenfold; if the thief does not have anything to give, he shall be killed (Roth, 1995: 82).

Thus, the trial verdicts were grounded on some facts confirmed during the trial as well as on appropriate law articles as premises to infer conclusions from them by applying logical inference rules. In Example 3.1.10, the conclusion is as follows:

1) The fact checked by the court: "5 sheep branded with a star that it belongs to Ištar of Uruk and Nanaya were led away by Bēl-iqīša, the son of Ṣillaya".
2) The article delivered from the Law Code: "If a man steals 5 sheep and it belongs to Ištar of Uruk and Nanaya, he shall give thirtyfold (150 sheep)."
3) From (1) and (2) by *modus ponens*: "Bēl-iqīša shall give 150 sheep."

Among the branded sheep there were 5 unblemished lambs which were not branded. It means that for them we cannot apply the law article to compensate thirtyfold. But we can apply another law article:

1) "If sheep (or other cattle) belongs to someone, its offspring belongs to him, as well."
2) The fact: "This sheep belonged to Ištar of Uruk and Nanaya."
3) From (1) and (2) by *modus ponens*: "Its offspring (5 unbranded lambs) belongs to Ištar of Uruk and Nanaya."

Example 3.1.10. Record of proceedings against Bēl-iqīša dated to 526 B.C.

Transliteration	Translation	Argumentative Reconstruction
1. [1-*en* UDU *pu-ḫal* 4 UDU U$_8$.MEŠ] NIGIN 5 *ṣe-e-nu šá* MUL-*tu*$_4$ *še-en-du* 2. [*ù*] [5 *par*]-*rat*.ME *ta-mi-ma-a-ta* NIGIN 10 *ṣe-e-nu* 3. NIG$_2$.GA ^{d}INNIN UNUG^{ki} *u* ^{d}*na-na-a šá qa-pu-ut-tu*$_4$ 4. *šá* ^{md}*a-nu*-LUGAL-URI$_3$ DUMU-*šú šá* ^{m}LUGAL-DU *šá ina* ITI APIN MU 2-*kám* 5. ^{m}*ka-am-bu-zi-ia* LUGAL TIN.TIR^{ki} LUGAL KUR.KUR ^{md}ENBA-*šá* 6. DUMU-*šú šá* ^{m}*ṣil-la-a ina* ŠU.2 ^{md}a-nu-LUGAL-URI$_3$ A-*šú šá* ^{m}LUGAL-DU *i-bu-ku-ma* 7. *ina* ITI ŠE MU 3-*kám* ^{m}*ri-mut u* ^{md}*ba-ú*-APIN-*eš* 8. ^{lu2}DI.KU$_5$.ME 150 *ṣe-e-nu ku-um ṣe-e-nu šá* ^{d}15 9. *šen-de-e-ti* 1-*en a-di* 30 *ù* 5 *parrat ta-mi-ma-a-ta* 10. NIGIN 155 *ṣe-e-nu a-na e-ṭe*$_3$-*ru šá* ^{d}INNIN UNUG^{ki} 11. *i-na ṭup-pi iš-ṭu-ru-ma e-li* ^{md}ENBA-*šá ú-kin-nu* 12. U$_4$ 25-*kám šá* ITI ŠE MU 3-*kám ṣe-e-nu a'* 155 ^{md}EN-BA-*šá* 13. DUMU-*šú šá* ^{m}*ṣil-la-a ibba-kám-ma ina* E$_2$.AN.NA *i-šim-mi-it-ma* 14. *a-na* NIG$_2$.GA E$_2$.AN.NA *i-namdin* ^{m}IR$_3$-^{d}U.GUR DUMU-*šú šá* ^{m}DU-A 15. DUMU ^{m}*e-gi-bi pu-ut e-ṭe*$_3$-*ru šá ṣe-e-nu-a'* 16. 155 *na-ši-i i-na ú-šu-uz-zu šá* ^{md}NA$_3$-DU-[IBILA] 17. ^{lu2}ŠA$_3$.TAM E$_2$.AN.NA DUMU-*šú šá* ^{m}*na-di-nu* DUMU ^{m}*da-bi-bi* 18. ^{md}NA$_3$-ŠEŠ-MU ^{lu2}SAG-LUGAL ^{lu2}EN *pi-qit-ti* E$_2$.AN.NA 19. ^{lu2}*mu-kin-nu* ^{m}IR$_3$-^{d}AMAR.UTU DUMU-*šú šá* ^{m}NUMUN-*ia* DUMU ^{m}*e-gi-bi* 20. ^{md}30-APIN-*eš* DUMU-*šú šá* ^{md}NA$_3$-MU-SI.SA$_2$ DUMU ^{m}DU$_3$-DINGIR 21. ^{md}EN-SUM-IBILA DUMU-*šú šá* ^{md}AMAR.UTU-MU-MU DUMU ^{md}EN-IBILA-URI$_3$	1.–6. [1 ram 4 ewes] total 5 sheep branded with a star and 5 unblemished lambs, a total of 10 sheep, property of Ištar of Uruk and Nanaya, from the pen of Anu-šarra-uṣur son of Šarrukīn, which in Araḫšamna, year 2 of Cambyses, king of Babylon, king of the lands, Bēl-iqīša son of Ṣillaya led away (in payment) from Anu-šarrauṣur son of Šarru-kīn. 7.–11. In Addaru, year 3, Rīmūt and Bau-ēreš, the judges, wrote in a tablet and determined for Bēl-iqīša to pay 150 sheep, thirtyfold for the sheep branded for Ištar and 5 unbranded lambs, a total of 155 sheep, for repayment to Ištar of Uruk. 12.–14. On 25 Addaru, year 3, Bēliqīša son of Ṣillaya shall bring these 155 sheep, brand them in the Eanna and give them to the property of the Eanna. 14.–16. Arad-Nergal son of Mukīnapli descendant of Egibi assumes responsibility for the repayment of these 155 sheep. 16.–17. In the presence of Nabûmukīn-apli, the *šatammu* of the Eanna, son of Nādinu descendant of Dābibī; 18. Nabû-aḫa-iddin, the royal official in charge of the Eanna. 19. Witnesses: Arad-Marduk, son of Zēriya descendant of Egibi; 20. Sîn-ēreš son of Nabû-šumu-līšir descendant of Ibni-ili; 21. Bēl-nādin-apli son of Mardukšuma-iddin descendant of Bēl-aplauṣur;	1.–6. The standpoint of the court: 5 sheep branded with a star and 5 unblemished lambs, a total of 10 sheep, is a property of Ištar of Uruk and Nanaya, but 5 sheep branded with a star were led away by Bēl-iqīša, the son of Ṣillaya. 7.–14. The first legal decision of the court: (*i*) Bēl-iqīša must pay 150 sheep, thirtyfold for the sheep branded for Ištar and (*ii*) 5 unbranded lambs, a total of 155 sheep, for repayment to Ištar of Uruk. The decision of *i* is inferred by *modus ponens* from the law article from the legal codes since Ur III that each stolen thing from the temple deserves the thirtyfold penalty for this thing: If a thing is stolen from the temple, the thief must repay the stolen thing thirtyfold. 5 sheep branded with a star. Bēl-iqīša, the thief, must bring 150 sheep (= 5 · 30) and give them to the property of the Eanna. The decision of *ii* is inferred by *modus ponens*, too, from the legal tradition that if sheep (or other cattle) belongs to someone, its offspring belongs to him, as well. This sheep belonged to Ištar of Uruk and Nanaya. Its offspring (5 unbranded lambs) belongs to Ištar of Uruk and Nanaya. 14.–16. The second legal decision of the court: (*iii*) Arad-Nergal son of Mukīnapli descendant of Egibi assumes responsibility for the repayment of these 155 sheep. There are the following two conclusions by *modus ponens*: (*a*) If Bēl-iqīša repays 155 sheep in time, Arad-Nergal is free. Bēl-iqīša repays 155 sheep in time. Arad-Nergal is free. (*b*) If Bēl-iqīša does not repay 155 sheep on schedule, Arad-Nergal is not free. Bēl-iqīša does not repay 155 sheep. Arad-Nergal must repay 155 sheep.

Example 3.1.10 contd. ...

...Example 3.1.10 contd.

Transliteration	Translation	Argumentative Reconstruction
22. ^{m}*na-di-nu* DUB.SAR DUMU ^{m}*e-gi-bi* 23. ^{m}IR3-^{d}[AMAR.UTU] DUB. SAR DUMU ^{md}EN-IBILA-URI$_3$ 24. UNUG^{ki} ITI ŠE U$_4$ 12-*kám* MU 3-*kám* ^{m}*kám-bu-zi-ia* 25. LUGAL TIN.TIR^{ki} LUGAL KUR.KUR	22. Nādinu, the scribe, descendant of Egibi; 23. Arad-Marduk, the scribe, descendant of Bēl-apla-uṣur. 24.–25. Uruk. 12 Addaru, year 3 of Cambyses, king of Babylon, king of the lands.	16.–18. The decision is signed by Nabûmukīn-apli, the *šatammu* of the Eanna, Nabû-aḫa-iddin, the royal official in charge of the Eanna. 19.–23. Witnesses who signed the decision: Arad-Marduk, Sîn-ēreš, Bēl-nādin-apli, Nādinu, Arad-Marduk. 24.–25. The place of legal decision: Uruk. The date: 12 Addaru, year 3 of Cambyses, king of Babylon.

The trial judgment is not ultimate, but conditional:

a) "If Bēl-iqīša repays 155 sheep in time, Arad-Nergal as a guarantor of Bēl-iqīša is free."

b) "If Bēl-iqīša does not repay 155 sheep on schedule, Arad-Nergal is not free (he must repay 155 sheep)."

It is worth noting that among other trial records of the Neo-Babylonian period, there are inference rules used implicitly or explicitly which can be reconstructed as *modus ponens*, introducing conjunction, introducing implication, and many other rules of present-day propositional logic. See Section 3.2.

3.1.6 Conclusions

The first foundations of logic and legal argumentation appeared in Mesopotamia at least 4,000 years ago. Within this tradition, the following strong argumentative/logical rules began to be strictly enforced:

- There are two parties of disputants: (i) a protagonist who formulates a standpoint by means of an assertive and (ii) an antagonist who disagrees with the protagonist's standpoint and formulates an alternative statement by means of another assertive.
- There is a rational judge represented by some community representatives delegated for the trial especially and the role of this judge consists in that he follows only logical conclusions from facts and law articles as premises.
- The protagonist (as well as the antagonist) has a right to defend his statements by arguments: oral testimonies of at least two witnesses or legal documents signed properly.
- The rational judge oversees the arguments put forward by the protagonist (as well as by the antagonist) and accepts them if and only if the standpoint or its negation follows from the arguments just logically by applying inference rules—first by *modus ponens*.

- A legal decision is taken by the rational judge based on the arguments accepted by him as justified.
- All rhetorical tropes such as rhetorical questions and *argumenta ad hominem* are completely avoided by the disputants and rational judge during the trial.

As we see, the standards of legal argumentation in Mesopotamia were really very strong.

3.2 Explication of Logical Inferences from Judicial Acts in Akkadian, Aramaic, and Ancient Greek

3.2.1 Problem Setting

Logeme structures presented in the texts written in the Sumerian and Akkadian languages lend themselves well to logical formalizations by means of modern propositional logic (Schumann, 2017b, 2019a, 2020, 2021a, 2021b; Schumann and Sazonov, 2021). The reconstructed correct reasoning of Babylonians contained at times a long chain of logical steps. In the meanwhile, articles from legal codes of Mesopotamia were sometimes used as premises for this reasoning. Let us remind that the code of Ur-Nammu (i.e., of the founder of Ur III) was the earliest surviving. This circumstance gives a strong structuralist argument in defense of the hypothesis of the existence of legislation in Mesopotamia: (i) the rationale in the records of court sessions is very complex, and this presupposes the perfect logical competence of their authors, although in such a sophisticated form, the competence is rather typical for modern logicians; hence, it can be assumed that this extraordinary competence was a part of legal hermeneutics; (ii) in the logical reasoning, the initial premises (assumptions, non-logical axioms) are found as some articles in the codes; thus, it can be assumed that some of the first premises (non-logical axioms) for making decisions in judicial opinions were taken from the corresponding articles in the Babylonian codes, so it demonstrates that these codes had the meaning of legislation, see Example 3.1.10.

On the island of Elephantine in Egypt, many papyri have been found containing legal texts in Coptic, Aramaic, Ancient Greek, and Latin. Meanwhile, the island's legal documents in Aramaic, and then in Ancient Greek, have much in common with the composition of the corresponding documents in Akkadian to the extent that some legal phrases were taken in the Aramaic language from the New Babylonian jurisprudence or were based on the terminology of the Mesopotamian legal system as such. This indicates that the Aramaic (Achaemenid) and then the Ancient Greek (Hellenistic) legal systems became a continuation of the Babylonian tradition. It is worth noting here that the only surviving codex in the dialect of the Ancient Greek language, found in Gortyn on the island of Crete and dating back to the 5th century B.C., was compiled in the legal tradition of Mesopotamia, e.g., many articles of the Gortynian laws repeat the code of Hammurabi, who ruled around 1792–1750 B.C.

So, this section shows that a strong similarity can be found between the structures of the logemes of legal documents in Akkadian, Aramaic, and Ancient Greek, which suggests that their authors had a close logical competence, and their logical culture had a single source, which is reconstructed as the earliest legal tradition in the world, originating for the first time in the Sumerian language.

The life of Akkadians was structured by two systems: (i) justice (*dīnu*) and (ii) omens (*alaktu*) (Gabbay, 2016; Lenzi, 2011). So, there were detailed lists of signs (*ittātu*/GISKIM.MEŠ) of two types: (i) articles of the law code; (ii) omens (forecasting) about everything with correlated events for divination. Each article of the code and each omen for divination has the same logical structure "If (Akkadian *šumma*) *S*, then *P*." The code was used for legal proceedings (to serve *dīnu*) and each court looked for an appropriate article of the code for a claim *S* to infer the trial decision *P* from the article "If *S*, then *P*" and the proved claim *S* just by *modus ponens*. The omens (as a part of *alaktu*) were used for prognoses and each scholar looked for an appropriate omen for an event *S* to infer the forecasting *P* from the rule "If *S*, then *P*" and the appearance of *S* just by *modus ponens*, as well (Rochberg, 2010). Hence, *modus ponens* was the most significant inference rule in the Sumerian-Akkadian culture. Also, we can find many examples of *modus tollens* in legal proceedings and forecastings.

In Greece, Chrysippus of Soli (Χρύσιππος ὁ Σολεύς, 279–206 B.C.) defined *modus ponens* and *modus tollens* for the first time correctly. His samples:

i) *modus ponens*: "If it is day, then it is light; but, in fact it is day; therefore, it is light" (Sextus Empiricus, *Against the Logicians* II: 224);
ii) *modus tollens*: "If it is day, then it is light; but it is not light; therefore, it is not day" (Sextus Empiricus, *Against the Logicians* II: 225).

We do not know whether Chrysippus applied his propositional logic in legal hermeneutics. But his prominent Roman follower, Marcus Tullius Cicero (106–43 B.C.) did it well in his *Topica* (Reinhardt, 2003). The Babylonians did it well too, but many and many centuries earlier, and not only in legal proceedings, but also in divination.

In this section, I propose a kind of formalization for *dīnu* and I show that legal proceedings reported in Neo-Babylonian trial records were based on an axiomatic system, where trial decisions were drawn automatically just by applying some inference rules to articles of a code or to previous trial decisions. It is a significant fact that Neo-Babylonian trial records and business contracts, including inheritance records, marriage contracts, and other documents concerning money, movable and immovable properties, consist of complex syllogisms without logical fallacies at all. These syllogisms are well formalizable within the contemporary propositional logic. Omens (forecastings), i.e., *alaktu*, can be formalized logically also, see Section 3.3.

The Talmud is a continuation of Aramaic legal tradition that became a continuation of Neo-Babylonian (Akkadian) law in turn. The Talmudic hermeneutics studied logically in (Abraham et al., 2015; Abraham et al., 2010; Abraham and Gabbay and Schild, 2010, 2013; Schumann, 2008, 2010a, 2010b, 2010c, 2010d, 2011a, 2011b, 2012a, 2012b, 2012c, 2013a, 2013b, 2013c, 2013d, 2015, 2016, 2017a, 2019b; Schumann and Kuznetsov, 2017) has the following two parts: (i) a *classical propositional logic* (implicitly used before the Talmud by the Assyrians and Neo-Babylonians for inferring court decisions on the basis of laws, Hebrew: *dîn*, Aramaic: *dīn'a*, a derivative word from the Akkadian *dīnu*); (ii) some *non-classical logics* for deducing new laws from the text of the Torah (Hebrew:

hălāḵāh, a derivative word from the Akkadian *alaktu*), see Subsection 1.2.2. Both parts have their roots in the Babylonian tradition.

Conventionally, Aristotle (384–322 B.C.) is considered a father of symbolic logic. Nevertheless, we can assume on the material of trial records (Holtz, 2009, 2014) that the Greek logic (rather the Stoic one) was based on a Sumerian-Akkadian legal hermeneutics, see Subsection 1.2.4. Notice that some Hebrew trial records in Aramaic made within a Babylonian tradition are found at Elephantine in Egypt (Botta, 2009). It is possible to show that judges knew some foundations of propositional logic well that was applied in deducing court decisions (Schumann, 2017b, 2019a, 2020). Hence, the origin of classical symbolic logic should have been connected to establishing a logical tradition of the Sumerian-Akkadian jurisprudence at first. The legal tradition of the Talmud is a direct continuation of the Babylonian tradition.

The structure of this section is based on the following conclusion drawn by applying the Mill's joint method of agreement and difference:

> The oldest fragments over the world consisting of complex syllogisms without logical fallacies (*x*) are presented by Mesopotamian legal documents (trial records and business contracts) (*A*);
>
> The Neo-Babylonian legal documents are made in Akkadian (*B*) within the Mesopotamian legal tradition (*A*);
>
> The Neo-Babylonian legal documents (*B*) are well formalizable as an axiomatic system extending the classical propositional logic to draw legal conclusions (*y*);
>
> Some Aramaic legal documents written in Aramaic (*C*) and excavated in Elephantine belong to the Jews (*D*);
>
> They (*C*) are made within the Aramaic legal tradition of the Achaemenid dynasty continuing the Akkadian legal tradition of Neo-Babylonians (*B*);
>
> The Aramaic legal documents of Elephantine (*C*) are well formalizable within the classical propositional logic of drawing legal conclusions and this logic is the same as of the Neo-Babylonian trial records (*y*);
>
> The Greek (then Byzantine) legal documents of Elephantine (*E*) are a direct continuation of the Aramaic legal tradition of Elephantine (*C*);
>
> These documents (*E*) are well formalizable within the classical propositional logic of drawing legal conclusions, too (*y*).
>
> ---
>
> Therefore, most probably, the first symbolic propositional logic (*x y*) was founded within the Mesopotamian legal tradition to draw legal conclusions (*A B*).
>
> Formally:
>
> *A* occurs together with *x*
> *A B* occur together with *x y*
> *A B C D E* occur together with *x y*
>
> ---
>
> Therefore, *A* is the cause, or the effect, or a part of the cause of *x* and *B* is the cause, or the effect, or a part of the cause of *y*

Hence, my reasoning follows Mill's joint method of agreement and difference to claim that the Sumerian-Akkadian jurisprudence (A) was a possible cause of using propositional logic correctly (x) and the Neo-Babylonian ways of making trial decisions (B) was a possible cause of using the axiomatic system extending the classical propositional logic to draw legal conclusions (y).

For the first time this idea was put forward in Schumann (2017b). In this section, I show a strong connection of symbolic logic to the ancient legality by proposing a logical formalization of legal reasoning belonging to Neo-Babylonians, Achaemenid Arameans, and Greeks of Hellenistic kingdoms to show that at the time of first Greek logicians there was a well-developed tradition of symbolic logic in legality rooted in the Neo-Babylonian and Achaemenid jurisprudence.

3.2.2 Logical Context for Studying Mesopotamian Texts

Main presuppositions of symbolic logic are formulated as follows: (i) each sentence (proposition or statement) is either true or false; (ii) a true sentence can be always verified by a reference to an appropriate fact described in the sentence and taking place indeed; (iii) there are logical inference rules which give only true conclusions from true premises; (iv) *modus ponens* and *modus tollens* are two basic inference rules. Let A and B be some propositions which can be either true or false. Then the rule of *modus ponens* is defined thus: Assume that "If A, then B" and A are true, then we can conclude that B is true, too. The rule of *modus tollens*: "If A, then B" is true, but B is false, then A is false, also.

For the first time, a symbolic logic with these presuppositions was explicitly introduced by Chrysippus (ca. 279–206 B.C.), the Stoic philosopher. There are no preserved texts authored by him, only some quotations. Thereby among all these quotations, the best illustration of his reasoning is given in the *Topica* written by Marcus Tullius Cicero (106 B.C.–43 B.C.). All the examples of Chrysippian conclusions in this treatise are taken from the legal hermeneutics, e.g.:

> If a slave has not been declared free either by the censor, or by the praetor's rod, or by the will of his master, he is not free: but none of those things is the case: therefore, he is not free (Yonge and London, 1891).

If S is not A ("a slave declared free by the censor") or S is not B ("a slave declared free by the praetor's rod") or S is not C ("a slave declared free by the will of his master"), then S is not D ('free'). S is not A. S is not B. S is not C.

Then S is not D

In this reasoning there are applied two inference rules: (i) *introducing disjunction* ("S is not A; S is not B; S is not C; then S is not A or S is not B or S is not C"); (ii) *modus ponens* ("If S is not A or S is not B or S is not C, then S is not D; S is not A or S is not B or S is not C; then S is not D"). Hence, we deal with the conclusions with two logical steps here. It is not so simple. But nevertheless, according to the *Topica*, even complicated logical conclusions of the Chrysippian symbolic logic were actively being applied in jurisprudence and in trial decisions at the time of Cicero.

It is important that the Chrysippian symbolic logic was being applied in trial records, business contracts, and testaments much earlier than Chrysippus lived, namely since some (not all) first legal documents of the third dynasty of Ur (ca. about 2111 B.C.–2003 B.C.).

Now the question is whether a symbolic logic can be used unintentionally (i.e., without studying logic as a system previously). Let us consider the following Lewis Carroll's logical puzzle. We have the four premises, and we should find out a conclusion from them:

a) None of the unnoticed things, met with at sea, are mermaids.
b) Things entered in the log, as met with at sea, are sure to be worth remembering.
c) I have never met with anything worth remembering, when on a voyage.
d) Things met with at sea, that are noticed, are sure to be recorded in the log.

Everybody who did not study a symbolic logic deeply cannot solve this task in any way. But the answer is very simple and made only by the three logical steps: "I have never met with a mermaid at sea. Another similar puzzle proposed by Lewis Carroll:

a) All babies are illogical.
b) Nobody is despised who can manage a crocodile.
c) Illogical persons are despised.

It can be solved now by two steps in inferring. The true answer is as follows: "Anyone who can manage a crocodile is not a baby." But it is impossible to be solved by anyone who did not study logic in advance, too.

Now let us take a task assuming only one step in inferring:

a) My gardener is well worth listening to on military subjects;
b) Nobody is worth listening to on military subjects, unless he can remember the battle of Waterloo.

How many people can conclude correctly from these two premises? Only a few respondents. But in Babylonian trial records, business contracts, and testaments there are correct conclusions with ten and even much more logical steps. And we face no logical fallacies. It cannot be occasionally. The point is that symbolic logic is not an innate knowledge, and it is not an innate ability (see Section 1.3). The Babylonian legal tradition was the only tradition in the human history before Aristotle and Chrysippus that had a perfect logical competence: very large and complicated logical conclusions without fallacies. Nevertheless, there is neither a logical textbook nor a logical treatise in Sumerian or Akkadian excavated till date (but we have some late logical parts of Talmudic books in Aramaic). But we have no excavated grammar of Akkadian or Sumerian, although there are known first dictionaries—Sumerian and Akkadian lexical lists ordered by topic, such as *ur*$_5$-*ra* = *ḫu-bul-lu*$_4$, see (Poebel, 1936).

3.2.3 Historical Context for Studying Mesopotamian Texts

There are the following cuneiform codices (Joannès, 2000; Westbrook, 2009): (i) the *Code of Ur-Nammu* (ca. 2048–2030 B.C. according to the short chronology) belonging to the third dynasty of Ur and written in Sumerian; (ii) the *Code of Lipit-Ištar* (ca. 1870–1860 B.C. by the short chronology) written also in Sumerian and belonging to the fifth king of the first dynasty of Isin; (iii) the *Laws of Ešnunna* (19th century B.C.) written in Akkadian and belonging to a king of Ešnunna, the city located in northern Mesopotamia on the Tigris River and becoming politically important after the fall of the third dynasty of Ur; (iv) the *Code of Hammurabi* (ca. 1728–1686 B.C. by the short chronology) written in Akkadian and belonging to Hammurabi, the sixth king of the first Babylonian dynasty; (v) the *Middle Assyrian Laws* (since the 12th century B.C.) written in the Assyrian dialect of Akkadian and much connected to Tiglath-Pileser I, the king of the Assyrian Empire; (vi) the *Hittite Laws* (ca. 1650–1100 B.C.) written in Hittite and containing the corpus of laws of the Hittite Empire; (vii) the *Neo-Babylonian Laws* (626 B.C.–539 B.C.) written in the Neo-Babylonian dialect of Akkadian and containing the corpus of laws of the Neo-Babylonian Empire. Aramaic was the everyday language at the time of Neo-Babylonians, but this language became the language of administration and culture only since the Achaemenid Empire (ca. 550 B.C.–330 B.C.), although cuneiform texts in Akkadian remained popular, also. Hence, (viii) the *Achaemenid Laws* were written in the Neo-Babylonian dialect of Akkadian as well as in Aramaic in the Aramaic script. Since establishing the Hellenistic Empire by Alexander the Great (334 B.C.–323 B.C.), (ix) the *Hellenistic Laws* were written mainly in Greek, but also Aramaic was used for the same purpose.

In Assyriology, there is quite a popular claim that these codes and laws did not function as a normative and binding legislation. For example, Raymond Westbrook puts forward a hypothesis that the codes were legal treatises, i.e., a secondary legal authority just to show possible legal lists with individual items formulated in casuistic style (Westbrook, 2009). There is even a more radical opinion that the codes were only a kind of ideology like the Bible was for Christians and it is till now. For instance, in Christianity different Biblical commandments such as "thou shalt not steal" or "thou shalt not covet" never played a basis of legislation. In Christianity, the Bible was never used in criminal and civil laws. It is explained by that since Constantine the Great (ca. 272 A.D.–337 A.D.) Christianity has been accepted in the Roman Empire only as ideology—the Roman laws are maintained to be applied as criminal and civil laws further. The idea that the Babylonian codes and laws were never a form of legislation is an unconscious 'Chrstianization' of Babylonians.

Among different arguments supporting this 'Christianization' of Babylonians the strongest is that we never find explicit references to these codes as sources of law. The problem is that many cuneiform tablets are not published yet. Nevertheless, in unpublished texts there are these references in fact. For example, in some fragments of trial and business documents of Old Assyrian period (ca. 2025 B.C.–1378 B.C.) there is an explicit expression "the words of the stele" (*awāt naruāim*), see for the details in Veenhof (1995), with the meaning to be such a reference. Also, we know that these references were not necessary. They were just assumed implicitly. So, there was excavated a very important Hebrew ostracon at Meṣad Hašavyahu, dated to from 640 B.C. to 609 B.C., the time of Josiah, the king of Judah (Albright, 1969: 568). The text of this ostracon is as follows:

ישמע אדני השר
את דבר עבדה. עבדך
קצר. היה. עבדך. בח
צר אסם. ויקצר עבדך
ויכל ואסם כימם. לפני שב
ת כאשר כל {ע}בדך את קצר וא
סם כימם ויבא הושעיהו בן שב
י. ויקח. את בגד עבדך כאשר כלת
את קצרי זה ימם לקח את בגד עבדך
וכל אחי. יענו. לי. הקצרם אתי בחם.
{ה}ש {מש} אחי. יענו. לי אמן נקתי מא
{שם}.........בגדי ואמלא. לשר להש
{יב}..........עב{דך}.....אלו. רח
{מם. והש}בת את {בגד. ע}בדך ולא תדהם נ

yšm' 'dny hsr
't dbr 'bdh. 'bdk
qṣr. hyh. 'bdk. bḥ
ṣr 'sm. wyqṣr 'bdk
wykl w'sm kymm. lpny šb
t k'šr kl {'}bdk 't qṣr w'
sm kymm wyb' hwš'yhw bn šb
y. wyqḥ. 't bgd 'bdk k'šr klt
't qṣry zh ymm lqḥ 't bgd 'bdk
wkl 'ḥy. y'nw. ly. hqṣrm 'ty bḥm.
{h}š {mš} 'ḥy. y'nw. ly 'mn nqty m'
{šm}.........bgdy w'ml'. lsr lhš
{yb}.......... 'b{dk}..... 'lw. rḥ
{mm. whš}bt 't {bgd. '}bdk wl' tdhm n

Its translation:

Let my lord, the governor, hear the word of his servant! Your servant is a reaper. Your servant was in Haṣar 'Asam, and your servant reaped, and he finished, and he was storing up (the grain) during these days before the Sabbath. When your servant had finished the harvest and was storing (the grain) during these days, Hošeyahu came, the son of Šobi, and he seized the garment of your servant, when I had finished my harvest. It (is already now some) days (since) he took the garment of your servant. And all my companions can bear witness for me—they who reaped with me in the heat of the harvest—yes, my companions can bear witness for me. Amen! I am innocent from guilt. And he stole my garment! It is for the governor to give back the garment of his servant. So, grant him mercy in that you return the garment of your servant and do not be displeased (Smelik, 1991: 96).

As we see, the author of the text makes a statement in court. He affirms that he has at least two witnesses who can support his claim—it is enough witnessing at a Jewish (Babylonian) court (Pritchard, 1969). In his claim he is appealing to the unjust act of the fortress governor who confiscated the garment on the eve of the Sabbath. Whether the fortress governor could take this garment? According to the Torah, he cannot:

> 26 If thou at all take thy neighbour's raiment to pledge, thou shalt deliver it unto him by that the sun goeth down:
> 27 For that is his covering only, it is his raiment for his skin: wherein shall he sleep? and it shall come to pass, when he crieth unto me, that I will hear; for I am gracious (*Exodus* 22: 26–27).

> 12 And if the man be poor, thou shalt not sleep with his pledge:
> 13 In any case thou shalt deliver him the pledge again when the sun goeth down, that he may sleep in his own raiment, and bless thee: and it shall be righteousness unto thee before the Lord thy God (*Deuteronomy* 24: 12–13).

Thus, the author refers to the Torah only implicitly. But it is enough, because only the Torah was accepted as a primary source of legislation at the time of Josiah, the king of Judah (ca. 648–609 B.C.). The same situation can hold for Neo-Babylonian trial decisions. We do not need explicit references there too.

The Hebrew ostracon is also interesting as it assumes that an appropriate trial decision should be drawn only by two logical steps through applying *modus ponens* and *substitution rule*:

a) "If thou at all take thy neighbour's raiment to pledge, thou shalt deliver it unto him by that the sun goeth down" is true (because it is an axiom from the Torah);
b) "The fortress's governor took the raiment of the claimant" is true (because it is verified by witnessing).

The fortress's governor should deliver the raiment unto the claimant back.

In the Neo-Babylonian trial records, many more logical steps occur in drawing trial decisions and always it is supposed that there are axioms taken from a code, although without a direct reference. For instance, for stealing one item belonging to the god, the thief should give the same item thirtyfold, see Example 3.1.10. According to the Neo-Babylonian trial records, it is a non-logical axiom. But we know that this statement is an article of the *Hammurabi Code* (Roth, 1995: 82). Thus, we can suggest that this statement was an axiom for courts since the time of Hammurabi.

In this section, I am based on a structuralist argument confirming the hypothesis of Babylonian legislation: (i) the symbolic-logical reasoning of trial records are very complicated; so, they assume a perfect logical competence of their authors; (ii) in this symbolic-logical reasoning there are assumed some non-logical axioms; many of them occur in some Babylonian codes; thus, we can suggest that all these axioms in trial conclusions were taken from appropriate articles in Babylonian codes.[3]

[3] There I have not addressed the existence of Akkadian legal decisions from earlier periods. About them, please see (Hunger, 2009) and Section 3.1.

3.2.4 Axiomatic System for Legal Proceedings

The Sumerian law codes were first over the world, e.g., *Ur-Nammu* (ca. 2048–2030 B.C.) and *Lipit-Ištar* (ca. 1870 B.C.–ca. 1860 B.C.). Later the Akkadians and Babylonians, the successors of Sumerians, continued this Sumerian legal tradition and established some of their own codes, such as *Hammurabi* (1728–1686 B.C.), see Figure 3.2.1.

Many trial records were excavated of the Neo-Babylonian period, i.e., of the time beginning with the rise of the Babylonian King Nabopolassar in 626 B.C. and lasting until the end of the reign of Nabonidus because of the Achaemenid conquest in 539 B.C. There are also many trial records thereafter, but they were written within the same unchanged cuneiform textual tradition of Neo-Babylonia (Holtz, 2009, 2014). Therefore, they can also be called Neo-Babylonian. All these trial records are especially interesting from the point of view of symbolic logic. The matter is that these texts demonstrate a high logical competence of their authors. Furthermore, we can claim that these documents obviously show an axiomatization of legal proceedings. The thing is that Neo-Babylonian trial records are well preserved and many of them are large enough to be reconstructed logically.

In the Neo-Babylonian trial records, suitable articles from law codes are assumed to be non-logical axioms. Facts established by the court give another set of non-logical axioms. And according to some logical inference rules, just automatically the court makes an appropriate decision from axioms, i.e., from facts and articles

Figure 3.2.1. The stele of the *Law Code of Hammurabi*, Louvre Museum. **By courtesy:** Vladimir Sazonov.

of law codes, in a logically correct way. We can assume that the same situation was before the Neo-Babylonian period, too: (i) the law codes of different periods have had many unchanged articles and the *Hammurabi Code* can be examined as a significant early sample of all these legal codifications; (ii) the court procedure has remained almost the same since the very beginning; (iii) there is much evidence that the law codes were considered as a set of non-logical axioms for logical inferring, e.g., in some fragments of trial and business documents of Old Assyrian period (that took place from 2025 B.C. to 1378 B.C.) we find out an expression "the words of the stele" (*awāt naruāim*), see for the details in Veenhof (1995), with the meaning to be a reference to an axiom for inferring from it.

Each article from codes is formulated as follows: "If (*šumma*) you do such-and-such, *then* a trial decision will be as follows". So, on the one hand, in the antecedent of implication we deal with some combinations of illegal and legal actions conducted before the court: 'steal', 'possess', 'strike', "causes the boat to sink", etc. and, on the other hand, in the succedent of implication we deal with some legal consequences established by the court: 'compensate', 'kill', etc. There are the following main classes of people in the *Law Code of Hammurabi* (Roth, 1995): (i) 'man' or "free person" (*awīlu*), to denote men, women, and minors; (ii) 'commoner' (*muškēnu*); (iii) "male slave" (*wardu*) and "female slave" (*amtu*). Some classes to denote familial relationships: (i) 'son' (*māru*); (ii) 'wife' (*aššatu*); (iii) "first-ranking wife" (*ḫīrtu*); (iv) 'widow' (*almattu*), etc. Also, there are some abstract entities such as 'god' (i.e., a temple as legal person) or 'palace' (i.e., a king as legal person) to denote a form of possession. And some professional groups: 'boatman'; 'merchant', etc.

Let us formalize the articles of the law in the following way. Suppose the small letters $a, b, c, \ldots, x, y, z, \ldots$ are constants and variables to denote things. Capital letters are to denote classes of natural and legal persons: A (*awīlu*-class); M (*muškēnu*-class); T (temple or palace as legal persons), B (boatmen), etc. The expression Ax means that y belongs to the *awīlu*-class—My means that y belongs to the *muškēnu*-class, and Tz means that z belongs to the category of legal persons.

First-order formulas are defined in the following standard manner: (i) each $a, b, c, \ldots, x, y, z, \ldots$ is a first-order formula; (ii) each $Ax, My, Tz, \ldots$ is a first-order formula; (iii) each Boolean combination of formulas defined in (i) and (ii) is a first-order formula (Boolean connectives: $\neg$ negation, & conjunction, $\vee$ disjunction, $\Rightarrow$ implication, $\oplus$ strong disjunction).

Let an action 'verb' in the antecedent or succedent of articles of the law be denoted by [[verb]]. Each action is an n-place relation among people of different classes and things. For example: [[steal (Ax, z)]] means that x of the *awīlu*-class steals z; [[strike (Ax, Ay)]] means that x of the *awīlu*-class strikes y of the same class; [[boat-to-sink (Bx, y)]] means that a boatman x causes the boat y to sink; [[possess (Ty, z)]] means that y is a legal person who possesses z; [[compensate (Ax, Ty, z)]] means that x of the *awīlu*-class must compensate the legal person y by paying z; [[killed (Ax)]] means that x of the *awīlu*-class is sentenced to be killed, etc.

Second-order formulas are defined thus: (i) each [[steal (Ax, z)]], [[strike (Ax, Ay)]], [[boat-to-sink (Bx, y)]], [[possess (Ty, z)]], [[compensate (Ax, Ty, z)]], [[killed (Ax)]], … is a second-order formula; (ii) each Boolean combination of formulas defined in (i) is a second-order formula (Boolean connectives: ¬ negation, & conjunction, ∨ disjunction, ⇒ implication, ⊕ strong disjunction).

Now, let us consider some articles from codes as non-logical axioms for legal proceedings. Some examples:

A1

If a man x of the *awīlu*-class steals a thing z and it belongs to a temple or king y, then x must replace z thirtyfold. If z belongs to a commoner y, he must replace it tenfold. If the thief does not have anything to give, he shall be killed:

$(([[\text{steal }(Ax, z)]] \ \& \ [[\text{possess }(Ty, z)]]) \Rightarrow [[\text{compensate }(Ax, Ty, z \cdot 30)]]) \ \&$
$(\neg\, [[\text{compensate }(Ax, Ty, z \cdot 30)]] \Rightarrow [[\text{killed }(Ax)]])$
&
$(([[\text{steal }(Ax, z)]] \ \& \ [[\text{possess }(Cy, z)]]) \Rightarrow [[\text{compensate }(Ax, Cy, z \cdot 10)]])$
$\& \ (\neg\, [[\text{compensate }(Ax, Cy, z \cdot 10)]] \Rightarrow [[\text{killed }(Ax)]])$

In the *Hammurabi Code*:

(vi 57–69) *šumma awīlum lu alpam lu immeram lu imēram lu šaḫâm ulu elippam išriq šumma ša ilim šumma ša ekallim adi* 30-*šu inaddin šumma ša muškēnim adi* 10-*šu iriab šumma šarrāqānum sa nadānim la išu iddâk.*

§8 If a man steals an ox, a sheep, a donkey, a pig, or a boat—if it belongs either to the god or to the palace, he shall give thirtyfold; if it belongs to a commoner, he shall replace it tenfold; if the thief does not have anything to give, he shall be killed (Roth, 1995: 82).

A2

If a man x of the *awīlu*-class strikes another man, he must pay 60 shekels of silver:

$[[\text{strike }(Ax, Ay)]] \Rightarrow [[\text{compensate }(Ax, Ay, 60 \text{ shekels})]]$

In the *Hammurabi Code*:

(xl 82–87) *šumma mār awīlim lēt mār awīlim ša kīma šuāti imtaḫaṣ* 1 *mana kaspam išaqqal.*

§203 If a member of the *awīlu*-class should strike the cheek of another member of the *awīlu*-class who is his equal, he shall weigh and deliver 60 shekels of silver (Roth, 1995: 121).

A3

If a boatman x causes the boat y to sink, he must compensate all things z contained there:

$([[\text{boat-to-sink }(Bx, y)]] \ \& \ [[\text{contain }(y, z)]] \ \& \ [[\text{possess }(Ay, z)]]) \Rightarrow [[\text{compensate }(Bx, Ay, z)]]$

In the *Laws of Ešnunna* (ca. 1790 B.C.):

(A i 25–26) *šumma malaḫḫum īgīma elippam uṭṭebbe mala uṭebbû umalla.*

§5 If the boatman is negligent and causes the boat to sink, he shall restore as much as he caused to sink (Roth, 1995: 60).

Hence, all the articles are formulated as a set of implications connected by conjunctions: $(A_1 \Rightarrow B_1)$ & $(A_2 \Rightarrow B_2)$ & … & $(A_n \Rightarrow B_n)$. In the meanwhile, the antecedents $A_1, A_2, \ldots, A_n$ of these implications are understood as particulars and their succedents $B_1, B_2, \ldots, B_n$ as appropriate generals.

One of the main features of law codes explicitly stated first by Cicero in his *Topica* (Reinhardt, 2003) (the work devoted to legal hermeneutics in the Stoic/Sumerian-Akkadian way) is a full enumeration of particulars $A_1, A_2, \ldots, A_n$, related to one general B, i.e., a full list of implications $A_1 \Rightarrow B, A_2 \Rightarrow B, \ldots, A_n \Rightarrow B$ with the same B, as it was supposed in any law code since *Hammurabi*. These $A_1, A_2, \ldots, A_n$ should be exclusive thereby. It means that they should be connected by strong disjunctions "either... or ..." (Akkadian: "*ūl* … *ūl*…", symbolically: "… ⊕..."). In this case there is the following equivalence: $(A_1 \oplus A_2 \oplus \ldots \oplus A_n) \Leftrightarrow B$. From this we can draw the following conclusion:

$$\frac{\dfrac{A_1 \Rightarrow B;\ A_2 \Rightarrow B;\ \ldots;\ A_n \Rightarrow B;\ C \Rightarrow \neg A_1;\ C \Rightarrow \neg A_2;\ \ldots;\ C \Rightarrow \neg A_n}{(A_1 \oplus A_2 \oplus \ldots \oplus A_n) \Leftrightarrow B; \qquad C \Rightarrow \neg (A_1 \oplus A_2 \oplus \ldots \oplus A_n)}}{C \Rightarrow \neg B.}$$

Cicero formulates this rule thus:

> Next, the enumeration of the parts (sc. of the whole), which is handled in the following way: If someone has not been freed by either having his name entered in the census-roll or by being touched with the rod or by a provision in a will, then he is not free. None of these applies to the individual in question. Therefore, he is not free (Reinhardt, 2003: 121).

This logical rule implemented in any code may be named a *completeness of legal information*. This completeness means that if we take any factual verified case C of an indictment, then for any general B from the code, each court can announce either a verdict $C \Rightarrow B$ or a verdict $C \Rightarrow \neg B$ inferred from the code just logically.

Each court has considered only cases supported by documents or couples of witnesses, see Example 3.1.9. Symbolically, we apply *modus tollens*:

1. If a man or woman brings a claim to present a case, he or she must support his or her claim about the case by documents or witnesses [the non-logical axiom supposed in any code].
2. There are no documents and witnesses for this case [a fact proved by the court].

Then, the case is not valid to be approved.

The case of the claim, C, is beyond all the n cases $A_1, A_2, \ldots, A_n$ given in a code (killing, stealing, etc.), which are usually considered by the court, i.e., we have: $C \Rightarrow \neg (A_1 \oplus A_2 \oplus \ldots \oplus A_n)$, where each case A_i is supported by appropriate documents and couples of witnesses. Then this case is out of any legal proceeding, B, i.e.: $C \Rightarrow \neg B$. In other words, if someone is C, he or she is guilty either in A_1 (killing) or in A_2 (stealing) or… in A_n. Formally: $C \Rightarrow (A_1 \oplus A_2 \oplus \ldots \oplus A_n)$. The court has considered all documents and witnesses in respect to the defendant and the case is not A_1 (killing) and it is not A_2 (stealing) and … it is not A_n: $(\neg A_1 \,\&\, \neg A_2 \,\&\, \dots \,\&\, \neg A_n)$. This expression is the same as $\neg (A_1 \vee A_2 \vee \dots \vee A_n)$. In turn, from the latter formula it follows that $\neg (A_1 \oplus A_2 \oplus \dots \oplus A_n)$, because there is a tautology: $\neg (A_1 \vee A_2 \vee \dots \vee A_n) \Rightarrow \neg (A_1 \oplus A_2 \oplus \dots \oplus A_n)$. Then, by *modus tollens* we deduce that he or she is not C (he or she is not guilty, i.e., not defendant):

$$\frac{C \Rightarrow (A_1 \oplus A_2 \oplus \dots \oplus A_n);\quad \neg (A_1 \oplus A_2 \oplus \dots \oplus A_n)}{\neg C}$$

Hence, not only articles of the law, such as **A1**, **A2**, and **A3**, were examined as non-logical axioms for legal proceedings, but also facts established by the court by appealing to signed documents or presented witnesses. Let us show how a document has supported a view of defendant partly. There is a signed document that ^{m}Iddin-Marduk possesses 480 *gur* of dates and the boatman ^{m}Amurru-natan son of ^{m}Ammaya is responsible to deliver them safely. Nevertheless, ^{m}Nergal-rēṣūa the slave of ^{m}Iddin-Marduk testimonies that 47 *gur* 1 *pi* are missing. But there is a contract that the boatman ^{m}Amurru-natan gave back 7 *gur* 1 *pi* of dates from missing items. The trial decision is that ^{m}Amurru-natan must pay 40 *gur* of dates, the missing amount of those dates, and assigned them to ^{m}Iddin-Marduk:

1. ^{md}U.GUR-*re-ṣu-u*$_2$*-a* ^{lu2}*qal-la ša*$_2$ ^{m}MU-^{d}AMAR.UTU
2. *a-na* ^{lu2}DI.KU$_5$.MEŠ *ša*$_2$ ^{md}NA$_3$IM.TUK LUGAL TIN.TIR^{ki}
3. *iq-bi um-ma* ^{m}MU-^{d}AMAR.UTU EN-*a*
4. 4 ME 80 GUR ZU$_2$.LUM.MA *e-pi-ru-tu*
5. *ul-tu* EDIN *a-na* ^{giš}MA$_2$.MEŠ *ša*$_2$ ^{md}KUR.GAL-*na-tan*
6. ^{lu2}MA$_2$.LAḪ$_5$ A-*šu*$_2$ *ša*$_2$ ^{m}*am-ma-a u*$_2$*-še-li-ma*
7. *pu-ut* EN.NUN-*tim ša*$_2$ ZU$_2$. LUM.MA *u*$_2$*-ša*$_2$*-aš*$_2$*-ši-iš*
8. ^{giš}MA$_2$.MEŠ *a-na* TIN.TIR^{ki} *u*$_2$*-še-la-am-ma*
9. *ši-pir-tu*$_4$ *ša*$_2$ ^{m}MU-^{d}AMAR. UTU *id-di-nam-ma*
10. 4 ME 80 GUR ZU$_2$.LUM. MA *ina lib*$_3$-[*bi-šu*$_2$] *ša*$_2$*-ṭi-ir*
11. *re-eš* ZU$_2$.LUM.MA *aš*$_2$*-ši-ma* 47 GUR 1 PI
12. *ina lib*$_3$*-bi ma-ṭu-*⌈*u*$_2$⌉ *a-na* UGU
13. *mi-ṭi-tu*$_4$ *ša*$_2$ ZU$_2$.LUM.MA *it-ti* ^{md}KUR.GAL-*na-tan*
14. *ar-gum*$_2$*-ma u*$_2$-ŠAR-X-RI *um-ma* ZU$_2$.[LUM.MA]-*ka*
15. *ul aš*$_2$*-ši ar*$_2^{?}$*-ki ba-ti-qu* X X X...
16. 4$^{!}$ GUR 1 PI ⌈ZU$_2$.⌉[LUM. MA]
17. *u*$_3$ *ku-tal-la ša*$_2$ ⌈^{giš}MA$_2$*-ni*⌉ X-X *u*$_2$
18. ZU$_2$.LUM.MA *šu*$_2$*-nu-tu*$_2$ *i-na* X-*šu*$_2$-[
19. *rik-su it-ti-šu*$_2$ *ni-iš-ku-us*
20. *um-ma* 7 GUR 1 PI ZU$_2$.LUM. MA

21. *md*KUR.GAL-*na-tan ina sar-tu iš-šu-u*$_2$
22. *ar*$_2$*-ki ri-ik-su šu-a-tu*2 ^{md}KUR. GAL-[*na-tan*]
23. *šuṭ-ur-ma a-di u*$_4$*-mu an-ni-i* X
24. *i-na-an-na i-na maḫ-ri-ku-nu ub-la-aš*$_2$
25. EŠ.BAR-*a-ni šuk-na* ^{lu2}DI.KU$_5$.MEŠ *dib-bi-šu*$_2$*-nu*
26. *iš-mu-u*$_2$ *rik-su šu-a-tu*$_2$ *u ši-pir-tu*$_4$
27. *ša*$_2$ ^{m}MU-^{d}AMAR.UTU *ša*$_2$ 4 ME 80 GUR ZU$_2$.LUM.MA
28. *ina lib*$_3$*-bi šaṭ-ru ša*$_2$ ^{md}U.GUR*re-ṣu-u*$_2$*-a ub-la*
29. *ma-ḫar-šu-nu iš-tas-su-u*$_2$ ^{md}KUR.GAL-*na-tan*
30. *i-ša*$_2$*-lu-ma na-šu-u*$_2$ *ša*$_2$ ZU$_2$. LUM.MA *ša*$_2$ *ina sar-tu*$_4$
31. *na-šu-u*$_2$ *e-li ra-ma-ni-šu*$_2$ *u*$_2$*-kin-ma*
32. 40 GUR ZU$_2$.LUM.MA *mi-ṭi-tu*$_4$ *ša*$_2$ ZU$_2$.LUM.MA *šu*$_2$*-nu-šu*$_2$
33. *e-li* [^{m}]^{d}KUR.GAL-*na-tan ip-rusu-ma*
34. *a-na* ^{md}U.GUR-*re-ṣu-u*$_2$*-a* ^{lu2}[*qal-la ša*$_2$] ^{m}MU-^{d}AMAR.UTU
35. *id-di-nu ina* EŠ.BAR ⌈*di-i-ni*⌉ *šu-a-tim*
36. ^{md}U.GUR-[GI ^{lu2}DI.KU$_5$] DUMU *ši-gu-u*$_2$*-a*
37. ^{md}NA$_3$-ŠEŠ.MEŠ-MU ^{lu2}DI. KU5 [DUMU]⌈*e-gi-bi*⌉
38. ^{md}NA$_3$-[MU-GI].NA ^{lu2}DI.KU$_5$ DUMU *ir-a-*[*ni*]
39. ^{md}EN-[ŠEŠ.MEŠ]-⌈MU⌉ ^{lu2}DI.KU$_5$ DUMU ^{md}ZALAG^{d}30
40. ^{md}EN-[KAR]-⌈*ir*⌉ ^{lu2}DI.KU$_5$ DUMU ^{md}30-*tab-ni*
41. ^{md}NA$_3$-MU-GAR-*un* DUB. SAR DUMU ^{lu2}GAL-DU$_3$
42. TIN.TIR^{ki} ITI ŠE U$_4$ 4-*kam*$_2$
43. MU 10-*kam*$_2$ ^{md}NA$_3$-IM.TUK LUGAL TIN.TIR^{ki}
    ^{na4}KIŠIB ^{md}U.GUR-GI ^{lu2}DI.KU$_5$
    ^{na4}KIŠIB ^{md}NA$_3$-ŠEŠ.MEŠ-MU [^{lu2}DI].KU$_5$
    ^{na4}KIŠIB ^{md}NA$_3$-MU-GI.NA ^{lu2}DI. KU$_5$
    ^{na4}KIŠIB ^{md}EN-ŠEŠ.MEŠ-MU ^{lu2}DI.KU$_5$
    ^{na4}[KIŠIB] ^{md}EN-KAR-[*ir*] ^{lu2}DI.KU$_5$

(1–3) ^{m}Nergal-rēṣūa the slave of ^{m}Iddin-Marduk said thus to the judges of Nabonidus, king of Babylon:

(3–6) "^{m}Iddin-Marduk, my master, loaded a shipment of 480 *kur* of dates for transport (?) from the hinterland on the boats belonging to ^{m}Amurru-natan, the boatman, son of ^{m}Ammaya."

(7) "He had him bear the responsibility for keeping the dates."

(8–10) "He brought the boats to Babylon, and he gave me ^{m}Iddin-Marduk's message. 480 Gur of dates was written i[n it]."

(11–12) "I took account of the dates, and 47 *gur* 1 *pi* were missing."

(12–14) I raised a claim against ^{m}Amurru-natan concerning the missing amount of the dates and … thus:

(14–15) "'I did not take your dates.'"

(15) "Afterwards, an informer …

(16) "'4 Gur 1 Pi of dates …

(17) "'and behind my boat …

(18) "'those dates in …

(19–20) "We contracted a contract stating thus: '^{m}Amurru-natan illegally took 7 *gur* 1 *pi* of dates.'"

(22–23) "After ^{m}Amurru-[natan] wrote this contract until today …

(24) Now, I have brought him before you."

(25) "Establish our decision!"

(25–26) The judges heard their arguments.

(26–29) They read before them that contract and ^{m}Iddin-Marduk's message in which 480 Gur of dates was written which ^{m}Nergalrēṣūa brought.

(29–30) They questioned ^{m}Amurru-natan.

(30–31) (Regarding) the taking of the dates, he established about himself that they were taken illegally.

(32–35) They decided that ^{m}Amurru-natan must pay 40 *gur* of dates, the missing amount of those dates, and assigned them to ^{m}Iddin-Marduk, [slave] of ^{m}Nergal-rēṣua.

(35) At the decision of this case:

(36) ^{m}Nergal-[ušallim, the judge], descendant of Šigûa;

(37) ^{m}Nabû-aḫḫē-iddin, the judge, [descendant of] Egibi;

(38) ^{m}Nabû-[šuma-ukī]n, the judge, descendant of Ir'an[ni];

(39) ^{m}Bēl-[aē]-iddin, the judge, descendant of Nūr-Sîn;

(40) ^{m}Bēl-ēṭir, the judge, descendant of Sîn-tabni;

[*Scribe*:]

(41) ^{m}Nabû-šuma-iškun, the scribe, descendant of Rāb-bānê.

[*Date*:]

(42–43) Babylon. 4 Addaru, year 10 of Nabonidus, king of Babylon.

[*Seals of authorities*]

[*Left edge*:]

Seal of ^{m}Nergal-ušallim, the judge;

Seal of ^{m}Nabû-aḫḫē-iddin [the jud]ge;

Seal of ^{m}Nabû-šuma-ukīn, the judge;

[*Right edge*:]

Seal of ^{m}Bēl-aḫḫē-iddin, the judge;

[*Seal*] of ^{m}Bēl-ēt[ir], the judge (Holtz, 2009: 28–32).

1. If a boatman has a contract to deliver some items safely, he is responsible for them and must repay them if they are missing [the non-logical axiom from the code, compare to **A3**];
2. The boatman ^{m}Amurru-natan son of ^{m}Ammaya has a contract to deliver 480 *gur* of dates belonging to ^{m}Iddin-Marduk [the fact established by the trial];
3. ^{m}Nergal-rēṣūa the slave of ^{m}Iddin-Marduk testimonies that 47 *gur* 1 *pi* are missing [the claim];
4. There is a contract that the boatman ^{m}Amurru-natan gave back 7 *gur* 1 *pi* of dates from missing items [the fact established by the trial];

Then, ^{m}Amurru-natan son of ^{m}Ammaya must repay 40 *gur* of dates to ^{m}Iddin-Marduk on 4 Addaru, year 10 of Nabonidus, king of Babylon.

This conclusion is perfect logically and we can even formalize it within an axiomatic system for the Neo-Babylonian legal proceedings (where articles from a code and facts confirmed by a trial are non-logical axioms and propositional logic is to deduce from the axioms some new conclusions). Assume that we have non-logical axioms of the following two types: (i) the articles of the law formulated *explicitly*, as well as the directly derivated propositions from the articles by using logical inference rules (the articles of the law formulated *implicitly*); (ii) the facts as signed documents or sworn testimonies checked by the court.

In this axiomatic system we apply some inference rules. The two basic rules for inferring from axioms A, $\neg B$, $A \Rightarrow B$ are as follows:

Modus ponens (**MP**):

$$\frac{A;\ A \Rightarrow B}{B}$$

Modus tollens (**MT**):

$$\frac{\neg B;\ A \Rightarrow B}{\neg A}$$

Some other inference rules:

Introducing conjunction (**I&**):

$$\frac{A;\ B}{A \,\&\, B}$$

Deleting conjunction (**D&**):

$$\frac{A \,\&\, B}{A;\ B}$$

Introducing disjunction (**I∨**):

$$\frac{A;\ B}{A \vee B}$$

Let S[*B*] mean that *B* is a variable *a*, *b*, *c*, …, *x*, *y*, *z*, … or *Ax*, *My*, *Tz*, … belonging to a formula S. Let *A* be a constant. An additional inference rule:

Substitution rule (**SR**):

$$\frac{S[B];\ A \Rightarrow B}{S[A]}$$

In other words, if $A \Rightarrow B$ holds, then we can substitute *A* for any occurrence of *B* in a formula S. For example, in expressions such as ([[steal (*Ax*, *z*)]] & [[possess (*Ty*, *z*)]]) ⇒ [[compensate (*Ax*, *Ty*, $z \cdot 30$)]] from a code, we can replace variables by constants: *Ax* by ‘ᵐAḫa-iddin’, *Ty* by “the Lady-of-Uruk and Nanaya”, and *z* by ‘ducks’, etc., if there are implications: ‘ᵐAḫa-iddin’ ⇒ *Ax*, “the Lady-of-Uruk and Nanaya” ⇒ *Ty*, and ‘ducks’ ⇒ *z*, validated by a court

Inference rules **MP**, **MT**, **I&**, **D&**, **I∨**, **SR** are enough for inferring trial decisions from the code and facts checked by the court, but to reduce a length of deduction we can introduce some new inference rules of the form:

$$\frac{A;\ B}{C}$$

if and only if the formula $(A \ \&\ B) \Rightarrow C$ is a tautology of propositional logic.

3.2.5 Logical Conclusions Drawn in Neo-Babylonian Trial Records

Each excavated Neo-Babylonian trial record can be examined as a conclusion drawn from non-logical axioms automatically. Let us consider Example 3.1.10 (Holtz, 2014: 178–181) again, but now formally. In this trial record, two judges determine that Bēl-iqīša, who led away 5 sheep belonging to “Ištar of Uruk and Nanaya” (a temple), must repay 155 sheep to the property of this temple, because 150 sheep is the **thirtyfold** penalty for 5 branded sheep and the 5 unbranded lambs are supposed to be born after stealing. This trial record is symbolized as an inference by *modus ponens* (**MP**) as follows:

1. If a man steals *X* sheep and it belongs to the god (to a temple), then he must replace it **thirtyfold** (i.e., the amount of $X \cdot 30$) [the non-logical axiom from the code, see **A1**];
2. Ištar of Uruk and Nanaya is a temple [it is a fact, because “a temple” is a generalization for the case of “Ištar of Uruk and Nanaya”];

3. Bēl-iqīša son of Ṣillaya led away 5 sheep belonging to Ištar of Uruk and Nanaya [the fact established by the trial];
4. But 5 unblemished lambs are a property of Ištar of Uruk and Nanaya too [the fact established by the trial];
5. Arad-Nergal son of Mukīnapli descendant of Egibi is a guarantor for the repayment of Bēl-iqīša son of Ṣillaya [the fact established by the document];

Then, Bēl-iqīša son of Ṣillaya must repay 155 sheep to Ištar of Uruk and Nanaya on 25 Addaru, year 3 of Cambyses, king of Babylon.

More formally step by step:

Axiom **A1**:

1) $([[\text{steal}\ (Ax, z)]]\ \&\ [[\text{possess}\ (Ty, z)]]) \Rightarrow [[\text{compensate}\ (Ax, Ty, z \cdot 30)]]$

   Facts proved by the trial:

2) $a \Rightarrow Ax$, where $a :=$ "Bēl-iqīša son of Ṣillaya".
3) $b \Rightarrow Ty$, where $b :=$ "Ištar of Uruk and Nanaya".
4) $c \Rightarrow z$, where $c :=$ "5 sheep".
5) $[[\text{steal}\ (a, c)]]$
6) $[[\text{possess}\ (b, c)]]$

Thus, (1)–(6) are axioms for the court and basing on them the court should draw a logical conclusion. According to the substitution rule, we can substitute a for Ax, b for Ty, and c for z in **A1**. Then we obtain by **SR**:

7) $([[\text{steal}\ (a, c)]]\ \&\ [[\text{possess}\ (b, c)]]) \Rightarrow [[\text{compensate}\ (a, b, c \cdot 30)]]$

   From (5) and (6) we deduce by **I&**:

8) $[[\text{steal}\ (a, c)]]\ \&\ [[\text{possess}\ (b, c)]]$

   Then, finally, by **MP** (*modus ponens*) we conclude from (7) and (8) that

9) $[[\text{compensate}\ (a, b, c \cdot 30)]]$

This (9) means that Bēl-iqīša, son of Ṣillaya, must compensate Ištar of Uruk and Nanaya by paying 150 sheep. Why must he repay additional 5 unblemished lambs? These 5 were born after stealing. Hence, on the one hand, they belong to Ištar of Uruk, too, but, on the other hand, they were not stolen. Therefore, they must be given back without any additional compensation. To deduce this statement, we should appeal to the property of the completeness of legal information. The 5 unblemished lambs were born to the 5 sheep after stealing. Hence, they belong to Ištar of Uruk, but they were not stolen. Let $A_1, A_2, \ldots, A_n$ be all the n cases related to the obligation B to compensate additionally. The case of the 5 lumbs, C, is beyond all the n cases $A_1, A_2, \ldots, A_n$, i.e., we have: $C \Rightarrow \neg (A_1 \oplus A_2 \oplus \ldots \oplus A_n)$. Then $C \Rightarrow \neg B$. So, they must be given back without compensation.

The next example about additional compensation:

1. 2 UZ.TUR^{mušen} *ša*$_{2}$ [^{d}GAŠAN] UNUG^{ki} *u*$_{3}$ ^{d}*na-na-*[*a ša*$_{2}$ *qa-puut-tu*$_{4}$]
2. *ša*$_{2}$ ^{m}*ni-din-tu*$_{4}$ *u*$_{3}$ ^{m}*gu-za-nu* DUMU.MEŠ *ša*$_{2}$ ^{md}*na-na-a* [MU... U$_{4}$ 11-*kam*$_{2}$ *ša*$_{2}$ ITI AB]
3. MU 2-*kam*$_{2}$ ^{m}*kam-bu-zi-ia* LUGAL TIN.TIR^{ki} LUGAL [KUR.KUR...
4. *ša*$_{2}$ KA$_{2}$.GAL ^{d}15 *di-i-ku-ma i-na ṭi-ṭu*$_{3}$ [*qit-bu-ru*...]
5. ^{m}ŠEŠ-SUM.NA *u*$_{3}$ ^{md}*a-nu*-ŠEŠ. MEŠ-TIN-[iṭ] [DUMU.MEŠ *ša*$_{2}$ ^{md}NA$_{3}$-KAD$_{2}$ ^{md}*na-na-a*-ŠEŠ-MU]
6. DUMU-*šu*$_{2}$ *ša*$_{2}$ ^{md}*na-na-a*-KAM$_{2}$ *u*$_{3}$ ^{m}ŠEŠ-SUM.NA [DUMU-*šu*$_{2}$ *ša*$_{2}$ ^{m}*ki-na-a*...] ID$_{2}$
7. *ša*$_{2}$ MUŠEN.MEŠ *i-na meš-ḫišu*$_{2}$*-nu di-i-ku-ma i-*[*na ṭi-ṭu*$_{3}$ *iqte-bi-ru*]
8. *a-na ma-ḫar* ^{md}NA$_{3}$-DUIBILA ^{lu2}ŠA$_{3}$.TAM E$_{2}$.AN. NA DUMU-*šu*$_{2}$ *ša*$_{2}$ ^{m}[*na-di-nu* DUMU *da-bi-bi*]
9. *u*$_{3}$ ^{md}NA$_{3}$-ŠEŠ-MU ^{lu2}SAG. LUGAL ^{lu2}EN *pi-qit-tu*$_{4}$ E$_{2}$.AN. NA [...]
10. *ina* UKKIN *iq-bu-u*$_{2}$ *um-ma* U$_{4}$ 11-*kam*$_{2}$ *ša*$_{2}$ ITI AB MU 2-*kam*$_{2}$ *ni-i-ni u*$_{3}$ ^{md}*na-na-a*-MU
11. DUMU-*šu*$_{2}$ *ša*$_{2}$ ^{md}*in-nin*-NUMUN-DU$_{3}$ *it-ti a-ḫa-meš ina ku-tal* BAD$_{3}$ ID$_{2}$ *ni-ḫi-*[*ir*]*-ru*
12. 2 UZ.TUR^{mušen}.ME NIG$_{2}$.GA ^{d}GAŠAN UNUG^{ki} *ša*$_{2}$ *qa-pu-uttu*$_{4}$ *ša*$_{2}$ ^{m}*ni-din-tu4 u3 mgu-za-nu*
13. DUMU.MEŠ *ša*$_{2}$ ^{md}*na-na-a*-MU *ki-i ni-du-ku i-na ṭi-ṭu*$_{3}$ *ni-iqte-bir*
14. [*pag*]*-ra-nu ša*$_{2}$ UZ.TUR^{mušen}*-a'* 2 *ša*$_{2}$ ^{m}ŠEŠ-MU *u*$_{3}$ ^{md}DIŠ-ŠEŠ. MEŠ-TIN-*iṭ* DUMU.MEŠ
15. *ša*$_{2}$ ^{md}NA$_{3}$-KEŠDA-*ir* ^{md}*na-naa*-ŠEŠ-MU DUMU-*šu*$_{2}$ *ša*$_{2}$ ^{md}*na-na-a*-APIN-*eš*
16. ^{m}ŠEŠ-SUM.NA DUMU-*šu*$_{2}$ *ša*$_{2}$ ^{m}*ki-na-a u* ^{md}*na-na-a*-MU DUMU-*šu*$_{2}$ *ša*$_{2}$ ^{md}*in-nin*-NUMUN-DU$_{3}$
17. *i-du-ku-ma ina ṭi-ṭu*$_{3}$ *iq-bi-ri i-na* UKKIN ^{lu2}*qi-pa-a-nu u* ^{lu2}DUMU DU$_{3}$*-i.*[MEŠ]
18. *in-nam-ru-ma ki-i pi-i* ^{lu2}*mu-kin-nu-tu ša*$_{2}$ ^{m}ŠEŠ-SUM.NA
19. ^{md}*a-nu*-ŠEŠ.MEŠ-TIN-*iṭ* ^{md}*na-na-a*-ŠEŠ-MU *u*$_{3}$ ^{m}ŠEŠMU *i-na* UKKIN *qi*$_{2}$*-pa-a-nu*
20. *u*$_{3}$ ^{lu2}DUMU DU$_{3}$*-i*.ME *e-li ram-ni-šu*$_{2}$*-nu* [*u*$_{3}$] [*ma-ḫar*] ^{md}NA$_{3}$-DU-IBILA
21. ^{lu2}ŠA$_{3}$.TAM E$_{2}$.AN.NA ^{md}NA$_{3}$-ŠEŠ-MU ^{lu2}SAGLUGAL ^{lu2}EN *pi-qit-tu*$_{4}$ E$_{2}$.AN.NA UKKIN
22. ^{lu2}TIN.TIR^{ki}.ME *u*$_{3}$ ^{lu2}UNUG^{ki}*-a-a ki* UZ.TUR^{mušen} 1-*en a-di* 30 *ku-um* UZ.TUR^{mušen}.ME-*a'*
23. 2 *e-li* ^{m}ŠEŠ-MU *u*$_{3}$ ^{md}DIŠ-PAP. ME-TIN-*iṭ* DUMU.MEŠ *ša*$_{2}$ ^{md}NA$_{3}$-KAD$_{2}$ ^{md}*na-na-a*-ŠEŠ-MU

24. DUMU-*šu*$_2$ *ša*$_2$ ^{md}*na-na-a*-KAM ^{m}ŠEŠ-MU DUMU-*šu*$_2$ *ša*$_2$ ^{m}*ki-na-a u* ^{md}*na-na-a*-MU DUMU-*šu*$_2$ *ša*$_2$ ^{md}INNIN.NANUMUN-DU$_3$
25. ^{lu2}EN *ar'-<ni> šu2-nu ša*$_2$ *la in-nam-ru a-na e-ṭe-ru a-na* NIG$_2$. GA E2.AN. NA *šul-lu-un-du*
26. *e-li-šu*$_2$*-nu ip-ru-su* UZ.TUR. MEŠ-*a* 60-*šu ib-ba-ku-nim-ma a-na* NIG$_2$.GA E$_2$.AN.NA
27. *i-nam-di-nu ina u*$_2$*-šu-uz-zu ša*$_2$ ^{md}NA$_3$DU-IBILA ^{lu2}ŠA$_3$.TAM E$_2$.AN.NA DUMU ^{m}*da-bi-bi*
28. ^{md}NA$_3$-ŠEŠ-MU ^{lu2}SAGLUGAL ^{lu2}EN SIG$_5$ E$_2$.AN.NA ^{lu2}*mu-kin-nu* ^{md}30-KAM$_2$ DUMU-*šu*$_2$ *ša*$_2$ ^{md}NA$_3$-MUSI.SA
29. DUMU ^{m}*ib-ni*-DINGIR ^{md}UTU-DU-IBILA DUMU*šu*$_2$ *ša*$_2$ ^{md}DI.KU$_5$-PAP. ME-MU DUMU ^{m}*ši-gu-u*$_2$*-a*
30. ^{m}*la-a-ba-ši*-^{d}AMAR.UTU DUMU-*šu*$_2$ *ša*$_2$ ^{m}IR3-dEN DUMU *me-gi-bi* ^{md}AMAR. UTU-MU-ŠEŠ DUMU-*šu*$_2$ *ša*$_2$ ^{md}EN-TIN-*iṭ*
31. DUMU ^{m}*bu-u*$_2$*-ṣu* $^{m<d>}$ENKAR-^{d}NA$_3$ ^{lu2}SAG ^{md}*a-nu*-MUDU$_3$ DUMU-*šu*$_2$ *ša*$_2$ ^{md}NA$_3$SUR DUMU ^{md}[PN]
32. ^{md}INNIN-ŠEŠ-MU DUMU*šu*$_2$ *ša*$_2$ ^{md}NA$_3$-DU$_3$-ŠEŠ DUMU ^{m}KUR-i ^{m}*lu-ṣa-ana*-ZALAG$_2$-^{d}UTU DUMU-*šu*$_2$ *ša*$_2$ ^{m}*šu-la-a*
33. DUMU ^{lu2}E$_2$.MAŠ-^{d}MAŠ ^{md}DIŠ-ŠEŠ-MU DUMU-*šu*$_2$ *ša*$_2$ ^{m}ŠU DUMU ^{m}KUR-*i*

33–34. ^{m}*na-di-nu* DUB.SAR DUMU-*šu*$_2$ *ša*$_2$ ^{md}EN-ŠEŠ. MEŠ-BA-*ša*$_2$ DUMU ^{m}*e-gi-bi* ^{m}IR$_3$-^{d}AMAR. UTU DUB.SAR DUMU-*šu*$_2$ *ša*$_2$ ^{m}[^{d}AMAR. UTU-MU-MU DUMU ^{md}EN-A-URI$_3$]

35. UNUG^{ki} ITI AB U$_4$ 12-*kam*$_2$ MU 2-kam$_2$ ^{m}*kam-bu-zi*-[*ia* LUGAL TIN.TIR^{ki} LUGAL KUR.KUR].

(1–4) 2 ducks, property of the Lady-of-Uruk and Nanaya [from the pen of] ^{m}Nidintu and ^{m}Guzānu, sons of ^{m}Nanaya-iddin [... on 11 Tebētu] year 2 of Cambyses king of Babylon king [of the lands] at the Ištar Gate, killed and [buried] in mud.

(5–10) ^{m}Aḫa-iddin and ^{m}Anuaḫḫē-bulliṭ, [sons of ^{m}Nabû-kāṣir, ^{m}Nanaya-aḫa-iddin] son of ^{m}Nanaya-ēreš and ^{m}Aḫa-iddin [son of Kīnaya...] in whose working-area the birds were killed [and buried in mud], said thus before ^{m}Nabû-mukīn-apli, *šatammu* of the Eanna, son of ^{m}[Nādinu descendant of Dābibī], and ^{m}Nabû-aḫa-iddin the *ša rēš šarri*, the administrator of the Eanna [...], in the assembly:

(10–11) "On 11 Ṭebētu, year 2, we were digging below the canal wall, together with ^{m}Nanaya-iddin son of ^{m}Innin-zēra-ibni."

(12–13) "When we killed 2 ducks, property of the Lady-of-Uruk, from the pen of ^{m}Nidintu and ^{m}Guzānu, sons of ^{m}Nanaya-iddin, we buried them in mud."

(14–18) The corpses of these 2 birds that ^{m}Aḫa-iddin and ^{m}Anu-aḫḫē-bulliṭ sons of ^{m}Nabûkāṣir, ^{m}Nanaya-aḫa-iddin son of ^{m}Nanaya-ēreš, ^{m}Aḫa-iddina son of ^{m}Kīnaya, and ^{m}Nanaya-iddin son of ^{m}Innin-zēra-ibni killed and buried in mud were inspected in the assembly of the *qīpu* officials and the *mār banî*.

(18–26) In accordance with the testimony of ^{m}Aḫa-iddin, ^{m}Anu-aḫḫē-bulliṭ, ^{m}Nanaya-aḫa-iddin and ^{m}Aḫa-iddin against themselves in the assembly of the *qīpu* officials and the *mār banî*, and [before] ^{m}Nabû-mukīn-apli, the *šatammu* of the Eanna, ^{m}Nabû-aḫa-iddin, the *ša rēš šarri*, administrator of the Eanna, the assembly of Babylonians and Urukians—they decided that ^{m}Aḫa-iddin and ^{m}Anu-aḫḫē-bulliṭ sons of ^{m}Nabû-kāṣir, ^{m}Nanaya-aḫa-iddin son of ^{m}Nanaya-ēreš, ^{m}Aḫa-iddin son of ^{m}Kīnaya, and ^{m}Nanaya-iddin son of ^{m}Inninzēra-ibni, their accomplice in crime who was not seen, must pay a **thirtyfold** restitution for the 2 ducks to the property of the Eanna.

(26–27) They shall bring and pay these 60 ducks to the property of the Eanna.

(27) In the presence of ^{m}Nabûmukīn-apli, the *šatammu* of the Eanna, descendant of Dābibī.

(28) ^{m}Nabû-aḫa-iddin, the *ša rēš šarri*, administrator of the Eanna.

Witnesses: ^{m}Sîn-ēreš son of ^{m}Nabû-šumu-līšir descendant of Ibni-ilī;

(29) ^{m}Šamaš-mukīn-apli son of ^{m}Madānu-aḫḫē-iddin descendant of Šigûa;

(30–31) ^{m}Lâbāši-Marduk son of ^{m}Arad-Bēl descendant of Egibi; ^{m}Marduk-šuma-uṣur son of ^{m}Bēluballiṭ descendant of ^{m}Būṣu;

(31) ^{m}Bēl-eṭēri-Nabû, *ša rēši*; ^{m}Anušuma-ibni son of ^{m}Nabû-ušēzib descendant of [PN];

(32–33) ^{m}Innin-aḫa-iddin son of ^{m}Nabû-bāni-aḫi descendant of Kurī; ^{m}Lūṣa-ana-nūr-Šamaš son of ^{m}Šulaya descendant of Šangû-Ninurta;

(33) ^{m}Anu-aḫa-iddin son of ^{m}Gimillu descendant of Kurī;

(33–34) ^{m}Nādinu, the scribe, son of ^{m}Bēl-aḫḫē-iqīša descendant of Egibi; ^{m}Arad-Marduk, the scribe, son of [^{m}Marduk-šuma-iddin descendant of Bēl-apla-uṣur]

(35) Uruk. 12 Ṭebētu, year 2 of Cambyses, king of Babylon, king of the lands (Holtz, 2009: 47–52).

Formally:

1. If a man steals X ducks and they belong to the god (to a temple), then he must replace them **thirtyfold** (i.e., the amount of $X \cdot 30$) [the non-logical axiom from the code, see **A1**];
2. The Lady-of-Uruk and Nanaya is a temple [it is a fact, because "a temple" is a generalization for the case of "the Lady-of-Uruk and Nanaya"];
3. ^{m}Aḫa-iddin and ^{m}Anu-aḫḫē-bulliṭ sons of ^{m}Nabû-kāṣir, ^{m}Nanaya-aḫa-iddin son of ^{m}Nanaya-ēreš, ^{m}Aḫa-iddin son of ^{m}Kīnaya, and ^{m}Nanaya-iddin son of ^{m}Inninzēra-ibni killed 2 ducks and buried them in mud [the fact established by the trial];
4. The 2 ducks are a property of the Lady-of-Uruk and Nanaya [the fact established by the trial];

Then, ^{m}Aḫa-iddin and ^{m}Anu-aḫḫē-bulliṭ sons of ^{m}Nabû-kāṣir, ^{m}Nanaya-aḫa-iddin son of ^{m}Nanaya-ēreš, ^{m}Aḫa-iddin son of ^{m}Kīnaya, and ^{m}Nanaya-iddin son of ^{m}Inninzēra-ibni must repay 60 ducks to the Lady-of-Uruk and Nanaya on 12 Ṭebētu, year 2 of Cambyses, king of Babylon.

Axiom **A1**:

1) ([[steal (Ax, z)]] & [[possess (Ty, z)]]) ⇒ [[compensate ($Ax, Ty, z \cdot 30$)]]

   Facts proved by the trial:

2) (a_1 & a_2 & a_3 & a_4 & a_5) ⇒ Ax, where (a_1 & a_2) := "[m]Aḫa-iddin and [m]Anu-aḫḫē-bulliṭ sons of [m]Nabû-kāṣir", a_3 := "[m]Nanaya-aḫa-iddin son of [m]Nanaya-ēreš", a_4 := "[m]Aḫa-iddin son of [m]Kīnaya", and a_5 := "[m]Nanaya-iddin son of [m]Inninzēra-ibni".
3) b ⇒ Ty, where b := "the Lady-of-Uruk and Nanaya".
4) c ⇒ z, where c := "2 ducks".
5) [[steal (a_1 & a_2 & a_3 & a_4 & a_5, c)]]
6) [[possess (b, c)]]

   Expressions (1)–(6) are axioms for the court. According to **SR**, we can substitute (a_1 & a_2 & a_3 & a_4 & a_5) for Ax, b for Ty, and c for z in **A1**. Then we obtain:

7) ([[steal (a_1 & a_2 & a_3 & a_4 & a_5, c)]] & [[possess (b, c)]]) ⇒ [[compensate (a_1 & a_2 & a_3 & a_4 & a_5, b, $c \cdot 30$)]]

   From (5) and (6) by **I&**:

8) [[steal (a_1 & a_2 & a_3 & a_4 & a_5, c)]] & [[possess (b, c)]]

   From (7) and (8) by **MP**:

9) [[compensate (a, b, $c \cdot 30$)]]

In Judaism, there is the same sufficient condition to define a case as theft. This condition is called an acquisition (קניין) in the form of lifting or pulling of a thing belonging to another man. So, according to Judaism, killing 2 ducks and buring them in mud is a case of theft, too.

Another example of trial record is denoted BM 46660, see Holtz (2014: 43–44) and tells us that Marduk-šarranu has accused Kīnaya of striking his son and, as a result, two siblings, a brother, and a sister, guarantee that Kīnaya will appear at the court. If Kīnaya escapes, then the two must pay compensation to Marduk-šarranu:

1'. [*u* [m]*ki-na-a* DUMU-*šú šá* [m]BA]-*šá a-na*
2'. ⌈[lu2]DUMU⌉ [DU$_3$ x x x] *it-ti a-ḫameš*
3'. *il-la-ku-ú-ma di-i-nu* [*šá*]
4'. [md]AMAR.UTU-LUGAL-*a-nu a-na* [m]*ki-na-*[*a*]
5'. *iq-bu-ú um-mu* DUMU-*u-*[*a*]
6'. *ta-an-da-ḫa-aṣ ina* IGI [lu2][…]
7'. *i-dab-bu-ub* [md]NA$_3$-[NUMUNMU]
8'. A-*šú šá* [m]ŠEŠ.MEŠ-*šá-iá u* [[f]*iṣṣur*-X]
9'. NIN-*šú pu-ut* [m][*ki-na-a*]
10'. A-*šú šá* [m]BA-*šá-a na-*[*šu-u ki-i*]
11'. [m]*ki-na-a iḫ-te-*[*li-qu*]

12’. ZI.MEŠ *šá* DUMU-*šú* sa_2 ^{md}[AMAR.UTU-LUGAL-*a-nu*]
13’. ^{md}NA$_3$-NUMUN-MU *u* ^{f}*iṣ*-[*ṣur*]-[
14’. *ú-šal-lim-mu* ^{lu2}*mu-kin-nu* ^{m}[PN].

(1’–3’) [… and Kīnaya son of Iq]īšaya will go to the *mār* [*banî*] together

(3’–7’) They (!) /He will argue the case [in which] Marduk-šarranu said thus to Kīnaya “You **struck** my son!” before the …

(7’–10’) Nabû-zēra-iddin son of Aḫḫūšaya and [Iṣṣur-X], his sister, assume responsibility for [Kīnaya] son of Iqīšaya.

(10’–14’) If Kīnaya escapes, Nabûzēra-iddin and Iṣṣur-[X] will pay compensation for the life of the son of Marduk-šarranu.

(14’–15’) Witnesses: PN
(Holtz, 2014: 43–44).

Symbolically:

1. If a man **strikes** somebody, then he must pay compensation [the non-logical axiom from the code, see **A2**];
2. If a man cannot pay, his guarantors must pay [the non-logical axiom from a code we can reconstruct];
3. Kīnaya struck Marduk-šarranu’s son [the proven fact];
4. Nabû-zēra-iddin son of Aḫḫūšaya and his sister Iṣṣur-X are guarantors for Kīnaya;
5. If Kīnaya appears at the trial, he must pay compensation to Marduk-šarranu [the first conditional verdict];
6. If Kīnaya escapes, his two guarantors (Nabû-zēra-iddin son of Aḫḫūšaya and his sister Iṣṣur-X) must pay compensation to Marduk-šarranu [the second conditional verdict];

Then, either Kīnaya or his two guarantors (Nabû-zēra-iddin son of Aḫḫūšaya and his sister Iṣṣur-X) must pay compensation to Marduk-šarranu.

Axiom **A2**:

1) [[strike (*Ax*, *Ay*)]] ⇒ [[compensate (*Ax*, *Ay*, 60 shekels)]]

   An axiom from a code which can be reconstructed:

2) (¬ [[pay (*Ax*, *Ay*, *n*)]] & [[guarantor (*Az*, *Ax*)]]) ⇒ [[pay (*Az*, *Ay*, *n*)]]

   which means that if a man *x* of the *awīlu*-class cannot pay *n* for *y* and there is a guarantor *z* of the same class, this guarantor must pay.

   Paying is a more general case than compensating (the direct conclusion just semantically):

3) [[compensate (*Ax*, *Ay*, *n*)]] ⇒ [[pay (*Ax*, *Ay*, *n*)]]

   Facts proved by the court:

4) [[strike (*a*, *b*)]]
5) $a \Rightarrow Ax$, where a := "Kīnaya son of Iqīšaya".
6) $b \Rightarrow Ay$, where b := "the son of Marduk-šarranu".
7) $(c \,\&\, d) \Rightarrow Az$, where $c \,\&\, d$:= "Nabû-zēra-iddin son of Aḫḫūšaya and his sister Iṣṣur-X".

Expressions (1)–(7) are non-logical axioms for the court. According to **SR**, we can substitute *a* for *Ax*, *b* for *Ay* in **A2**. Then we obtain:

8) [[strike (*a*, *b*)]] ⇒ [[compensate (*a*, *b*, 60 shekels)]]

   From (8) and (4) we have by **MP**:

9) [[compensate (*a*, *b*, 60 shekels)]]

Now we can substitute [[compensate (*Ax*, *Ay*, *n*)]] for [[pay (*Ax*, *Ay*, *n*)]] in (2). Then we substitute *a* for *Ax*, *b* for *Ay*, *c* & *d* for *Az*, and "60 shekels" for *n* in (2). So, we have from (2) by **SR**:

10) (¬ [[compensate (*a*, *b*, 60 shekels)]] & [[guarantor (*c* & *d*, *a*)]]) ⇒ [[compensate (*c* & *d*, *b*, 60 shekels)]]

    Let us introduce a new inference rule:

$$\frac{A;\ (\neg A \,\&\, B) \Rightarrow C}{A \vee C},$$

because $(A \,\&\, ((\neg A \,\&\, B) \Rightarrow C)) \Rightarrow (A \vee C)$ is a tautology of propositional logic. Then from (9) and (10) by this rule we infer:

11) [[compensate (*a*, *b*, 60 shekels)]] ∨ [[compensate (*c* & *d*, *b*, 60 shekels)]]

Thus, we have deduced the trial decision denoted by (11) just automatically.

The next instance of conditional verdicts is taken from the text denoted by BM 31162, found in Opis, and dated to 23.VIII.40 of Nebuchadnezzar, the King of Babylon (5 November, 565 B.C.), see Holtz (2014: 45–47). In this trial record, Gudaya, the guarantor of a grain loan to Katimu', testifies that he presented Katimu' to Bau-ēreš (the creditor) to repay the debt. Bau-ēreš has pressed the charges that he has not been repaid by Katimu'. Gudaya must present two witnesses now. If Gudaya finds these witnesses for his claim, then he is clear. If Gudaya does not support his statement by witnessing, then Gudaya must repay the barley and the interest to Bau-ēreš:

1. *a-di* U_4 1-*kám šá* ITI GAN [m]*guda-a*
2. A-*šú šá* [m]*ḫi-in-ni*-DINGIR.MEŠ 2 [lu2]DUMU-DU_3.MEŠ
3. [lu2]*mu-kin-ne-e-šú a-na* [uru]*ú-pi-ia ib-ba-kám-ma*
4. *a-na* [md]KA_2-KAM_2 A-*šú šá* [md]NA_3-DU_3-ŠEŠ
5. *ú-kan-ni šá* [m]*ka-ti-mu-*ʾ A-*šú šá*
6. [m]*ḫa-gu-ru šá pu-ut še-pi-šú ina* ŠU.2
7. [md]KA_2-KAM_2 *iš-šu-ú ina a-danni-šú*
8. [m]g[*u-d*]*a-a i-bu-ka-šim-*[*ma*]

9. [*a-na* ^{m}]^{d}KA$_2$-KAM$_2$ *id-di-nu*
10. *ki-i uk-tin-nu-uš za-ki*
11. *ki-i la uk-tin-nu-uš a-ki-i ú-ìl-tim*
12. ŠE.BAR *u* HAR.RA-*šú a-na* ^{md}KA$_2$-KAM$_2$ *it-ta-din*
13. ^{lu2}*mu-kin-nu* ^{m}*si-lim*-^{d}EN A-*šú šá*
14. ^{m}*ba-la-ṭu* ^{m}MU-^{d}AMAR.UTU A-*šú šá*
15. ^{md}NA3-KI-*ia u* ^{lu2}UMBISAG ^{md}NA$_3$-ŠEŠ.MEŠ-MU
16. A-*šú šá* ^{m}*šu-la-a* A ^{m}*e-gi-bi* ^{uru}*úpi-ia*
17. ITI APIN U$_4$ 23-*kám* MU 40-*kám*
18. ^{d}NA$_3$-NIG$_2$.DU-URI$_3$ LUGAL TIN.TIR^{ki}.

(1–9) By 1 Kislīmu, Gudaya son of Ḫinni-ilī shall bring two *mār banî* (as) his witnesses to Opis and establish, against Bau-ēreš son of Nabû-bāniaḫi, that, at the time (of the termination of the loan), Gudaya brought Katimu' son of Ḫagūru—for whose presence he (Gudaya) assumed guarantee to Bau-ēreš—to him (Bau-ēreš) and handed (Katimu') over to Bau-ēreš.

(10) If he (Gudaya) establishes (the case) against him (Bau-ēreš), he (Gudaya) is clear.

(11–12) If he (Gudaya) does not establish (the case) against him (Bauēreš), then he (Gudaya) shall pay Bauēreš barley and its interest according to the debt-note.

(13–14) Witnesses: Silim-Bēl son of Balāṭu;

(14–15) Iddin-Marduk son of Nabûittiya;

(15–16) and the scribe: Nabû-aḫḫēiddin son of Šulaya descendant of Egibi.

(16–18) Opis. 23 Araḫšamna, year 40 of Nebuchadnezzar, king of Babylon (Holtz, 2014: 45–47).

Formally:

1. If a man takes a loan, he must repay the debt according to the debt-note in the presence of a guarantor [the non-logical axiom from a code we can reconstruct];
2. If a man cannot pay, his guarantors must pay [the non-logical axiom from a code we can reconstruct];
3. Gudaya son of Ḫinni-ilī was a guarantor that Katimu' took a loan from Bau-ēreš [the documented fact];
4. If Gudaya has two witnesses that he presented Katimu' to Bau-ēreš to repay the debt, Gudaya is free [the first conditional verdict];
5. If Gudaya has no witnesses that he presented Katimu' to Bau-ēreš, Gudaya must pay Bauēreš barley and its interest according to the debt-note [the second conditional verdict];

Then, either Gudaya is free, or he must pay on 23 Araḫšamna, year 40 of Nebuchadnezzar, king of Babylon.

Non-logical axioms from a code:

1) ([[loan (*Ax*, *Ay*, *n*)]] & [[guarantor (*Az*, *Ax*)]]) ⇒ [[repay-debt (*Ax*, *Ay*, *n*)]]

Formula (1) means that if a man *x* of the *awīlu*-class takes a loan from *y* of the same class and *x* has a guarantor *z* of the same class, *x* must repay the debt according to the debt-note in the presence of a guarantor *z*.

2) (¬ [[pay (*Ax*, *Ay*, *n*)]] & [[guarantor (*Az*, *Ax*)]]) ⇒ [[pay (*Az*, *Ay*, *n*)]]

3) [[repay-debt (*Ax*, *Ay*, *n*)]] ⇒ [[pay (*Ax*, *Ay*, *n*)]]

Indeed, [[pay (*Ax*, *Ay*, *n*)]] is a generalization for [[repay-debt (*Ax*, *Ay*, *n*)]]. It is proved just semantically.

4) ([[pay (*Ax*, *Ay*, *n*)]] & [[witness (*Ap*, *Ax*)]] & [[witness (*Aq*, *Ax*)]]) ⇒ [[free (*Ax*, *n*)]]

Formula (4) means that a man *x* of the *awīlu*-class pays *n* for *y* of the same class in the presence of two witnesses *p* and *q* of the *awīlu*-class, then *x* is free for paying *n*.

5) (¬ ([[pay (*Ax*, *Ay*, *n*)]] & [[witness (*Ap*, *Ax*)]] & [[witness (*Aq*, *Ax*)]]) & [[guarantor (*Az*, *Ax*)]]) ⇒ [[pay (*Az*, *Ay*, *n*)]]

Formula (5) means that if a man *x* of the *awīlu*-class does not pay *n* for *y* of the same class in the presence of two witnesses *p* and *q* of the *awīlu*-class and there is a guarantor *z* for *x*, then *z* must pay.

Facts proved by the court:

6) *a* ⇒ *Ax*, where *a* := "Katimu' son of Ḫagūru".

7) *b* ⇒ *Ay*, where *b* := "Bau-ērеš son of Nabû-bāniaḫi".

8) *c* ⇒ *Az*, where *c* := "Gudaya son of Ḫinni-ilī".

9) [[loan (*a*, *b*, *n*)]]

10) [[guarantor (*c*, *a*)]]

Expressions (1)–(10) are axioms for the court. According to **SR**, we can substitute *a* for *Ax*, *b* for *Ay*, *c* for *Az* in (1) to obtain the following expression:

11) ([[loan (*a*, *b*, *n*)]] & [[guarantor (*c*, *a*)]]) ⇒ [[repay-debt (*a*, *b*, *n*)]]

Now, according to **SR**, we can substitute [[repay-debt (*Ax*, *Ay*, *n*)]] for [[pay (*Ax*, *Ay*, *n*)]] and then *a* for *Ax*, *b* for *Ay*, *c* for *Az* in (2):

12) (¬ [[repay-debt (*a*, *b*, *n*)]] & [[guarantor (*c*, *a*)]]) ⇒ [[repay-debt (*c*, *b*, *n*)]]

By substituting [[repay-debt (*Ax*, *Ay*, *n*)]] for [[pay (*Ax*, *Ay*, *n*)]] and then *c* for *Ax*, *b* for *Ay* in (4) we have by **SR**:

13) ([[repay-debt (*c*, *b*, *n*)]] & [[witness (*Ap*, *c*)]] & [[witness (*Aq*, *c*)]]) ⇒ [[free (*c*, *n*)]]

By substituting [[repay-debt (*Ax*, *Ay*, *n*)]] for [[pay (*Ax*, *Ay*, *n*)]] and then *a* for *Ax*, *b* for *Ay*, *c* for *Az* in (5) we have by **SR**:

14) (¬ ([[repay-debt (*a*, *b*, *n*)]] & [[witness (*Ap*, *a*)]] & [[witness (*Aq*, *a*)]]) & [[guarantor (*c*, *a*)]]) ⇒ [[repay-debt (*c*, *b*, *n*)]]
Let us introduce a new inference rule:

$$\frac{A;\ (A \,\&\, \neg\, B) \Rightarrow C}{\neg B \Rightarrow C},$$

because $(A \,\&\, ((A \,\&\, \neg\, B) \Rightarrow C)) \Rightarrow (\neg B \Rightarrow C)$ is a tautology of propositional logic.
Then from (10) and (14) we deduce:

15) ¬ ([[repay-debt (*a*, *b*, *n*)]] & [[witness (*Ap*, *a*)]] & [[witness (*Aq*, *a*)]]) ⇒ [[repay-debt (*c*, *b*, *n*)]]

From (9), (10) by **I&**:

16) [[loan (*a*, *b*, *n*)]] & [[guarantor (*c*, *a*)]]

Then from (16) and (11) by **MP**:

17) [[repay-debt (*a*, *b*, *n*)]]

Then we apply for (15) and (17) the following inference rule:

$$\frac{A;\ \neg(A \,\&\, B) \Rightarrow C}{\neg B \Rightarrow C},$$

because $(A \,\&\, (\neg(A \,\&\, B) \Rightarrow C)) \Rightarrow (\neg B \Rightarrow C)$ is a tautology of propositional logic.
As a result, we obtain:

18) ¬ ([[witness (*Ap*, *a*)]] & [[witness (*Aq*, *a*)]]) ⇒ [[repay-debt (*c*, *b*, *n*)]]

From (18) and (13) by inference rule **I**∨:

19) (¬ ([[witness (*Ap*, *a*)]] & [[witness (*Aq*, *a*)]]) ⇒ [[repay-debt (*c*, *b*, *n*)]]) ∨ (([[repay-debt (*c*, *b*, *n*)]] & ([[witness (*Ap*, *c*)]] & [[witness (*Aq*, *c*)]])) ⇒ [[free (*c*, *n*)]])

This (19) is just a verdict of the trial. So, it is obtained automatically, also.

Usually, any relationship between creditors and debtors was regulated by a legal proceeding that may be formalized as follows:

1. The creditor (*C*) has pressed the charges that the debtor (*D*) has not given back the *X* shekels taken from him.
2. This *D* is testifying at the trial: "The *X* shekels of *C* which I owed, I have paid to him in the presence of two witnesses: W_1 and W_2." In accordance with the words of the stele, it means that *D* is free.
3. If his witnesses W_1 and W_2 are confirming: "*D* has repaid the *X* shekels to *C*," then *D* must swear together with his witnesses and *D* is free and *C* forfeits his claims.
4. And if *D*'s witnesses do not confirm *D*'s statement, *C* must swear together with his witnesses W_3 and W_4 that *D* has taken the *X* shekels from *C* in the presence of W_3 and W_4 and *D* must pay *C*'s money back."

This legal proceeding has the following logical structure:

1. If a man takes a loan, he must do it in the presence of two witnesses W_3 and W_4 [the non-logical axiom from a code];
2. If a man took a loan in the presence of two witnesses W_3 and W_4, he must repay the debt [the non-logical axiom from a code];
3. If a man repays the debt, he must do it in the presence of two witnesses W_1 and W_2 [the non-logical axiom from a code];
4. If a man repays the debt in the presence of two witnesses W_1 and W_2, he is free [the non-logical axiom from a code];
5. There are two witnesses W_3 and W_4 that a debtor took a loan from a creditor [a documented fact];

Then, either the debtor must repay the debt or if he repaid it in the presence of two witnesses W_1 and W_2, then he is free.

To sum up, the Neo-Babylonian trial records reconstructed in Holtz (2009, 2014) assume an axiomatization of justice to infer verdicts automatically. In some cases, we see even quite long deductions. All these deductions are perfect logically and they are proofs that the Neo-Babylonian justice was grounded logically in facts. Thus, the first (proto-) logical axiomatic system in all the world was implicitly proposed by the Neo-Babylonians at least or their predecessors. But it is very probable that the same system existed much earlier and it was established by the Sumerians from the very beginning of Sumerian-Akkadian legal culture, see Section 3.1 (Schumann, 2017b, 2019a, 2020).

3.2.6 Aramaic and Greek Legal Documents of Elephantine

In Elephantine (island on the Nile in Southern Egypt) there were excavated many papyri containing legal texts and compiled in Coptic, Aramaic, Greek, and Latin. The legal documents of Elephantine and Syene in Aramaic have many similarities between the Aramaic and Old Babylonian legal formulae (Cowley, 1923). For instance, the structure of documents and many legal phrases were directly taken from the (Neo-) Babylonian jurisprudence or were based on the Mesopotamian legal system (Cowley, 1923). This indicates that the Aramaic legal system exposed in the Achaemenid Empire is, in fact, a continuation of the (Neo-) Babylonian tradition.

Let us consider a Jewish business contract in Aramaic dated to 12 September, 471 B.C. This document was drawn up between Konaiah son of Zadak and Mahseiah son of Jedaniah. According to the contract, Konaiah has access to Mahseiah's gateway to build there a wall to continue all along the common wall between their two properties (lines 3–4): [[build ('Konaiah', 'wall')]]. The wall is regarded as the property of Mahseiah (lines 4–5): [[possess ('Mahseiah', 'wall')]]. Konaiah and his heirs agree that Mahseiah and his heirs can build on that wall and will have a free access through the gateway and if Konaiah and his heirs deny these rights, then they will incur a penalty of 5 karsh (lines 6–14): [[claim ('Konaiah' & "his heirs", ¬ [[possess ('Mahseiah' & "his heirs", 'wall')]])]] ⇒ [[compensate ('Konaiah' &

"his heirs", 'Mahseiah' & "his heirs", "5 karsh of pure silver")]]. The text of this contract is as follows:

> [*Date*:]
> [1]On the 18th of Elul, that is day 28 of Pachons, year 15 of Xerxes the king,
>
> [*Parties*:]
> said [2]Konaiah son of Zadak, an Aramean of Syene of the detachment of Varyazata, [2]to Mahseiah son of Jenadiah, an Aramean of Syene [3]of the detachment of Varyazata, saying:
>
> [*Building Rights*:]
> I came to you, and you gave me the gateway of the house of yours to build [4]a wall there.
>
> [*Investiture*:]
> That wall is yours—(the wall) which adjoins the house of mine at its corner which is above. [5]That wall shall adjoin the side of my house from the ground upwards, from the corner of my house which is above to the house of Zechariah.
>
> [*Restraint Waiver I*:]
> [6]Tomorrow or the next day, I shall not be able to restrain you from building upon that wall of yours.
>
> [*Penalty I*:]
> [7]If I restrain you, I shall give you silver, 5 karsh by the stone(-weight)s of the king, pure silver,
>
> [*Reaffirmation I*:]
> and that wall [8]is likewise (yours).
>
> [*Restraint Waiver II*:]
> And if Konaiah die tomorrow or the next day, son or daughter, brother, or sister, [9]near or far, member of a detachment or town [8]shall not be able [9]to restrain Mahsah or a son of his from building upon [10]that wall of his.
>
> [*Penalty II*:]
> Whoever shall restrain (one) of them shall give him the silver which is written above
>
> [*Reaffirmation II*:]
> and the wall [11]is yours likewise and you have right to build upon it upwards.
>
> [*Restraint Waiver III*:]
> And I, Konaiah, shall not be able [12]to say to Mahsah, saying: "(ERASURE: Not) That gateway is not yours and you shall not go out into the street which is [13]between us and between the house of Peftuauneit the boatman."
>
> [*Penalty III*:]
> If I restrain you, I shall give you the silver which is written above
>
> [*Reaffirmation III*:]
> [14]and you have right to open that gateway and to go out into the street which is between us (and Pefṭuauneit).

[*Scribe*:]
[15]Wrote Pelatiah son of Ahio this document at the instruction of Konaiah.

[*Witnesses*:]
The witnesses herein: [16](2nd hand) witness Mahsah son of Isaiah; (3rd hand) witness Shatibarzana son of *'trly*; (3rd hand) witness Shatibarzana son of *'trly*; [17](4th hand) witness Shemaiah son of Hosea; (5th hand) witness Phrathanjana son of Artakarana; (6tn hand) [18]witness Bagadata son of Nabukudurri; (7th hand) *Ynbwly* son of Darga; (8tn hand) [19]witness Baniteresh son of Wahpre; (9tn hand) witness Shillem son of Hoshaiah.

[*Endorsement*:]
[20]Document (sealing) of the wall which is built which Konaiah wrote for Mahsah (Porten, 1996: 152–157).

As we see, Konaiah and Mahseiah, although they are Jews, are called Arameans in the contract. This reflects the double-identity of Jews in Elephantine as well as in other parts of Achaemenid Empire: both Aramean and Jew. Among the eight witnesses, only three are Jews and the others display a mixed onomasticon: Persian, Caspian, Babylonian, and Egyptian (lines 16–19). All this attests to good neighbor relationships among different peoples. The contract is fomalizable in the following way:

1) [[build (a, b)]] $\Rightarrow$ [[possess (c, b)]]
2) [[claim $(a \,\&\, d, \neg$ [[possess $(c \,\&\, d, b)$]])]] $\Rightarrow$ [[compensate $(a \,\&\, d, c \,\&\, d, n)$]]
3) "Konaiah son of Zadak" $\Rightarrow a$
4) 'wall' $\Rightarrow b$
5) "Mahseiah son of Jedaniah" $\Rightarrow c$
6) "his heirs" $\Rightarrow d$
7) "5 karsh of pure silver" $\Rightarrow n$

Non-logical axioms (1)–(7) given in the contract allow a possible court to draw the following two conclusions just by *modus ponens* (**MP**) and substitution rule (**SR**):

1. Konaiah son of Zadak built up the wall. Then from (1), (3), (4), (5) it follows that it is a property of Mahseiah son of Jedaniah.
2. Konaiah and his heirs denied that Mahseiah and his heirs can build on that wall and have a free access through the gateway. Then from (2), (3), (4), (5), (6), (7) it is concluded that Konaiah and his heirs should incur a penalty of five karsh.

We see the similar set of logemes involved into the court judgment as it was used in the Neo-Babylonian period, see the previous subsection.

The next document compiled in Aramaic and dated to 2 January, 464 B.C. contains a judicial settlement of disputes between Dargamana son of Khvarshaina and Mahseiah son of Jedaniah. Dargamana complained that Mahseiah took his land, but neither party could produce a document of title who is landowner in fact. The court, headed by the Persian Damidata, settled the dispute as follows. Mahseiah with his wife and son swore by YHW, perhaps in the Elephantine Jewish Temple, that the land does not belong to Dargamana (lines 4–7): $\neg$ [[possess ('Dargamana', 'land')]].

As a consequence, Dargamana was satisfied by this oath (lines 11–12) and drew up the withdrawal that he or any child or sibling in his name will incur a penalty of 20 karsh, if they dispute this court decision (lines 12–16): [[claim ('Dargamana', [[possess ('Dargamana', 'land')]])]] ⇒ [[compensate ('Dargamana' & "his heirs", 'Mahseiah' & "his heirs", "20 karsh of pure silver")]]. The document:

RECTO

[*Date*:]
[1]On the 18th of Kislev, that is d[ay 13+]4 (= 17) of Thoth, year 21 (of Xerxes the king), the beginning of the reign when [2]Artaxerxes the king sat on his throne,

[*Parties*:]
said Dargamana son of Khvarshaina, a Khwarezmian whose place [3]is made in Elephantine the fortress of the detachment of Artabanu, [3]to Mahseiah son of Jedaniah, a Jew who is in the fortress of Elephantine [4]of the detachment of Varyazata, saying:

[*Complaint*:]
You swore to me by YHW the God in Elephantine the fortress, you, and your wife [5]and your son, all (told)[3], about the land of mine on account of which I complained against you before [6]Damidata and his colleagues the judges,

[*Oath I*:]
and they imposed upon you for me the oath to swear by YHW on account of [7]that [6]land, [7]that it was not land of Dargamana, mine, behold I.

[*Boundaries*:]
Moreover, behold the boundaries of that land [8]which you swore to me on account of it: my house, Dargamana is to the east of it; and the house of Konaiah son of Zadak, [9]a Jew of the detachment of Atropharna, is to the west of it; and the house of [Jeza]niah son of Uriah, [10]a Jew of the detachment of Varyazata, is below it; and the house of Espemet son of Peftuauneit, [11]a boatman of the rough waters, is above it.

[*Oath II*:]
You swore to me by YHW

[*Satisfaction*:]
and satisfied [12]my heart about that land.

[*Waiver of Suit*:]
I shall not be able to institute against you suit or process—I, or son of mine or daughter [13]of mine, brother, or sister of mine near or far—about that land (against) you, or son of yours or daughter of yours, brother, or sister of yours, near or far.

[*Penalty*:]
[14]Whoever shall institute against you (suit) in my name about that land shall give you silver, 20, that is twenty, karsh by the stone(-weight)s of [15]the king, silver 2 q(uarters) to the ten,

[*Affirmation of Investiture*:]
and that land is likewise yours and you are withdrawn from [16]any suit (in) which they shall complain against you on account of that land.

[*Scribe and Place*:]
Wrote Itu son of Abah [17]this [16]document [17]in Syene the fortress at the instruction of Dargamana.

[*Witnesses*:]
(2nd hand) Witness Hosea, son of Peṭekhnum; (3rd hand) witness [18]Gaddul son of Igdal; (4th hand) witness Gemariah son of Ahio; (5th hand) Meshullam son of Hosea; (6th hand) [19]Sinkishir son of Nabusumiskun; (7th hand) witness Hadadnuri the Babylonian; (8th hand) [20]witness Gedaliah son of Ananiah; (9th hand) [21]witness Aryaicha son of Arvastahmara.

VERSO

[*Endorsement*:]
[22]Document (sealing) of withdrawal which [Dargama]na son of Khvarshaina wrote for Mahseiah (Porten, 1996: 158–162).

Formally:

1) ¬ [[possess (*a*, *b*)]].
2) [[claim (*a*, [[possess (*a*, *b*)]])]] ⇒ [[compensate (*a* & *c*, *d* & *c*, *n*)]]
3) "Dargamana son of Khvarshaina" ⇒ *a*
4) 'land' ⇒ *b*
5) "his heirs" ⇒ *c*
6) "Mahseiah son of Jedaniah" ⇒ *d*
7) "20 karsh of pure silver" ⇒ *n*

It is worth noting that Mahseiah is now called Jew, because for the court decision it was necessary to have his oath and his religious identity played a significant role there. Among the eight witnesses, five are Jews and the others are Babylonian and Persian (lines 19–21).

In accordance with axioms (1)–(7) each next court can conclude by **MP** and **SR**:

> Dargamana, son of Khvarshaina, brought a lawsuit against Mahseiah son of Jedaniah that the land belongs to Dargamana. Then from (2), (3), (4), (5), (6), (7) it follows that Dargamana and his heirs should incur a penalty of 20 karsh.

This text is based on the same set of logemes as it was in the Akkadian legal culture, e.g., see Example 3.1.2.

The next document is compiled in Greek and dated to 29 June–28 July, 284 B.C. It is a testament and voluntary disposition of property belonging to the spouses Dionysios, a Temnian, and Kallista, a Temnian. According to this testament, if anything happens to Dionysios, then he leaves all his own belongings to Kallista: [[die ('Dionysios')]] ⇒ [[inherit ('Kallista', 'Dionysios', "all his own belongings")]].

Dionysios also inherits all from the deceased spouse (lines 3–5): [[die (‘Kallista’)]] ⇒ [[inherit (‘Dionysios’, ‘Kallista’, “all her own belongings”)]]. If then, anything happens to Dionysios after he inherited the Kallista’s property, then he leaves the belongings to all his own sons: ([[inherit (‘Dionysios’, ‘Kallista’, “all her own belongings”)]] & [[die (‘Dionysios’)]]) ⇒ [[inherit (“sons of Dionysios”, ‘Dionysios’, “all her own belongings”)]]. If then, anything happens to Kallista after she inherited the Dionysios’ property, then she leaves the belongings to all her own sons (lines 6–8): ([[inherit (‘Kallista’, ‘Dionysios’, “all her own belongings”)]] & [[die (‘Kallista’)]]) ⇒ [[inherit (“sons of Kallista”, ‘Kallista’, “all her own belongings”)]]. Nevertheless, there is the following exception in the inheritance awarded to sons: Bakchios, Herakleides, and Metrodoros may receive something from Dionysios and Kallista for their labor while their father and mother are alive, but if they are married, the belongings of Dionysios and Kallista will be shared in common by all the sons (lines 8–10): ¬ [[married (‘Bakchios’)]] ⇒ [[pay (‘Dionysios’ & ‘Kallista’, ‘Bakchios’, “shares for the labor”)]]; ¬ [[married (‘Herakleides’)]] ⇒ [[pay (‘Dionysios’ & ‘Kallista’, ‘Herakleides’, “shares for the labor”)]]; ¬ [[married (‘Metrodoros’)]] ⇒ [[pay (‘Dionysios’ & ‘Kallista’, ‘Metrodoros’, “shares for the labor”)]]. If Dionysios or Kallista owe a debt, all the sons in common should repay their debts: [[loan (‘Dionysios’ ∨ ‘Kallista’, *Ay*, *n*)]] ⇒ [[repay-debt (“sons of Dionysios and Kallista”, *Ay*, *n*)]]. If any one of the sons denies in repaying their debts, he should repay one thousand drachmas of silver (lines 12–13): ([[loan (‘Dionysios’ ∨ ‘Kallista’, *Ay*, *n*)]] & ¬ [[repay-debt (“son of Dionysios and Kallista”, *Ay*, *n*)]]) ⇒ [[compensate (“son of Dionysios and Kallista”, ‘Dionysios’ & ‘Kallista’, “1000 drachmas”)]]. Please see:

RECTO

[*Date*:]

[1]In the 40th year of the reign of Ptolemy, in the month of Gorpiaios, Menelaos son of Lagos being priest.

[*Title*:]

Con[2]tract and acknowledgment.

[*Parties*:]

Dionysios, a Temnian, has composed this with his wi[3]fe [2]Kallista, a Temnian.

[*Testament I*:]

[3]If anything should happen to Dionysios, he leaves all his own belongings to Kallista, and she is in control of [4]all the belongings as long as she lives. If anything should happen to Kallista while Dionysios is alive, [5]Dionysios is in control of the belongings.

[*Testament II*:]

And if anything (then) happens to Dionysios, let him leave the belongings [6]to all his own sons. In the same way, let Kallista, if anything should (then) happen to her, leave the [7]belongings to all the sons, except for the shares which [8]Bakchios, Herakleides, and Metrodoros [7]may receive from Dionysios and Kallista for their labor [8]while their father and mother are alive. [9]But if Bakchios, Herakleides, and Metrodoros are [8]married and registered, let the belongings of Dionysios and Kallista be (shared) [10]in common by all the sons.

[*Obligation of Heirs*:]
If Dionysios or Kallista, while alive, should be in need or owe a debt [11]let all the sons in common feed them and all join in repaying their debts.

[*Penalty*:]
If any one of them [12]should not be willing to support them or join in repaying their debts or should not join in burying them, let him repay one thousand drachmas of silver, [13]and let there be requisition from the one who is insubordinate and does not act in accordance with what is written.

[*Release of Obligation*:]
If [14]Dionysios or Kallista should leave any debt, let it be permitted to the sons not to enter into (possession of the inheritance) if they do not wish to [15]when Dionysios and Kallista have died.

[*Validity*:]
Let this contract be valid in every respect everywhere, [16]wherever it may be brought, as if the covenant had been made there.

[*Guardian of document*:]
They have deposited the contract willingly [17]with the contract keeper, Herakleitos.

[*Witnesses*:]
Witnesses:
Polykrates an Arcadian; Androsthenes a Coan; [18]Noumenios a Cretan; Simonides a Maronean; Lysis and Herakleitos Temnians.

VERSO
of Dio of Bakchios of [Poly] of Noume of Lysis nysios (sealings) of Kallista (sealings) c[rates] nios (sealings) of Metrodoros of Simonides of Herakleides of Herakletos of Androsthenes (Porten, 1996: 412–413).

Symbolically:

1) [[die (a)]] ⇒ [[inherit (b, a, c)]]
2) [[die (b)]] ⇒ [[inherit (a, b, c)]]
3) ([[inherit (b, a, c)]] & [[die (b)]]) ⇒ [[inherit (d, b, c)]]
4) ([[inherit (a, b, c)]] & [[die (a)]]) ⇒ [[inherit (d', a, c)]]
5) ¬ [[married (e)]] ⇒ [[pay (a & b, e, f)]]
6) [[loan ($a \vee b$, Ay, n)]] ⇒ [[repay-debt (d & d', Ay, n)]]
7) ([[loan ($a \vee b$, Ay, n)]] & ¬ [[repay-debt (g, Ay, n)]]) ⇒ [[compensate (g, a & b, m)]]
8) 'Dionysios' ⇒ a
9) "shares for the labor" ⇒ f
10) 'Kallista' ⇒ b
11) "all own belongings" ⇒ c
12) "sons of Kallista" ⇒ d

13) "sons of Dionysios" ⇒ *d'*

14) 'Bakchios' or 'Herakleides' or 'Metrodoros' ⇒ *e*

15) "one of the sons of Dionysios or Kallista" ⇒ *g*

16) "1000 drachmas" ⇒ *m*

From axioms (1)–(16) each next court can conclude by **MP** and **SR**:

1. Dionysios died. Then Kallista inherits.
2. Kallista died. Then Dionysios inherits.
3. Kallista inherited and died. Then her sons inherit.
4. Dionysios inherited and died. Then his sons inherit.
5. Bakchios or Herakleides or Metrodoros are not married. Then each of them before his marriage can receive something from Dionysios and Kallista for his labor.
6. Dionysios or Kallista owed a debt. Then their sons should repay the debt.
7. Dionysios or Kallista owed a debt and one of the sons denied repaying the debt. Then he should compensate 1000 drachmas.

It is possible also to draw more complex conclusions by applying more complex inference rules. They are applied in the way of Babylonians showing that we deal with the same set of logemes.

There are 52 Greek papyri found in Elephantine and dated from 310 B.C. to 613 C.E. If we say about legal documents like the testament cited above, then their structure is taken and borrowed from the Aramaic documents which were much earlier there: date, parties, the crux of the matter by means of implications, witnesses (συγγραφοφιλαξ, "keeper of contracts"), and sealings.

3.2.7 Conclusions

Logic as a kind of social practice was invented within the Semitic legal culture. For instance, the *Law Code of Gortyn* (Willetts, 1967), the only preserved code of the Greeks, was written in the way of the *Code of Hammurabi* with many citations from Semitic codes (Schumann, 2017b). Hence, we can assume that the Greek legal tradition is a continuation of the Semitic one. In fact, we know only five logical systems of the Ancient Greeks: (i) the Aristotelian syllogistic (Aristotle's *Prior Analytics*); (ii) the Aristotelian modal logic (Aristotle's *On Interpretation*); (iii) the Stoic propositional logic (Hans Von Arnim's *Stoicorum Veterum Fragmenta*); (iv) the Stoic modal logic (Cicero's *On the Nature of the Gods*, *On Divination*, *On Fate*); (v) the Epicuren inductive logic (Philodemus' *On Phenomena and Inferences*). These logical systems can be formalized within the modern symbolic logic. In the Babylonian trial records we face lengthy deductions within an axiomatic form we can reconstruct. It is direct evidence for the following: (i) from Ur III there existed a (proto-) axiomatic thinking formalized in this section, and this system was implicitly used from the very beginning of the Sumerian-Akkadian legality; (ii) this (proto-) axiomatic system was implicitly implemented in justice to make it transparent.

Recently, the European Commission has formulated e-justice as a promising way to develop an open society, see Chapter 5. But the Babylonians and Assyrians more than 3,500 years ago implemented an analogue of e-justice very effectively. Let us remember what e-justice is. It uses technology, information, and communication to improve citizens' access to justice and improve the efficiency of judicial action. In other words, e-justice is a logical formalization of justice as such, allowing us to implement different expert systems to make justice efficient and transparent.

In this section, I have considered the logical practice of Babylonians as closer to their trial records as possible, but their modes of logical reasoning can be formalized differently, including the form of sequent calculus. For instance, we can define the following sequent calculus for the trial deductions of Babylonians. Let $S_1.S_2.S_3.\ldots.S_n \to P_1.P_2.P_3.\ldots.P_n$ mean that $(S_1 \,\&\, S_2 \,\&\, S_3 \,\&\, \ldots \,\&\, S_n) \Rightarrow (P_1 \vee P_2 \vee P_3 \vee \ldots \vee P_n)$ is true, where $S_1, S_2, S_3, \ldots, S_n, P_1, P_2, P_3, \ldots, P_n$ are metavariables running over formulas defined in Subsection 3.2.4 for formalyzing the Babylonian legal proceedings. The only logical axiom of this sequent calculus is $S \to S$. We can use a standard system of natural deduction with the following additional inference rules:

$$\frac{A \Rightarrow B.\mathrm{S}[B] \to \mathrm{P}[B]}{\mathrm{S}[A] \to \mathrm{P}[A]} \qquad \frac{\neg A \Rightarrow B.\mathrm{S}[B] \to \mathrm{P}[B]}{\mathrm{S}[\neg A] \to \mathrm{P}[\neg A]}$$

Thus, the perfect logical competence existed before Aristotle and the Stoics and it was invented within the Sumerian-Akkadian legal culture (Schumann, 2017b, 2019a, 2020).

3.3 Explication of Boolean Algebra from Divination Texts in Akkadian

3.3.1 Problem Setting

Texts compiled for divination in Akkadian consist of long lists, containing numerous implications of the form: "If a sign (omen) *A* is observed, then an event *B* will occur." At the same time, among the collections of omens, there was a special group of texts that were compiled as codes covering all possible combinations of identifiable pairs of signs *A* and not-*A* for the prediction of *B*, where each sign *A* has its opposite, not-*A*. Such codes have been around since the 18th century B.C.

As we remember, conventionally, there were considered only two 'independent' logical traditions, see Subsection 1.2.4: (i) the ancient Greek logic founded by Aristotle (384–322 B.C.) and Chrysippus (ca. 279–ca. 206 B.C.) and their successors (Orthodox-Byzantian, Catholic-Scholastic, Islamic, etc.); (ii) the ancient Indian logic founded in Gandhāra in the 2nd century A.D. at the time, when the Greek language was still official for edicts, business, and duties for about four hundred years in this region, and the successors of the 'Gandhāran' logic (*nyāya*, Buddhist logic, *navya-nyāya*, and so on). Nevertheless, recently there are many works about the logical hermeneutics (*middôt*, מידות) in Judaism (Abraham et al., 2015; Abraham et al., 2010; Abraham et al., 2010, 2013; Schumann, 2008, 2010a, 2010b, 2010c, 2010d, 2011a, 2011b, 2012a, 2012b, 2012c, 2013a, 2013b, 2013c, 2013d,

2015, 2016, 2017a, 2019b; Schumann and Kuznetsov, 2017), see Subsection 1.2.2. These works about "Talmudic symbolic logic" mean that there is at least one new symbolic-logical tradition beyond (i) and (ii). In turn, it means that we did not know the true history of logic and what its origin was. The point is that the good logical tradition of the Talmud would mean that the Greek and Talmudic logics were concurrent and may have a common root in the Akkadian culture.

This section is devoted to a logical reconstruction of algebraic aspects of Babylonian divination in Akkadian. Whether this reconstruction is correct can easily be verified—I have given all the passages in both Akkadian and English translations so that the reader can verify this. The purpose of this section is to show the perfect logical competence of the authors of divination in Akkadian, mainly a series of omens reconstructed by De Zorzi (2009, 2011). It is not trivial, because we do not face such a perfect competence in other cultures of the Bronze Age. In Greece, the good logical competence of some authors is well visible rather after Aristotle (i.e., since the 4th century B.C.) and in India the good logical competence of some authors is well visible rather after Nāgārjuna (according to some Chinese chronicles, he was born in Kashmir and was a contemporary of Kaṇiṣka the Great, who ruled in the 1st–2nd century A.D., and was invited by Kaṇiṣka to organize the Buddhist Council, i.e., the perfect logical competence of some Indian authors can be observed rather after the 2nd century A.D.). Hence, in Mesopotamia, we observe the perfect logical competence much earlier than in Ancient Greece (4th century B.C.) and in Ancient India (2nd century A.D.).

In Babylonian divinations, each individual forecasting is presented strongly as the conditional: "If a sign (omen), then an event". Nevertheless, omen series consisting of many conditionals and used for divinations are very different, according to their content and composition rules. So, their possible content can be as follows: astrology, physiognomy, interpretation of dreams, extispicy, oil interpretation, interpretation of aromatic smoke, and many others (Manetti, 1993: 6). There are the following three main types of composition rules for conditionals involved into forecasting:

1) The earliest rule for composing omens is denoted now by scholars as "divinatory empiricism" (Manetti, 1993: 7). It consists of recording events which really occurred one after another in the past: "If a sign (omen) [was], then an event [was]". In the instances of it we detect a logical fallacy *post hoc ergo propter hoc* very often. This rule of composing omens is typical for "historical oracles", where the apodosis of the conditional occurs rather in the past tense.
2) The second rule of composing involves a "chain of association between elements of the protasis and elements of the apodosis" of conditionals (Manetti, 1993: 7). Here we deal either with a phonetic association between the signifiers or with a tropic association between the signifieds.
3) The third rule of composing omens offers "codes which cover a finite series of completely identifiable cases" (Manetti, 1993: 7). This type of structure of omen series has appeared since the second quarter of the second millennium B.C. Here we find codes or systematic collections of very detailed divinatory signs presented in all possible combinations. For instance, let us take the following

four signs: “the threshold of the Door of the Palace”, “the middle of the Door of the Palace”, “on the right”, “on the left” (Manetti, 1993: 11). So, we have the four signs (omens), where there are the two possible locations: “the threshold of the Door of the Palace”, “the middle of the Door of the Palace”, with the two possible additional details: “on the right”, “on the left”. Consequently, we have 2^2 of their possible combinations: (i) “the threshold of the Door of the Palace” & “on the right”; (ii) “the threshold of the Door of the Palace” & “on the left”; (iii) “the middle of the Door of the Palace” & “on the right”; (iv) “the middle of the Door of the Palace” & “on the left”. Hence, the algorithm of composing omens of this type is as follows. Let us take $2k$ signs of the first order, $2l$ signs of the second order, $2m$ signs of the third order, etc., where $k, l, m \geq 1$. The point is that there are considered only binary oppositions so that each sign has an opposition within a couple. Then we have $2^{(k+l+m+\dots)}$ of all possible combinations of the protasis of conditionals.[4] In the meanwhile, each possible combination is presented as an item in long lists of divinations containing all the combinations. As a result, we deal with a code of conditionals: (i) **if** “the threshold of the Door of the Palace” & “on the right”, **then**…; (ii) **if** “the threshold of the Door of the Palace” & “on the left”, **then**…; (iii) **if** “the middle of the Door of the Palace” & “on the right”, **then**…; (iv) **if** “the middle of the Door of the Palace” & “on the left”, **then**…, where for $2k$ signs of the first order, $2l$ signs of the second order, $2m$ signs of the third order, etc., where $k, l, m \geq 1$, we always have $2^{(k+l+m+\dots)}$ conditionals for foreseeing. In the meanwhile, in this foreseeing any logical inconsistency or contradiction is avoided. First, at the place of the protasis of conditionals, we have different $2^{(k+l+m+\dots)}$ combinations without any contradiction and it holds true by definition and construction—each opposite combination of signs (omens) is contained in another conditional. Second, at the place of the apodosis of conditionals for each combination from $2^{(k+l+m+\dots)}$ possibilities used in the protasis, we have one of the appropriate signs from the list of not more than $2^{(k'+l'+m'+\dots)}$ signs used in the apodosis, where each k', l', m', … corresponds to k, l, m, …, respectively, so that each sign from the apodosis has an opposition within a couple, too, which is contained in another conditional. So, we deal with not more than $2^{(k'+l'+m'+\dots)}$ signs in the apodosis of conditionals and they are distributed in the way to avoid any logical inconsistency or contradiction, as well. Each opposite sign is contained in another apodosis of another conditional. Indeed, for different oppositions at the place of protasis we observe different oppositions at the place of apodosis. Consequently, we obtain a rule-based expert system to name within the modern AI and symbolic logic. And in this system, any inconsistency (contradiction) is excluded absolutely. Furthermore, this rule-based expert system assumes a binary logic of always opposite signs. Therefore, by applying logical inference rules we cannot draw a contradiction here. It is very nontrivial that we face such a logical system since the second quarter of the second millennium B.C.

[4] In composing the divination lists, the Babylonians used only finite sets $2^{(k+l+\dots+m)}$ of signs, but we can expand their approach to put infinite sets $2^{(k+l+m+\dots)}$ of signs to produce an infinite Boolean algebra.

In any case, it is the first known example of rule-based expert system in its explicit form in the world history. The Stoic sign-inference (σημείωσις) is only a reflection of this doctrine, see Subsection 1.2.4.

In this section, I focus on the omen series possessing the structure described in (3). My claim is that in these omen series we can reconstruct an implicit usage of Boolean algebra—a very advanced logical-algebraic technique.

Usually, in Assyriology, omen series are analyzed for reconstructing some cultural contexts of Babylonians, such as social (Hunger, 2009) or even economic one (how astrologically they can have predicted market rates) (Ossendrijver, 2019). But a reconstruction of Babylonian science based on the omen lists is of special interest—first, I mean a reconstruction of their exact astrology with advanced mathematical calculations (Brack-Bernsen and Hunger, 2002; Brack-Bernsen, 2010).

A good philosophical reflection on conditionals used in the omen series of type (3) is proposed in Rochberg (2009, 2010), where F. Rochberg paid attention to the fact that we can treat conditionals of type (3) even truth-functionally. Then she noted that in the rule-based expert systems presented in the omen series of (3), we can detect some combinations at the place of protasis, which are physically impossible. From this fact, she concluded that in the implications "If *P*, then *Q*" of divinations, the relation between the antecedent *P* and the consequent *Q* "need not be physical, causal, or even divinely produced for *P* to indicate *Q*" (Rochberg, 2009: 11). So, we can predict from the implications "If *P*, then *Q*", as she remarked, only being based on the antecedent *P* that is true now. But if *P* is physically impossible, then such a prediction is impossible, too. It is right. However, there are other combinations of signs which are always physically possible. Then in the list of $2^{(k+l+m+\ldots)}$ combinations for binary signs (omens) $2k$, $2l$, $2m$, ..., there are ever possible some predictions made by *modus ponens*, since all the possible combinations are given for us (some of them must be realized at any moment).

Then F. Rochberg assumed that we can say that a kind of logic did not exist in Mesopotamia: "The question is not whether Mesopotamian omens reflect a conscious interest in the nature of conditionals or in the logic of conditional statements. They do not" (Rochberg, 2009: 8). However, this assumption is quite contentious. Yes, some antecedents in divinations can be physically impossible, but they are ever possible logically and this property is a determining factor for our reconstruction of Mesopotamian logic. Furthermore, she did not pay attention that in the omen series of (3), we deal with the rule-based expert systems which are always complete and inconsistent for binary signs (omens) $2k$, $2l$, $2m$, ... So, in contrast, we can draw logical conclusions without contradictions for ever. It is a logic, surely.

Let us refer to Kolmogorov's probability theory. In this theory, we do not consider events whether they are physically possible or not, but we consider events, whether they are logically possible or not. For instance, events such as "a self-assembling Boeing" or "a flying crocodile" seem to be physically impossible for us at least now, but they are always logically possible, therefore they cannot be treated as a zero percent chance. Only a contradictory proposition such as "in a coin toss we obtain a head and a tail" is probabilistically equal to a zero percent chance, because this proposition has a logical form of contradiction: "a head and not a head" or "a tail and not a tail" (A & $\neg A$). In the list of $2^{(k+l+m+\dots)}$ combinations for binary signs (omens) $2k$, $2l$, $2m$, …, there are no contradictions at all, therefore, all combinations are logically possible. This fact tells us that each complete divination list (code) was composed rather as a logical system, which is reconstructed in this section. Furthermore, the omen lists of type (3) are a knowledge base which can be treated probabilistically in the meaning of Kolmogorov's theory.

Methodologically, my approach to formalize logical ideas of omen series (3) is close to the approach presented by Brack-Bernsen (2010). She noted that divinations were a sacral knowledge taught orally and we can reconstruct this knowledge by assuming that there is an inner structure in composing conditionals. For example, astrology presented by a list of conditionals was regarded as a secret doctrine in Mesopotamia (BM 42282+):[5]

> 1. Tablet of the secret of heaven, the hidden thing of the great gods. He must not give it out of hand; let him teach (it) to his son whom he loves.
> 2. To teach (it) to a non-citizen of Babylon or a non-citizen of Borsippa or anyone who is not learned, is a taboo of Nabû and Nisaba (Brack-Bernsen, 2010: 278).

As a result, we find some lists of astrological conditionals fixing some strong relations among numerals without any instructions how it did work in fact. But the assumption that there is an inner mathematical structure allowed L. Brack-Bernsen to reconstruct a very advanced mathematical technique used in the analyzed conditionals implicitly for astrological predictions. To the same extent, I've assumed that in the omen series of (3), there exists an inner algebraic structure which can be formalized by logicians. As it turned out, it is true. The implicit usage of advanced astrological calculations (Brack-Bernsen and Hunger, 2002; Brack-Bernsen, 2010) is a good proof that the Babylonians knew this system in a way. On the one hand, the thesis that theoretical knowledge follows from practical usage is a fallacy in general case. For example, beavers build dams according to some advanced building technology. However, they do not 'know' how they do it. But, on the other hand, we are talking about a kind of human exact knowledge and not about unconscious practices of swarms (such as colonies of beavers). In the case of human knowledge, there is always a form of its self-reflection, and the question is how far this self-reflection is developed. From this point of view, it is impossible to know an advanced mathematics for astronomy/astrology without its knowledge with a form of its reflection which can be even not mathematical in the narrow sense of modern mathematics. To the same extent, it is excluded to use (even implicitly) the Boolean algebra (very advanced logical-algebraic technique) without a form of its reflection. There is no problem to point somebody who reasons according to all the rules of propositional logic without knowing them in the sense of modern logicians. But, if he or she does it in all examples, then we can assume a form of logical reflection. Moreover, if there is a group of people who do it in their social practice, then we

[5] The astrology of Babylonians has many features of astronomy, including good predictions of planetary rotations, but in general it does not satisfy Galileo's paradigm, then it cannot be called astronomy in the narrow sense.

can assume that they follow a logical tradition in this practice. The matter is that, according to experimental data of cognitive science, logic (as well as mathematics) is not an innate ability and even scientists are cognitively biased and make logical fallacies (Kern et al., 1983; Sjoberg, 2017), although they especially learnt logic. In other words, logic is a knowledge that is possible only within a permanent reflection which consciously eliminates cognitive biases and logical fallacies. The divination lists of (3) are consciously composed in accordance with the rules which have some logical meanings. It does not mean that it is a logic in the meaning of modern symbolic logic, but it is a knowledge containing some logical algorithms which were understood by them in a way.

In composing the omen lists of type (3), the Babylonians put forward the following logical-philosophical presuppositions (they hold true anyway for these texts; therefore, they are readily reconstructed based on the appropriate cuneiform fragments):

(O_1) For each event *Q* there is an appropriate combination *P* of signs (omens) such that the implication "If *P*, then *Q*" holds true.

(O_2) Each individual proposition *A* about an event (even in the future) is either true or false: $A \vee \neg A$.

(O_3) Everything happens in accordance with omens.

We know that the first system of propositional logic was created by Chrysippus, the Stoic philosopher. His main logical-philosophical presuppositions are as follows:

(S_1) For each event *Q* there is a reason *P* (this reason is not only a cause, but it can be a sign (σημεῖον), too) such that the implication "if *P*, then *Q*" holds true. In the Stoic terms: "Nothing happens without a cause", *nihil enim fieri sine causa potest* (Cicero, *De Divinatione* II: 61); "No motion is without a cause" *motum nullum esse sine causa* (Cicero, *De Fato* 23).

(S_2) Each individual proposition (ἀξιώματα) *A* (even about future) is either true or false: $A \vee \neg A$.

(S_3) Each individual proposition (even about future) *A* is necessarily true, or it is necessarily false: $\Box A \vee \Box\neg A$.

(S_4) Each (past, present, future) event *A* is actual and then necessary: $A \Rightarrow \Box A$.

(S_5) Everything happens in accordance with fate (τὰ πάντα καθ' εἱμαρμένην γίνεται).

We see that the structure of logemes (O_1)–(O_3) has an injective mapping into the structure of logemes (S_1)–(S_5), with preserving the syntagmatic and paradigmatic order. Namely, we are dealing with the same propositional logic, only the logic of Chrysippus has an additional modal part. So, there is a strong similarity between the logemes of (O_1)–(O_3) and the logemes of (S_1)–(S_5), and the Stoic presuppositions can be evaluated as a continuation and further elaboration of the Babylonian assumptions. Indeed, the Stoic propositional logic assumes that each individual proposition can be considered as a succedent of true implication—to the same extent as it was understood in the Mesopotamian (proto-) logic from the beginning. Hence, the ideas of logical determinism were first expressed in an implicit way not by the Stoic philosophers, but by the Babylonian authors of divinations of type (3).

The influence of Babylonian logical ideas presented in the divinations of (3) on the Stoic logic is well traced. The point is that the Babylonian art of divination (called "the Chaldean art" at the time of Old Stoa) was regarded by the Stoics as an important science included by them into their doctrine. The Stoic philosophical teaching on divinations was founded by Zeno of Citium (ca. 334–262 B.C.) and Cleanthes (ca. 330–230 B.C), see Cicero's *De Divinatione* I: 6. In turn, Chrysippus "discussed the whole theory of divination in two books" (*Chrysippus, qui totam de divinatione duobus libris explicavit sententiam*), and "wrote one book on oracles and another on dreams" (*uno praeterea de oraculis, uno de somniis*), see Cicero's *De Divinatione* I: 6. Evidently, the Stoics were inspired by divinations in their modal-logical investigations.

The idea that implications from the divination lists of type (3) have a pure logical meaning is quite old (Rochberg, 2009, 2010). Some basic arguments for this claim are collected in Subsection 3.3.2. De Zorzi (2009, 2011) has shown that there is an algebraic intuition in composing omen lists of type (3). I explicate this intuition formally in Subsection 3.3.3 and demonstrate that all reasoning of omen series reconstructed by De Zorzi are correct within the Boolean algebra so that each conditional divination is a tautology of propositional logic with the following two values: positive sign (+1) and negative sign (–1). We can take other omen series, such as (Gehlken, 2012), and check that the prophecy logic is the same. In Subsection 3.3.4, I show some Akkadian roots for the Talmudic terminology: *hălāḵāh*, [*tôrāh še-*]*bbəʿal-peh, middāh, bāraytāʾ*, which were first used in Akkadian commentaries to divination lists.

3.3.2 Divination as Logical Inference

Since the Old Babylonian period (ca. 1800 B.C.) each forecasting was formulated in the form of implication: "If (*šumma*) *P*, [then] *Q*" ($P \Rightarrow Q$). The same form of implication was engaged in formulating articles in the Mesopotamian law codes since the earliest ones (ca. 2100 B.C.) (Roth, 1995): "If someone does thus and such, he shall be sentenced so". Hence, we can suppose that the idea of using implication in formulating divinations might have been loaned from legal documents.

While in the Old Babylonian period omens usually started with the lexeme BE-*ma* with the obvious reading *šum*$_4$*-ma* 'if', in the Neo-Assyrian period (from 911 B.C. to 609 B.C.) another lexeme DIŠ was used, just to designate 'item' or 'entry', although this form '*P*:*Q*' of divinations had the meaning of implication, too:

> DIŠ MUL SAG.ME.GAR *ana* UL Á.MUŠEN *iṭeḫḫi*.
> **If** (or, 'given', or, "on the condition that") Jupiter approaches the Eagle Constellation (Aquila)
> *miqittim būlim išakkan murṣānu ina mati ibašši zunnu arḫi* 1.KAM *ul izannun*.
>
> there will be (meaning "there is as a result") disease of the cattle herds, sickness in the land (and) no rain for one month (Rochberg, 2010: 380).

Usually, it was assumed that each divination $P \Rightarrow Q$ was first based on an empirical foundation for the association of P and Q, see Gehlken (2012). In other words, it was supposed that omens were an erroneous attempt to formulate empirical/physical laws. Nevertheless, there are many textual proofs that divinations do not concern physics at all, but logical descriptions of the world which should consider all events $P_1, P_2, \ldots, P_n$ from impossible to necessary for inferring all conclusions $Q_1, Q_2, \ldots, Q_n$ from the set of implications $P_1 \Rightarrow Q_1, P_2 \Rightarrow Q_2, \ldots, P_n \Rightarrow Q_n$:

> The relation between protasis and apodosis, or antecedent and consequent, is the working dynamic of Mesopotamian divination, and although to us that connection, at least semantically, is not readily understood, it is subject to certain relational (propositional logical) rules in the same way as are the casuistic statements of the cuneiform laws. I contend that the relation of antecedent and consequent in the omens can be understood as a matter of conditional logic, which applies independently of any semantic, causal, or empirical connection there may be (or not be) between the statements P and Q so related (Rochberg, 2010: 379).

The matter is that many divinations introduce the so-called 'impossible' phenomena which cannot appear as such:

> That is, the list of statements (P) constitute data, or knowledge, on the basis of which the diviner makes judgements and draws conclusions about what will happen. The use of the terms possible and impossible are, among other things, relative to one's accepted knowledge of how and what things are (Rochberg, 2010: 389).

For example, in some divinations there are examined the eclipses that occur on days of the month when eclipses are not possible at all or there is regarded the sun appearing at midnight that is impossible, too. The point is that the main goal of divinations is to consider all the combinations of phenomena—from the phenomena, possible in the world, and up to the phenomena, which cannot occur at any time. For instance, there are only three watches, and it is known that the sun appears only in the morning watch. But there are the following three combinations: (i) "the sun appears in the evening watch"; (ii) "the sun appears in the middle watch"; (iii) "the sun appears in the morning watch". All these combinations are considered in the following Old Babylonian text:

> ^{d}*Utu ina bararti* IGI.DU$_8$ *šaḫluqti maš nīši.*
>
> If the sun appears in the evening watch: destruction of the herds of the people.
>
> ^{d}*Utu ina qablīti* IGI.DU$_8$ *bartum ana šarri.*
>
> If the sun appears in the middle watch: revolt against the king.
>
> ^{d}*Utu ina šaturri* IGI.DU$_8$ *ina ali šuātu šarru šanumma ibašši.*
>
> If the sun appears in the morning watch: in that city there will be another king (Rochberg, 2010: 389).

Thus, the divinations $P \Rightarrow Q$ are to consider all the combinations of events P, Q for predicting their consequences Q based on the omens P. Each divination as an implication "[If] P, Q" tells us about *purussû* "[divine] decision" or 'verdict' Q inferred by *modus ponens*:

Divination: $P \Rightarrow Q$;
Observer: P

Diviner: Q

more precisely by *transitivity of implication*:

Divination: $P \Rightarrow Q$;
Observer: an event E that is a particular case of P (i.e., $E \Rightarrow P$)

Diviner: $E \Rightarrow Q$

Thus, the role of diviner was like the role of judge—both should have applied logical inference rules in formulating appropriate divine decisions concerning our future in divinations or our presence in trial verdicts. While the diviner applied *modus ponens* to articles of omen series, the judge applied *modus ponens* to articles of law codes (Schumann, 2017b, 2019a, 2020, 2021a, 2021b; Schumann and Sazonov, 2021). For example, in the subscript to a late astrological text from Uruk we read: BE-*ma* EŠ.BAR 3,20 *ana* IGI-*ka*, "In order for you to see a divine decision concerning the king..." (TU 11 rev. 37) (Rochberg, 2010: 416).

As we see, the divine decisions should be inferred by *modus ponens* or *transitivity of implication*. In the meanwhile, all the possible combinations are fixed in divinations to introduce a full enumeration of particulars $P_1, P_2, \ldots, P_n$, appearing in the antecedent of the divinations and related to one general Q, appearing in the succedent, i.e., a full list of implications $P_1 \Rightarrow Q, P_2 \Rightarrow Q, \ldots, P_n \Rightarrow Q$ with the same Q. Those $P_1, P_2, \ldots, P_n$ should be exclusive—they should be connected by strong disjunctions "either... or ..." (Akkadian: "*ūl* ... *ūl*...", symbolically: "...⊕..."). In this case there is the following equivalence: $(P_1 \oplus P_2 \oplus \ldots \oplus P_n) \Leftrightarrow Q$. From this we can draw the following conclusion:

$$P_1 \Rightarrow Q; P_2 \Rightarrow Q; \ldots; P_n \Rightarrow Q; E \Rightarrow \neg P_1; E \Rightarrow \neg P_2; \ldots; E \Rightarrow \neg P_n$$

$$(P_1 \oplus P_2 \oplus \ldots \oplus P_n) \Leftrightarrow Q; \qquad E \Rightarrow \neg (P_1 \oplus P_2 \oplus \ldots \oplus P_n)$$

$$E \Rightarrow \neg Q.$$

This property of divinations can be called a *completeness of forecasting information*. This completeness means that if we take any empirical event E, then for any general ("divine decision") Q from the divination list, each diviner can announce either a "divine verdict" $E \Rightarrow Q$ or a "divine verdict" $E \Rightarrow \neg Q$ inferred from the divination list just logically. The same completeness of information was assumed in collecting the Babylonian law code (Schumann, 2017b, 2019a, 2020, 2021a, 2021b; Schumann and Sazonov, 2021), see Subsection 3.2.4.

3.3.3 Logical Reconstruction of Divination Lists

Let us reconstruct a logical intuition guessed in composing divination lists. In this reconstruction, we propose a semi-formal system with the following alphabet. Assume that small letters *a*, *b*, *c*, …, *x*, *y*, *z*, … are signs to denote individuals. Meanwhile, the letters *a*, *b*, *c*, … are constants, and *x*, *y*, *z*, … are variables. First-order formulas are defined as follows: (i) each *a*, *b*, *c*, …, *x*, *y*, *z*, … is a first-order formula; (ii) each Boolean combination of formulas defined in (i) is a first-order formula (Boolean connectives: ¬ negation, & conjunction, ∨ disjunction, ⇒ implication, ⊕ strong disjunction).

Let an action 'verb' in the antecedent or succedent of divinations be denoted by [[verb]]. Each action is an *n*-place relation among things. For example: [[twinkle-from-right-to-left (*x*, *y*)]] means that *x* twinkles from the right of *y* to the left of *y*; [[recover (*x*)]] means that *x* is recovering; [[wail (*x*)]] means that *x* wails; [[get-up (*x*)]] means that *x* gets up, etc. Let an opposite action to the action 'verb' be denoted by [[$\underline{\text{verb}}$]]. For example, [[$\underline{\text{twinkle-from-left-to-right}}$ (*x*, *y*)]] is equal to [[twinkle-from-right-to-left (*x*, *y*)]] or [[$\underline{\text{lower-cheek-to-right}}$ (*x*)]] is equal to [[lower-cheek-to-left (*x*, *y*)]]. Second-order formulas are defined as follows: (i) each [[twinkle-from-right-to-left (*x*, *y*)]], [[recover (*x*)]], [[wail (*x*)]], [[get-up (*x*)]], [[$\underline{\text{twinkle-from-right-to-left}}$ (*x*, *y*)]], [[$\underline{\text{recover}}$ (*x*)]], [[$\underline{\text{wail}}$ (*x*)]], [[$\underline{\text{get-up}}$ (*x*)]], … is a second-order formula; (ii) each Boolean combination of formulas defined in (i) is a second-order formula (Boolean connectives: ¬ negation, & conjunction, ∨ disjunction, ⇒ implication, ⊕ strong disjunction).

Now, let us formulate divinations as statements of this semi-formal system. In (Zorzi, 2009), it was detected that there is an asymmetry in omens with opposite verbs:

1. *šumma*-DIŠ *amēlu*-NA *ana ṣibûtīšu*-Á.ÁŠ^{šú} *itebbīma* ZI^{ma} *kakkabu*-MUL *ištu*-TA *imitti*-15 *amēli*-NA *ana šumēl*-GÙB *amēli*-NA *iṣarriḫ*-SUR *damqu*-SIG$_5$
2. *šumma*-DIŠ *ištu*-TA *šumēli*-150 *ana imitti*-15 *iṣarriḫ*-SUR *aḫītu*-BAR
3. *šumma*-DIŠ *kakkabu*-MUL *ina ku-tál amēli*-NA *ištu*-TA *imitti*-15 *ana šumēli*-150 *iṣarriḫ*-SUR *aḫītu*-BAR
4. *šumma*-DIŠ *kakkabu*-MUL *ina ku-tál amēli*-NA *ištu*-TA *šumēli*-150 *ana imitti*-15 *iṣarriḫ*-SUR *damqu*-SIG$_5$.

1. If a man leaves (in order to achieve) his purpose and a star twinkles **from the right** of the man **to the left** of the man—favourable.
2. If (it) twinkles **from the left to the right**—unfavourable.
3. If a star twinkles **on the back** of the man **from the right to the left**—unfavourable.
4. If a star twinkles **on the back** of the man **from the left to the right**—favourable (Zorzi, 2009: 97).

Symbolically:

i) ([[leave (*y*)]] & [[twinkle-in-front (*x*, *y*)]] & [[twinkle-from-right-to-left (*x*, *y*)]]) ⇒ [[favourable (*y*)]]

ii) ([[leave (y)]] & [[twinkle-in-front (x, y)]] & [[twinkle-from-left-to-right (x, y)]]) ⇒ [[$\underline{\text{favourable}}$ (y)]]

iii) ([[leave (y)]] & [[twinkle-on-back (x, y)]] & [[twinkle-from-right-to-left (x, y)]]) ⇒ [[$\underline{\text{favourable}}$ (y)]]

iv) ([[leave (y)]] & [[twinkle-on-back (x, y)]] & [[twinkle-from-left-to-right (x, y)]]) ⇒ [[favourable (y)]]

We have the four opposite combinations: (i) [[twinkle-in-front (x, y)]] & [[twinkle-from-right-to-left (x, y)]] "x twinkles from the right of y to the left of y in front of y"; (ii) [[twinkle-in-front (x, y)]] & [[twinkle-from-left-to-right (x, y)]] "x twinkles from the left of y to the right of y in front of y"; (iii) [[twinkle-on-back (x, y)]] & [[twinkle-from-right-to-left (x, y)]] "x twinkles from the right of y to the left of y on the back of y"; (iv) [[twinkle-on-back (x, y)]] & [[twinkle-from-left-to-right (x, y)]] "x twinkles from the left of y to the right of y on the back of y".

Each combination from (i) to (iv) gives either the positive sign [[favourable (y)]] or negative sign [[$\underline{\text{favourable}}$ (y)]]. We know that "to be in front" is a positive sign in the omens and "to be on the back" is a negative sign. At the same time, "to move from right to left" is a positive sign and "to move from left to right" is a negative sign. And 'leave' is neutral and can be ignored. Let us denote a positive sign by +1 and a negative sign by –1. Then we have:

i) (+1 & +1) ⇒ +1;
ii) (+1 & –1) ⇒ –1;
iii) (–1 & +1) ⇒ –1;
iv) (–1 & –1) ⇒ +1.

Logically, we can reconstruct the following truth tables for ¬, &, and ⇒, respectively:

Truth table for ¬

[[verb]]	[[$\underline{\text{verb}}$]]	¬ [[verb]]	¬ [[$\underline{\text{verb}}$]]
+1	–1	–1	+1
–1	+1	+1	–1

Truth table for &

[[verb$_1$]]	[[verb$_2$]]	[[verb$_1$]] & [[verb$_2$]]
+1	+1	+1
+1	–1	–1
–1	+1	–1
–1	–1	–1

Truth table for ⇒

[[verb$_1$]]	[[verb$_2$]]	[[verb$_1$]] ⇒ [[verb$_2$]]
+1	+1	+1
+1	–1	–1
–1	+1	+1
–1	–1	+1

Indeed, according to these truth tables, we have:

i) ((+1 & +1) ⇒ +1) = (+1 ⇒ +1) = +1;
ii) ((+1 & –1) ⇒ –1) = (–1 ⇒ –1) = +1;
iii) ((–1 & +1) ⇒ –1) = (–1 ⇒ –1) = +1;
iv) ((–1 & –1) ⇒ +1) = (–1 ⇒ +1) = +1.

To sum up, in the divination list of Zorzi (2009: 97) we see an intuition of Boolean algebra ({–1, +1}, ¬, &, ⇒). In the meanwhile, all the implications from the analyzed divination list are considered logical axioms, i.e., they have always the truth value +1.

Let us regard another example, whether its implications are logical axioms according to reconstructed truth tables:

> 122. *šum*$_4$*-ma alpu*-GUD *is-su-us-ma it-bi kašād*-KUR^{ad} *ṣibûti*-ÁŠ
> 123. *šum*$_4$*-ma alpu*-GUD *is-su-us-ma ul*-NU *it-bi lā*-NU *kašād*-KUR^{ad} *ṣibûti*-ÁŠ
> 124. *šum*$_4$*-ma alpu*-GUD *itbīma*-ZI^{ma} *lēssu*-TE^{su} *ana imittīšu*-15^{šú} *iddi*-ŠUB *lā*-NU *kašād*-KUR^{ad} *ṣibûti*-ÁŠ
> 125. *šum*$_4$*-ma alpu*-GUD *itbīma*-ZI^{ma} *lēssu*-TE^{su} *ana šumēlīšu*-150^{šú} *iddi*-ŠUB *kašād*-KUR^{ad} *ṣibûti*-ÁŠ.
>
> 122. If an ox **wailed and (then) got up**—attaining of the desire.
> 123. If an ox **wailed and (then) didn't get up**—not attaining of the desire.
> 124. If an ox **got up and lowered his cheek to the right**—not attaining of the desire.
> 125. If an ox **got up and lowered his cheek to the left**—attaining of the desire (Zorzi, 2009: 103–104).

Symbolically:

v) ([[wail (*x*)]] & [[get-up(*x*)]]) ⇒ [[favourable (*y*)]]
vi) ([[wail (*x*)]] & [[$\underline{\text{get-up}}$(*x*)]]) ⇒ [[$\underline{\text{favourable}}$ (*y*)]]
vii) ([[wail (*x*)]] & [[get-up(*x*)]] & [[lower-cheek-to-right (*x*)]]) ⇒ [[$\underline{\text{favourable}}$ (*y*)]]
viii) ([[wail (*x*)]] & [[get-up(*x*)]] & [[lower-cheek-to-left (*x*)]]) ⇒ [[favourable (*y*)]]

We see that [[wail (*x*)]] and [[get-up(*x*)]] and [[lower-cheek-to-left (*x*)]] are positive signs, and [[$\underline{\text{get-up}}$(*x*)]] and [[lower-cheek-to-right (*x*)]] are negative signs. Therefore, we formalize these conditional statements in the following manner, according to the truth tables:

v) ((+1 & +1) ⇒ +1) = +1;
vi) ((+1 & –1) ⇒ –1) = +1;
vii) ((+1 & +1 & –1) ⇒ –1) = +1;
viii) ((+1 & +1 & +1) ⇒ +1) = +1.

Thus, all the conditional statements from (v) to (viii) are logical axioms, too. One more example:

63. *šumma*-DIŠ *kakkabu*-MUL *ištu*-TA *imitti*-15 *amēli*-NA *ana šumēl*-150 *amēli*-NA *iṣruḫ*-SUR^{uḫ} *damqu*-SIG$_{15}$ *ana marṣi*-GIG *imât*-BA.ÚŠ
64. *šumma*-DIŠ *kakkabu*-MUL *ištu*-TA *šumēl*-150 *amēli*-NA *ana imitti*-15 *amēli*-NA *iṣruḫ*-SUR^{uḫ} *aḫītu*-BAR *ana marṣi*-GIG *iballuṭ*-AL.TI
65. *šumma*-DIŠ *kakkabu*-MUL *ana pān*-IGI *amēli*-NA *iṣruḫ*-SUR^{uḫ} *kašād*-KUR^{ád} *ṣibûti*-AŠ *ana marṣi*-GIG *imât*-BA.ÚŠ
66. *šumma*-DIŠ *mi-iḫ-ra ana pān*-IGI *amēli*-NA *iṣruḫma*-SUR$^{\text{uḫ-ma}}$ *eli*-UGU? *ītiq*-DIB^{iq} *lā*-NU *kašād*-KUR^{ád} *ṣibûti*-AŠ *ana marṣi*-GIG *iballuṭ*-AL.TI.

63. If a star twinkled from the right of the man to the left of the **man—favourable**; the **patient: he will die**.
64. If a star twinkled from the left of the man to the man right of the **man—unfavourable**; the **patient: he will recover**.
65. If a star twinkled in front of the man—**attaining of the desire**; the **patient: he will die**.
66. If (a star) twinkled directly in front of a man and passes by him?—**not attaining of the desire**; the **patient: he will recover** (Zorzi, 2009: 105).

Formally:

ix) $([[\underline{\text{ill}}\ (y)]]\ \&\ [[\text{twinkle-from-right-to-left}\ (x, y)]]) \Rightarrow [[\text{favourable}\ (y)]]$

x) $([[\text{ill}\ (y)]]\ \&\ [[\text{twinkle-from-right-to-left}\ (x, y)]]) \Rightarrow [[\underline{\text{favourable}}\ (y)]]$

xi) $([[\underline{\text{ill}}\ (y)]]\ \&\ [[\text{twinkle-from-left-to-right}\ (x, y)]]) \Rightarrow [[\underline{\text{favourable}}\ (y)]]$

xii) $([[\text{ill}\ (y)]]\ \&\ [[\text{twinkle-from-left-to-right}\ (x, y)]]) \Rightarrow [[\text{favourable}\ (y)]]$

xiii) $([[\underline{\text{ill}}\ (y)]]\ \&\ [[\text{twinkle-in-front}\ (x, y)]]) \Rightarrow [[\text{favourable}\ (y)]]$

xiv) $([[\text{ill}\ (y)]]\ \&\ [[\text{twinkle-in-front}\ (x, y)]]) \Rightarrow [[\underline{\text{favourable}}\ (y)]]$

xv) $([[\underline{\text{ill}}\ (y)]]\ \&\ [[\text{twinkle-in-front-pass}\ (x, y)]]) \Rightarrow [[\underline{\text{favourable}}\ (y)]]$

xvi) $([[\text{ill}\ (y)]]\ \&\ [[\text{twinkle-in-front-pass}\ (x, y)]]) \Rightarrow [[\text{favourable}\ (y)]]$

In these axioms, there are two positions of a man y: he is ill (i.e., $[[\text{ill}\ (y)]]$) or not (i.e., $[[\underline{\text{ill}}\ (y)]]$), that is, he has a positive or negative status—he is himself a positive or negative sign. Hence, there are the following positive signs in these omens: $[[\underline{\text{ill}}\ (y)]]$, $[[\text{twinkle-from-right-to-left}\ (x, y)]]$, $[[\text{twinkle-in-front}\ (x, y)]]$. The negative signs: $[[\text{ill}\ (y)]]$, $[[\text{twinkle-from-left-to-right}\ (x, y)]]$, $[[\text{twinkle-in-front-pass}\ (x, y)]]$. So, the same, according to the truth tables:

ix) $((+1\ \&\ +1) \Rightarrow +1) = (+1 \Rightarrow +1) = +1;$

x) $((-1\ \&\ +1) \Rightarrow -1 = (-1 \Rightarrow -1) = +1;$

xi) $((+1\ \&\ -1) \Rightarrow -1) = (-1 \Rightarrow -1) = +1;$

xii) $((-1\ \&\ -1) \Rightarrow +1) = (-1 \Rightarrow +1) = +1;$

xiii) $((+1\ \&\ +1) \Rightarrow +1) = (+1 \Rightarrow +1) = +1;$

xiv) $((-1\ \&\ +1) \Rightarrow -1) = (-1 \Rightarrow -1) = +1;$

xv) $((+1 \,\&\, -1) \Rightarrow -1) = (-1 \Rightarrow -1) = +1$;

xvi) $((-1 \,\&\, -1) \Rightarrow +1) = (-1 \Rightarrow +1) = +1$.

Hence, each atomic divination (conditional statement for forecasting) is a logical axiom (always true statement, according to the truth tables). The informal rules for composing the divination lists can be formulated as follows:

M1 *A succedent of conditional divination gives a positive sign if and only if all signs (more than one) in the antecedent are positive or all signs (more than one) in the antecedent are negative.*

M2 *A succedent of conditional divination gives a negative sign if and only if there are positive and negative signs in the antecedent simultaneously.*

Let F be an n-place function that maps from the set of signs $\{-1, +1\}^n$ in the antecedent to the set of signs $\{-1, +1\}$ in the succedent in the conditional divinations. Then, according to **M1** and **M2**, F is a logical biconditional $\Leftrightarrow$ with the following truth table:

Truth table for $\Leftrightarrow$

$[[verb_1]]$	$[[verb_2]]$	$[[verb_1]] \Leftrightarrow [[verb_2]]$
+1	+1	+1
+1	–1	–1
–1	+1	–1
–1	–1	+1

Each atomic divination has the following form:

Axiom

$([[verb_1]] \,\&\, [[verb_2]] \,\&\, \ldots \,\&\, [[verb_n]]) \Rightarrow F([[verb_1]], [[verb_2]], \ldots, [[verb_n]])$, i.e.,
$([[verb_1]] \,\&\, [[verb_2]] \,\&\, \ldots \,\&\, [[verb_n]]) \Rightarrow ([[verb_1]] \Leftrightarrow [[verb_2]] \Leftrightarrow \ldots \Leftrightarrow [[verb_n]])$.

In the Boolean algebra, this statement is an axiom (it is true for any valuation). In the divination lists, this statement has only the value +1. As we see, each atomic divination is a tautology of two-valued propositional logic. It cannot be occasionally that the Babylonians formulated the divination lists with the perfect logical competence—they should have known a kind of symbolic logic or at least some logical algorithms.

Rules **M1** and **M2** are verified in some passages from the Bible, too:

> And he showed me Joshua the high priest standing **before** the **angel of the Lord**, and **Satan** standing at his **right hand** to accuse him (*Zechariah* 3: 1)

Let x be Joshua. "An angel" ([[angel (x)]]) is a positive sign. "To be before a man" ([[before (x)]]) is a positive sign. The combination of both positive signs "an angel is before a man" gives a positive sign for this man. Nevertheless, 'Satan' ([[Satan (x)]]) is a negative sign. "To be at the right hand" ([[at-the-right-hand (x)]]) is a positive sign. The combination of negative and positive signs "Satan is at his right hand" is a negative sign. The resulting combination of all four signs "an angel is before a man and Satan is at his right hand" gives a negative sign:

xvii) $([[\text{angel}\ (x)]]\ \&\ [[\text{before}\ (x)]]\ \&\ [[\text{at-the-right-hand}\ (x)]]\ \&\ [[\text{Satan}\ (x)]]) \Rightarrow [[\underline{\text{favourable}}\ (y)]]$,

where y is a man who sees this vision. According to **M1** and **M2**, we have the always true proposition:

xvii) $((+1\ \&\ +1\ \&\ +1\ \&\ -1) \Rightarrow (+1 \Leftrightarrow +1 \Leftrightarrow +1 \Leftrightarrow -1)) = (-1 \Rightarrow -1) = +1.$

However, the Lord decided to help the man:

וַיֹּאמֶר יְהוָה אֶל־הַשָּׂטָן יִגְעַר יְהוָה בְּךָ הַשָּׂטָן וְיִגְעַר יְהוָה בְּךָ הַבֹּחֵר בִּירוּשָׁלִָם הֲלוֹא זֶה אוּד מֻצָּל מֵאֵשׁ׃

wayyōʾmer ʾăḏōnāi ʾel-haśśāṭān yiḡʿar ʾăḏōnāi bəḵā haśśāṭān wəyiḡʿar ʾăḏōnāi bəḵā habbōḥēr bîrûšālāim hălôʾ zeh ʾûḏ muṣṣāl mēʾēš.

And the Lord said unto Satan: "The Lord rebuke thee, O Satan, yea, the Lord that hath chosen Jerusalem rebuke thee; is not this man a brand plucked out of the fire?" (*Zechariah* 3: 2)

As a consequence, the Lord changed the omen:

וִיהוֹשֻׁעַ הָיָה לָבֻשׁ בְּגָדִים צוֹאִים וְעֹמֵד לִפְנֵי הַמַּלְאָךְ׃

wîhôšuaʿ hāyāh lāḇuš bəḡāḏîm ṣôʾîm wəʿōmēḏ lip̱nê hammalʾāḵ.

Now Joshua was **clothed with filthy garments** and stood **before** the **angel** (*Zechariah* 3: 3).

xviii) $([[\text{angel}\ (x)]]\ \&\ [[\text{before}\ (x)]]\ \&\ [[\text{clothed-with-filthy-garments}\ (x)]]) \Rightarrow [[\text{favourable}\ (y)]]$

In truth values:

xviii) $((+1\ \&\ +1\ \&\ +1) \Rightarrow (+1 \Leftrightarrow +1 \Leftrightarrow +1)) = (+1 \Rightarrow +1) = +1.$

This proposition is a tautology, as well.

The fact that rules **M1** and **M2** might be verified in the Bible means that the Biblical prophets had some rudiments from the tradition of the Akkadian divinations. In this tradition, divinations as axioms of Babylonian prophecy logic are formulated due to **M1** and **M2**.

The axiomatic system of Babylonian prophecy logic contains some inference rules which can be reconstructed by us. Assume that A, $\neg B$, $A \Rightarrow B$ are axioms. Then the basic inference rules are as follows:

Modus ponens (**MP**):

$$\frac{A;\ A \Rightarrow B}{B}$$

Modus tollens (**MT**):

$$\frac{\neg B;\ A \Rightarrow B}{\neg A}$$

Some other inference rules:

Introducing conjunction (**I&**):

$A; B$

———

$A \& B$

Deleting conjunction (**D&**):

$A \& B$

———

$A; B$

Introducing disjunction (**I∨**):

$A; B$

———

$A \vee B$

Let S[*B*] mean that *B* is a sign *a*, *b*, *c*, …, *x*, *y*, *z*, … belonging to a formula S. Let *A* be a constant. An additional inference rule:

Substitution rule (**SR**):

$S[B]; A \Rightarrow B$

———

$S[A]$

In other words, if $A \Rightarrow B$ holds, then we can substitute *A* for any occurrence of *B* in a formula S.

Inference rules **MP**, **MT**, **I&**, **D&**, **I∨**, **SR** are enough for inferring "divine decisions" from the divination lists and facts, but to reduce a length of deduction we can introduce some new inference rules of the form:

$A; B$

——

C

if and only if the formula $(A \& B) \Rightarrow C$ is a tautology of propositional logic.

Let us show an example of drawing conclusions by Babylonian diviners.

> If a woman gives birth and (the foetus) has two ears on the right and none on the left—the **gods (who were) angry will return to the land and the land will live in peace.**
> If a woman gives birth and (the foetus) has two ears on the left and none on the right—**the advice of the land will be unheeded.**
> If a malformed foetus has a second ear inside its right ear—**the prince will have advisers.**
> If a malformed foetus has a second ear inside its left ear—**the advisers of the prince will advise him badly** (Zorzi, 2011: 53).

Assume that "the foetus has two ears on the left and none on the right" is a true event now. We should infer a forecasting in our semi-formal system. The axiom for the forecasting is as follows:

1) ([[foetus (x_1)]] & [[two-ears-on-right (x_1)]] & [[two-ears-on-left (x_1)]]) ⇒ [[peace (y)]] &
   ([[foetus (x_2)]] & [[two-ears-on-right (x_2)]] & [[two-ears-on-left (x_2)]]) ⇒ [[unheeded (y)]] &
   ([[foetus (x_3)]] & [[second-ear-inside-right-ear (x_3)]]) ⇒ [[advisers (z)]] &
   ([[foetus (x_4)]] & [[second-ear-inside-left-ear (x_4)]]) ⇒ [[advisers (z)]]

The empirical facts:

2) $a \Rightarrow x_2$, where a := "a concrete malformed foetus".
3) $b \Rightarrow y$, where b := "a concrete land".
4) [[foetus (a)]]
5) [[two-ears-on-right (a)]]
6) [[two-ears-on-left (a)]]

From (1) we deduce by **D&**:

7) ([[foetus (x_2)]] & [[two-ears-on-right (x_2)]] & [[two-ears-on-left (x_2)]]) ⇒ [[unheeded (y)]]

Then by **SR** we obtain from (7) the following statement:

8) ([[foetus (a)]] & [[two-ears-on-right (a)]] & [[two-ears-on-left (a)]]) ⇒ [[unheeded (b)]]

From (4), (5), (6) we can infer by **I&**:

9) [[foetus (a)]] & [[two-ears-on-right (a)]] & [[two-ears-on-left (a)]]

Now by *modus ponens* from (8) and (9):

10) [[unheeded (b)]].

Thus, just automatically the diviner concludes that [[unheeded (b)]], i.e., "the advice of land b will be unheeded".

Rules **M1** and **M2** allowed the Babylonians to formulate deterministic forecastings for all events with all possible combinations of positive and negative signs. It gives a main feature of their prophecy logic. It is totally deterministic: plausible events are impossible; all events are necessary.

3.3.4 Akkadian Roots of Some Talmudic Hermeneutic Terminology

As it is known from (Abraham et al., 2015; Abraham et al., 2010; Abraham et al., 2010, 2013; Schumann, 2008, 2010a, 2010b, 2010c, 2010d, 2011a, 2011b, 2012a, 2012b, 2012c, 2013a, 2013b, 2013c, 2013d, 2015, 2016, 2017a, 2019b; Schumann and Kuznetsov, 2017), the Talmud introduces a specific logical hermeneutics consisting of the following two parts: (i) a *classical propositional logic* that may be reconstructed for inferring court decisions on the basis of Torah's articles and wise men orders; (ii) some *non-classical logics* for deducing new laws of wise men from the text of the Torah (the so-called *middôṯ* in Hebrew), see Subsection 1.2.2. At least the first part has its roots in the Mesopotamian tradition (Schumann, 2017b, 2019a, 2020, 2021a, 2021b; Schumann and Sazonov, 2021): (1) a kind of classical propositional logic (more exactly, its practice) was borrowed by Jewish wise men not from the Greeks, but from the Babylonians (the Akkadians and then the Arameans) to deduce trial decisions by *modus pones*, *modus tollens*, substitution rule, etc. from the law codes whose articles are formulated in the casuistic form: "If/when (Akkadian: *šumma*) this or that occurs, this or that must be done" in the same way as it is formulated in the Bible, e.g., "If [*w 'im*] he has not been redeemed in any of those ways, he and his children with him shall go free in the jubilee year" (*Leviticus* 25: 54); (2) a kind of classical propositional logic, such as prophecy logic with rules **M1** and **M2**, was implicitly used to induce new omen lists for forecasting; meanwhile each divination was formulated in the same casuistic form: "If (*šumma*) *P*, [then] *Q*", therefore each forecasting was inferred by *modus pones* or *modus tollens*, as well (Rochberg, 2010).

A Mesopotamian origin of Talmudic logic is well visible due to the hermeneutic terminology which has Akkadian roots quite often (Gabbay, 2016). Let us start with the following prayer in Akkadian:

13. *i-zi-za-nim-ma* DINGIR.MEŠ GAL.MEŠ *ši-ma-a da-ba-bi*
14. *di-ni di-na a-lak-ti lim-da.*

13. Stand by me, O great gods, and give heed to my suit,
14. Judge my case and grant me an (oracular) decision! (Lenzi, 2011).

The term *dīnu* is *dîn* (דִּין) in Hebrew and the term *alaktu* is *hălāḵāh* (הֲלָכָה) in Hebrew, see also (Abusch, 1987). So, this short phrase may have the following expanded meaning:

[O great gods,] let [the human] trial against me be a trial [of righteous people], show me the [righteous] way by your omens.

In this section, I have just shown that symbolic-logical tools were applied by the Akkadians not only for (i) justice (*dīnu*), but also for (ii) composing omens (*alaktu*). In other words, there were detailed lists of signs (*ittātu*/GISKIM.MEŠ) [the Stoic σημεῖον] of two types: (i) articles of the law code; (ii) omens for divination. Each article of the code and each divination has the same logical structure "If (Akkadian *šumma*) *S*, then *P*." The code was used for legal proceedings (as a part of *dīnu*), while the omens (as a part of *alaktu*) were used for prognoses. In both cases, *modus ponens* and *modus tollens* were actively engaged as inference rules. Let us remember that in Ancient Greece, Chrysippus (ca. 279–ca. 206 B.C.) defined *modus ponens* and *modus tollens* for the first time correctly (Reinhardt, 2003).

The Sumerian writing of the word GISKIM (Akkadian *ittu*) consists of the graphic element IGI 'eye'. It means that *dīnu* and *alaktu* were two fundamental tools to see everything. The Hebrew word *'ot* came from the Akkadian *ittu*. So, Kabbalistic aspects in interpreting Hebrew 'letters' (*'otiot*, אותיות) as starting points for everything have Akkadian philosophical roots in fact. According to the Akkadians, *ittu* is a basic element participating in all the combinations (lists of law articles or omens) creating the world. And these combinations are represented in both law codifications (*dīnu*) and omens (*alaktu*). Both systems are a revelation of the heaven. And our task is only to read the 'word' (Akkadian *amatu*) as a message from gods:

> [d]SAG.ME.GAR *ina* EGIR 30 *i-ti-ti-zi an-ni-u pi-šìr-š*[*ú*]/DIŠ MUL.SAG. ME.GAR *ina* [EGIR [d]30] GUB-*iz* MÍ.KÚR *ina* KUR [GÁL-*ši*]/LUGAL *be-lí* ⸢*a*⸣-[*bu-tú*] *in-nu-*⸢*u*⸣ [*ši-i*].

> Jupiter stood behind the moon; this is its interpretation: "If Jupiter stands [behind the Moon]—there will be hostility in the land." Oh king, my lord, [this is] a **w[ord]** concerning us! (Gabbay, 2016: 283).

Later, the Akkadian *alaktu* (in the meaning to follow omens) might have been transformed into the Hebrew *hălāḵāh* (in the meaning to follow commandments). The point is that some Akkadian omens had the following reading: "If you perform something bad in context *X*, then there will be evil for you in context *X*" and "If you perform something good in context *X*, then there will be good for you in context *X*". And this reading is quite similar to the Jewish commandments.

Thus, quite probably, the commandments of the Torah were divinations at the beginning.

For the first time, the phenomenon of canonical corpus of texts appeared in Mesopotamia:

> Almost all of the texts that have attracted commentaries belong to the canonical corpus of first millennium B.C.E. Mesopotamian literature. I use the term 'canonical' to refer to texts that meet four criteria: (1) they belong to a relatively closed group of texts that circulated in various localities and in different periods during the first millennium B.C.E.; (2) their manuscripts are standardized and display a minimum number of variants; (3) they have been grouped into corpora, often as segments of an individual composition or (especially in Nineveh) as a series of compositions arranged in a standard sequence; and (4) they are attributed to a figure of great authority, often a divinity. Although the notion of canonization is not universally accepted in studies of Mesopotamian literature, I prefer to treat texts that meet these four criteria as members of a canon, even if there are limits and exceptions—as is the case with any canonical corpus (Gabbay, 2016: 4).

This canonical corpus mainly consisted of legal codes and omens. In the meanwhile, there existed a tradition of "oral lore"—how to interpret this corpus (i.e., the written Scripture) correctly. In Akkadian the oral lore is denoted as *ša pî* ("of the mouth"), *ša pī ummâni* ("according to a scholar"), and *šūt pî* ("those of the mouth"). They are very similar to the Hebrew expression [*tôrāh še-*]*bbə*ʿ*al-peh*, תורה שבעל פה ("[Scripture that is] upon the mouth").

A logical rule for interpreting the written Scripture of omens was called in Akkadian *nindanu ša bārûti* ("measurement of the *bārûtu* lore") at one commentary (Gabbay, 2016: 300). The term *nindanu* is interesting because it has a meaning closer to the Hebrew *middāh*, מִדָּה ('measurement', a logical rule to interpret the Torah) of the same root. In Bactrian μιδο ('measure') is perhaps a derivative from the Semitic *middāh* (Sims-Williams, 2000: 204). The Akkadian term *bārûtu* means divination or lore of the diviner. There is a closer Hebrew term *bāraytā*ʾ of the same root in the meaning "lore of the Torah". It is worth noting that the Judaic logical rules (*middô*<u>*t*</u>) were well-defined explicitly in the ***Bāraytā***ʾ *Dərabbî Yišmāʿēʾl* (ברייתא דרבי ישמעאל) for the first time, see Subsection 1.2.2. Also, the Judaic gematria as a way of hermeneutic interpretation has its origin in Assyria (Lieberman, 1987).

3.3.5 Conclusions

Logical rules (*middô*<u>*t*</u>, מידות) of Judaism were formalized in several logical papers and books (Abraham et al., 2015; Abraham et al., 2010; Abraham et al., 2010, 2013; Schumann, 2008, 2010a, 2010b, 2010c, 2010d, 2011a, 2011b, 2012a, 2012b, 2012c, 2013a, 2013b, 2013c, 2013d, 2015, 2016, 2017a, 2019b; Schumann and Kuznetsov, 2017). The main conclusion of all these works is that (i) the *middô*<u>*t*</u> of Judaism were concurrent to the ancient Greek logic, on the one hand, and (ii) the *middô*<u>*t*</u> are a kind of symbolic logic (see Subsection 1.2.2), because they can be formalized logically as well as the ancient Greek logic was formalized, on the other hand.

In this section, I propose a logical reconstruction for *alaktu* (composing omens) which were implicitly used by the Babylonians. It was a prophecy logic, where the Boolean algebra was used implicitly.

It is possible to show that both Greek logic and Talmudic logical hermeneutics have their common roots in the Sumerian-Akkadian culture (Schumann, 2017b, 2019a, 2020, 2021a, 2021b; Schumann and Sazonov, 2021). Both traditions have developed concurrently, but both had had the same origin in Mesopotamia (Schumann, 2017b). Hence, the ancient Greek logic was not the first, obviously. We can assume that logical reasoning was invented within the Semitic legal culture as a kind of practice (Reinhardt, 2003; Schumann, 2017b, 2019a, 2020, 2021a, 2021b; Schumann and Sazonov, 2021).

In Mesopotamia there are two (proto-)logical systems (Schumann, 2017b, 2019a, 2020, 2021a, 2021b; Schumann and Sazonov, 2021): (i) the Babylonian prophecy logic (for composing divination lists in Akkadian of different periods, first of all the Assyrian ones, ca. the first millennium B.C.); (ii) the Babylonian propositional logic reconstructed by us for the legal proceedings (trial records, first of all the trial records of the Neo-Babylonian period—from 626 B.C. to 539 B.C. (Holtz, 2009, 2014)). It is direct evidence for the fact that in Mesopotamia there was a good logical competence.

References

Abraham, M., Gabbay, D., Hazut, G., Maruvka, Y.E. and Schild, U. 2010. *The Textual Inference Rules Klal uPrat How the Talmud Defines Sets*. London: College Publications.

Abraham, M., Gabbay, D. and Schild, U. 2010. *Non-deductive Inferences in the Talmud*. London: College Publications.

Abraham, M., Gabbay, D. and Schild, U. 2013. *Principles of Talmudic Logic*. London: College Publications.

Abraham, M., Belfer, I., Gabbay, D. and Schild, U. 2015. *Fuzzy Logic, and Quantum States in Talmudic Reasoning*. London: College Publications.

Abusch, I.T. 1987. Alaktu and Halakhah Oracular decision, divine revelation. *The Harvard Theological Review* 80(1): 15–42.

Albright, W.F. 1969. Palestinian inscriptions: a letter from the time of Josiah. *In*: *Ancient Near Eastern Texts Relating to the Old Testament*. Princeton: Princeton University Press.

Botta, A.F. 2009. *The Aramaic and Egyptian Legal traditions at Elephantine: An Egyptological Approach*. London; New York: T&T Clark.

Brack-Bernsen, L. and Hunger, H. 2002. TU 11, A collection of rules for the prediction of lunar phases and of month lenghts. *SCIAMVS* 3: 3–90.

Brack-Bernsen, L. 2010. Methods for understanding and reconstructing Babylonian predicting rules. pp. 277–297. *In*: Annette Imhausen and Tanja Pommerening (eds.). *Writings of Early Scholars in the Ancient Near East, Egypt, Rome, and Greece*. Beiträge zur Altertumskunde, Vol. 286. Berlin: de Gruyter.

Cole, S.W. 1996. *Nippur IV. The Early Neo-Babylonian Governor's Archive from Nippur*. Oriental Institute Publications. Vol. 114. Chicago: The Oriental Institute of the University of Chicago.

Cowley, A.E. 1923. *Aramaic Papyri of the Fifth Century B.C.* Oxford: Clarendon Press.

Culbertson, L.E. 2009. *Dispute Resolution in the Provincial Courts of the Third Dynasty of Ur.* A dissertation submitted in partial fulfillment of the requirements for the degree of Doctor of Philosophy (Near Eastern Studies) in the University of Michigan.

Culbertson, L. 2015. Local Courts in Centralizing States: The Case of Ur III Mesopotamia. pp. 185–202. *In*: Emberling, G. (ed.). *Social Theory in Archaeology and Ancient History*. doi:10.1017/cbo9781107282056.010.

Eemeren van, F.H. and Grootendorst, R. 2004. *A Systematic Theory of Argumentation. The Pragma-dialectical Approach*. Cambridge: Cambridge University Press.

Eemeren van, F.H., Garssen, B., Krabbe, E.C.W., Francisca, A., Henkemans, S., Verheij, B. and Wagemans, J.H.M. 2014. *Handbook of Argumentation Theory*. Dordrecht Heidelberg, New York, London: Springer.

Friberg, J. 2007. *A Remarkable Collection of Babylonian Mathematical Texts*. Manuscripts in the Schøyen Collection Cuneiform Texts I. New York: Springer.

Gabbay, U. 2016. *The Exegetical Terminology of Akkadian Commentaries*. Series: Culture and History of the Ancient Near East. Vol. 8. Brill: Leiden, Boston.

Gehlken, E. 2012. *Weather Omens of Enuma Anu Enlil: Thunderstorms, Wind and Rain (tablets 44–49)*. (Cuneiform monographs; 43). Brill: Leiden, Boston.

Holtz, Sh.E. 2009. *Neo-Babylonian Court Procedure*. Series: Cuneiform monographs. Brill: Leiden, Boston.

Holtz, Sh.E. 2014. *Neo-Babylonian Trial Records*. Society of Biblical Literature Atlanta.

Hubbell, H.M. (tr.). 2006. *Cicero. On Invention, The Best Kind of Orator, Topics* (Loeb Classical Library 386). Cambridge, MA: Harvard University Press.

Hunger, H. 2009. The relation of Babylonian astronomy to its culture and society. pp. 62–73. *In*: *Proceedings of the International Astronomical Union*. Vol. 5. doi: 10.1017/S1743921311002158.

Joannès, F. (éd.). 2000. *Rendre la justice en Mésopotamie. Archives judiciaires du Proche-Orient ancien (IIIe-Ier millénaires avant J.-C.)*. Saint-Denis: Presses Universitaires de Vincennes, "Temps et espaces".

Kern, L.H., Mirels, H.L. and Hinshaw, V.G. 1983. Scientists' understanding of propositional logic: An experimental investigation. *Social Studies of Science* 13(1): 131–146.

Kienast, B. and Volk, K. 1995. *Die Sumerischen und Akkadischen Briefe des III Jahrtausends aus der Zeit vor der III Dynastie von Ur*. Freiburg: F. Steiner Verlag.

Lafont, B. 2000. Les texts judiciaries sumériens. pp. 35–68. *In*: Joannes (ed.). *Rendre la justice en Mesopotamie*. Presses Universitaires de Vincennes.

Lenzi, A. 2011. *Akkadian Prayers and Hymns: A Reader*. Series: Ancient near east monographs. Atlanta: Society of Biblical Literature.

Lieberman, S.J. 1987. A Mesopotamian background for the so-called Aggadic 'Measures' of Biblical Hermeneutics? *Hebrew Union College Annual* 58: 157–225.

Manetti, G. 1993. *Theories of the Sign in Classical Antiquity*. Indiana University Press.

Mercer, S.A.B. 1913. The oath in cuneiform inscriptions. *Journal of the American Oriental Society* 33: 33–50.

Michalowski, P. and Walker, C.B. 1989. A new sumerian "law code". *In*: Behrens, H. et al. (eds.). *Dumu-e2-dub-ba-a: Studies in Honor of Ake W. Sjoberg*. Philadelphia.

Molina, M. 2000. *La ley mas Antigua: Textes legales sumerios*. Madrid.

Molina, M. 2008. New Ur III court records concerning slavery. pp. 125–143. *In*: Michalowski, P. (ed.). *On the Third Dynasty of Ur: Studies in Honor of Marcel Sigrist. Journal of Cuneiform Studies Supplementary Series 1*. ASOR.

Ossendrijver, M. 2019. Babylonian market predictions. pp. 53–78. *In*: *Keeping Watch in Babylon*. Leiden, the Netherlands: Brill.

Poebel, A. 1936. The beginning of the fourteenth tablet of Harra Hubullu. *The American Journal of Semitic Languages and Literatures* 52(2): 111–114.

Porten, B. 1996. *The Elephantine Papyri in English. Three Millennia of Cross-Cultural Continuity and Change*. Leiden, the Netherlands: Brill.

Pritchard, J.B. (ed.). 1969. *Ancient Near Eastern Texts Relating to the Old Testament with Supplement*. Princeton: Princeton University Press.

Reinhardt, T. (tr.). 2003. *Ciceronis. Topica*. Oxford University Press.

Rochberg, F. 2009. Conditionals, inference, and possibility in Ancient Mesopotamian Science. *Science in Context* 22(1): 5–25.

Rochberg, F. 2010. *In the Path of the Moon: Babylonian Celestial Divination and Its Legacy*. Series: Studies in ancient magic and divination. Leiden, Boston: Brill.

Roth, M. 2001. Reading Mesopotamian Law Cases PBS 5 100: A Question of Filiation. *Journal of the Economic and Social History of the Orient* 44/3: 243–292.

Roth, M.T. 1995. *Law Collections from Mesopotamia and Asia Minor*. With a contribution by Harry A. Hoffner, Ir.; edited by Piotr Michalowski. Scholars Press Atlanta, Georgia.

Schumann, A. 2008. Non-well-foundedness in Judaic logic. *Studies in Logic, Grammar, and Rhetoric* 13(26): 41–60.

Schumann, A. (ed.). 2010a. *Judaic Logic*. Gorgias Press.

Schumann, A. 2010b. Preface, Introduction. pp. 7–8. *In*: Schumann, A. (ed.). *Judaic Logic*. Book series: Judaism in Context, 8. Gorgias Press, 11–33.

Schumann, A. 2010c. Judaic syllogistics: The Baba Qama from the logical point of view. pp. 239–254. *In*: Schumann, A. (ed.). *Judaic Logic*. Book series: Judaism in Context, 8. Gorgias Press.

Schumann, A. 2010d. Preface: Religious logic as a part of philosophical logic. pp. 10–19. *In*: Schumann, A. (ed.). *Logic in Religious Discourse*. Ontos Verlag.

Schumann, A. 2011a. Preface. *History and Philosophy of Logic* 32(1): 1–8.

Schumann, A. 2011b. Qal wa-Ḥomer and theory of massive-parallel proofs. *History and Philosophy of Logic* 32(1): 71–83.

Schumann, A. 2012a. *Talmudic Logic*. London: College Publications.

Schumann, A. 2012b. Kabbalistic-Leibnizian automata for simulating the universe. pp. 259–280. *In*: Evgeny Katz (ed.). *Molecular and Supramolecular Information Processing. From Molecular Switches to Logic Systems*. Wiley-VCH, Weinheim.

Schumann, A. 2012c. Cellular automaton of the Lord. *Polish Journal of Philosophy* 6(2): 103–110.

Schumann, A. 2013a. Logical cornestones of judaic argumentation theory. *Argumentation* 27(3): 305–326.

Schumann, A. (ed.). 2013b. *Pragmatic Studies in Judaism*. Gorgias Press.

Schumann, A. 2013c. Preface, Introduction. pp. 11–12. *In*: Schumann, A. (ed.). *Pragmatic Studies in Judaism*. Book series: Judaism in Context, 14. USA: Gorgias Press, 15–24.

Schumann, A. 2013d. Pragmatic limits of halakhic logic. pp. 67–98. *In*: Schumann, A. (ed.). *Pragmatic Studies in Judaism*. Book series: Judaism in Context, 14. USA: Gorgias Press.

Schumann, A. 2015. *Logika Talmudu*. Kraków: Copernicus Center Press.

Schumann, A. 2016. Physarum polycephalum syllogistic L-systems and Judaic roots of unconventional computing. *Studies in Logic, Grammar, and Rhetoric* 44(57): 181–201.

Schumann, A. and Kuznetsov, A.V. 2017. Talmudic foundations of mathematics. pp. 67–74. *BICT 2017 – 10th EAI International Conference on Bio-Inspired Information and Communications Technologies*.
Schumann, A. (ed.). 2017a. *Philosophy and History of Talmudic Logic*. London: College Publications.
Schumann, A. 2017b. On the Babylonian origin of symbolic logic. *Studia Humana* 6(2): 126–154.
Schumann, A. 2018. On a logical reconstruction of the Kathavatthu. pp. 1509–1512. *In*: *Proceedings of the 3rd International Conference on Contemporary Education, Social Sciences and Humanities* (ICCESSH 2018).
Schumann, A. 2019a. Did the Neo-Babylonians construct a Symbolic Logic for legal proceedings? *Journal of Applied Logics – IfCoLog Journal of Logics and Their Applications* 6(1): 31–82.
Schumann, A. 2019b. Do ant colonies obey the Talmud? pp. 261–272. *In*: Adamatzky, A., Akl, S. and Sirakoulis, G.Ch. (eds.). *From Parallel to Emergent Computing*, CRC Press.
Schumann, A. 2019c. On the origin of Indian Logic from the viewpoint of the Pāli Canon. *Logica Universalis* 13: 347–393.
Schumann, A. 2020. Legal argumentation in Mesopotamia since Ur III. *Journal of Argumentation in Context* 9(2): 126–154.
Schumann, A. 2021a. Logical reasoning for forecasting in Mesopotamia. *Studies in Logic, Grammar, and Rhetoric* (accepted).
Schumann, A. 2021b. On the origin of Logical Determinism in Babylonia. *Logica Universalis* 15: 331–357.
Schumann, A. and Sazonov, V. 2021. Omens as Logic. *NABU: Nouvelles Assyriologiques Brèves et Utilitaires* 3: 192–195.
Sims-Williams, N. 2000. *Bactrian Documents from Northern Afghanistan I: Legal and Economic Documents*. Oxford: Oxford University Press.
Sjoberg, E. 2017. Logical fallacies in animal model research. *Behavioral and Brain Functions* 13(1): 3.
Smelik, K. 1991. *Writings from Ancient Israel*. Westminster: John Knox Press.
Veenhof, K.R. 1995. In accordance with the words of the stele: Evidence for Old Assyrian legislation. *Chicago-Kent Law Review* 70(4): 1717–1744.
Walton, D. 1995. *Argumentation Schemes for Presumptive Reasoning*. Series: Studies in Argumentation Theory. Routledge.
Weisberg, D.B. 2003. *Neo-Nabylonian Texts in the Oriental Institute Collection*. Oriental Institute Publications. Vol. 114. Chicago: The Oriental Institute of the University of Chicago.
Westbrook, R. 2009. *Law from the Tigris to the Tiber: The Writings of Raymond Westbrook./Raymond Westbrook*; edited by Bruce Wells and Rachel Magdalene. Vol. 1 *The Shared Tradition*. Indiana: Eisenbrauns Winona Lake.
Willetts, R.F. 1967. *The Law Code of Gortyn*. De Gruyter, pp. 37–50.
Yonge, C.D. and London, B.A. (trs.). 1891. *The Orations of Marcus Tullius Cicero*. George Bell & Sons, York Street, Covent Garden.
Zorzi, N.De. 2009. Bird divination in Mesopotamia, new evidence from BM 108874. *KASKAL Rivista di storia, ambienti e culture del Vicino Oriente Antico*. Vol. 6.
Zorzi, N.De. 2011. The omen series *Šumma Izbu* – internal structure and hermeneutic strategies. *KASKAL. Rivista di storia, ambienti e culture del Vicino Oriente Antico*. Vol. 8.

Chapter 4

Directions of the Development of Logical Competence in the Hellenistic Period, which Covered the Cultures of Greece, the Middle East, and North India

The development of international trade, starting from the earliest empires of Mesopotamia, such as the III dynasty of Ur, and up to the transnational states of Central Asia ruled by the dynasties of Achaemenids, Seleucids, Indo-Greeks, Indo-Scythians, Kuṣāṇas, and some others, contributed to the development of international trade law within some unified standards.

In the legal discourse of the ancient world, the following two logical concepts became fundamental: 'particular' (it represents a species) and 'general' (it represents a genus). The 'particular' *A* is considered as a case of corresponding 'general' *B*, so that this general *B* is inferred from this *A* as its property. For example, the following implication is always true: "If this is *Dalbergia sisso* (*A*), then this is a tree (*B*)" ("Each *Dalbergia sisso* (*A*) is a tree (*B*)"), where "*Dalbergia sisso*" is a particular and 'tree' is a general. Therefore, if the implication "If *A*, then *B*" is semantically correct (true), then this means that *A* is a specific case of *B*, and *B* is a common characteristic for *A*.

The idea of how to distinguish between the particular and the general, and how to use this distinction to build true consequences, first appeared in the Sumerian-Akkadian culture within the framework of legal discourse (see Subsection 3.2.4). This idea also emerged in the classical Chinese culture in Mozi's hermeneutics (see Subsection 1.2.3) but without any deductive logic. So, in Chinese it is grammatically correct to say: "A horse is a horse, two horses are a horse" (*yī mǎ mǎ e, èr mǎ mǎ e*). Thus, we do not distinguish here (i) 'horse' as a real horse, which can serve as a specific example (subject) for our reasoning, and (ii) 'horse' as a general concept, which can be a general characteristic (predicate) for real items. The lack of such a distinction led to the fact that in the Chinese classical philosophy there was no independent logical doctrine, although some Buddhist logical treatises were translated into Chinese.

Among all the logical teachings of the ancient world, in which the whole spectrum of logical relations between the particular and the general was thoroughly analyzed, only three independent traditions existed: (1) the syllogistic of Aristotle (384–322 B.C.); (2) the logical rules (*middôṯ*) of the Talmudic hermeneutics of Hillel (ca. 75 B.C.–ca. 10 A.D.), Rabbi Ishmael (ca. 70–135 A.D.), and Rabbi Akiva (ca. 17–137 A.D.); (3) the Indian logic of the Yogācāra and Nyāya schools (since the 2nd century A.D.) with their inferences about analyticity (*svabhāva*).

Among these three traditions, only Aristotle's syllogistic lends itself to an unambiguous axiomatization. The fact is that it is built as a theory of categorical judgments which are either true or false. For these judgments, the following three types of relations are set between the particular (subject) and the general (predicate): the inclusion of the particular into the general, the exclusion of the particular from the general, the intersection of the particular and the general. Thus, any concept with any context can be analyzed from the standpoint of such a teaching. This provides a certain universal applicability of Aristotle's syllogistic and makes it relatively easy to be axiomatized.

The situation is completely different regarding the Talmudic hermeneutics and the logical teaching of the Yogācāra and Nyāya schools.

In the Judaic hermeneutics, the particular and the general are taken only from the text of the Pentateuch. What is a particular and what is a general there follows only from the context of Holy Scripture. For example, in *Genesis* 44: 8, the word 'gold' means "theft of various things, including gold". Hence, here 'gold' is understood as a particular of theft, but not gold as such. Consequently, in the Talmudic hermeneutics, we are not dealing with judgments that are true or false, but with expressions that are particular or general, and only in the context of the Scripture we are able to understand whether a given concept or expression is general or particular for finding appropriate consequences. Logical conclusions in the Talmudic hermeneutics are drawn specifically for the derivation of legislative acts from the text of the Pentateuch. This complicates the task of formalizing the logical rules of the Talmud—every time we must remain in the context of the Holy Scripture, and here the distinction between true and false judgments is inapplicable. This is explained by the fact that the judgments of the Torah are considered all equally true, but they differ, at the same time, as particular or general or exclusive.

Similar difficulty in formalizing (axiomatizing) the Indian logic of Yogācāra and Nyāya takes place since judgments here are not distinguished as either true or false in the sense of the ancient Greek philosophers. Even the very concept of judgment as a true or false description of facts was absent in Indian logical treatises of the 2nd–5th centuries A.D. The Indian logic arose, most likely, within the framework of the discourse of *abhidharma*—the Buddhist teaching on the classification of all *dharma*s (mental elements of our consciousness). This discourse began to develop around the 3rd century B.C. and was the philosophical quintessence of Buddhism. In this discourse, the need arose to study the logical relationship between *dharmin* ("bearer of dharma", understood as a particular or a subject) and *dharma* (understood as a general or a predicate). It is assumed that these relationships can be expressed in a language in different ways, but it is necessary to focus on their pre-linguistic, mental aspect. As we can see, the logical teaching in India arose from the need for (Buddhist) *yoga* with a logically correct study of the *dharma*s as different mental states of the *yogi*.

Thus, in the syllogistic of Aristotle, in the logical hermeneutics of the Talmud, and in the logic of Yogācāra and Nyāya about *svabhāva*, the equally important logical concepts of particular and general have a very different cultural and even religious context: (i) in the ancient Greek culture, they are used to construct logically correct reasoning in rhetoric; (ii) in the Talmud, they are to summarize legal cases based on the sayings of the Pentateuch; and (iii) in the Buddhist culture, they are to generalize mental states, including states given through the *yoga*. Hence, the corresponding logeme structures of all the three traditions are similar, but their illocutionary and perlocutionary complexes of philosophical speeches are different. But, meanwhile, they have a common core—the Sumerian-Akkadian definition of the particular and the general, which is correct from the standpoint of modern logic.

All three traditions arose during the Hellenistic and then Roman periods; for example, the Indian logic appeared in Gandhāra in the 2nd century A.D.—in a region that was controlled by Hellenistic dynasties or dynasties with a strong Hellenistic influence from the 4th–3rd centuries B.C. to the end of the 4th century A.D. All the three traditions were created within a certain literary genre: (i) Aristotle's syllogistic within the genre of a philosophical treatise; (ii) the logical hermeneutics of Judaism within the Talmudic genre with dialogues and commentaries to them; and (iii) the logic of Yogācāra and Nyāya within the genre of philosophical *sūtra*s, which students had to learn by heart. All three genres are different and represent different ways of manifesting the developed philosophical reflection that appeared during the Hellenistic and Roman periods.

This chapter traces the genesis of these three traditions and shows their common root in the form of the Sumerian-Akkadian logical culture (considered in Chapter 3), and the direct or indirect influence of the logical ideas of the Early Stoa on the doctrine of Yogācāra and Nyāya.

4.1 Emergence of New Literary Genre of Philosophical Treatise in Ancient Greece in the Context of the History of Logical Competence

4.1.1 Problem Setting

The main emphasis in this section is placed on the emergence of ancient Greek philosophical prose as a new literary genre that played an exceptional role in the development of philosophical reflection of European nations. To do this, the following three literary genres are compared from the point of view of logical discourse: the historical prose in the meaning of Herodotus (Subsection 4.1.3), histories (Hebrew: *ʾaggādôṯ*) in Judaism (Subsection 4.1.4), and omens in different dialects of Akkadian (Subsection 4.1.5). I am planning to show that all the three genres assume different logical discourses. So, in the historical prose, the following presupposition is implicitly applied by its authors: there exist contingent events and any kind of logical determinism is excluded. In contrast, in the *ʾaggādôṯ*, there is supposed the logical determinism as such: each event can be described as an omen or a fulfillment of prophecy. And some events are collected as cautionary tales in these *ʾaggāḏôṯ.* The same idea of logical determinism is detected in the genre of divinations as codes in Akkadian. Some rudiments of this genre can be found even in the Bible and the Talmud. In this genre, long lists of predictions are drawn up so that each divination is formulated as a conditional proposition to which the logical rules *modus ponens* and *modus tollens* are to apply: "If there is an omen *A*, then an event *B* appears."

Thus, the history in the meaning of Herodotus (ca. 484–425 B.C.) was impossible for the early Rabbinical Judaism as well as for the Akkadian culture. The reason is that the existence of contingency was not accepted there—they did not allow for unforeseen circumstances. In other words, the structure of the logemes of the Akkadian books on divination and the structure of the logemes of the books of the Haggadah (*ʾaggāḏāh*) have a strong similarity—both narratives admit the logical determinism. On the other hand, both structures have no similarity to the structure of logemes of ancient Greek prose (first, to the structure of logemes of historical prose in the sense of Herodotus, but also including the philosophical prose).

Therefore, the ancient Semitic philosophy and logic cannot be expressed in the form of prosaic works. Hence, a logical handbook in Akkadian is possible, but it needs a new way of reconstructions—it should have been written within a genre still unknown. In this section, I show that the ideas of logical determinism (expressed out in Stoicism at first), in fact, have their true origin in Mesopotamia. So, the perfect logical competence was not accidental there.

The ancient Greek logic founded by Aristotle (384–322 B.C.) and Chrysippus (ca. 279–206 B.C.) is considered the first piece of symbolic logic in the world. Nevertheless, the Judaic hermeneutical (logical) tradition [*middôṯ*, מידות] (Abraham et al., 2010a, 2010b, 2013, 2015; Schumann, 2010, 2011, 2012, 2013, 2017a.) has its roots in the Babylonian legal culture (Gabbay, 2016; Schumann, 2017b, 2019a, 2020a). This culture is much older than the ancient Greek logic.

In many fragments in Akkadian, especially in trial records and business correspondence dated from the 24th century B.C. to the 6th century B.C., we observe a perfect logical competence of their authors. It is not trivial—in some texts we see lengthy consequences of correct logical conclusions by applying some logical inference rules step by step including *modus ponens*, see (Schumann, 2017a, 2019a, 2020a), see Chapter 3. There are no other cultures in the ancient world of that time where something similar can be detected—only some later exceptions such as Aramaic legal texts since the 6th century B.C., Greek philosophical and legal texts since the 5th–4th centuries B.C., some Indian philosophical texts since the 2nd century A.D., and a few others. Furthermore, we can trace back how Greek legal documents containing perfect logical reasoning, also, were a continuation of legal tradition in Aramaic of the Achaemenid Empire and the latter tradition was a continuation of legal tradition of Neo-Babylonians with their origin from the Akkadian culture in turn, see Schumann (2019a, 2020a).

Now that is the question—whether a logical handbook could have been written in Akkadian? Here there is possibly a great mistake of Westernization of ancient cultures of the Near East—our expectation to find out a Greek philosophical prose in Babylonia before Hellenism. Evidently, it is absolutely excluded. But it is not because philosophy or logic was not possible there. The true explanation is that in Mesopotamia some literary genres were excluded because of specific ways of thinking. For instance, the genre of historical prose and then of philosophical prose cannot have existed indeed. It is worth noting that this prosaic genre did not exist in the Rabbinic Judaism at the time of the Talmud as well and for the same reason—because of a special way of Judaic thinking.

The logical tradition established in Mesopotamia was used for expressing a transparency, inevitability, and necessity of divine and royal power in the human society (logic for legal proceedings) and in the nature (logic of omens). This power was supposed to be eternal. Therefore, each divine or royal decision was forever obligatory. Under these conditions, it was assumed that a logical determinism holds true, i.e., each statement about past, actual, or future events is either necessarily true or necessarily false. This intuition from modal logic became the first version of *eternalism*—a philosophical belief that all existence in time is equally real.

So, in this section, I am going to reconstruct modal logical and eternalistic ideas of Babylonians who were the first in introducing logical determinism as a version of eternalism.

During the Hellenistic and Roman periods, the following three genres, expressing philosophical reflections on logical competence, were created independently of each other (Subsection 4.1.6): (i) ancient Greek philosophical treatises (prose) devoted to logical reasoning (e.g., written by Aristotle and Chrysippus) since the 4th century B.C.; (ii) collections of thematic dialogues in Hebrew or Aramaic taken from the Talmud or from other books such as *Bāraytā*ʾ (ברייתא) since the 1st century B.C., in which some authoritative conversations concerned *middôṯ*—some logical rules of Judaic hermeneutics for drawing conclusions from the Torah; (iii) philosophical *sūtra*s in Sanskrit[1] (collections of thematic aphorisms used to express a school doctrine) composed since the 2nd century A.D., belonging to different schools of Indian tradition (such as Yogācāra and Nyāya schools), and dedicated to *pramāṇa*—sources of correct knowledge. In this section, the emphasis is placed, first on the emergence of ancient Greek prose as a new literary genre, which played an exceptional role in the development of historical and philosophical reflections in Europe.

The philosophical prose in Ancient Greece is one of the most significant manifestations of Greek prose as such. As we will see, philosophy and logic in Semitic cultures cannot have been expressed in the form of prosaic works. Therefore, to reconstruct the logical discourse in Akkadian, Aramaic, and Hebrew languages, it is necessary to consider the circumstances that the prosaic genre, in the meaning of Herodotus or Aristotle, simply did not exist in the traditional Semitic cultures. But there was a genre of thematic dialogues, which was continued by the Jerusalem and Babylonian Talmuds.

The aim of the section is to show that the ideas of logical determinism originated in Mesopotamia and then they were philosophically reflected in some treatises of the

[1] Most probably, the first logical works in India have appeared in Gāndhārī—one of the Prakrits.

Early Stoa, whose philosophers were inspired by the Babylonian art of divination, as they themselves stated.

4.1.2 Modal Logical Preliminaries

Let us consider some basic notions of modal logic, which will be used for reconstructing modal logical intuitions in the ancient Greek historical prose (Subsection 4.1.3), in the Hebrew and Aramaic *ʾaggādôṯ* (Subsection 4.1.4), and in the Akkadian codes of omens (Subsection 4.1.5). The alphabet of propositional modal logic consists of the following signs:

- propositional letters (atoms): $p_0, p_1, \ldots$;
- propositional connectives: negation $\neg$ ("not…"), disjunction $\vee$ ("… or …"), conjunction & ("… and …"), implication $\Rightarrow$ ("if…, then, …"), equivalence $\Leftrightarrow$ ("… if and only if…");
- two modal operators: necessity $\Box$ ("… is necessarily") and possibility $\Diamond$ ("… is possibly").

A well-formed formula of modal logic is defined as follows:

- each propositional letter is a formula;
- if A is a formula, then $\neg A$ is a formula;
- if A and B are formulas, then $A \vee B$, $A \& B$, $A \Rightarrow B$, $A \Leftrightarrow B$ are also formulas;
- if A is a formula, then $\Box A$ is a formula, too;
- if A is a formula, then $\Diamond A$ is a formula, too.

Axioms of propositional modal logic are formed from the following statements:

- all propositional axioms such as the rule of excluded middle (*tertium non datur*) $A \vee \neg A$ or the rule of non-contradiction $\neg(A \& \neg A)$, etc.;
- all instances of some modal schemata such as $\Box(A \Rightarrow B) \Rightarrow (\Box A \Rightarrow \Box B)$.

The set of axioms is closed under the following two inference rules:

- *modus ponens*: from $A \Rightarrow B$ and A it follows that B;
- *necessitation rule*: if A is an axiom, then $\Box A$ is an axiom, too.

As we see, the propositional modal logic is an extension of the standard propositional logic by adding some modal schemata and the necessitation rule. The most important additional modal schemata are as follows, for more details see (Garson, 2006; Bull and Segerberg, 1984):

(K) $\Box(A \Rightarrow B) \Rightarrow (\Box A \Rightarrow \Box B)$
(D) $\Box A \Rightarrow \Diamond A$
(T) $\Box A \Rightarrow A$
(CD) $\Diamond A \Rightarrow \Box A$
(=) $A \Rightarrow \Box A$

They are used in different modal logics for expressing either logical contingency or logical determinism—these systems are given in Table 4.1.1.

Table 4.1.1. Different systems of modal logics.

	Modal Logic	**Axioms**
Basic modal logic	**K**	propositional axioms together with (K)
Modal logics for expressing logical contingency	**D**	propositional axioms together with (K) and (D)
	T	propositional axioms together with (K) and (T)
Modal logics for expressing logical determinism	**CD**	propositional axioms together with (K) and (CD)
	K=	propositional axioms together with (K) and (=)

An appropriate semantics for a propositional modal logic is presented by a Kripke model $M = (X, a, R, V)$, where

- X is a set of indices;
- a is a distinguished index;
- R is a binary relation on the indices;
- V is a function assigning a valuation $V(x)$ to each index x and a truth-value $V(x)(A)$ to each index x and atom A.

The distinguished index a is to show an actual time. The relation R is to be said a possibility since a. Let us take x from X such that aRx. Then aRx shows a possibility since a at an index x.

A true valuation of formulas at an index x in a model M is defined thus:

1) for atomic A, A is true at x in M Iff $V(x)(A) = \mathrm{T}$, where T means truth;
2) $\neg A$ is true at x in M Iff A is not true at x in M;
3) $A \vee B$ is true at x in M Iff A is true at x in M or B is true at x in M;
4) $A \& B$ is true at x in M Iff A is true at x in M and B is true at x in M;
5) $A \Rightarrow B$ is true at x in M Iff if A is true at x in M, then B is true at x in M;
6) $A \Leftrightarrow B$ is true at x in M Iff A is true at x in M if and only if B is true at x in M;
7) $\Box A$ is true at x in M Iff for all y with xRy, A is true at y in M;
8) $\Diamond A$ is true at x in M Iff for some y with xRy, A is true at y in M.

A formula A is valid if A is true in all models, and A is satisfiable if A is true in some model.

Different modal logics are sound and complete in different Kripke models with different properties of the relation R, see Table 4.1.2.

Let us assume that R is serial and $\Box A$ holds true at a in M. It means that for all y with aRy, A is true at y in M, but the seriality of R gives some x such that aRx. Hence, it guarantees that (D) is true in all Kripke models with the serial R. Suppose, now that R is reflexive and $\Box A$ holds true at a in M. Then, for all y with aRy, A is true at y in M, but due to the reflexivity of R we have aRa. From this it follows that (T) is true in all Kripke models with the reflexive R. We can check that if R is reflexive, then R is serial, but not vice versa. Then **T** contains all the theorems of **D**, but not vice versa.

Table 4.1.2. Different modal logics in Kripke models with different R.

	Modal Logic	Properties of R
Basic modal logic	**K**	R can be any relation
Modal logics for expressing logical contingency	**D**	R is serial: there exist y such that xRy
	T	R is reflexive: xRx
Modal logics for expressing logical determinism	**CD**	R is unique: $(zRx \,\&\, zRy) \Rightarrow x = y$
	K=	R is bisimilar: $xRy \Rightarrow x = y$

Let R be unique and $\Diamond A$ holds true at a in M. Then, for some y with aRy, A is true at y in M, but the uniqueness of R shows that aRy for all y. So, (CD) is true in all Kripke models with the unique R. Let R be bisimilar and A holds true at a in M. Then, from aRy we infer $a = y$, therefore, A is true at y in M for all y. Thus, (=) is true in all Kripke models with the bisimilar R. Let us note that if R is bisimilar, then R is unique, but not vice versa. Then **K=** contains all the theorems of **CD**, but not vice versa.

4.1.3 Prose as a Literary Genre in Ancient Greece for Expressing Contingency

For the first time, historical narrative (prose) as a new literary genre had arisen during the early classical period in Ancient Greece. Hecataeus of Miletus (Ἑκαταῖος ὁ Μιλήσιος; ca. from 550 B.C to 476 B.C.) is regarded as a founder of this genre. He had written the following two works which survived only in short fragments: the *Survey of the Earth* or *Periodos gēs* (Περίοδος γῆς) and the *Genealogies* (Γενεαλογίαι), also cited as the *History* (Ἱστορία) or *Hērōologia* (Ἡρωολογία). The *Periodos gēs* had been divided into two sections: the first called the *Europe* (Εὐρώπη) and devoted to the geography and customs of the western part of the Achaemenid Empire and the second called the *Asia* (Ἀσία) and containing the data on the regions and the nations in the eastern part of the Achaemenid Empire. The book *Genealogies* was the first attempt to classify the mythical era genealogically.

The new genre was continued in the *Histories* (Ἱστορίαι) consisting of IX books, which survived till now and were written in the Ionic dialect of classical Greek by Herodotus Halicarnassus (Ἡρόδοτος Ἁλικαρνᾱσσεύς; ca. from 484 B.C. to 425 B.C.). Mainly, this work was devoted to the Greco-Persian wars. Factually, the books of Herodotus became a sample of historical prose for all the succeeding generations. Some significant features of this literary genre are as follows: (i) a prosaic consideration of some important facts about different geographical locations with mentioning the practice and customs of their communities; (ii) a prosaic description of different events as results of human actions caused by a conflict of interests; and (iii) a consideration of different opinions on the conflict of interest that was behind the event. Hence, in the historical prose established in Ancient Greece two types of facts were introduced which should have been examined necessarily: (a) geographical facts including the general data on some regions and nations, and (b) historical facts including a description of conflicts of interest causing appropriate events.

Let us consider a fragment (VI: 137, 1–VI: 137, 4) of this narrative from the *Histories* to show features mentioned above. In this fragment (Godley, 1920), Herodotus described how Miltiades, son of Cimon (Μιλτιάδης; ca. 550–489 B.C.), took possession of Lemnos, the island in the northern part of the Aegean Sea, and then he ceded it to the Athenians. It is a historical fact. An appropriate conflict of interest consisted in that "the Pelasgians were driven out of Attica by the Athenians" (Πελασγοὶ ἐπείτε ἐκ τῆς Ἀττικῆς ὑπὸ Ἀθηναίων ἐξεβλήθησαν) and "the Athenians gave to the Pelasgians Lemnos as a dwelling-place in reward for the wall that had once been built around the acropolis" (ἐπείτε γὰρ ἰδεῖν τοὺς Ἀθηναίους τὴν χώρην, τὴν σφίσι αὐτοῖσι ὑπὸ τὸν Ὑμησσὸν ἐοῦσαν ἔδοσαν Πελασγοῖσι οἰκῆσαι μισθὸν τοῦ τείχεος τοῦ περὶ τὴν ἀκρόπολιν κοτὲ ἐληλαμένου) (see *Ibid.*). Then the Athenians decided to take Lemnos back. There are two opinions about this conflict that "the act of the Athenians was just or unjust" (εἴτε ὦν δὴ δικαίως εἴτε ἀδίκως) (see *Ibid.*). "Hecataeus the son of Hegesandrus declared in his history that the act was unjust" (Ἑκαταῖος μὲν ὁ Ἡγησάνδρου ἔφησε ἐν τοῖσι λόγοισι λέγων ἀδίκως), because Lemnos was in the possession of the Pelasgians at the time of war (see *Ibid.*). Nevertheless, Herodotus declared that the Athenians themselves say that their reason for expelling the Pelasgians from the island was just and he considered their reason (see *Ibid.*).

Aristotle (Ἀριστοτέλης; 384–322 B.C.) was one of the first philosophers who defined the features of the genre of historical prose. He emphasized in his *Poetics* (Ποιητική/*Poetica*) 1451b (Fyfe, 1932) that "the writings of Herodotus could be put into verse and yet would still be a kind of history" (εἴη γὰρ ἂν τὰ Ἡροδότου εἰς μέτρα τεθῆναι καὶ οὐδὲν ἧττον ἂν εἴη ἱστορία). Hence, a prosaic narration is not its main feature. But the real difference from poetry consists in that the author in the historical genre "tells us what happened" and the author in poetry "what might happen" (ἀλλὰ τούτῳ διαφέρει, τῷ τὸν μὲν τὰ γενόμενα λέγειν, τὸν δὲ οἷα ἂν γένοιτο), see *Ibid.* Therefore, "poetry is something more philosophical and serious than history" (διὸ καὶ φιλοσοφώτερον καὶ σπουδαιότερον ποίησις ἱστορίας ἐστίν), because "poetry tends to give something general" (καθόλου), namely one plot with a continuous action, while "history gives particular opinions" (ἕκαστον λέγει) about many plots with many different actions (see *Ibid*). Aristotle explained this difference as follows. Poetry describes something general, because it assumes that "a certain type of man will do or say either probably or necessarily" (ἔστιν δὲ καθόλου μέν, τῷ ποίῳ τὰ ποῖα ἄττα συμβαίνει λέγειν ἢ πράττειν κατὰ τὸ εἰκὸς ἢ τὸ ἀναγκαῖον), i.e., there is only one plot, where all events are connected among themselves by chance (τὸ εἰκὸς) or by a causal relationship (τὸ ἀναγκαῖον) (see *Ibid.*).

In tragedies, one plot should be around a man "who is not pre-eminently virtuous and just, and yet it is through no badness or villainy of his own that he falls into the fortune, but rather through some flaw in him, he being one of those who are in high station and good fortune, like Oedipus and Thyestes" (ὁ μήτε ἀρετῇ διαφέρων καὶ δικαιοσύνῃ μήτε διὰ κακίαν καὶ μοχθηρίαν μεταβάλλων εἰς τὴν δυστυχίαν ἀλλὰ δι' ἁμαρτίαν τινά, τῶν ἐν μεγάλῃ δόξῃ ὄντων καὶ εὐτυχίᾳ, οἷον Οἰδίπους καὶ Θυέστης), *Poethics* 1453a9–1453a11 (Fyfe, 1932). Thus, in tragedies the plot should be devoted to the passing of a common man from good fortune to misfortune (ἐξ εὐτυχίας εἰς δυστυχίαν). In histories we cannot have a general plot at all.

To sum up, according to Aristotle, the historical events cannot be collected within one plot. It means that for the historical events there are no connections by chance (τὸ εἰκὸς) or by a causal relationship (τὸ ἀναγκαῖον), see Aristotle's *Poetics* 1451b. So, Aristotle treated historical events as something contingent. Let R show a possibility since a. Then, in accordance with Aristotle, for necessary events, R is serial: there exist y such that aRy, see Table 4.1.2; and for contingent events, R is not unique—it is false that $(zRx \,\&\, zRy) \Rightarrow x = y$.

A sea battle became an often mentioned historical event at the time of Aristotle. For instance, Thucydides (Θουκυδίδης; ca. from 460 B.C. to 400 B.C.), the Athenian historian and general, wrote the *Histories* (Ἱστορίαι) devoted to the Peloponnesian war, where we find out many descriptions of sea battles such as the following narration: "Meanwhile Leon and Diomedon with the Athenian fleet from Lesbos issuing from the OeLacedaenussae, the isles off Chios, and from their forts of Sidussa and Pteleum in the Erythraeid, and from Lesbos, carried on the war against the Chians from the ships, having on board heavy infantry from the rolls pressed to serve as marines" (καὶ Λέων καὶ Διομέδων ἔχοντες τὰς ἐκ Λέσβου Ἀθηναίων ναῦς, ἔκ τε Οἰνουσσῶν τῶν πρὸ Χίου νήσων καὶ ἐκ Σιδούσσης καὶ ἐκ Πτελεοῦ, ἃ ἐν τῇ Ἐρυθραίᾳ εἶχον τείχη, καὶ ἐκ τῆς Λέσβου ὁρμώμενοι τὸν πρὸς τοὺς Χίους πόλεμον ἀπὸ τῶν νεῶν ἐποιοῦντο· εἶχον δ' ἐπιβάτας τῶν ὁπλιτῶν ἐκ καταλόγου ἀναγκαστούς), *Histories* VIII: 24, 2 (Thucydides, 1910). Also, sea battles are often mentioned even in comedies—e.g., in the *Frogs* (Βάτραχοι) written by Aristophanes (Ἀριστοφάνης; ca. from 446 B.C. to 386 B.C.): "If there were a sea battle, and then they had bottles of vinegar, they could squirt them in the eyes of enemies" (εἰ ναυμαχοῖεν κᾆτ' ἔχοντες ὀξίδας ῥαίνοιεν ἐς τὰ βλέφαρα τῶν ἐναντίων), *Frogs* 1440 (tr. by Matthew Dillon).

In the book, *On Interpretation* (Περί ερμηνείας/*De Interpretatione*), Aristotle defined a modal logic for the first time—a formal logic with alethic modalities for any proposition p: possibility ("It is possible that p", formally: $\Diamond p$) and necessity ("It is necessary that p", formally: $\Box p$). He considered some modal propositions as always true statements which are, in fact, axioms of the present-day's modal logic **T**, see Tables 4.1.1 and 4.1.2. In this system **T**, there are assumed contingent events—the events p which possibly appear and possibly disappear ("It is possible that p and it is possible that not-p simultaneously"). And a sea battle is the Aristotelian classical example of such an event p: "A sea battle must either take place tomorrow or not, but it is not necessary that it should take place tomorrow, neither it is necessary that it should not take place, yet it is necessary that it either should or should not take place tomorrow" (Δὲ οἷον ἀνάγκη μὲν ἔσεσθαι ναυμαχίαν αὔριον ἢ μὴ ἔσεσθαι, οὐ μέντοι γενέσθαι αὔριον ναυμαχίαν ἀναγκαῖον οὐδὲ μὴ γενέσθαι γενέσθαι μέντοι ἢ μὴ γενέσθαι ἀναγκαῖον, Aristotle, *On Interpretation* IX: 19a). Hence, by Aristotle, we can state: "Either a sea battle appears tomorrow, or it does not appear tomorrow" (either p or not-p), because it is a logical law of excluded middle. But we cannot state modally: "Either it is necessary that a sea battle appears tomorrow, or it is necessary that it does not appear tomorrow" (either $\Box p$ or $\Box$not-p). The negation of the statement "Either $\Box p$ or $\Box$not-p" is represented by the statement "$\Diamond p$ and $\Diamond$not-p". The statement "Either $\Box p$ or $\Box$not-p" tells us about a logical determinism that everything appears necessarily or disappears necessarily. Its negation "$\Diamond p$ and $\Diamond$not-p" tells us about a logical contingency that there exists something that appears possibly and disappears possibly. And historical events such as a sea battle belong to the class of this contingency (i.e., we have the negation that *R* is unique).

Thus, poetry is "more philosophical and serious than history" (διὸ καὶ φιλοσοφώτερον καὶ σπουδαιότερον ποίησις ἱστορίας ἐστίν), see Aristotle's *Poetics* 1451b, because history concerns contingent events such as sea battles and these events cannot be collected as one plot. In this way we can draw a conclusion that, according to Aristotle, history is a literary genre where some contingent events presenting the past of some people are examined. It means that the historical narrative could arise only due to the assumption that there exist contingent events which cannot be collected through a causal relationship, but they can be just mentioned and then described in detail.

The appearance of historical prose in Ancient Greece also caused the appearance of philosophical prose as an additional literary genre of prose—it became a new genre of meta-narration where all possible kinds of narration including prose and poetry can be analyzed theoretically within the three disciplines: dialectics (our reasoning), physics (natural forces), and ethics (our customs and characters). While in history there are possible different opinions explaining conflicts of interest, in philosophy there are possible different opinions in meta-narration. So, these opinions can be collected, as well, and Diogenes Laërtius (Διογένης Λαέρτιος; ca. from 180 A.D. to 240 A.D.) wrote the *Lives and Opinions of Eminent Philosophers* (Βίοι καὶ γνῶμαι τῶν ἐν φιλοσοφίᾳ εὐδοκιμησάντων/*Vitae Philosophorum*) where there are considered biographies and views of the most prominent Greek philosophers.

4.1.4 'Aggādāh as a Literary Genre in Judaism for Expressing Logical Determinism

At first glance, *'aggādāh* (אגדה) including non-legalistic exegetical texts recorded in the Talmud and Midraš (hence, compiled over the period of late antiquity—from the 3rd to the 5th centuries A.D.) and containing different histories (first, cautionary tales) can be treated as a historical prose, too, close to the historical narrative founded by Hecataeus of Miletus. To show some really significant differences from Greek history, let us consider just one story taken from the *'Ên Ya'ăqōḇ* (עין יעקב)—a big compilation of different *'aggāḏôṯ* (Hebrew: 'histories') from the Talmud and Midraš edited and issued by Ya'aqov ben Šlomoh 'ibn Ḥabib (יעקב בן שלמה אבן חביב; ca. from 1460 to 1516 A.D.), the rabbi born at Zamora, Spain.

This *'aggādic* story (*'Ên Ya'ăqōḇ, Giṭṭîn* V: 5), we are going to explore, describes the destruction of the Second Temple—the most tragic historical event for the Jews just before their exile from Israel. But in this story, there is neither any important geographical information nor a consideration of the conflict of interest causing the Judaic war against Rome—the things needed in the Greek history, surely. Instead of that we read a tale about the following woman—"Martha, the daughter of Beithus, who came from one of the richest families in Jerusalem" (מרתא בת בייתוס עתירתא דירושלים) (See *Ibid.*). During the siege of Jerusalem by the Roman army, she was hungry and sent a messenger to buy a "fine flour" (סמידא). But it was sold out and he came back without the fine flour. Then she said to him to bring a "usual flour" (חיורתא), but it was sold, too. After that she sent him to buy a "dark flour" (גושקרא), but it was sold, also. At the end, she asked him to bring a "barley flour" (קמחא), and again it happened that this was sold out. She put her shoes on and said she herself would try to find something to eat. Nevertheless, "while walking in the street dirt infected her foot and it caused her death" (אי משכחנא מידי למיכל איתיב לה פרתא בכרעה ומתה) (see *Ibid.*). It is the end of the story.

In this tale, there occur the following six events with Martha: (i) the fruitless attempt to buy the most expansive flour (סמידא) by the messenger for her; (ii) the unsuccessful attempt to buy the expansive, but quite cheaper flour (חיורתא) for her; (iii) the failed attempt to buy the cheaper flour (גושקרא); (iv) at the end, the fruitless attempt to buy the cheapest flour (קמחא); (v) the decision of the lady to buy herself something to eat; (vi) the death of the lady because of the dirt on her foot. We observe a cascade of worsening events from the easy misfortune of (i) to the scariest misfortune of (vi). So, the plot of this story embodies the growth of misfortunes from the light to the worst.

According to the *'aggādāh* (*'Ên Ya'ăqōb, Giṭṭîn* V: 5), there were two eyewitnesses of the described misfortunes of the lady: Rabban Yoḥanan ben Zakk'ai (רבן יוחנן בן זכאי; the 1st century A.D.) and Rabbi Ṣadoq (רבי צדוק; the 1st century A.D.). Both defined this event as a clear omen for the prompt destruction of Jerusalem and the Second Temple. Meanwhile, Rabban Yoḥanan ben Zakk'ai applied to her the following passage from the Torah (*Deuteronomy* 28: 56) (see *Ibid.*), to tell the inhabitants of Jerusalem about fulfilling a prophecy right now: "And she who is most tender and dainty among you, so tender and dainty that she would never venture to set a foot on the ground, shall begrudge the husband of her bosom, and her son and her daughter." When Rabbi Ṣadoq saw how before the death Martha "took all the gold and silver and threw it in the streets, saying: 'What do I need this for?'" (ניחא נפשה אפיקתא לכל דהבא וכספא שדיתיה בשוקא אמרה האי למאי), he applied the following passage from the Prophet (*Ezekiel* 7: 19) (see *Ibid.*), to warn all the inhabitants of Jerusalem about the prompt destruction: "They shall throw their silver into the streets, and their gold shall be treated as something unclean. Their silver and gold shall not avail to save them in the day of the Lord's wrath—to satisfy their hunger or to fill their stomachs. Because they made them stumble into guilt."

To sum up, the story of Martha was examined by the rabbis (two witnesses), on the one hand, as an omen for the demolition of the Second Temple and, on the other hand, as a fulfilment of the old prophecies.

It is worth noting that each misfortune from (i) to (v) was a personal omen (warning) for Martha to stop her actions. Nevertheless, she ignored these evident signs and continued to go to her own end. Why did she deserve the terrible end of (vi)? There is a good *'aggādic* explanation (*'Ên Ya'ăqōb, Yəḇāmôṯ* VI: 1) for this historical fact, also: Yehošu'a ben Gaml'a (יהושע בן גמלא), the Jewish high priest in about 64–65 A.D., betrothed the widow Martha, the daughter of Boethus, and was appointed by the King as high priest, to whom a widow is always prohibited, he nevertheless married her (מתני' מעשה ביהושע בן גמלא שקידש למרתא בת בייתוס ומינהו המלך להיות כהן גדול וכנסה וכו' (דף ס"א)). So, at the beginning she committed a great sin to become the wife of the high priest being a widow, then she ignored personal omens and, therefore, she deserved to become an omen for everybody in Jerusalem that the city is doomed.

Let us come up with a conclusion how far the *ʾaggāḏôṯ* ('histories') in Judaism differ from the ἱστορίαι ('histories') in Ancient Greece. Firstly, contingent events as such are impossible in the Judaic narrations (thinking)—each event can be examined as both an omen and fulfilling a prophecy simultaneously. Consequently, any compilation of historical facts as contingent events cannot be accepted in Judaism at all. Contingent events exist only in the mind of unjust people. In reality any event is an omen for us and a result of prophecy coming from some old sages. In other words, the relation R is always unique or bisimilar, see Table 4.1.2. In the meantime, the bisimilarity of R means that all events are actual: if aRy, then $a = y$. For each event a there are at least two witnesses (such as Rabban Yoḥanan ben Zakk'ai and Rabbi Ṣadoq) who can consider a because of a past prophecy and as a new omen for the next prophecy. It is a kind of eternalism—all historical events are actual and connected among themselves by a unique or bisimilar R.

Notice that Flavius Josephus (Φλάβιος Ἰώσηπος; ca. 37–100 A.D.) wrote the *Antiquities of the Jews* (*Antiquitates Judaicae*/Ἰουδαϊκὴ ἀρχαιολογία) in the way of Greek chronics, but in general this way of history with many contingent events was pushed back by the rabbis. The sacral history of the Jewish nation cannot be treated as a history written by Herodotus Halicarnassus and other Greek authors. For the same reason, the *Septuagint* (*Interpretatio Septuaginta Seniorum*/Ἡ μετάφρασις τῶν Ἑβδομήκοντα), the Greek translation of the Hebrew Bible in the 3rd century B.C., was considered by the rabbis as a profanation of the sacral *ʾaggāḏôṯ* and their wrong interpretations. So, the tenth day of the Hebrew month of Ṭevet (טֵבֵת) is celebrated as a fasting day marking the beginning of the siege of Jerusalem by Nebuchadnezzar II of Babylonia ([d]Nabû-kudurri-uṣur; נְבוּכַדְנֶאצַּר, ca. from 605 B.C. to 562 B.C.). And the same day of the Jewish calendar marks the completion of the *Septuagint* to show that the idea of this translation was wrong from its beginning.

Thus, a history in the meaning of Herodotus was impossible in Judaism. The *ʾaggāḏôṯ* cannot become the ἱστορίαι forever. The Judaic thinking categorically avoids the intuition of the modal logic **T** which assumes the existence of contingent events p such that "$\Diamond p$ and $\Diamond$not-p". The story of Martha as a tale about the destruction of the Second Temple demonstrates that in Judaism the intuition of logical determinism is widely accepted—each event p is logically determined to be an omen or a prophecy fulfillment: "Either $\Box p$ or $\Box$not-p", which is an axiom of logics **CD** and **K=**, see Tables 4.1.1 and 4.1.2.

4.1.5 Omens as a Literary Genre in Akkadian for Expressing Logical Determinism and the Religious Roots of Stoic Modal Logic

As we know, Aristotle in his *On Interpretation* (Περί ερμηνείας/*De Interpretatione*) introduced some modal statements as axioms which correspond to the present-day logic **T**. In this logic there are possible contingent events p such that "$\Diamond p$ and $\Diamond$not-p". The intuition of this logic was shared by a lot of Greek philosophers with one exception—the Stoics who supported the idea of logical determinism and completely avoided any contingency. This intuition of the Stoics was fulfilled recently in the modal logic **K=** where there is logical determinism and contingent events are impossible in principle, see (Garson, 2006).

According to the Stoics, "nothing happens without a cause" (*nihil enim fieri sine causa potest*), see the *On Divination* (*De Divinatione*) II: 61 written by Marcus Tullius Cicero (106–43 B.C.). As a result, each individual proposition (ἀξιώματα) p (even about future) is necessarily true, or it is necessarily false: "Either $\Box p$ or $\Box$not-p". For example, a sea battle tomorrow appears necessarily, or it does not appear necessarily. Hence, the world is subordinated by necessity (ἀνάγκη) and fate (εἱμαρμένη). That divine force is also called by them destiny (πεπρωμένη), pneuma (πνεῦμα), cause (αιτία), reason (λόγος), Zeus, will (βούλησις) of Zeus. Due to the fate, each event has its cause and then it is fully determined.

Let us assume that p is a cause for q. Then p determines q: "If p, then q", and p explains q: "q because of p". On the other hand, q can be regarded as a sign for p: "If there is a sign q, then an event p takes place". Some examples of reasoning based on signs which are provided by the Stoics: "If this one has milk in her breast, then she gave birth"; "If sweat flows through the surface, then there are imperceptible pores"; "If this one has a scar, then he has had a wound," see the *Outlines of Pyrrhonism* (Πυρρώνειοι ὑποτυπώσεις/*Pyrrhoniae Hypotyposes*) II: 106 and the *Against the Logicians* (Πρὸς λογικούς/*Adversus Logicos*) II: 252; II: 254–255 written by Sextus Empiricus (Σέξτος Ἐμπειρικός; ca. 160–210 A.D.).

In accordance with the Stoics, implications (conditional statements) by appealing to signs such as "If there is a sign q, then an event p takes place" are used in divinations: "If there is an omen q, then an event p occurs." So, according to the Stoics, omens are not causes of events, but they are their signs, while events might be real causes of omens sometimes.

So, for the Stoics, any proposition is either necessarily true or necessarily false only because an event described by an appropriate proposition always has a natural cause. This claim is used for proving the existence of fate by *modus tollens*:

> Si est motus sine causa, non omnis enuntiatio, quod ἀξίωμα dialectici appellant, aut vera aut falsa erit; causas enim efficientis quod non habebit, id nec verum nec falsum erit; omnis autem enuntiatio aut vera aut falsa est; motus ergo sine causa nullus est. Quod si ita est, omnia, quae fiunt, causis fiunt antegressis; id si ita est, fato omnia fiunt; efficitur igitur fato fieri, quaecumque fiant (Cicero, *De Fato* X: 20–21).

> If uncaused motion exists, it will not be the case that every proposition (termed by the logicians an *axioma*) is either true or false, for a thing not possessing efficient causes will be neither true nor false; but every proposition is either true or false; therefore, uncaused motion does not exist. Precedent causes for all things means fate and causal determinism. It is better to give up the truth/falsity of propositions. If this is so, all things that take place take place by precedent causes; if this is so, all take place by fate; it therefore follows that all things that take place take place by fate (tr. by H. Rackham).

In the formal notation: If *p* ("uncaused motion"), then not-*q* ("not the case that every proposition is either true or false"); *q*; thus, not-*p*. Then there is fate. And according to fate, each actual event occurs necessarily:

> Quia futura vera, inquit, non possunt esse ea, quae causas, cur futura sint, non habent; habeant igitur causas necesse est ea, quae vera sunt; ita, cum evenerint, fato evenerint (Cicero, *De Fato* XI: 26).

> The reason is, says he, that future things that have not got causes of *why* they will be in the future cannot be true; therefore, those that are true must necessarily have causes; accordingly, when they have occurred, they will have occurred by fate (tr. by H. Rackham).

Even the gods have no foreknowledge of events without knowing sufficient causes:

> Ita ne praeterita quidem ea, quorum nulla signa tamquam vestigia extarent, Apollini nota esse censebat; quo minus futura! causis enim efficientibus quamque rem cognitis posse denique sciri, quid futurum esset. Ergo nec de Oedipode potuisse Apollinem praedicere nullis in rerum natura causis praepositis, cur ab eo patrem interfici necesse esset, nec quicquam eius modi (Cicero, *De Fato* XIV: 32–33).

> Therefore Carneades held the view that Apollo had no knowledge even of these past events which had left behind them no trace of their passage—how much less had he knowledge of future events, for only by knowing the efficient causes of all things was it possible to know the future; therefore it was impossible for Apollo to foretell the fate of Oedipus when there were no causes fore-ordained in the nature of things making it necessary for him to murder his father, nor could he foretell anything of the sort (tr. by H. Rackham).

The Stoics, including his father, Zeno of Citium (Ζήνων ὁ Κιτιεύς; ca. 334–262 B.C.) who founded the Stoic teaching on divinations, and his prominent pupil, Cleanthes (Κλεάνθης; ca. 330–230 B.C.) who described this teaching in detail, all them defended nearly every sort of divination (*sed cum Stoici omnia fere illa defenderent, quod et Zeno in suis commentariis quasi semina quaedam sparsisset et ea Cleanthes paulo uberiora fecisset, accessit acerrimo vir ingenio*), see Cicero's *De Divinatione* I: 6. After that Chrysippus "discussed the whole theory of divination in two books" (*Chrysippus, qui totam de divinatione duobus libris explicavit sententiam*), and "wrote one book on oracles and another on dreams" (*uno praeterea de oraculis, uno de somniis*) (see *Ibid.*). Other Stoics published many books on similar subjects too, such as Diogenes of Babylon (Διογένης Βαβυλώνιος; ca. 230–150 B.C.) and Posidonius (Ποσειδώνιος; ca. 135–51 B.C.), the friend of Cicero (see *Ibid.*). The Stoics also supported the astrology developed by the Chaldeans as an especial art of divination:

> Principio Assyrii, ut ab ultimis auctoritatem repetam, propter planitiam magnitudinemque regionum quas incolebant, cum caelum ex omni parte patens atque apertum intuerentur, traiectiones motusque stellarum observitaverunt, quibus notati, quid cuique significaretur memoriae prodiderunt. Qua in natione Chaldaei, non ex artis sed ex gentis vocabulo nominati, diuturna observatione siderum scientiam putantur effecisse, ut praedici posset, quid cuique eventurum et quo quisque fato natus esset (Cicero, *De Divinatione* I: 2).

> First of all to seek authority from the most distant sources—the Assyrians, on account of the vast plains inhabited by them, and because of the open and unobstructed view of the heavens presented to them on every side, took observations of the paths and movements of the stars, and having made note of them, transmitted to posterity what significance they had for each person. And in that same nation the Chaldeans—a name which they derived not from their art but their race—have, it is thought, by means of long-continued observation of the constellations, perfected a science which enables them to foretell what any man's lot will be and for what fate he was born (Falconer, 1923).

The special perspective of the Stoics on logical determinism and their support of all sorts of divination can be explained by some Semitic roots of many Stoic philosophers—e.g., their founder, Zeno, was a Phoenician. In the meanwhile, the Stoics had strongly negative attitudes towards ἱστορίαι ('histories') in the meaning of Herodotus, because they did not accept the existence of contingency.

In Semitic cultures of the Achaemenid Empire rooted mainly from the Akkadian one, a divination was usually formulated by means of the following implications, as we remember: "If [Akkadian: *šumma*] we observe an omen *p*, an event *q* should take place". This implication tells us about "[divine] decision" or 'verdict' (Akkadian: *purussû*) inferred by *modus ponens*:

Divination: "If *p*, then *q*";
Observer: "There is an omen *p*"

Diviner: "There is a forecasting of *q*"

The Stoic instance of *modus ponens* taken by them from an astrological divination made by the Chaldeans, see Cicero's *On Fate* (*De Fato*) VI: 12 is as follows: (i) The first premise: "If somebody was born at the rise of Sirius, he never dies at sea" (*si quis oriente Canicula natus est, in mari non morietur*). (ii) The second premise: "Fabius was born at the rise of Sirius" (*Fabius oriente Canicula natus est*). (iii) Then, according to the inference rule of *modus ponens*, we draw the following conclusion: "Fabius never dies at sea" (*Fabius in mari non morietur*).

We can assume that the Stoic propositional logic grounded on implications (conditional statements) and *modus ponens* is a kind of their philosophical reflection on the tradition of Semitic divinations based on conditional statements with *modus ponens*, too. The point is that this kind of formulation for divinations (forecasting) is well traced back since the Old Babylonian period (i.e., since ca. 1800 B.C.): "If (*šumma*) [an omen] *p*, [then a prognosis] *q*." The same form of implication was used in formulating articles in the Mesopotamian law codes since the earliest ones at the time of Ur III: "If [someone committed] *p*, [then he or she shall be sentenced to] *q*." Then the court sentence is delivered by *modus ponens*: he or she committed *p*, then the decision of court is *q*.

In the Old Babylonian period divinations usually started with the lexeme BE-*ma* with the reading *šum*$_4$-*ma* 'if'. In the Neo-Assyrian period another lexeme DIŠ was used, just to designate 'item' or 'entry' with the same meaning of implication. Divinations formulated by the implications in Akkadian were composed in long lists where all logical combinations for different contexts of occurring *p* and *q* are analyzed.

As we remember, see Section 3.3, these lists of omens were composed sometimes as codes, where all entities are introduced within a binary logic—each sign or event is examined as either positive or negative (see Subsection 3.3.3). For instance, let us take the following four signs: 'falcon', 'raven', "on the right", "on the left". So, we obtain the four omens as 2^2 possible combinations of four signs: (i) 'falcon' & "on the right"; (ii) 'falcon' & "on the left"; (iii) 'raven' & "on the right"; (iv) 'raven' & "on the left". Hence, the algorithm of composing omens is as follows. Let us take $2k$ signs of the first order, $2l$ signs of the second order, $2m$ signs of the third order, etc., where $k, l, m \geq 1$. The point is that there are considered only binary oppositions so that each sign has an opposition within a couple. Then we have $2^{(k+l+m+\ldots)}$ of all possible combinations giving different antecedents of conditionals. Therefore, taking the four signs, we deal with a code of four conditionals: (i) if 'falcon' & "on the right", then…; (ii) if 'falcon' & "on the left", then…; (iii) if 'raven' & "on the right", then…; (iv) if 'raven' & "on the left", then… Taking $2k$ signs of the first order, $2l$ signs of the second order, $2m$ signs of the third order, etc., where $k, l, m \geq 1$, we always have $2^{(k+l+m+\ldots)}$ conditionals for prediction.

In these omen codes, any logical inconsistency is avoided. For instance, 'falcon' is a positive sign, but 'raven' is not; "on the right" is a positive sign, "on the left" is not. Therefore, the antecedents of codes give a Boolean matrix (truth table): (i) if "positive (falcon)" & "positive (on the right)", then …; (ii) if "positive (falcon)" & "negative (on the left)", then …; (iii) if "negative (raven)" & "positive (on the right)", then …; (iv) if "negative (raven)" & "negative (on the left)", then … For $2k$ signs, we obtain 2^k antecedents in the form of Boolean matrix. In the consequents of conditionals, each opposite sign is contained in another conditional. So, for different oppositions at the place of antecedent, we have different oppositions at the place of consequent. This type of composing conditionals can be named a *rule-based expert system* in the modern meaning. And any logical inconsistency (contradiction) is excluded there.

There are lists of divinations in Akkadian devoted to different subjects: from astrological to even quite ethical (e.g., omens about a city), see (Freedman, 2017; Koch(-Westenholz), 2015; Jeyes, 1989; Maul, 2007; Reiner, 1998; Winitzer, 2017;

Zorzi, 2009, 2011). In this way, some Jewish commandments can be regarded as rudiments of some ethical divinations in Hebrew formulated by means of the conditional statements at first:

> עַל שָׁלֹשׁ עֲבֵרוֹת נָשִׁים מֵתוֹת בִּשְׁעַת לֵדָתָן, עַל שֶׁאֵינָן זְהִירוֹת בַּנִּדָּה וּבַחַלָּה וּבְהַדְלָקַת הַנֵּר
>
> *ʿal šālōš ʿăb̲ērôt̲ nāšîm mēt̲ôt̲ bišʿat̲ lēd̲āt̲ān, ʿal šeʾênān zəhîrôt̲ banniddāh ûb̲aḥallāh ûb̲əhad̲lāqat̲ hannēr.*
>
> Women die in childbirth for the following three transgressions: Because [עַל] they are not careful with [obeying the laws] of menstruation; and because they are not careful [to separate some] batter [when baking to give to the priest]; and because they are not careful with the lighting of the [Sabbatical] lamp (*Mišnah Šabbat* II: 5).

In this passage, the breaking of the three Jewish commandments: ‘menstruation’ (*nniddāh*), ‘batter’ (*ḥallāh*), and “lighting the Sabbatical lamp” (*had̲lāqat̲ hannēr*), is represented as omens for the death of woman in her childbirth. Hence, usually the Akkadian divinations with an ethical dimension were expressed out in the form of implications: “If you perform something that is treated as a bad omen, then an evil event happens to you” and “If you perform something that is treated as a good omen, then a favourable event happens to you.” Nevertheless, this formulation as conditionals was preserved in Judaism only as a rudiment in the Talmudic commentaries.

Based on the analysis of different lists of divinations, presented as codes, we can formulate the following four general rules for composing omens (Zorzi, 2009, 2011):

a) The consequent of divination gives a positive forecasting (event) if and only if a singular omen in the antecedent is positive.
b) The consequent of divination gives a negative forecasting (event) if and only if a singular omen in the antecedent is negative.
c) Let us assume that there is more than one omen in the antecedent. The consequent of divination gives a negative forecasting (event) if and only if there are positive and negative omens in the antecedent simultaneously.
d) Let us assume that there is more than one omen in the antecedent. Then the consequent of divination gives a positive forecasting (event) if and only if all omens in the antecedent are positive or all omens in the antecedent are negative.

Let us exemplify these rules by some fragments in Akkadian to formalize them logically:

> 1. DIŠ NA *ana* Á.ÁŠ-*šú* ZI-*ma* ŠÚR.DÙ^{MUŠEN} TA 15 NA *ana* 2,30 NA *i-ti-iq* Á.ÁŠ-*su* KUR-*ad*
> 2. DIŠ KIMIN ŠÚR.DÙ^{MUŠEN} 15 NA *iz-ziz-ma ta-ḫe-e* NA DU-*ak* NA.BI KI DU-*ku* Á.TUG TUK-*ši*
> 3. DIŠ KIMIN Ú.NÁG.GA^{MUŠEN} *iz-zi-iz-ma* GÙB NA GÙ-*si* NA.BI KI IGI.MEŠ-*šú* GAR-*nu* DU-*ma* ḪA.LA KÚ
> 6. DIŠ KIMIN KASKAL ZI-*ma* Ú.NÁG.GA^{MUŠEN} *ina* 15 NA GUB-*ma* GÙ-*si* NA.BI KI IGI.MEŠ-*šú* GAR-*nu* NU DU-*ak* ŠÀ.BI NU DÙG.GA

24. DIŠ NA ŠE.NUMUN È-*ma* Ú.NÁG.GA[MUŠEN] *ina* UGU-*šú* DU-*ma ina* 2,30 NA GÙ-*si* AB.SÍN GUN-*sà ut-tar*

1. If a man goes off on his errand, and a falcon crosses from the right of the man to the left of the man—he will attain his desire.
2. If ditto a falcon stays on the right of the man and proceeds alongside the man—that man: wher(ever) he goes, he will have gain.
3. If ditto a raven stays and caws to the left of the man—that man: he will go where he decides and he will enjoy a profit.
6. If ditto goes on a journey and a raven proceeds to the right of the man and caws—that man: he won't go where he decides, and his heart won't be satisfied.
24. If a man throws the seed and a raven goes upon it and caws to the left of the man—the furrow will increase its yield (Zorzi, 2009: 92–94, 120).[2]

My logical commentaries to these omens are as follows:

1. The omen 'falcon' (f) is a positive sign as well as "crossing from the right of the man to the left of the man" (r) is positive as all other omens associated with the right side and moving from the right. Hence, in verse 1, there are two positive omens f & r in the antecedent, then, according to (d), it gives a favourable forecasting (p): $(f \,\&\, r) \Rightarrow p$.
2. The omen 'falcon' (f) is positive, the omen "staying on the right of the man" (r) is positive, and the omen "proceeding alongside the man" (a) is positive. So, all the omens f & r & a are positive simultaneously and, by (d), we obtain a favourable conclusion again (p): $(f \,\&\, r \,\&\, a) \Rightarrow p$.
3. The omen 'raven' (the opposite to 'falcon': $\neg f$) is a negative sign, the omen "[raven] caws" ($\neg c$) is negative, too, the omen "to the left of the man" (the opposite to 'right': $\neg r$) is negative, also, to the same extent as all other omens connected to the left side. Thus, all the omens in the antecedent $\neg f$ & $\neg c$ & $\neg r$ are negative simultaneously. In accordance with (d), it gives a positive prediction (p): $(\neg f \,\&\, \neg c \,\&\, \neg r) \Rightarrow p$.
6. The omen 'raven' ($\neg f$) is negative, the omen "[raven] caws" ($\neg c$) is negative, but the omen "proceeding to the right of the man" (r) is positive. We have a mix of negative and positive signs: $\neg f$ & $\neg c$ & r. It means that, according to (c), the prediction should be negative (the opposite to the positive forecasting: $\neg p$) and it is so: $(\neg f \,\&\, \neg c \,\&\, r) \Rightarrow \neg p$.
24. The omen "raven goes upon it" ($\neg f$) is negative, the omen "[raven] caws" ($\neg c$) is negative, the omen "to the left of the man" ($\neg r$) is negative. By (d), we obtain a positive prediction (p): $(\neg f \,\&\, \neg c \,\&\, \neg r) \Rightarrow p$.

[2] In this fragment, we see a commutativity of conjunction—the omen "to the left/right" is located at different places in two verses: "[raven] stays and caws to the left of the man" (verse 3) and "raven proceeds to the right of the man and caws" (verse 6) but changing places does not change the meaning of conjunction. From other fragments in Akkadian, we see that conjunction and disjunction were understood as commutative and associative operations and implication had the property of transitivity.

Let +1 be a positive meaning of an appropriate omen and –1 be a negative meaning of omen (see Subsection 3.3.3). Let [[*p*]] be a meaning of *p* that is either positive or negative. Then the example above can be presented as the following symbolic expressions:

1. ((+1 & +1) ⇒ +1) = (([[*f*]] & [[*r*]]) ⇒ [[*p*]]).
2. ((+1 & +1 & +1) ⇒ +1) = (([[*f*]] & [[*r*]] & [[*a*]]) ⇒ [[*p*]]).
3. ((–1 & –1 & –1) ⇒ +1) = (([[¬*f*]] & [[¬*c*]] & [[¬*r*]]) ⇒ [[*p*]]).
6. ((–1 & –1 & +1) ⇒ –1) = (([[¬*f*]] & [[¬*c*]] & [[*r*]]) ⇒ [[¬*p*]]).
24. ((–1 & –1 & –1) ⇒ +1) = (([[¬*f*]] & [[¬*c*]] & [[¬*r*]]) ⇒ [[*p*]]).

Now, we can reformulate rules (a), (b), (c), (d) symbolically:

a) +1 ⇒ +1.

b) –1 ⇒ –1.

c) (–1 & +1 & …) ⇒ –1.

d) (+1 & +1 & …) ⇒ +1, where there are only +1 in the antecedent;
(–1 & –1 & …) ⇒ +1, where there are only –1 in the antecedent.

In this way, we can reconstruct semantics for logical compositions of atomic propositions involved into predictions. So, we can check that in verses 1, 2, 3, 6, 24, as well as in other Akkadian fragments, the following truth tables are implicitly used (see Subsection 3.3.3):

Truth table for ¬ (negation):

[[omen/event]]	[[¬ omen/event]]
+1	–1
–1	+1

Truth table for & (conjunction):

[[omen 1/event 1]]	[[omen 2/event 2]]	[[omen 1/event 1]] & [[omen 2/event 2]]
+1	+1	+1
+1	–1	–1
–1	+1	–1
–1	–1	–1

Truth table for ⇒ (implication):

[[omen 1/event 1]]	[[omen 2/event 2]]	[[omen 1/event 1]] ⇒ [[omen 2/event 2]]
+1	+1	+1
+1	–1	–1
–1	+1	+1
–1	–1	+1

According to these truth tables, we have that in the considered example in Akkadian, the divination list contains only propositional axioms (tautologies):

1. $((+1 \,\&\, +1) \Rightarrow +1) = +1$.
2. $((+1 \,\&\, +1 \,\&\, +1) \Rightarrow +1) = +1$.
3. $((-1 \,\&\, -1 \,\&\, -1) \Rightarrow +1) = +1$.
6. $((-1 \,\&\, -1 \,\&\, +1) \Rightarrow -1) = +1$.
24. $((-1 \,\&\, -1 \,\&\, -1) \Rightarrow +1) = +1$.

Furthermore, rules (a), (b), (c), (d) define how these logical axioms should be formulated (see Axiom of Subsection 3.3.3):

a) $(+1 \Rightarrow +1) = +1$.

b) $(-1 \Rightarrow -1) = +1$.

c) $((-1 \,\&\, +1 \,\&\, \ldots) \Rightarrow -1) = +1$.

d) $((+1 \,\&\, +1 \,\&\, \ldots) \Rightarrow +1) = +1$;

$((-1 \,\&\, -1 \,\&\, \ldots) \Rightarrow +1) = +1$.

Let us consider another Akkadian fragment with an appropriate divination.

48′. DIŠ KA_5.A *ina* E_2 [NA TU]-*ub* E_2 BI ŠU KUR-*su* : ŠUB-*di*
49′. DIŠ KA_5.A UR.GI_7 *iṭ-ru-ud* TUR URU LAL-*ṭi* KI.LAM
50′. DIŠ KA_5.A *ana* UR.GI_7 TE-*ḫi* ZA$Ḫ_2$ KUR BI
51′. DIŠ KA_5.A *ana* UDU.$NITA_2$ TE-*ḫi* BIR-*aḫ* KUR BI
52′. DIŠ KA_5.A *ina* KASKAL TA [ZAG LU_2] *ana* GUB_3 LU_2 DIB-*iq* LU_2 BI A_2.$AŠ_2$-*su* KUR-*ad*$_2$
53′. DIŠ KA_5.A *ina* KASKAL TA GUB_3 LU_2 *ana* ZAG LU_2 DIB-*iq* LU_2 BI A_2.$AŠ_2$-*su* NU KUR-ad$_2$
54′. DIŠ KA_5.A *ana* IGI LU_2 *i-šur-ma ina* GIR_3.2-*šu*$_2$ SAḪAR.ḪI.A *is-lu* LU_2 BI *bu-ti-iq-ta*$_5$ IGI-*mar*
55′. DIŠ KA_5.A *ana* ZAG LU_2 DIB NU KUR-*ad*$_2$ A_2.$AŠ_2$
56′. DIŠ KA_5.A *ana* GUB_3 LU_2 DIB KUR-*ad*$_2$ A_2.$AŠ_2$
57′. DIŠ LU_2 *ana* A_2.$AŠ_2$-*šu*$_2$ ZI-*ma* KA_5.A TA 15 LU_2 *ana* 150 LU_2 DIB-*iq* LU_2 BI KI DU INIM *ḫa-di-e* TE-*šu*$_2$
58′. DIŠ LU_2 *ana* A_2.$AŠ_2$-*šu*$_2$ ZI-*ma* KA_5.A TA 150 LU_2 *ana* 15 LU_2 DIB-*iq* LU_2 BI I.BI_2.ZA IGI-*mar*
59′. DIŠ LU_2 *ana* A_2.$AŠ_2$-*šu*$_2$ ZI-*ma* KA_5.A *ana* 15 LU_2 GU_3-*si-ma ta-ḫi-e* LU_2 $KAŠ_4$-*um* NA BI KI ZI-*u*$_2$ *ṣal-ti i-ger-ru-šu*$_2$ NIG_2.SIG_5 IGI

48′. If a fox en[ters a man’s] house, the hand (of a god) will affect that house/it will be abandoned.
49′. If a fox chases away a dog—diminution of the city, reduction of prices.
50′. If a fox approaches a dog (sexually)—destruction of that land.
51′. If a fox approaches a sheep (sexually)—dispersal of that land.
52′. If a fox on the road crosses from [a man’s right] to a man’s left, that man will attain his wish.

> 53′. If a fox on the road crosses from a man's left to a man's right, that man will not attain his wish.
> 54′. If a fox circles around in front of a man (and) scratches up dirt with its feet, that man will see a flood.
> 55′. If a fox crosses toward a man's right—no attainment of a wish.
> 56′. If a fox crosses toward a man's left—attainment of a wish.
> 57′. If a man is going about his business, and a fox crosses from the man's right to the man's left, joyful news will come to that man wherever he goes.
> 58′. If a man is going about his business, and a fox crosses from the man's left to the man's right, that man will experience losses.
> 59′. If a man is going about his business, and a fox cries out on the man's right and runs toward the man, they will file suit against that man wherever he goes on his business, (but) he will experience good fortune (Freedman, 2017: 36–37).

Symbolically:

48′. "Fox entering a man's house" is negative, then it gives an unfavourable forecasting: $(-1 \Rightarrow -1) = +1$.
49′. "Fox chasing away a dog" is negative, therefore the conclusion is negative, too: $(-1 \Rightarrow -1) = +1$.
50′. "Fox approaching a dog" is negative, then the conclusion is negative: $(-1 \Rightarrow -1) = +1$.
51′. "Fox approaching a seep" is negative, then the prediction is negative: $(-1 \Rightarrow -1) = +1$.
52′. "Fox crossing from a man's right to a man's left" is positive, from which we infer a positive prediction: $(+1 \Rightarrow +1) = +1$.
53′. "Fox crossing from a man's left to a man's right" is negative, then we deduce a negative event: $(-1 \Rightarrow -1) = +1$.
54′. "Fox circling around in front of a man" is positive, but "fox scratching up dirt with its feet" is negative, that we summarize that the expected event should be negative: $((+1 \;\&\; -1) \Rightarrow -1) = +1$.
55′. "Fox crossing toward a man's right [from left]" is negative and the prediction is negative: $(-1 \Rightarrow -1) = +1$.
56′. "Fox crossing toward a man's left [from right]" is positive and the prediction is positive: $(+1 \Rightarrow +1) = +1$.
57′. "Fox crossing from the man's right to the man's left" is positive and the forecasting is positive: $(+1 \Rightarrow +1) = +1$.
58′. "Fox crossing from the man's left to the man's right" is negative and the conclusion is negative: $(-1 \Rightarrow -1) = +1$.
59′. "Fox crying out on the man's right" is positive and "fax running toward the man" is positive, then the prediction concerns "good fortune": $((+1 \;\&\; +1) \Rightarrow +1) = +1$.

The binary thinking based on the Boolean algebra is universal for many other omen series, e.g., for some tablets 1–21 (OPKF 17). Let us show it. From the context, we see that "white fungus" (*katarru peṣû*) is a negative sign. Their possible combinations with other signs are as follows:

5. DIŠ KA.TAR BABBAR *ina ne-rib* KÁ T[ÙR...] SAG.PA.LAGAB É NA [...].
15. DIŠ KA.TAR *ina É iš-pik-ki* GAR É BI *i-ḫar-*[*ru-ub*]
16. DIŠ KA.TAR BABBAR *ina qi-rib* É [NA GAR] É BI *ip-pe-eḫ-*[*ḫi*?].
17. DIŠ KA.TAR BABBAR *ina šu-bat* NA ZAG GAR BIR-*ah* É BI.
18. DIŠ KA.TAR BABBAR *ina šu-bat* NA GÙB GAR *sa-dir* É BI.

5. If white fungus [...] in the entryway of the gate of the court[yard?...] grief in that house [...].
15. If there is <white> fungus in a warehouse, that house will be devas[tated].
16. [If there is] white fungus in the middle of [a man's] house that house will be closed up.
17. If there is white fungus on the right in a man's residential quarter—dispersal of that house.
18. If there is white fungus on the left in a man's residential quarter that house will be normal (Freedman, 1998: 192–193).

The logical meaning of these verses is as follows:

5. If –1 ("white fungus") & +1 ("in the entryway of the gate of the courtyard"), then –1 ("grief in that house"). It is equal to +1.
15. If –1 ("white fungus") & +1 ('warehouse'), then –1 ("that house will be devastated"). It is equal to +1.
16. If –1 ("white fungus") & +1 ("in the middle of [a man's] house"), then –1 ("that house will be closed up"). It is equal to +1.
17. If –1 ("white fungus") & +1 ("on the right in a man's residential quarter"), then –1 ("dispersal of that house"). It is equal to +1.
18. If –1 ("white fungus") & –1 ("on the left in a man's residential quarter"), then +1 ("that house will be normal"). It is equal to +1.

On the other hand, "black fungus" (*katarru ṣalmu*) is a positive sign that along with all other signs gives only positive prognoses:

49. DIŠ KA.TAR GE$_6$ *ina tal-lak-ti* É NA GAR *ka-liš* AL.SA$_6$.
50. DIŠ KA.TAR GE$_6$ *ina* GÚ.ḪAŠ *šu-bat* É LÚ *it-tab-ši* SUḪUS É BI GI.NA.
51. DIŠ KA.TAR GE$_6$ *ina šu-bat* LÚ ZAG *it-tab-ši lu* ÚKU NÌ.TUK.
52. DIŠ KA.TAR GE$_6$ *ina šu-bat* LÚ GÙB *it-tab-ši* LÚ *ú-wa-at-tar*.
54. DIŠ KA.TAR GE$_6$. EŠ ŠÀ É NA *ma-lu-ú* EN É NÌ.TUK SUḪUŠ.BI GI.NA.

49. If there is black fungus on the path to a man's house it will be entirely favourable.
50. If black fungus appears on the back of the residential quarter of a man's house, the foundation of that house will be secure.

51. If black fungus appears on the right in a man's residential quarter if he is poor, he will become rich.
52. If black fungus appears on the left in a man's residential quarter the man will become prominent.
54. If black fungi fill the interior of a man's house the owner of that house will become rich; its foundation will become secure (Freedman, 1998: 194–197).

Formally:

49. If +1 ("black fungus") & +1 ("on the path to a man's house"), then +1 ("it will be entirely favourable"). It is equal to +1.
50. If +1 ("black fungus") & –1 ("on the back of the residential quarter of a man's house", then +1 ("the foundation of that house will be secure"). Hence, we have "if (+1 & –1), then +1", but it is equal to +1 in the Boolean algebra.
51. If +1 ("black fungus") & +1 ("on the right in a man's residential quarter if he is poor"), then +1 ("he will become rich"). It is equal to +1.
52. If +1 ("black fungus") & –1 ("on the left in a man's residential quarter"), then +1 ("the man will become prominent"). It is equal to +1.
54. If +1 ("black fungus") & +1 ("the interior of a man's house"), then +1 ("the owner of that house will become rich; its foundation will become secure"). It is equal to +1.

The sign "red fungus" (*katarru sāmu*) is negative and gives only an unfavourable forecasting together with all other signs:

60. DIŠ KA.TAR SI.A *ša mi-iq-tu₄* MU.NI *ina* É LÚ ZAG GAR ŠUB-*di* É
61. DIŠ KA.TAR SI.A *ša mi-iq-tu₄* MU.NI *ina* É LÚ GÙB GAR ŠUB-*di* É

60. If red fungus named *miqtu* appears on the right in a man's house—abandonment of the house.
61. If red fungus named *miqtu* appears on the left in a man's house—abandonment of the house (Freedman, 1998: 196–197).

Formally:

60. If –1 ("red fungus named *miqtu*") & +1 ("on the right in a man's house"), then –1 ("abandonment of the house"). It is equal to +1.
61. If –1 ("red fungus named *miqtu*") & –1 ("on the left in a man's house"), then –1 ("abandonment of the house"). It is equal to +1.

Thus, different omen series are composed in the way to be axioms of Boolean algebra (that is, to have only the value +1) and, therefore, we can reconstruct the same Boolean algebra ({–1, +1}, ¬, &, ⇒) in different omen series, where each singular prediction is composed as a tautology (always true expression, i.e., with the value +1) of this algebra. So, sometimes rules from (a) to (d) are applied in composing omens and sometimes not, but in any case each omen is formulated as a logical axiom. It shows that the Mesopotamians not only possessed the perfect logical competence, but also used some logical-algebraic constructions.

In Biblical verses we find out the same understanding of right and left hands as positive and negative omens respectively. The normal situation is that the Lord (“a wise man’s heart”) is at the right hand and Satan (“a fool’s heart”) at the left:

> לֵב חָכָם לִימִינוֹ וְלֵב כְּסִיל לִשְׂמֹאלוֹ׃
>
> *lēḇ ḥāḵām lîmînô wəlēḇ kəsîl liśmō ʾlô.*
>
> A wise man’s heart is **at his right hand**; but a fool’s heart **at his left** (*Ecclesiastes* 10: 2).
>
> יְהוָה שֹׁמְרֶךָ יְהוָה צִלְּךָ עַל־יַד יְמִינֶךָ׃
>
> *ʾăḏōnāi šōmərəḵā ʾăḏōnāi ṣilləḵā ʿal-yaḏ yəmîneḵā.*
>
> The Lord is your Keeper. **The Lord** is your shade **on your right hand** (*Psalm* 121: 5).

“Being before” and “being on the back” are two additional Biblical omens with the same treatment as in the Akkadian divinations—as positive and negative omens respectively. A natural place for good forces (angels and the Lord) is to be before us and a natural place for evil forces (Satan) is to be on the back of us:

> שִׁוִּיתִי יְהוָה לְנֶגְדִּי תָמִיד כִּי מִימִינִי בַּל־אֶמּוֹט׃
>
> *šiwwîṯî ʾăḏōnāi lənegdî ṯāmîḏ kî mîmînî bal-ʾemmôṭ.*
>
> I have set the Lord always before me: because he is at my right hand, I shall not be moved (*Psalm* 16: 8).

If Satan stands at the right hand of a man, it means that it is unnatural (e.g., it is a combination of positive and negative singular omens at once) and this presages a strongly unfavourable event for this man:

> הַפְקֵד עָלָיו רָשָׁע וְשָׂטָן יַעֲמֹד עַל־יְמִינוֹ׃
>
> *hap̱qēḏ ʿālāyw rāšāʿ wəśāṭān yaʿămōḏ ʿal-yəmînô.*
>
> Set thou a wicked man over him: and let Satan stand at his right hand (*Psalm* 109: 6).

Let us consider two predictions in one passage:

> שְׂמֹאלוֹ תַּחַת לְרֹאשִׁי וִימִינוֹ תְּחַבְּקֵנִי׃
>
> *śəmōʾlōw taḥaṯ lərōʾšî wîmînōw təḥabbəqēnî.*
>
> His **left hand** [is] **under** my head, and his **right hand** doth **embrace** me (*Song of Songs* 2: 6).

In the first prediction (*śəmōʾlōw taḥaṯ lərōʾšî*) we have a combination of two negative omens which should give a favourable forecasting:

$$(([[\text{“his left hand”}]] \ \& \ [[\text{“under my head”}]]) \Rightarrow [[p]]) = ((-1 \ \& \ -1) \Rightarrow +1) = +1.$$

In the second prediction (*ymînōw təḥabbəqēnî*) we have a combination of two positive omens which should give a favourable conclusion, also:

$$(([[\text{“his right hand”}]] \ \& \ [[\text{“embracing me”}]]) \Rightarrow [[p]]) = ((+1 \ \& \ +1) \Rightarrow +1) = +1.$$

In the Rabbinical Judaism as well as in the much older Akkadian culture, we must be vigilant to omens which are scattered at all steps before us. Let us consider an example of this reflection on omens, taken from the Talmud (*Šabbat* 32a). A rabbi must have crossed a river in a ferry in which a gentile sat. Rav (רב; ca. from 175 A.D. to 247 A.D.) said that he would not cross a river under this condition, because a judgment can be reckoned with the gentile, and Rav will be caught together with him when the gentile is punished indeed (רב לא עבר במברא דיתיב ביה עכו"ם אמר דילמא מיפקיד ליה דינא עליה ומתפיסנא בהדיה). Whereas Šmu'el (שמואל; ca. from 165 A.D. to 254 A.D.) said that he would cross a river in a ferry with the gentile, because "Satan does not have dominion over two nations at once: he settles his accounts with people from each nationality separately" (שמואל לא עבר אלא במברא דאית ביה עכו"ם אמר שטנא בתרי אומי לא שליט) (see *Ibid.*). The reasoning of Šmu'el is especially interesting. He assumes that for each event (such as a punishment) there exists a separate omen and different nations cannot be judged by the same omens. It is a kind of logical determinism: nothing happens without its own omen.

Thus, the Stoic philosophers are increasingly voicing their concern about the Semitic way of divinations formulated by means of conditional statements from which an appropriate forecasting is inferred by *modus ponens*. This thinking was rooted from the Akkadian art of divination through conditionals. Many rudiments of this thinking are contained in the Bible and the Talmud, as well. Therefore, the logical determinism, expressed well explicitly by the Stoics at first, has its origin in the Akkadian divinations in fact, where this determinism was represented as an intuition that each event has its own omen and, therefore, it can be explained completely by conditionals.

Indeed, according to the Babylonians, for each event q there is a suitable omen p such that $p \Rightarrow q$ is an always true proposition. In the Stoic words, "nothing happens without a cause" (*nihil enim fieri sine causa potest*), see the *On Divination* (*De Divinatione*) II: 61. Therefore, we can apply *modus ponens* to infer q from $p \Rightarrow q$. So, each actual event q is inferable from some axioms. Then we can apply the necessitation rule and infer $\Box q$. As we see, logical determinism and eternalism were first established in Babylonia in fact. Each actual event q is necessary.

As we know, conditional propositions for the Stoics express cause-and-effect or sign-and-prediction relations, for example "if it is day, then it is light". In the *Introduction to Syllogisms*, Chrysippus proposed "five non-demonstrable arguments (modes)" (inference rules) for inferring from causes/effects or signs/predictions:

- *Modus ponendo ponens* (*or modus ponens*): "If it is day, then it is light; it is day now; hence it is light" (Sextus Empiricus, *Adversus Logicos* II: 224, 431; *Pyrrhoniae Hypotyposes* II: 157); if p, then q; p; then q.
- *Modus tollendo tollens* (*or modus tollens*): "If it is day, then it is light; it is not light; hence it is not day" (Sextus Empiricus, *Adversus Logicos* II: 225, 433; *Pyrrhoniae Hypotyposes* II: 157); if p, then q; not-q; then not-p.
- *Modus ponendo tollens*: "It cannot take place that it is both day and night; it is day now; hence it is not night" (Sextus Empiricus, *Adversus Logicos* II: 225; *Pyrrhoniae Hypotyposes* II: 158); $\neg(p \,\&\, q)$; p; then $\neg q$.

- *Modus tollendo ponens* I: "It is either light or dark; but it is light; hence it is not dark" (Sextus Empiricus, *Adversus Logicos* II: 225; *Pyrrhoniae Hypotyposes* II: 158); $p \oplus q$; p; then $\neg q$.
- *Modus tollendo ponens* II: "It is either light or dark; but it is not dark; hence it is light" (Sextus Empiricus, *Pyrrhoniae Hypotyposes* II: 158); $p \oplus q$; $\neg q$; then p.

According to the Stoics, the implication possesses the property of transitivity: "When we have in a deduction combined premises of any conclusion, we potentially have in them as well this conclusion though it is not expressed explicitly" (Sextus Empiricus, *Adversus Logicos* II: 231).

By using the five non-demonstrable arguments, it is possible to check whether our reasoning is logically correct. For instance, let us check the following conclusion:

> If phenomena similarly are all being in an identical state, signs are phenomena, then signs similarly are all being in an identical state. Nevertheless, signs anyway are not similarly all being in an identical state. And phenomena similarly are all being in an identical state, therefore, signs are not phenomena (Sextus Empiricus, *Adversus Logicos* II: 234).

In the form of the Stoic way of formalization by using numerals: "If it is the first and the second, then it is the third [*A.Sch.*: the transitivity of implication]; it is not the third, but it is the first; therefore, it is not the second" (Sextus Empiricus, *Adversus Logicos* II: 235). Formally: $(p \,\&\, q) \Rightarrow r$; $\neg r \,\&\, p$; then $\neg q$. From here it is obvious that the given proof consists of the 2nd and 3rd modes, namely, first, $(p \,\&\, q) \Rightarrow r$; $\neg r$; then $\neg(p \,\&\, q)$; second, $\neg(p \,\&\, q)$; p; then $\neg q$.

Each antecedent in proof trees is called a 'sign' by the Stoics: "A sign is a premise, being a basis in a correct conclusion from true to true, and it opens the conclusion and permanently it appears as a present sign for available" (Sextus Empiricus, *Adversus Logicos* II: 256). Therefore signs (omens) logically determine conclusions irrespective of the fact of whether implication takes place now, in the past or in the future.

So, in the Stoic logic there is no place for contingency, i.e., the statement "both it is possible to be, and it is possible not to be" is always false. In the meanwhile, it means that debates have no meaning. Indeed, from two points of view only one is true, and any compromises therefore are eliminated—they are not possible logically:

> Someone who said: "Never take decisions without having listened to both sides," was objected to by Zeno by means of the following reasoning. Either, the first put forward exhaustive arguments (then there is no need to listen to the second because the required is found out). Or, he did not put them forward (he was not on call of court or, answering, was not intelligible, it is all the same). Hence, either he proved his correctness or did not prove. So, in any case there is no need to listen to the second (Plutarch, *De Stoicorum repugnantiis* VIII: 1034e).

To sum up, the Stoic doctrine about destiny is a specific theory of modal logic in which modal operators 'possible' and 'necessary' satisfy the logic **K=**, see Tables 4.1.1

and 4.1.2. But this approach was nothing more than a reflective conceptualization of Babylonian logical teaching about divination.

4.1.6 Conclusions

In this section, I have logically compared the three literary genres: the historical prose of Ancient Greece (causing the emergence of philosophical prose here), the *ʾaggāḏôṯ* from the Talmud, and the omens in Akkadian. It was pointed out that the Greek historical prose was grounded on the idea of existence of contingent events. This idea is fully rejected in Judaism and in the Akkadian divinations. As a result, the phenomenon of historical prose in the meaning of Herodotus was excluded in these Semitic cultures—it was out of their ways of thinking. But instead, in the *ʾaggāḏôṯ* and Akkadian divinations we observe an intuition of logical determinism: everything can be explained by omens; therefore, everything appears necessarily, or it disappears necessarily. This intuition was well expressed later in the Stoic philosophy: everything can be explained by causes or signs (omens). Consequently, many correct forms of applying sophisticated logical inference rules in the texts in Akkadian are not accidental—this derives from the intuition of logical determinism intrinsic to some Semitic cultures rooted from the Akkadian one.

It is worth noting that the Stoic logical approach to conditionals causing their logical determinism is supported in the Indian logic, too—in the schools of Yogācāra and Nyāya (Schumann, 2018b, 2021c), see Subsection 4.3.6.

Hence, the historical prose in Ancient Greece arose within a specific thinking assuming the existence of contingent events—this phenomenon has taken place since the 6th century B.C. This trend was excluded in Semitic cultures of that time. The ancient Greek thinking, admitting of contingency, was emancipated thereby from religious dogmatism. In turn, this influenced the emergence of philosophical prose as a new genre of meta-narration where instead of religious canons and commentaries to them, their own worldview, based on some reasoning, was developed. This worldview could vary and consisted of three disciplines: dialectics (our reasoning), physics (natural forces), and ethics (our customs and characters), treated differently by different philosophers. As a result, various approaches to dialectics, physics, and ethics could be collected in some surveys such as the *Lives and Opinions of Eminent Philosophers* (Βίοι καὶ γνῶμαι τῶν ἐν φιλοσοφίᾳ εὐδοκιμησάντων/*Vitae Philosophorum*), composed by Diogenes Laërtius.

Nevertheless, we cannot say that logic can be taught only within the philosophical prose (i.e., within philosophical treatises which appeared in Ancient Greece). There were three genres supporting the logical discourse in the ancient world: (1) ancient Greek philosophical treatises as a kind of prose; (2) thematic dialogues of the Jerusalem and Babylonian Talmuds; (3) either *sūtra*s (doctrinal texts to memorize) or thematic dialogues in Indian schools such as Madhyamaka, Yogācāra, and Nyāya. In the very next section, I shall consider the earliest logical treatise in India—the *Milindapañha* which was composed as a thematic dialogue devoted to understanding the Buddhist doctrine on *abhidharma*—a classification of all mental states (*dharma*s).

4.2 Role of the Logical Treatise *Milindapañha* in the Genesis of Indian Logic

4.2.1 Problem Setting

Logical treatises in Ancient India were compositions of two types: thematic dialogues or philosophical *sūtras*—but, at the same time, comments on the *sūtras* were often composed as thematic dialogues. The earliest logical treatise in Ancient India, whose author possesses the perfect logical competence and a developed reflection on this competence, is presented by *Milindapañha*. This section provides an argument supporting the claim that the authors of the Pāli Canon and *Milindapañha* did not know about the Nyāya school of logic and knew nothing about syllogisms defined in the *Nyāya-sūtra* (Subsection 4.2.3). The methodology for this conclusion is discussed in Subsection 4.2.2.

The argument is focused on the following claims (Schumann, 2019c, 2020b): (i) the terms denoting logic (such as *nyāya*) and occurring in the Pāli Canon were not used, in accordance with their contextual meanings, to denote the school of logic or the *Nyāya-sūtra* (Subsection 4.2.3); (ii) in some early Pāli *sutta*s some terms denoting logical reasoning have sometimes negative connotations (Subsection 4.2.3), nevertheless in the Theravāda tradition there are many explanations why we need logic still; (iii) in the Pāli Canon, the terms like *nyāya* have the meaning of the method of Buddhists distinguishing them from non-Buddhists (Subsection 4.2.3); (iv) it seems that in the Pāli texts there are a few logical syllogisms used for the logical purpose, but the majority of syllogisms are used without understanding of their logical nature, but they are rather used for the purpose of rhetoric, e.g., syllogisms often play a rhetorical role in the *Kathāvatthu* (Subsection 4.2.4); (v) so, the author of the *Kathāvatthu* did not have good competence in symbolic (systematic) logic, because there are a lot of sophisms and other fallacies among correct syllogisms (Subsection 4.2.4); (vi) in the *Kathāvatthu* and in other *Abhidhamma* texts there is no requirement to give examples in syllogisms—but it is one of the principal requirements of the Nyāya tradition and the Yogācāra logic (Subsection 4.2.4); (vii) the author(s) of the *Milindapañha* (its historical context is considered in Subsection 4.2.5) did have a good competence in symbolic logic, e.g., there are no sophisms and we face a requirement here to give examples for verifying statements—this requirement is explained in the text in a more primitive way terminologically than it was done in the *Nyāya-sūtra*, hence this text preceded the Nyāya philosophy and could have had an impact on the latter (Subsection 4.2.6).

Accounting for the genesis of logical knowledge in the early Pāli literature from (i) to (vii), we can assume that the Gandhāran (Greco-Buddhist) influence on the origin of Indian systematic logic is highly possible and at least not excluded *if* we concentrate just on the Pāli texts. To infer this statement, *Mill's joint method of agreement and difference* is applied:[3]

On the one hand, the *Milindapañha* is the only early Pāli source in which we deal with a proto-Nyāya logic—the main feature of this text is to be logical among all other canonical texts, and, on the other hand, the *Milindapañha* was written in Gandhāra, the region where the Greek language was official for 300–400 years (e.g., it was used for edicts, business documentation, courts, and taxation as *lingua franca*), and the Hellenistic influences on social life here were evident. Thus, the early Buddhist knowledge of systematic logic is not supposed to be connected to the *Nyāya-sūtra*, but could have been established under a Hellenistic (Gandhāran) influence.

Formally:

> The canonical texts, such as the *Yamaka*, the *Kalahavivāda-sutta* (*Sutta-nipāta* IV: 11), the *Vijñānakāya* (it is not a Pāli text, but a Sanskrit one—one of the seven Sarvāstivāda Abhidharma texts), the *Kathāvatthu*, and the *Milindapañha*, deal with syllogisms;
>
> The *Yamaka* deals with a converse of implications (reversing its two parts);
>
> The *Kalahavivāda-sutta* deals with a transitivity of implications (combining several implications having a joint part);
>
> The *Vijñānakāya* deals with *modus ponens* and *modus tollens*;
>
> The *Kathāvatthu* deals with *modus ponens* and *modus tollens*;
>
> The *Milindapañha* deals with *modus ponens* and *modus tollens*;
>
> The *Yamaka*, the *Kalahavivāda-sutta*, the *Vijñānakāya*, and the *Kathāvatthu* do not provide syllogisms with a Nyāya or proto-Nyāya way of verifying premises;
>
> The *Milindapañha* provides syllogisms with a proto-Nyāya way of verifying premises;
>
> The *Yamaka*, the *Kalahavivāda-sutta*, the *Vijñānakāya*, and the *Kathāvatthu* use sophisms and combine correct and incorrect syllogisms;

3 The following five methods of induction were introduced by philosopher John Stuart Mill in the book *A System of Logic* (1843): direct method of agreement, method of difference, joint method of agreement and difference, method of residue, and method of concomitant variations. These methods give us the best way of modelling the historical reality based on the scattered archaeological and textological data. In this section, my conclusions follow these methods. One of the possible schemata for the joint method of agreement and difference is as follows:

A, B, C occur together with *x, y, z*;
A, D, E occur together with *x, v, w*;
B, C occur with *y, z*;

Therefore, *A* is the cause, or the effect, or a part of the cause of *x*.

The *Milindapañha* does not use sophisms and applies only correct syllogisms;

The *Milindapañha* is, perhaps, the only early Pāli text written in Gandhāra by, one expects, a direct Hellenistic or Gandhāran (Greco-Buddhist) influence;

Therefore, probably, the correct application of inference rules in the early Buddhist logic is explained by a Hellenistic or Gandhāran (Greco-Buddhist) influence.[4]

Notice that Mill's joint method of agreement and difference that I have applied in this section is not deductive, but plausible with a high probability.

Hence, I have performed an experiment to check the logical culture of ways of using syllogisms in the Pāli Canon and, consequently I am probabilistically concluding that the authors of the early Pāli texts did not know the *Nyāya-sūtra.* It means that based on the corpus of these texts we can claim that at the time of the authors of the early Pāli texts (until the 1st century A.D. or even later) the Nyāya school of logic did not exist. (According to some other data, the *Nyāya-sūtra* is dated not earlier than to the 2nd century A.D., e.g., there are quotings in this *sūtra* from some early Madhyamaka and Yogācāra texts dated to this century or later and written probably in Gandhāra, too—their early fragments in Gāndhārī are excavated in this area).

The following presuppositions are the most principal for any system of symbolic logic:

i) Each proposition should be factual and then, it is either true or false. It is true if it correctly describes an appropriate fact, otherwise it is false. For example, the proposition "it's raining" is true if it's raining now indeed in a specified place.
ii) Into our reasoning we can involve only true propositions.
iii) There are logical schemata that are called inference rules and they infer only true propositions from true premises. Hence, our conclusions are ever true if we apply inference rules in relation to true premises.

---

[4] Formally, a very short version:

Y (*Yamaka*), *Kl* (*Kalahavivāda-sutta*), *V* (*Vijñānakāya*), *Kt* (*Kathāvatthu*), *M* (*Milindapañha*) occur together with *x* (correct applications of *inference rules*) and *y* (incorrect applications of *inference rules*); *M* as the only Pāli text from Gandhāra occur together with *x* (correct applications of *inference rules*) and *z* (proto-*nyāya* doctrine on verification premises) and without *y* (incorrect applications of *inference rules*);
Y (*Yamaka*), *Kl* (*Kalahavivāda-sutta*), *V* (*Vijñānakāya*), *Kt* (*Kathāvatthu*) occurs with *y* (incorrect applications of *inference rules*) and without *z* (proto-*nyāya* doctrine on verification premises);

Therefore, *M* as the only Pāli text from Gandhāra is the cause, or the effect, or a part of the cause of *x* and *z*. In other words, to be Gandhāran is the cause, or the effect, or a part of the cause of *x* and *z*.

As we remember, in symbolic logic the following two inference rules are fundamental:

1) *Modus ponens*. Let A and B be two factual propositions. Assume that "A implies B" and A are both asserted to be true. Then we can draw the conclusion that B must be true, too. Symbolically:

   $A \Rightarrow B$ is true; A is true.

   Then B is true, also.

   $$\frac{(A \Rightarrow B);\ A}{B.}$$

2) *Modus tollens*. Let A and B be two factual propositions again. Suppose "A implies B" is considered true, but it is not the case that B. Then we can draw the conclusion that it is not the case that A, too. Symbolically:

   $A \Rightarrow B$ is true; B is false.

   Then A is false, also.

   $$\frac{(A \Rightarrow B);\ \neg B}{\neg A.}$$

In the old Greek philosophy, the above presuppositions (i)–(iii) were widely accepted due to the logical works written by Aristotle (384–322 B.C.) and Chrysippus (ca. 279–ca. 206 B.C.). In the meanwhile, Aristotle proposed a modification of *modus ponens* and *modus tollens* for categorical propositions and Chrysippus formulated *modus ponens* and *modus tollens* conventionally in the way as said above.

Hence, if we can observe that someone understands presuppositions (i)–(iii) and can follow them in his/her reasoning, then he or she possesses a good logical competence. This competence is detected in the Nyāya as well as in the Madhyamaka and Yogācāra texts. Certainly, the *Nyāya-sūtra* was one of the first Indian documents, whose author(s) possessed a relatively good logical competence; particularly, the author(s) of the *Nyāya-sūtra* accepted presuppositions (i)–(iii) of symbolic logic. He explicitly pointed out that we cannot cast doubt, first, on factual propositions which are successfully verified (*pratyakṣa*) and second, on conclusions drawn correctly by means of inference rules from true premises (*anumāna*):

> *dṛṣṭānumitānāṃ hi niyogapratiṣedhānupapattiḥ* (*Nyāya-sūtra* III: 1, 51).
>
> It is incongruous to attribute or deny what [has already become] the subject of perception or inference.

In the *Nyāya-sūtra*, we face many examples of correct applications of *modus ponens* and *modus tollens*. For instance, *modus tollens* is applied here, although this reasoning is not accepted by the author(s) of the *Nyāya-sūtra*:

> *īśvaraḥ kāraṇam, puruṣakarmāphalyadarśanāt* ||19||
> *na, puruṣakarmābhāve phlāniṣpatteḥ* ||20||

tatkāritatvād ahetuḥ ||21||
(*Nyāya-sūtra* IV: 1, 19–21).

Īśvara is a reason for observing that human actions are fruitless.
Wrong, because without human actions there is no 'fruiting'.
Not an argument—due to the conditionality of the latter.

Formally: If the Lord (*īśvara*) is a cause (*A*), then human actions are without fruits (*B*). Nevertheless, they are ever with fruits ($\neg B$). Then the Lord is not a cause ($\neg A$):[5]

$$\frac{(A \Rightarrow B);\ \neg B}{\neg A.}$$

The *Nyāya-sūtra* is dated very differently by different scholars: from the 6th century B.C. to the 2nd century A.D. (Potter, 1977: 4). Let us notice that the dating from the 6th to the 5th century B.C. is quite improbable from the *archaeological* point of view. The matter is that at that time the Painted Grey Ware culture existed which is characterized by a very low-scale urbanization at the Ganga-Yamuna valley. For instance, the settlements of this culture could be quite large, but they had no town planning and consisted of buildings made from bamboo and loam which can be compared to today's slums of Mumbai—an unstructured housing and lack of infrastructure. There were no states and no cities in the strict sense. Therefore, it is obvious why there was neither a writing system nor money. The large-scale urbanization began only since ca. 400 B.C. The *śramaṇa* movement was a spiritual way to resist this urbanization—since that time it was just the very beginning of philosophical reflection of India in the pure meaning. Hence, we cannot date the *Nyāya-sūtra* from the 6th to the 5th century B.C. certainly. It was a late-Vedic period without any *sūtra*s.

In any case, among different date patterns, the 2nd century A.D. is the most reasonable dating of the *Nyāya-sūtra* from the standpoint of archaeology. At that time in the Kuṣāṇa Empire there was observed a flowering of sciences and arts in India, the first big fruit yielded by the large-scale urbanization started from ca. 400 B.C. Textologically, there are also many arguments supporting this dating. So, these arguments are collected by Vidyabhusana (1921) and they are as follows: (i) the *Nyāya-sūtra* contains many quotes from some early Madhyamaka and Yogācāra texts which are dated to the 2nd century A.D. or even later; (ii) the *Nyāya-sūtra* can be examined as a systematization of logical pieces from the *Caraka-saṃhitā* dated to the same century.

i) For instance, there are proofs that the author(s) of the *Nyāya-sūtra* accept *kṣaṇikavāda* (teaching on momentariness), a key doctrine of Yogācāra (as well as of the earlier Sautrāntika thought), e.g.:

sphaṭike api aparāparotpatteḥ ***kṣaṇikatvāt*** *vyaktīnām ahetuḥ* (*Nyāya-sūtra* III: 2, 10).

[5] It is worth noting that this phrase is so close to the Madhyamaka manner of refutation of God's existence, and it can be taken from the Madhyamaka texts such as *Īśvara-kartṛtva-nirākṛtiḥ-viṣṇoḥ-ekakartṛtva-nirākaraṇa* ascribed to Nāgārjuna, see Subsection 4.3.7.

> There is no reason [to deny the difference] in the crystal, because, due to momentariness of [all], the particles [of the crystal] are updated one after the other.

There are many quotings from the *Madhyamakaśāstra* of Nāgārjuna. The verse:

> *na svabhāvasiddhiḥ, āpekṣikatvāt* (*Nyāya-sūtra* IV: 1, 39).

is used to express the Madhyamaka doctrine and it is close to:

> *na hi svabhāvo bhāvānāṃ pratyayādiṣu vidyate |*
> *avidyamāne svabhāve parabhāvo na vidyate* ||5|| (*Madhyamakaśāstra* I: 5).

The next passage to show the Madhyamaka doctrine is as follows:

> *nāsanna sanna sadasat, sadasatorvaidharmayāt* (*Nyāya-sūtra* IV: 1, 48).

It is like the following verse:

> *na sannāsanna sadasan dharmo nirvartate yadā |*
> *kathaṃ nirvartako heturevaṃ sati hi yujyate* ||9|| (*Madhyamakaśāstra* I: 9).

The doctrine of Madhyamaka and Yogācāra is exemplified as follows:

> *māyāgandharvanagaramṛgatṛṣṇikāvadvā* (*Nyāya-sūtra* IV: 2, 32).

This example is taken from the following verse:

> *yathā māyā yathā svapno gandharvanagaraṃ yathā |*
> *tathotpādastathā sthānaṃ tathā bhaṅga udāhṛtam* ||34|| (*Madhyamakaśāstra* VII: 34).

Another Madhyamaka doctrine objected to by the author(s) of the *Nyāya-sūtra*:

> *vartamānābhāvaḥ, patataḥ patitapatitavyakālopapatteḥ* (*Nyāya-sūtra* II: 1, 40).

Please compare this to the passage:

> *gataṃ na gamyate tāvadagataṃ naiva gamyate |*
> *gatāgatavinirmuktaṃ gamyamānaṃ na gamyate* ||1|| (*Madhyamakaśāstra* II: 1).

Thus, the *Nyāya-sūtra* cannot be written earlier than the first texts of Madhyamaka and Yogācāra dated not earlier than the 2nd century A.D.

ii) The *Caraka-saṃhitā*, a medical treatise, includes many passages about *pramāṇa*, the sources of knowledge, like the Nyāya doctrine, e.g., some sources are as follows:

> *Pratyakṣa—Pratyakṣa* (perception) is the knowledge which is directly received by the self and the sense organs. Self-perceived are pleasure, pain, desire, aversion, etc., while sound, etc. are percieved by the sense organs.
> *Anumāna—Anumāna* (inference) is the reasoning supported by invariable concomittance such as the knowledge of *agni* by the power of digestion, that of strength by the power of exercise, auditory organ, etc. by the perception of sound, etc.

Aitihya—*Aitihya* (tradition) is the traditional authoritative source of knowledge such as *veda,* etc.
Aupamya—*Aupamya* (analogy) is the statement of similarity between things such is—analogy of *daṇḍaka* with *daṇḍa* (staff), that of *dhanuḥsthmbha* with bow and that of the provider of health with the archer (Sharma, 2014: 363).

The authorship or at least the deep editorship of this treatise is ascribed to Caraka, the great physician who was the medical attendant of Kaniṣka (2nd century A.D.), the Emperor of the Kuṣāṇas. So, in the Chinese text of the *Saṃyukta-ratna-piṭaka-sūtra* it is stated that Kaniṣka had the following three friends: his prime minister Mathara, the physician Caraka, and the poet Aśvaghoṣa (Levi, 1896). As we thus see, Caraka is considered a resident of Gandhāra, the Greco-Buddhist region.

Let us notice that there are many proofs that the Madhyamaka was founded in the same Gandhāra, too. First, there is archaeological evidence. The fragments of the *Aṣṭasāhasrikā Prajñāpāramitā,* the proto-Madhyamaka book, written in the Kharoṣṭhī script in the Gāndhārī language and dated to ca. 75 A.D. by the radiocarbon analysis were reconstructed by Harry Falk and Seishi Karashima (Falk and Karashima, 2012). They found out that these fragments are supposed to be a source text for the first Chinese translation of the *Aṣṭasāhasrikā Prajñāpāramitā* by Lokakṣema (ca. 179 A.D.). Also, they proved linguistically that the standard Sanskrit text can be a translation from Gāndhāri, because in Sanskrit there are many expanded phrases that are not present in the Gāndhārī source at all: "the language of the original text was Gāndhāri, just as was assumed on the basis of some expressions in Lokakṣema's translation which presuppose sound changes only found in Gāndhāri, and not in other Indian vernaculars of the time" (Falk and Karashima, 2012).

Some other fragments of Mahāyāna texts excavated in Gandhāra are: (i) the Gāndhārī manuscripts SC1 (PP-G, similar to the ASP 40), BC2 (*Bajaur-mahāyāna-sūtra* with parallels to the *Akṣobhya-vyūha*) in Kharoṣṭhī dated to the first or second century A.D.; (ii) the *Sucitti-sūtra* (NC2, similar to the *Vimalakīrtinirdeśa-sūtra*) with parallels to three Chinese translations (T477–479) and the *Pratyutpannabu ddhasaṃmukhāvasthitasamādhi-sūtra,* both written in Kharoṣṭhī and dated to the 1st or 2nd century A.D.; (iii) several small palm leaf fragments from Bamiyan with text passages familiar from the *Bodhisattvapiṭaka-sūtra,* the *Sarvapuṇyasamuccaya samādhi-sūtra,* and the *Bhadrakalpika-sūtra,* written in Kharoṣṭhī and dated to the 3rd–4th century A.D., etc. (Schlosser, 2016).

We know that the *prajñāpāramitā* teaching was very popular in the north-west of India in the Kuṣāṇa period. In the *Mañjuśrīmūlakalpa* (LIII v. 575) it is affirmed that under Kaniṣka the *prajñāpāramitā* was 'established' (*pratiṣṭhitā*) in the north-west. This doctrine with the *śūnyatā* concept served as a background for the fragments BC4 and BC11. In BC4, we read "benefit of freedom from all passions" (*vairāga-ănuśaṃsa*) and in BC11 "benefit of release" (*avasarga-ănuśaṃsa*) in practising the bodhisattva path started with understanding (*parijñā*) the origins of suffering and finished at abandoning (*prahāṇa*) these origins and realising the emptiness of all *dharma*s within a direct realization of the unconditioned (*lokottara-bhūta-jñāna*) (Schlosser, 2016).

There are some textological proofs that Nāgārjuna was a resident of the Kuṣāṇa Empire. It is stated in the *Rājataraṃgiṇī* dating from the 12th century that Nāgārjuna was a lord of the earth (*bhūmīśvaro*) in Kāśmīr under the rule of the following three Kuṣāṇa Emperors: Huṣka, Juṣka, Kaniṣka:

> *athābhavan svanāmankapuratrayavidhāyinaḥ* |
> ***huṣkajuṣkakaniṣk****ākhyās trayas tatraiva pārthivaḥ* ||168||
> *sa vihārasya nirmātā juṣko juṣkapurasya yaḥ* |
> *jayasvāmipurasyāpi śuddhadhīḥ saṃvidhāyakaḥ* ||169||
> *te turuṣkānvayodbhūtā api punyāśrayā nṛpāḥ* |
> *shuṣkaletrādideshеṣu maṭhacaityādi cakrire* ||170||
> *prājye rājyakṣane teṣāṃ prāyaḥ* ***kāśmīra****maṇḍalam* |
> *bhojyam aste sma bauddhānāṃ pravrajyorjitatejasām* ||171||
> *tadā bhagavataḥ śākyasiṃhasya paranirvṛteḥ* |
> *asmin mahilokadhātau sārdham varṣaśataṃ hy agāt* ||172||
> *bodhisattvaś ca deshe 'sminn eko bhūmīśvaro 'bhavat* |
> *sa ca* ***nāgārjunaḥ*** *srhīmān sadārhadvanasaṃsrhayi* ||173||
> (*Rājataraṃgiṇī* 168–173).

Hiouen Thsang, who visited India in 645 A.D., mentions Aśvaghoṣa, Deva, Nāgārjuna, and Kumāralabdha, as the contemporaries of Kaniṣka and "as the four suns which illumine the world". Aśvaghoṣa is named also the spiritual advisor of Kaniṣka (Beal, 1884/1981: 302–303).

Hence, to prove that the *Nyāya-sūtra* was created in Gandhāra in the 2nd century A.D. (or later) indeed, we should trace the proto-Nyāya teaching (first the *pramāṇa* doctrine) in the Buddhist sources before the Mahāyāna and show that this teaching can have been established in Gandhāra. This would give a new argument supporting the dating of the *Nyāya-sūtra* to the 2nd century A.D.

So, the logical treatise *Milindapañha* sets the earliest structure of logemes of Indian logic, which was then expanded and refined in the structures of the logemes of Yogācāra and Nyāya.

4.2.2 Discussion on Methodology of Studying the Indian Logic

According to some recent results in experimental psychology and cognitive science, systematic (or symbolic) logic is not an 'innate' knowledge of human beings. To know logic, we should especially study it and, first, study the ways how it can be applied in different situations: public discourse, science, or private strategy. Consequently, we can use strategic, creative, or even critical thinking without any logical competence even if we did not study systematic logic before, and even if we have studied it, we usually do not follow it in life. At first, psychologists Amos Tversky and Daniel Kahneman showed that even experts take current decisions without using systematic logic. Therefore, they established a research programme in cognitive science to study cognitive heuristics and biases, i.e., ways of making decisions in life: decisions in risk situations, prompt decisions, creative decisions and so on (Tversky, 1974).

In the section, I try to find out the origin of the emergence of logic in India according to the Pāli texts.

Logic is based on distinguishing particulars and generals. Let us consider an example from the Pāli Canon—how particulars and generals are examined in the *Yamaka*. In Sanskrit, *yamaka* means 'twin' or 'pair' and it is used to denote a rhetoric trope with a repetition of words after their permutation. In the *Nāṭyaśāstra* (dated to between 200 B.C. and 200 A.D.) there are enumerated excellent points (*laksaṇa*) of a good dramatic composition (*kāvya*) and *yamaka* belongs to one of the four "rhetorical figures of speech" (*alaṃkāra*), used when composing dramas. These four figures are as follows: simile (*upamā*), metaphor (*rūpaka*), condensed expression (*dīpaka*), and *yamaka*.

In the Pāli treatise, *Yamaka* the repetition of two words *A* and *B* after their permutation is used to show what a general is from *A* and *B* and what a particular is. So, this book considers many different pairs of *dhamma*s *A* and *B* by questions: "Is *A B*? But is *B A*?" Answering both questions allows us to define an inclusion relation respectively: "Is *A* a subset of *B*" or "Is *B* a subset of *A*?" There are possibly four answers: (i) *A* is a subset of *B* and *B* is a subset of *A* ($A = B$, i.e., *A* and *B* are of the same generality); (ii) *A* is a subset of *B* and *B* is not a subset of *A* ($A \subset B$, i.e., *A* is particular, and *B* is general); (iii) *A* is not a subset of *B* and *B* is a subset of *A* ($B \subset A$, i.e., *A* is general and *B* is particular); (iv) *A* is not a subset of *B* and *B* is not a subset of *A* (*A* and *B* are not comparable).

These four possible answers to the question "Is *A B*? But is *B A*?" can be represented as the following four converses of the universal affirmative syllogistic proposition "All *A* are *B*": (i) "All *A* are *B*" and "All *B* are *A*"; (ii) "All *A* are *B*" and "Not all *B* are *A*" (i.e., "Some *B* are not *A*"); (iii) "Not all *A* are *B*" (i.e., "Some *A* are not *B*") and "All *B* are *A*"; (iv) "Not all *A* are *B*" (i.e., "Some *A* are not *B*") and "Not all *B* are *A*"' (i.e., "Some *B* are not *A*").

For instance:

> (*ka*) *ye keci kusalā dhammā,* ***sabbe*** *te kusalamūlā*?
> (*kha*) *ye vā pana kusalamūlā,* ***sabbe*** *te dhammā kusalā*?
> (*Yamakapāḷi*, 1 *Mūlayamakaṃ* I: 1).

> All faultless states (are present). Are they **all** faultless roots?
> These faultless roots (are present). Are they **all** faultless states? (Nārada, 1998: 22).

Thus, among all the converses from (i) to (iv) for all pairs considered in the *Yamaka* we see just a correct declaration of what a general is and what a particular is and, at the same time, we cannot find out any logical inference rule, even in respect to converse or inverse. For example, we know that from "All *A* are *B*" it follows logically that "Some *B* are *A*" (conversation). Nevertheless, the author(s) of the *Yamaka* does not know this rule, see also (Jayatilleke, 1963: 306–310; Watanabe, 1983: 152) as well as others. This fact is apologetically explained as follows:

> The *Yamaka* does not consist of a set of logical exercises and is not a textbook on applied logic at all. The members of the pairs of statements do not stand to each other in the logical relation of one being an immediate inference of the other. To conceive them as such is wholly to misunderstand

> the purpose of the book, which is not an exercise in logical gymnastics, but is intended to convey to the reader the exact logical boundaries of important concepts in the light of their actual technical usage (Jayatilleke, 1963: 309).

So, on the one hand, the *Yamaka* demonstrates a well-developed philosophical discourse with distinguishing particulars and generals. But, on the other hand, its author does not express any knowledge of logical rules for inferring.

This situation with the *Yamaka* is an example of adopting the *first method* applied in this section, called a *structural analysis of logical competence*. We can always detect this competence or its deficit by textual analysis: whether there are some validations of inference rules which are correct from the standpoint of symbolic logic. Even if the author demonstrates a philosophical discourse with some logical notions, but she or he also often uses sophisms or does not apply inference rules at all, this means that she or he does not have the true logical competence.

Logic is a part of algebra and logical competence means that the author can combine some lexemes algebraically, e.g., she or he can draw true conclusions from true premises mechanically by means of some algebraic tools. In other words, the author can use some logemes which are mapped into a set of logemes of modern logic.

In fragments of trial records and omens written in Akkadian we can detect some algebraic tools used for trial decisions and forecasting (see Sections 3.2 and 3.3). Hence, the authors of these fragments possessed the good logical competence. In India, for example, this good competence is detected in the texts written by representatives of the Hindu schools of Nyāya and Vaiśeṣika and by representatives of the Buddhist schools of Madhyamaka and Yogācāra.

The *Kalahavivāda-sutta* (*Sutta-nipāta* IV: 11) is another example from the Pāli Canon in which the author involves difficult logical notions without any logical inference rules. In this discourse, for rhetorical purpose the author uses a transitivity of implications, when several implications with a joint part can be combined among themselves to build sorites (Watanabe, 1983: 130–136). Nevertheless, the author implements semantically true transitions some of which are not correct formally.

The Pāli Canon is a unique Indian text corpus because we have a chance to observe how the logical competence of its authors continuously grew up from zero in analyzing the same subject of *abhidhamma*. The first definitions of *dhamma*s and the first logical divisions of their concepts contained many fallacies because of "mutually intercrossing, over-lapping or partially coinciding notions", see an appropriate apologetics for this fact in (Jayatilleke, 1963: 294). But later they became correct and there were two ways of division: (i) by choosing contradictory terms, e.g., in classifying some *duka*s (pairs), see (Watanabe, 1983: 49); (ii) and by choosing contrary terms with a neutral between them, e.g., in classifying some *tika*s (triplets): *sukha* (happiness), *dukkha* (unhappiness), and *adukkhamasukha* (neither unhappiness nor happiness), see (Watanabe, 1983: 52).

In reconstructing the history of logical competence of the authors of the Pāli Canon, I appeal to the *second method, historical reconstructive hermeneutics*. This hermeneutics allows us to examine texts diachronically and it is grounded on the motto of Ludwig Wittgenstein: "use as meaning" according to which we should reconstruct meanings of words based on the understanding their contextual use.

This hermeneutics is the opposite of the *philosophical synchronic hermeneutics*. The latter examines texts synchronically for the sake of philosophical inspirations. It can be productive from the point of view of philosophy, although it is not scientific. For instance, sometimes the *Abhidhamma* is analyzed by Ronkin (2005) by the philosophical synchronic hermeneutics, e.g., when she compares *abhidhamma* and the *vaiśeṣika* way of classifying categories. On the one hand, this comparison is interesting philosophically, but, on the other hand, it ignores the fact that the Vaiśeṣika classification of categories demonstrates a good logical competence, while the Pāli classifications were proposed at a different time and some of them are not perfect logically. So, from the standpoint of historical reconstructive hermeneutics the latter fact would mean that some Pāli classifications of *dhamma*s were created earlier than the Vaiśeṣika classification. Historically, the Vaiśeṣika one can be compared to the *Abhidharmakośakārikā* of Vasubandhu, because their logical competences are comparable and perfect simultaneously.

Some versions of philosophical synchronic hermeneutics can be even absurd historically. For instance, according to the Mīmāṃsā hermeneutics, the Nyāya school of logic existed at the time of Kṛṣṇa and Arjuna, as the word of *nyāya* is mentioned in the *Mahābhārata*:

> The sciences called ***nyāya***, orthoephy and treatment of diseases; <...> a description of places of pilgrimage and other holy places of rivers, mountains, forests, the ocean, of heavenly cities and the *kalpa*s; the art of war <...> (*Mahābhārata* I: 1, 52).

The historical reconstructive hermeneutics allows us to trace back the logical competence of the authors of the Pāli Canon. In (Jayatilleke, 1963; Watanabe, 1983) there is proposed a well-grounded reconstruction of logical discourse in the early Pāli texts, although the first book is quite subjective, because its author tries to be so apologetic for Theravāda. In both books, their authors note that from the very beginning, the *abhidhamma* as well as other texts was often taught in the form of questions and answers and, as a result, some (proto-) logical techniques were thought up for the purpose of rhetoric.

For example, in the *Nikāya*s and *Āgama*s the following four kinds of explanations of questions were proposed (Jayatilleke, 1963: 281; Watanabe, 1983; 72–73): (i) *ekaṃsa-vyākaraṇīyo*, a categorical explanation when a questioner demands 'Yes' or 'No' from an answerer; (ii) *paṭipucchā-vyākaraṇīyo*, an answer by a counter-question when an answerer returns a reply in the form of a new question; (iii) *ṭhapanīyo*, when a questioner suggests some reply to an answerer but all such suggestions are set aside as inapplicable; (4) *vibhajja-vyākaraṇīyo*, an analytical explanation when a questioner requests some explanation of an answerer.

One of the first logical tools represented in the Pāli texts is to classify things according to the following fourfold assertions: (i) *S* is *P*; (ii) *S* is not *P*; (iii) *S* is and is not *P*; (iv) *S* neither is nor is not *P*. For instance, the Buddha engages these four-fold assertions to show that none of them "fit the case" (*upeti*). Thus, he says that when an enlightened person dies: (i) "he is reborn... does not fit the case" (*upapajjatī ti... na upeti*); (ii) "he is not reborn... does not fit the case" (*na upapajjatī ti... na upeti*);

(iii) "he is and is not reborn... does not fit the case" (*upapajjati ca na ca upapajjatī ti... na upeti*), (iv) "he is neither reborn nor not reborn... does not fit the case" (*n'eva upapajjati na na upapajjatī ti ... upeti*) (*Majjhima-nikāya* I: 486; Jayatilleke, 1963: 289). This type of answer corresponds to the following strategy in questioning: (i) "Is *S P*?"; (ii) "If not, is *S* not *P*?" (iii) "If not, is *S* both *P* and not *P*?" (iv) "If not, is *S* neither *P* nor not *P*?"

As we see, the first logical techniques, such as the fourfold assertions, which were invented in the Pāli Canon, were first used for some rhetorical purposes.

Thus, in this section, I propose the structuralist analysis of logical competence, how it is expressed in the Pāli Canon, by means of the historical reconstructive hermeneutics. My aim is to show that the *Milindapañha* is a unique Pāli text close to the true original point of logic's emergence in India.

4.2.3 Some Occurrences of the Terms 'Logic', 'Logical' in the Pāli Canon

The *Milindapañha* consists of many short dialogues and each dialogue can be examined as (or reduced to) a syllogism with the following four steps:

a) Is *A B*? (question or thesis);
b) *A* is (not) *B*, because… (argumentation or inference, *anumāna*);
c) The illustration for inference (b) is as follows: … (example, *opamma*);
d) Accepting or denying (a) because of (b).

This structure of syllogism proposed in the *Milindapañha* holds even for definitions:

a) "Venerable Nāgasena, what is the differentia (distinguishing characteristic) of *A*?"
   "*bhante nāgasena kiṃlakkhaṇā A*?" *ti.*
b) "The differentia of *A*, your majesty, is *B*."
   "*B -lakkhaṇā mahārāja A*" *ti.*
c) "Give me an illustration."
   "*opammaṃ karohī*" *ti.*
   Then the Nāgasena illustration follows.
d) Accepting the definition put forward by Nāgasena.
   "*kallo'si bhante nāgasenā*" *ti.*

Let us notice that this kind of definition is very close to the Aristotelian model of genus-differentia definitions: the item is defined through its differentia, but the difference from Aristotle is that Nāgasena always verifies his definitions by examples in the way of the Epicureans (see Subsection 1.2.4).

In the *Milindapañha* there is no name for the four-step syllogisms. The word *ñāya*, the Pāli analogue for the Sanskrit *nyāya*, occurs in the meaning "the method, the *dhamma*" *ñāya dhamma* (*Milindapañha* VI: 1, 3), i.e., as a method, distinguishing Buddhists from non-Buddhists. Let us emphasize that *ñāya* occurs very often in the

Pāli Canon, but never in the meaning of the Hindu school of logic. In most cases it means an analytic method or even a Buddhist method of cognition distinguishing Buddhists from non-Buddhists. For example, there is an expression "the noble method" (*ariya cassa ñāya*) in the *Gahapativagga* of the *Saṃyutta-nikāya*:

> *sāvatthiyaṃ viharati. atha kho anāthapiṇḍiko gahapati yena bhagavā tenupasaṅkami; upasaṅkamitvā bhagavantaṃ abhivādetvā ekamantaṃ nisīdi. ekamantaṃ nisinnaṃ kho anāthapiṇḍikaṃ gahapatiṃ bhagavā etadavoca:*
> *"yato kho, gahapati, ariyasāvakassa pañca bhayāni verāni vūpasantāni honti, catūhi ca sotāpattiyaṅgehi samannāgato hoti,* ***ariyo cassa ñāyo*** *paññāya sudiṭṭho hoti suppaṭividdho, so ākaṅkhamāno attanāva attānaṃ byākareyya: 'khīṇanirayomhi khīṇatiracchānayoni khīṇapettivisayo khīṇāpāyaduggativinipāto, sotāpannohamasmi avinipātadhammo niyato sambodhiparāyano'ti'"* (*Saṃyuttanikāya, Pañcaverabhayasutta* 41–42 [*Gahapativagga*]).

> At Savatthī. Then the householder Anathapiṇḍika approached the Blessed One, paid homage to him, and sat down to one side. The Blessed One then said to him:
> "Householder, when five fearful animosities have subsided in a noble disciple, and he possesses the four factors of stream-entry, and he has clearly seen and thoroughly penetrated with wisdom the **noble method**, if he wishes he could by himself declare of himself: 'I am one finished with hell, finished with the animal realm, finished with the domain of ghosts, finished with the plane of misery, the bad destinations, the nether world. I am a stream-enterer, no longer bound to the nether world, fixed in destiny, with enlightenment as my destination'" (Bodhi, 2000, vol. 1: 578).

Also, there is an expression "the method, the *dhamma*" (*ñāya dhamma*) in other texts:

> *sāvatthinidānaṃ. "gihino vāhaṃ, bhikkhave, pabbajitassa vā micchāpaṭipadaṃ na vaṇṇemi. gihi vā, bhikkhave, pabbajito vā micchāpaṭipanno micchāpaṭipattādhikaraṇahetu nārādhako hoti* ***ñāyaṃ dhammaṃ*** *kusalaṃ"* (*Saṃyuttanikāya, Mahāvaggapāḷi* I: 24 [*Dutiyapaṭipadāsutta*]).

> At Savatthī. "Bhikkhus, whether for a layperson or one gone forth, I do not praise the wrong way. Whether it is a layperson, or one gone forth who is practising wrongly, because of undertaking the wrong way of practice he does not attain the **method**, the **Dhamma** that is wholesome" (Bodhi, 2000, vol. 2: 1536).

Contextually, only the *Milindapañha* among other Pāli texts assumes that *ñāya* should include 'logic'—by assuming the four-step syllogism as a tool of true cognitions. As we see, the *Milindapañha* is unique not only because of its historical context (the only Pāli book directly connected to Gandhāra), but also due to its respect for syllogisms as a part of the method of Buddhists. The point is that the doctrinal difference between the Theravāda and the Mahāyāna teaching is significant, indeed, but the most intriguing difference holds in respect to logic. While in the Mahāyāna there are many logical treatises and logic is regarded as one of the most important Buddhist sciences and arts, in the Theravāda there is no interest in logic as such, there are no logical treatises in the strict sense. That fact is in line with the Buddha's words concerning logical matters in the earliest *sūtra*s (Pāli: *sutta*s). Let us refer to the *Kālāma-sutta* contained in the *Aṅguttara-nikāya* of the *Tipiṭaka*, the Pāli Canon. In this *sutta*, the Buddha sounds a note of caution about what should be avoided in Buddhism:

> *etha tumhe kālāmā mā anussavena, mā paramparāya, mā itikirāya, mā piṭakasampadānena, mā* ***takkahetu****, mā* ***nayahetu****, mā* ***ākāraparivitakkena****, mā diṭṭhinijjhānakkhantiyā, mā bhabbarūpatāya, mā samaṇo no garū'ti. yadā tumhe kālāmā attanā'va jāneyyātha, ime dhammā kusalā, ime dhammā anavajjā, ime dhammā viññuppasatthā, ime dhammā samattā samādinnā hitāya sukhāya saṃvattantī'ti. atha tumhe kālāmā upasampajja vihareyyātha* (*Tikanipātapāḷi* 66 [*Kālāmasutta* I: 189]).

> Come, Kālāmas. Do not go upon what has been acquired by repeated hearing; nor upon tradition; nor upon rumour; nor upon what is in a scripture; nor upon **surmise**; nor upon an **axiom**; nor upon **specious reasoning**; nor upon a bias towards a notion that has been pondered over; nor upon another's seeming ability; nor upon the consideration, "The monk is our teacher." Kālāmas, when you yourselves know: "These things are good; these things are not blameable; these things are praised by the wise; undertaken and observed, these things lead to benefit and happiness", enter on and abide in them (Thera, 1981: 5).

In this short-quoted text, the Buddha talks about things which are like the idols of the mind (*idola mentis*) proposed by Francis Bacon (1561–1626). For example, the Buddha warns us against traditional visions and dogmas including "repeated hearing" (*anussava*), 'tradition' (*paramparā*), 'rumor' (*itikirā*), "following a scripture" (*piṭaka-sampadāna*), "following an authority" (*bhabba-rūpatāya* and *samaṇo no garū*). Meanwhile, he warns us against any logical reasoning, too. Namely, first, he talks against 'surmise' (*takkahetu*), which is better to be translated as "because of (deductive) reasoning". It means that any truth of *dhamma* (the Buddha's teaching) cannot be proven by inferring from premises. Second, the Buddha warns us against "an axiom" (*nayahetu*), assuming it will be verified later. Third, he avoids "specious reasoning" (*ākāraparivitakka*), accepting something after considering its reasons. Fourth, he criticizes "a bias towards a notion that has been pondered over" (*diṭṭhinijjhānakkhantiyā*), starting from some view or opinion (*diṭṭhi*).

Hence, according to the *Kālāma-sutta*, logical matters (e.g., the art of debates) are excessive for Buddhism and logical reasoning is not enough for our liberation. Another commentary to this *sutta* is by Watanabe (1983: 105). We should just analyze our own action (*kamma*) and its consequences or effects (*phala*). The same attitude towards logical subjects is seen in the Theravāda (i.e., relatively early) Buddhism, as well.

In many other early *sutta*s, such as some in the *Sutta-nipāta*, the Buddha expresses the same sceptical view of logical reasoning, e.g.:

> *na heva saccāni bahūni nānā,*
> *aññatra saññāya niccāni loke;*
> *takkañca diṭṭhīsu pakappayitvā,*
> *saccaṃ musāti dvayadhammamāhu.*
> *Diṭṭhe sute sīlavate mute vā,*
> *ete ca nissāya vimānadassī;*
> *vinicchaye ṭhatvā pahassamāno,*
> *bālo paro akkusaloti cāha*
> (*Suttanipātapāḷi* 892–893 [*Cūḷabyūhasutta*]).

> [Buddha]:
> Indeed, there are not many and varied truths
> differing from perception of the ever-true in the world;
> but they work upon their views with logic: [*A.Sch.*: the author uses here the term *dhamma*, not logic]
> "Truth! Falsehood!" So, they speak in dualities.
> Based on what is seen, heard,
> On precepts and vows, or what is cognized,
> They look down on others.
> Convinced of their own theories,
> pleased with themselves,
> They say, "My opponent is a fool, no expert."
> (Mill, 2015: 279).

Thus, the word *ñāya*, the Pāli analogue for the Sanskrit *nyāya*, never occurs in the Pāli Canon in the meaning of the school of logic or the *Nyāya-sūtra*. The only case that seems to be a mention of the school of logic is as follows. In the *Milindapañha*, Menander or Milinda was regarded as an excellent expert in all the 19 Hindu sciences and arts:

> Of these two the novice became the King called Milinda in the city of Sagala in India. He was wise, experienced, clever, able; he was one who acted conscientiously at the time of doing all the (magic) devices, ceremonies and observances concerning things past, future, and present. Many were the arts he had mastered, that is to say: the revealed tradition, secular lore, the Sankhya, Yoga, Nyāya, and Vaiśeśika systems, accountancy, music, medicine, the four Vedas, the Purāṇas, the oral traditions, astronomy, conjuring, logic, spells, fighting, poetry, reckoning on the fingers, in a word, the nineteen (arts) (Horner, 1969, vol. 1: 5).

So, the Greek king was represented as a scholar even in the Hindu philosophy, including the doctrines of the Sāṃkhya, Yoga, Nyāya, and Vaiśeṣika schools. The problem is that T.W. Rhys Davids and then I.B. Horner have not translated this fragment correctly. Let us quote the same text in Pāli:

> *tesu sāmaṇero jambudīpe sāgalanagare milindo nāma rājā ahosi. paṇḍito byatto medhāvī paṭibalo atītānāgatapaccuppannānaṃ mantayogavidhānakiriyānaṃ, karaṇakāle nisammakārī hoti, bahūni cassa satthāni uggahitāni honti. seyyathidaṃ—suti sammuti* ***saṅkhyā yogā nīti visesikā*** *gaṇikā gandhabbā tikicchā catubbedā purāṇā itihāsā jotisā māyā ketu mantanā yuddhā chandasā buddhavacanena ekūnavīsati, vitaṇḍavādī durāsado duppasaho puthutitthakarānaṃ aggamakkhāyati, sakalajambudīpe milindena raññā samo koci nāhosi yadidaṃ thāmena javena sūrena paññāya, aḍḍho mahaddhano mahābhogo anantabalavāhano* (*Milindapañha* I: 1, 4).

We are reading here just *saṅkhyā yogā nīti visesikā*, i.e., the Sāṃkhya, Yoga, Nīti, and Vaiśeṣika schools, not Nyāya. The fact that the word *nyāya* in the meaning of the Hindu doctrine or school of logic does not occur in the Pāli Canon at all is significant evidence for us that Nyāya did not exist before the 1st century A.D., at a time when approximately, but not earlier, the Pāli Canon was finally edited. The translation of *nīti* as *nyāya* is a kind of historical falsification. First, *nyāya* is a teaching concerning *pramāṇa* how to verify or falsify logical reasoning. And, as we try to prove, this teaching did not exist before the 1st century A.D.

Let us examine the 19 Hindu sciences and arts for which Menander was said to be a great scholar: (1) *suti veda*—the Hindu holy texts presented by the Vedas; (2) *sammuti*—moral codes and sage advices such as the *Manusmṛti* or *Manu's Code of Law* finally edited after 400 A.D.; (3) *saṅkhyā*—the *Sāṃkhya* philosophy founded by Kapila, its earliest surviving text is the *Sāṁkhyakārikā* written by Iśvarakṛṣṇa from the 3rd to the 4th century A.D.; (4) *yogā*—the yoga philosophy founded by Patañjali, its basic text is the *Yoga-sūtra* dated from the 2nd to the 4th century A.D.; (5) *nīti*—the *nīti* philosophy or political philosophy including several arts: diplomacy and statecraft (Pāli: *rajanīti*); economics (Pāli: *atthanīti*); morality (Pāli: *lokanīti*); ethics (Pāli: *dhammanīti*); the Buddhist ethics that included didactic stories and maxims on numerous everyday subjects (Pāli: *vaddhananīti*); (6) *visesikā*—the Vaiśeṣika philosophy founded by Kaṇāda, its basic text is the *Vaiśeṣika-sūtra* cited, e.g., the *Jñānaprasthāna-śāstra*, one of the seven books of the Sarvāstivāda *Abhidharma* written from the 1st to the 2nd century A.D.; thus, the *abhidharma* (Pāli: *abhidhamma*) is a Buddhist alternative to *vaiśeṣika*; (7) *gaṇikā*—arithmetics; (8) *gandhabbā*—the Hindu literature on music; (9) *tikicchā*—the Hindu medicine; (10) *catubbedā*—the art of archery; (11) *purāṇā* – the corpus of histories and ancient tales; (12) *itihāsā*—the Hindu chronics saying "it happened thus", such as the *Mahābhārata*; (13) *jotisā*—the Hindu astrology that was influenced by the Greek astrology in the higher measure, because the *Yavanajātaka* (or the *Greek Jātaka*), a book on astrology translated from Greek in the 2nd century A.D., was one of the earliest sources of the Hindu astrology; (14) *māyā*—the knowledge of stratagem; (15) *ketu* that is traditionally read by the Theravāda as *hetu*—the art of weighing and analyzing the *pro et con* of the matter in question, in the way it was demonstrated in the *Kathāvatthu*; however, if it is *ketu* indeed, then it means Hindu omens; (16) *mantanā*—the Hindu art of incantations and sacrificial formulas; (17) *yuddhā*—the art of warfare; (18) *chandasā*—the Hindu art of reciting and composing hymns and poetry; (19) *buddhavacanena*—all the words of the Buddha, including the Pāli Canon.

For the first time, *ñāya* was used with the same kind of meaning of the Hindu school of logic, most probably, of Buddhist logic, just in the *Daṭhavansa* or the *History of the Tooth-Relic Gotama Buddha* written not earlier than in the 5th century A.D.:

> Then priests, wise and skilled in the Tipiṭaka, Jātaka, **Logic** [*A.Sch.*: *ñāya*], Agāma, and the like, and citizens who had their sole refuge in the three treasures, assembled there instantly through curiosity (Swamy, 1874: 72).

However, the meaning "the method of Buddhists" was also preserved:

> Who, again, as the hog Tundila, satisfied the people with the taste of the nectar of the Law, and as a sage, composing a **treatise on logic** [*A.Sch.*: *ñāya*], made his own Law prevail for a long time (Swamy, 1874: 51).

For the meaning of "logical inference" there is another Pāli analogue of *nyāya*, presented by *naya*. In this meaning the term of *naya* occurs often together with the term *hetu* in the same way as in the *Kālāma-sutta* quoted above, for example in the *Jātaka*:

> "*pucchassu maṃ rāja tavesa attho, yaṃ saṃsayaṃ kuruse bhūmipāla; ahaṃ taṃ nissaṃsayataṃ gamemi,* ***nayehi ñāyehi*** *ca* ***hetubhī*** *ca*" (*Jātaka*, 544, *Mahānipāto* 1298 [*Mahānāradakassapajātaka*]).

> Narada replied:
> "Ask me, O king; this is thy business; this doubt of thine which thou feelest, I will assuredly solve it for thee by **reasoning**, by **logic**, and by **proofs**."

The next instance of the same occurrence is taken from the *Mahāvagga* of the *Aṅguttara-nikāya*, where the Buddha also expresses a kind of scepticism about any logical reasoning:

> *etha tumhe, sāḷhā, mā anussavena, mā paramparāya, mā itikirāya, mā piṭakasampadānena, mā* ***takkahetu****, mā* ***nayahetu****, mā ākāraparivitakkena, mā diṭṭhinijjhānakkhantiyā, mā bhabbarūpatāya, mā samaṇo no garūti* (*Aṅguttaranikāya* I: 193, *Tikanipātapāḷi* 66 [*Sāḷhasutta*]).

> "Come, Salha, do not be satisfied with hearsay or with tradition or with legendary lore or with what has come down in scriptures or with conjecture or with **logical inference** or with **weighing evidence** or with a liking for a view after pondering it or with someone else's ability or with the thought 'The monk is our teacher'" (Thera, 2013).

Thus, the term *nyāya* in the meaning of the Hindu school of logic or the *Nyāya-sūtra* does not occur in the Pāli Canon, actually. Instead of that the Pāli term *ñāya* had the meaning of one or other aspect of the wisest Buddhist method and, according to the *Milindapañha*, the four-step syllogisms are an important part of cognitions within this method.

4.2.4 Logical Reconstructions of Some Conclusions in the *Kathāvatthu*

The *Milindapañha* is organized as a compendium of four-step syllogisms explaining *abhidhamma*. There is only one other similar treatise, *pakaraṇa*, written especially for the purpose of debates with non-Theravādins (more precisely historically, with non-Vibhajjavādins) for teaching the *abhidhamma*. This compendium of logical reasoning for different debates is called the *Kathāvatthu*, it is contained in the *Abhidhamma-piṭaka* of the Pāli Canon. In this subsection, I will try to show that its author(s) had no competence in logic because of many fallacies, although a lot of syllogisms of the *Kathāvatthu* are correct and difficult (Schumann, 2018a, 2019c). This is the main difference of that book from the *Milindapañha*, where there are no logical fallacies at all. For another logical reconstruction of the *Kathāvatthu*, please see Jayatilleke (1963: 305–368) and Watanabe (1983: 122–126, 154–174).

In this treatise we find many correct complex syllogisms, such as *modus tollens*: "If *A* is *B*, then *C* is *D*. But *C* is not *D*. Therefore, *A* is not *B*" (Aung and Davids, 1915: xlviii):

> Adherent: Is *A B*? (*ṭhapanā*)[6]
> Opponent: Yes.
> Adh.: Is *C D*? (*pāpanā*)
> Opp.: No.
> Adh.: But if *A* be *B*, then [you should have said] *C* is *D*. That *B* can be affirmed of *A*, but not *D* of *C*, is false. Hence your first answer is refuted. (*ropanā*).

Formally:

$$\frac{(A \Rightarrow B) \Rightarrow (C \Rightarrow D); \neg (C \Rightarrow D)}{\neg (A \Rightarrow B).}$$

Another example of correct syllogism as a modification of *modus tollens* is logically formulated in the *Kathāvatthu* as follows: "If *D* be denied of *C*, then *B* should have been denied of *A*. But you affirmed *B* of *A*. Therefore, that *B* can be affirmed of *A*, but not *D* of *C*, is wrong," or in the simpler way: "If *C* is not *D*, then *A* is not *B*. But *A* is *B*. Therefore, *C* is *D*" (Aung and Davids, 1915: xlviii). Formally:

$$\frac{\neg (C \Rightarrow D) \Rightarrow \neg (A \Rightarrow B); (A \Rightarrow B)}{(C \Rightarrow D).}$$

6 Let us notice that in symbolic logic the proposition "*A* is *B*" always has the formal meaning of implication: "If *A*, then *B*" ($A \Rightarrow B$) or "If something is *A*, then it is *B*, too" ($A \Rightarrow B$). This formal treatment of affirmative propositions in the way of implications was well known by Indian logicians such as Dharmakīrti (he was a representative of Yogācāra school). He exemplifies this relationship as follows: "*Dalbergia sisso* is a tree." Consequently, it means that if something is a *Dalbergia sisso*, then we can conclude that it is a tree, too, but not vice versa.

One of the most interesting items of evidence for the genuine role that logic plays in Theravāda Buddhism is contained in the first chapter of this text, called the *Puggalakathā*; the latter describes a debate between a Theravādin (more precisely historically, Vibhajjavādins), who is considered an orthodox Buddhist in the text, and a Puggalavādin, another Buddhist who believes in the existence of a soul-like personal entity (*puggalo*). The point is that the reasoning involved in the debate from the opposite sides shows that the Theravādin as well as the Puggalavādin do not understand the subject of logic as ultimate inferring, although they use correct syllogisms sometimes.

Let us introduce some symbolic notations to make their debate more transparent:

> *A* is *B* := "'The person' (*puggalo*) is known in the sense of a real and ultimate fact."
>
> *puggalo upalabbhati saccikaṭṭhaparamatthenāti* (*Kathāvatthu* 1).
>
> *A* is *C* := "Is 'the person' known in the same way as a real and ultimate fact is known?"
>
> *yo saccikaṭṭho paramattho, tato so puggalo upalabbhati saccikaṭṭhaparamatthenāti*? (*Kathāvatthu* 1).

Then their debate is taking place in the following manner:

> Theravādin: Is *A B*?
> Puggalavādin: Yes.
> Ther.: Is *A C*?
> Pugg.: No.
> Ther.: However, "if *A* is *B*, then *A* is *C*." Then that which you say here is wrong, because you state that "*A* is *B*" is true, but "*A* is *C*" is false. But if "*A* is *C*" is false, then "*A* is *B*" is false.

Symbolically:

$$\frac{(A \Rightarrow B) \Rightarrow (A \Rightarrow C);\ \neg (A \Rightarrow C)}{\neg (A \Rightarrow B).}$$

It is an *ad absurdum*, because $A \Rightarrow B$ is held by the Puggalavādin to be true. The Theravādin holds that this should mean that $A \Rightarrow C$ is true, too. Hence, we see that the final refutation is logically correct, according to the Theravādin's understanding of the terms *A* and *B*, here: "If *A* is *B*, then *A* is *C*." So, if "*A* is *B*" is true, "*A* is *C*" should be true, too. The Puggalavādin maintains that "*A* is *C*" is false. However, it means, as the Theravādin truly claims, according to their analysis that "*A* is *B*" should be false, also. This syllogism is a classical *modus tollens*. Hence, the Theravādin has just refuted the Puggalavādin's opinion. But let us look at the continuation of this dialogue:

> Puggalavādin: Is *A* not *B*?
> Theravādin: Yes, it is not.
> Pugg.: Is *A* not *C*?

Ther.: No, it is.
Pugg.: However, "if *A* is not *B*, then *A* is not *C*." Then that which you say here is wrong, namely, that "*A* is not *B*" is true, but "*A* is not *C*" is false. But if "*A* is not *C*" is false, then "*A* is not *B*" is false, also. Thus, you are wrong.
Pugg.: So, if "*A* is not *B*" is true, then "*A* is not *C*" is true. Now we, who admitted these propositions, do not consider ourselves to have been refuted. You say you have refuted us; anyway, we are not well refuted.

Symbolically:

$$\frac{\neg (A \Rightarrow B) \Rightarrow \neg (A \Rightarrow C);\ \neg\neg (A \Rightarrow C)}{\neg\neg (A \Rightarrow B).}$$

It is an *ad absurdum*, as well, because $A \Rightarrow C$ is true for the Puggalavādin. From this it follows that $A \Rightarrow B$ is true, also. So, the Puggalavādin puts forward another implication, namely: "if *A* is not *B*, then *A* is not *C*" $[\neg (A \Rightarrow B) \Rightarrow \neg (A \Rightarrow C)]$. The Theravādin states that "*A* is not *C*" is false. From this it should follow according to the same *modus tollens*, as the Puggalavādin notes now, that "*A* is not *B*" is false. It means that the Puggalavādin has just refuted the Theravādin's opinion.

Thus, we have the following opposite sides:

Theravādin:	"*A* is *B*" is false, "*A* is not *B*" is true; "*A* is *C*" is true, "*A* is not *C*" is false; if "*A* is *B*", then "*A* is *C*."
Puggalavādin:	"*A* is *B*" is true, "*A* is not *B*" is false; "*A* is *C*" is false, "*A* is not *C*" is true; if "*A* is not *B*", then "*A* is not *C*."

To apply the same *modus tollens*, the Theravādin appeals to the implication "if *A* is *B*, then *A* is *C*" as the first premise of his syllogism and the Puggalavādin to the same implication, but with negations "if *A* is not *B*, then *A* is not *C*" as the first premise of his syllogism. Who is right? Nobody! The problem is that the Theravādin as well as the Puggalavādin cannot agree on the first premise of their reasoning. Their dialogue looks like a logical paradox: the same propositions are true and false at the same time. One opposite side puts forward one implication to prove a contradictory statement. Another side puts forward the same implication, but with negations to prove another statement. Such a dialogue can become interminable. Indeed, we face many modifications of the first dialogue in the *Puggalakathā*.

Formally:

Theravādin:	if "*A* is *B*" is true by the Puggalavādin, then "*A* is *C*" should be true by the Puggalavādin also, but it is not.
Puggalavādin:	if "*A* is not *B*" is true by the Theravādin, then "*A* is not *C*" should be true by the Theravādin also, but it is not.

The problem is that the author of the *Kathāvatthu* does not know how the implication $A \Rightarrow B$ can be verified. In the Nyāya and Yogācāra logic, there are the following three ways of verifying the implication: (i) the (Aristotelian) way by showing that B is a general (genus) for A; (ii) the (Stoic or Chryssipus) way by checking that A is a cause for B; (iii) the (Stoic or Chryssipus) way by checking that B is a sign for A if A is a cause for B. Hence, the sentence $A \Rightarrow B$ means, according to Nyāya and Yogācāra, that A implies B as a genus for A or a causal consequence from A. There is the third possibility as well that there can be the sentence $B \Rightarrow A$ telling us that the sign B exists for occurring its cause A. This semantics for conditional sentences is unknown for the author of the *Kathāvatthu*. But without a verified implication, *modus tollens* plays just a rhetorical role.

Thus, on the one hand, the Theravādin as well as the Puggalavādin apply the formally correct modifications of *modus tollens* mentioned above, but, on the other hand, they do not give true inferences, but sophisms in fact, because the Theravādin uses the implication "if A is B is true, then A is C is true" where the antecedent occurs he considers false and the Puggalavādin uses the implication "if A is not B is true, then A is not C is true" where there is the antecedent he examines as false, too. However, we cannot infer from the false premises! This significant fact that *modus tollens* is a sophism because of the unverified implications is ignored by Jayatilleke (1963) and Watanabe (1983). A complicated reasoning with many formulas does not mean immediately that its author is a logician. The matter is that this reasoning should be correct formally with, necessarily, correct verifications of all premises.

Let us introduce the following new notations:

> A is B := "'The person' (*puggalo*) is known in the sense of a real and ultimate fact."
> C is B := "Material quality[7] is known in the sense of a real and ultimate fact."

Then we have the following next dialogue:

> Theravādin: Is A B, and is C B?
> Puggalavādin. Yes.
> Ther.: Is C one thing and A another?

[7] Then they have used the same reasoning where for "material quality" they have substituted the following new items: feeling; perception; coefficients (*saṅkhāras*); consciousness; the organ of sight; the organ of hearing; the organ of smell; the organ of taste; the organ of touch; visible object; sound; odour; taste; tangible object; mind (*sensis communis*); cognizable object; eye as subjective element; sights as subjective element; visual cognition as subjective element; ear as subjective element; sounds as subjective element; auditory cognition as subjective element; nose as subjective element; odours as subjective element; olfactory cognition as subjective element; tongue as subjective element; tastes as subjective element; gustatory cognition as subjective element; body as subjective element; touches as subjective element; tactile cognition as subjective element; mind as subjective element; mind-cognizing as subjective element; cognizables as objective element; eye as controlling power; ear as controlling power; nose as controlling power; tongue as controlling power; body as controlling power; mind as controlling power; female sex as controlling power; male sex as controlling power; life as controlling power; pleasure as controlling power; pain as controlling power; joy as controlling power; grief as controlling power; hedonic indifference as controlling power; faith as controlling power; energy as controlling power; mindfulness as controlling power; *samādhi* as controlling power; understanding as controlling power; the thought: "I shall come to know the unknown" as controlling power; the coming to know as controlling power; the having known as controlling power.

Pugg.: No, that cannot truly be said.
Ther.: However, if "*A* is *B*, and *C* is *B*", then "*A* and *C* are distinct things." You are wrong to admit "*A* is *B*, and *C* is *B*" and not "*A* and *C* are distinct things." If the latter is false, then the first is false.
Pugg.: Is *A B*?
Ther.: It is not.
Pugg.: Is *C B*?
Ther.: Yes.
Pugg.: Is *C* one thing and *A* another?
Ther.: No, that cannot be truly said.
Pugg.: If "*C* is *B*", then you should also have admitted that "*A* and *C* are distinct things." You are wrong in admitting the truth of "*C* is *B*" while you deny that of "*A* is *B*." If *A* and *C* are not distinct things, then *A* is *B*. Thus, your position is false.

Symbolically:

Theravādin:

$$\frac{((A \Rightarrow B) \,\&\, (C \Rightarrow B)) \Rightarrow (A \vee C);\ \neg\,(A \vee C)}{\neg\,(A \Rightarrow B) \,\&\, \neg\,(A \Rightarrow C).}$$

Nevertheless, it is an incorrect form. The logically corrected form is thus:

$$\frac{((A \Rightarrow B) \,\&\, (C \Rightarrow B)) \Rightarrow (A \vee C);\ \neg\,(A \vee C)}{\neg(A \Rightarrow B) \vee \neg\,(A \Rightarrow C).}$$

Puggalavādin:

$$\frac{(\neg\,(A \Rightarrow B) \,\&\, (C \Rightarrow B)) \Rightarrow (A \vee C);\ \neg\,(A \vee C)}{(A \Rightarrow B) \,\&\, \neg\,(C \Rightarrow B).}$$

It is an incorrect form, also. The logically corrected form is as follows:

$$\frac{(\neg\,(A \Rightarrow B) \,\&\, (C \Rightarrow B)) \Rightarrow (A \vee C);\ \neg\,(A \vee C)}{(A \Rightarrow B) \vee \neg\,(C \Rightarrow B).}$$

We deal here with two modifications of *modus tollens* again and in the same manner that the Theravādin and the Puggalavādin demonstrate that they do not know how implication can be verified. So, they apply different implications to infer contradictions, since they do not have a procedure for verifying conditional propositions as well as other propositions at all. The Theravādin is based on the scheme: "If *A* is *B* and *C* is *B*, then *A* and *C* are distinct things." If it is false that "*A* and *C* are distinct things," then it is false that "*A* is *B* and *C* is *B*." The Puggalavādin offers the following scheme: "If *A* is not *B* and *C* is *B*, then *A* and *C* are distinct things." If it is false that "*A* and *C* are distinct things," then it is false that "*A* is not *B* and *C* is *B*."

The main problem of the author(s) of the *Kathāvatthu* is that its author(s) does not know what the subject of logic is, but its subject is to infer automatically from premises which are verified as true sentences (Schumann, 2018a, 2019c). In the Hindu terms, they do not know what *pramāṇa* (Sanskrit: "means of knowledge") is—how we can verify sentences. Notice that an appropriate Pāli word *pamāṇa* occurs several times in the Pāli Canon, but never in the meaning of "means of knowledge". The teaching on *pramāṇa* appeared in India much later than all the texts of the Pāli Canon were composed. In this teaching all the sources of the true knowledge are classified: *pratyakṣa* (Pāli: *paccakkha*; 'evidence', 'axioms' or "underlying things", ὑποκείμενον in the Aristotelian meaning), *anumāna* (Pāli: *anumāna*; 'inference'), *upamāna* (Pāli: *upamāna*; 'comparison', 'analogy'), *arthāpatti* (Sanskrit: "postulation, derivation from circumstances", this word is not there in the Pāli Canon), *anupalabdhi* (Sanskrit: "non-perception, negative proof"; this is not a common word in the Pāli Canon, but in the *Milindapañha* (I: 138) it is claimed that a real self is *anupalabbhamāne*—it is not being apprehended) and *śabda* (Pāli: *sadda*; "word, testimony of past or present reliable experts"). In the European logic, *pramāṇa* is a logical semantics and a logical epistemology, i.e., the rules of how to ascribe meanings to logical propositions.

The Theravādin as well as the Puggalavādin are not familiar with any logical semantics. Therefore, they cannot agree on using premises. They do not know how to verify or falsify atomic propositions and how to build up true composite propositions based on the atomic ones. Although they know some correct syllogisms, they have no idea of how these syllogisms can be verified or falsified.

Hence, the *Kathāvatthu* cannot be evaluated as a logical treatise in fact. There is not even a hint of *pramāṇa* in this text. Meanwhile, there are many sophisms presented as true inference rules, such as:

> *vedanaṃ upādāya puggalassa paññattīti? āmantā. kusalaṃ vedanaṃ upādāya kusalassa puggalassa paññattīti? na hevaṃ vattabbe ... pe ...* (*Kathāvatthu* 192).

> Theravādin: Is the concept of soul derived from feeling?
> Puggalavādin: Yes.
> Ther.: Is the concept of good soul derived from good feeling?
> Pugg.: Nay, that cannot truly be said (Aung and Davids, 1915: 33–34).

> *vedanaṃ upādāya puggalassa paññattīti? āmantā. akusalaṃ vedanaṃ upādāya akusalassa puggalassa paññattīti? na hevaṃ vattabbe ... pe ...* (*Kathāvatthu* 193).

> Ther.: If the concept of soul is derived from feeling, is the concept of bad soul derived from bad feeling?
> Pugg.: Nay, that cannot truly be said (Aung and Davids, 1915: 33–34).

This text also contains a lot of references to authority (*śabda* of the *nyāya*) as an ultimate argument: "it was not said by the Exalted One [*A.Sch.*: i.e., by the Buddha]" and "it was said by the Exalted One".

In the *Kathāvatthu* there is a dispute with, probably, a follower of an idea according to that all real things are momentary. In other words, the Sautrāntikas ("those who rely upon the *sūtra*s [*A.Sch.*: and avoid the *Abhidhamma*]") think that items can exist for only one instantaneous moment:

> *ekacittakkhaṇikā sabbe dhammāti? āmantā. citte mahāpathavī saṇṭhāti, mahāsamuddo saṇṭhāti, sinerupabbatarājā saṇṭhāti, āpo saṇṭhāti, tejo saṇṭhāti, vāyo saṇṭhāti, tiṇakaṭṭhavanappatayo saṇṭhahantīti? na hevaṃ vattabbe ... pe ...* (*Kathāvatthu* 906).

> Controverted Point: That all things are momentary conscious units.
> Theravādin: Do you imply that a mountain, the ocean, Sumeru chief of mountains, the cohesive, fiery, and mobile elements, grass, twigs, trees, all last only so long in consciousness? You deny… (Aung and Davids, 1915: 363).

This dispute is attributed to the 'Sautrāntikas' by the Pāli commentary, but it is unusual and unexpected, since the Sautrāntikas existed from the 2nd to the 3rd century A.D. in the Hellenized region of Gandhāra and they represented a north-western branch of the Sarvāstivāda School whose ideas are contextually mentioned in the *Kathāvatthu*, also. The Sautrāntikas proposed the doctrine of momentariness (*kṣaṇikavāda*) mentioned in the verse above. After the 4th century A.D. the Sautrāntikas were transformed into the Yogācāra School—the most influential school of Gandhāran Buddhism. The idea of extreme momentariness was then adopted by the greatest Buddhist logicians and epistemologists, such as Dignāga (5th century A.D.) and Dharmakīrti (ca. 600–660).

The disputation with, possibly, an intended Vaibhāśika-Sarvāstivādin is as follows:

> *atītena samannāgatoti? āmantā. nanu atītaṃ niruddhaṃ vigataṃ vipariṇataṃ atthaṅgataṃ abbhatthaṅgatanti? āmantā. hañci atītaṃ niruddhaṃ vigataṃ vipariṇataṃ atthaṅgataṃ abbhatthaṅgataṃ, no ca vata re vattabbe—"atītena samannāgato"ti* (*Kathāvatthu* 568).

> Controverted Point: That a past or future experience is actually possessed.
> Theravādin: But is not the past extinct, departed, changed, come to an end, finished? And is not the future unborn, not yet become, not come into being, not produced, not brought to pass, not manifested? How then can you call eithersomething that is actually possessed? (Aung and Davids, 1915: 242).

The Vaibhāśikas represented a north-eastern branch of the Sarvāstivāda School that took root in Kashmir from the 2nd to the 3rd century A.D.

Hence, the *Kathāvatthu* as one of the most important texts of the *Abhidhamma* of the Pāli Canon contained some disputes with the intended Sautrāntikas who had avoided the *Abhidhamma* as such and the Vaibhāśikas who had proposed another approach to the *Abhidhamma* which differs a lot from the Theravāda approach. Later, the Vaibhāśika ideas on the *Abhidhamma* were partly used in the *Abhidharmakośa-bhāṣya*—the greatest work written by Vasubandhu (ca. from the 4th to the 5th century A.D.), who went on to be one of the most famous representatives of the Yogācāra School. Let us notice that Dignāga and Vasubandhu's texts are contained recently in the *Tengyur*, serving as the Tibetan Buddhist Canon.

To sum up, the logical fragments of the *Kathāvatthu* are not connected to the *pramāṇa* doctrine as a whole and then, most probably, they were transposed from disputes with some representatives of Northern Buddhism (e.g., the Gandhāran Buddhism), because the logic is applied in the *Kathāvatthu* mechanically, without understanding logical semantics or logical foundations. This feature distinguishes the *Kathāvatthu* from the *Milindapañha*, which proposes the four-step syllogisms, where the third step is used especially for verifying premises and inferences.

4.2.5 Historical Context of the Milindapañha

The *Milindapañha* text is organized as a recorded conversation between the Buddhist monk Nāgasena and the Greek king Menander I Soter (Pāli: *Milinda*; Greek: Μένανδρος Α′ ὁ Σωτήρ; 165/155–130 B.C.), the ruler of Arachosia, Gandhāra, Punjab, and Mathura (today's Afghanistan, Pakistan, Tajikistan, and some northern states of India) (Tarn, 1951). It is worth noting that Greek clans/dynasties had ruled Bactria since the beginning of Alexander the Great's Indian campaign, i.e., since 326 B.C. First, they ruled this land as satraps of Seleucid kings. But then around 250 B.C., Diodotus I Soter (Διόδοτος Α' ὁ Σωτήρ), the governor of the Seleucid province of Bactria, proclaimed independence. Since that time, Greek kings advanced from Bactria in the east and south, taking control over Arachosia, Gandhāra, Punjab, and Mathura. Since ca. 130 B.C. the Greek clans mainly left Bactria, their homeland, and concentrated mostly in Gandhāra, where there was the capital of their Empire, Taxila. At the end, the Greek clans controlled only eastern Punjab and, probably, Strato III (Greek: Στράτων Γ′) was the last Greek king who ruled from ca. 25 B.C. to 10 C.E. So, the time of Menander was a culmination point of the Greek power in India. To sum up, from the time of Alexander the Great, the Greek rule in some parts of India took over ca. 336 years.

The people of the Greek dynasties (Pāli: *yona*; Sanskrit: *yavana*) were replaced by Indo-Scythians or Śakas (Sanskrit: *śaka*) who at first continued the Greek Hellenization in India—at the beginning they continued to use the Greek language as official and to worship some Greek deities (Heracles, Zeus, Athena, Apollo and so on) (Bopearachchi, 1993; Hoover, 2013). At first, the Śakas occupied Sogdiana and Greek Bactria, then Arachosia, Gandhāra, Sindh, Kashmir, Punjab, Haryana, Uttar Pradesh, Rajasthan, Gujarat, and Maharashtra. There were different Indo-Scythian clans—which were, to a great extent, Hellenized. The following two dynasties were the most powerful and, at the same time, the most Hellenized of the Indo-Scythian clans: (i) the Western Kṣatrapas (ca. 35–400 A.D.) who ruled the western and central part of India (modern states of Gujarat, Maharashtra, Rajasthan, and Madhya Pradesh); (ii) the Kuṣāṇas (Bactrian: Κοþανο; Sanskrit: *Kuṣāṇa*) (ca. 35–375 A.D.) who at their peak ruled present-day Afghanistan, Pakistan, Tajikistan, south of Uzbekistan, and some northern parts of India up to Varanasi. But at the end of their power, the Kuṣāṇas controlled only eastern Punjab, while the Greeks controlled the same region at the time of the decline of their power 350 years previously.

Thus, the date of ca. 400 C.E. was an end of all Hellenized dynasties in India. The Western Kṣatrapas and the last Kuṣāṇas in eastern Punjab finally fell and their territories were invaded by the Gupta Empire. Later the Kidarites and Hephthalites expelled the Guptas and other clans from the territory once controlled by the Kuṣāṇas in their greatest period. The Hephthalites, according to numismatics, were primarily Zoroastrian in the Sassanian meaning—certainly neither Hindu, nor Buddhist.

Greek was used as an official language in Afghanistan, Pakistan, and the northern parts of India for at least three centuries; first of all it was used for edicts, trading, and receiving taxes. One of the taxable documents was found at Aï Khanum (today's Afghanistan) and it is dated to the 2nd century B.C. The writing appears only on one side and it is in black ink, presumably carbon-based. The writing material is prepared from skin. And this document is one of the oldest examples of texts on skin found until now in India. The text is as follows:

> In the reign of God *Antimachus* (Ἀντίμαχος) and *Eumenes* (Εὐμένης) and *Antimachus* (Ἀντίμαχος)…. year 4, month of *Olöus*, in *Asangorna*(?), when NN was guardian of the law. *Menodotus*, tax-gatherer, in the presence of NN, who was sent out likewise by *Demonax* the former …, and of *Simus*(?), who was … by agency of *Diodorus*, controller of revenues, acknowledges receipt from(?) NN the son(?) of *Dataes*(?), … of the payments due in respect of the purchase … (Rea, 1994).

"God *Antimachus*" mentioned in the inscription is a Greco-Bactrian king, named Θεός Ἀντίμαχος, whose rule generally dated from around 185 B.C.–170 B.C. On the Indian coins such as silver tetradrachms, he was depicted by the diademed bust of king on the obverse. On the reverse of these coins there was the following inscription: ΒΑΣΙΛΕΩΣ ΘΕΟΥ // ΑΝΤΙΜΑΧΟΥ, round the Greek deity Poseidon, standing to front, holding trident in the right hand and palm in the left hand (Lahiri, 1965; Mitchiner, 1975–1976; Sear, 1979).

The second name '*Antimachus*' perhaps belongs to Ἀντίμαχος Β′ or Νικηφόρος, the son of Θεός Ἀντίμαχος. That Antimachus was the next Greco-Bactrian king, who ruled on a vast territory from the Hindu-Kush to the Punjab around 170 B.C.

Since the territorial conquest of Arachosia, Gandhāra, Punjab, and Mathura by the Greek kings, there have been used the following two official languages: (i) Greek for edicts, trading, taxes, and all other things related to secular and political matters; (ii) Gāndhārī (one of the Prakrits) in the Kharoṣṭhī script (this script is obviously of the Aramaic origin) just for religious matters: hymns and philosophy (Tarn, 1951). This was particularly visible on the Indo-Greek coins including the coins of Menander. On obverse, we usually see a king portrait, a royal sign or a Greek deity with a Greek legend around it. On the reverse, we observe a Greek deity or a royal sign with a Gāndhārī legend in the Kharoṣṭhī script (Lahiri, 1965; Mitchiner, 1975–1976; Sear, 1979).

Please see the two examples of Menander's coin in Figures 4.2.1 and 4.2.2. On the Indo-Greek coins we can sometimes also find some Buddhist symbols, such as *dharmacakra* (Pāli: *dhammacakka*). In Gandhāra, many Buddhist *stūpa*s and temples were built by the Greeks: Dharmarajika *stūpa* and monastery (Figure 4.2.3), the Double-Headed Eagle *stūpa*, and the Apsidal Temple at Sirkap, etc. The Buddha was often depicted as a character accompanied by some Greek

Figure 4.2.1. Menander I Soter square hemi-obol (ca. 160–130 B.C.). *Obverse*: head of elephant right wearing bell around neck, Greek legend around: ΒΑΣΙΛΕΩΣ/ΣΩΤΗΡΟΣ/ΜΕΝΑΝΔΡΟΥ. *Reverse*: club of Heracles, monogram at right, Kharoṣṭhī legend around: *Maharajasa/tratarasa/Menamdrasa*. *Material*: bronze with the dark green patina. *Weight*: 1.7 gr. *Size*: 15 × 15 mm. The square coin reflects the preference of their Indian subjects for this shape which is similar to the punchmarked coins of the Mauryan Empire.

Figure 4.2.2. Menander I Soter silver drachm (ca. 160–130 B.C.). *Obverse*: diademed bust of king left, seen from behind, holding spear in thrust position in right hand, aegis on left shoulder, Greek legend around: ΒΑΣΙΛΕΩΣ ΣΩΤΗΡΟΣ/ΜΕΝΑΝΔΡΟΥ. *Reverse*: Athena Alkidemos standing left, holding sloping shield on outstretched left arm, hurling thunderbolt with right hand, monogram at right, Kharoṣṭhī legend around: *Maharajasa tratarasa/Menamdrasa*. *Material*: silver. *Weight*: 2.4 gr. *Size*: 17 mm.

Figure 4.2.3. General view of the Dharmarajika *stūpa*. This is one of the oldest Buddhist *stūpa*s in the world, around the 2nd century B.C., i.e., the time of the reign of Menander I Soter. **By courtesy:** Sadeed Arif.

deities: first of all, by Heracles and Erotes (Cupids), see (i) the fragment of a panel showing Vajrapāṇi (Heracles) and other figures attending the Buddha in British Museum, number: 1970,0718.1, with the date from the 2nd century to the 3rd century A.D.; (ii) the winged Cupids holding a wreath over the Buddha in Musée Guimet, number: MG21810, with the date about the 3rd century A.D.; (iii) the stele with scenes from the life of the Buddha and playing Cupids in Lahore Museum, number: G-109, with the date about the 3rd century A.D. All these facts testify to accepting the Buddhist doctrine by the Greeks before the growth of their Empire.

Hence, from the archaeological point of view we can, in fact, conclude that the Greek King Menander might have been a Buddhist follower. The talking of Nāgasena to him was possible, indeed. The only problem is that this conversation should have been in Gāndhārī, not Pāli. Nevertheless, some texts were excavated in Gāndhārī, such as *Dhammapada*, which are known to be written in Pāli, too. Therefore, the *Milindapañha* theoretically can be first in Gāndhārī, in the sacral language of Greco-Buddhists, and then translated into Pāli.

The Bactrian Greeks were well educated and have contributed to a philosophical conversation on a Buddhist subject. So, in Aï Khanum was found a philosophical dialogue (written on skin), close to some Buddhist matters:

col. II.

[———]/[——— οὐ μόνον]/[——— τῶν ἰδεῶν φάμ]εγ/[——— τὰ] αἰσθητὰ ἀλλὰ [κ]αὶ τὰς ἰδέας αὐ-/τὰς ἀλλήλων—φαμέγ/γὰρ εἶπεν—οὐκοῦν [ὡς]/ [αὐ]τὸ αἴτιον τῶ[ν α]ὐ-/[τῶ]ν ου[...] μετίσχει τῶν ὄντων τ[ὰ]ς ἰδέας/ὅπερ καὶ τ[ο]ῦ μετέχειν/τἄλλ[α.]ε τ[ο]ύτ[ων] αἴτια/[..]νο[..]ω[...]/[αἴ]τιον [..... —] [..]ς ἑτέρ[αις] καθ' ἑ-/[κά]στην [ἰδέα]ις εἰ μ[..]/[......] πρὸς τ[..]/[.....] α[....]λα/[......]νειδ[.]ων [.... ἑκά]στην [...]/[......]ω[....]/[.......]ότητος/[.........]πο[..]/[...... αἰσ]θητῶν [.......]α[.] εἰδῶν/[........]νους

col. III.

[———]/[———]/[———] ὥ[στ]ε/διὰ [τούτων τ]ῶν [αὐ]τῶν αἰτίω[ν ———]ον α-/ναγκαῖον ε[ἶναι] τὸ τῆς/μεθέξεως αἴ[τι]ον ἀκί-/νητον γὰρ ἕκαστον/τῶν εἰδῶν διὰ ταῦτά τε καὶ τὸ τὴγ γένεσιν/εἶναι καὶ τὴν φθορὰν/ἀΐδιον τὴν τῶν αἰσθη-/τῶν ἀναγκαῖον — εἶπεν —/ἀλλὰ μὴγ καὶ κυριώ-τατόγ γε καὶ πρῶτον/τῶν αἰτίων δόξειεν/ἂν τοῦ[το] δικαίως [—]/τοῦτο μὲγ γὰρ [αἴτι]ον/πᾶσι καὶ πάσαις ταῖς ἰδέαις [..] ἀ[λλ]ήλων/[...]ω[...]ωι[...]/[.]ε οὐθὲν οὐθενὸς τ[.]/[...]ει[.] αρ[......]/[.....] αὐτω[.——] [.]ε[.]ειν τω [...——]/[...]ων κ[.——]/[.]...[.. ——]/[.. ———]

> col. IV.
> [.]εε[...... — —]/τ[...]αλλ[.]λ[.— —]/[...]μερ[.]νω[..— —]/[..]νε[.]ταξ[.] με[..— —]
> γονωστε κα[..]ε[...]/μανθάνεις γάρ [— — —]/πάνυ γε εἶπε[ν — — —]/ μὴν εἰ γε [.... με]-/τέχει τω[.— — —]
> ἔσται πρω[.— — —]/μὴ μετέχοι [— — —]/ἄλλω [..— — —]/λ[..— — —]/εἶπ[εν — — —]
> τ[— — —]
> (Rapin, 1992).

This text contains some abstract notions such as 'ideas' (ἰδέαις), 'reason' (αιτία), "each individually" (καῦ' ἑκάστην), 'equalities' (ἰσότητος), "the first reason of reasons" (πρῶτον τῶν αἰτίων), etc. Also, there is the following Buddhist description: γένεσιν εἶναι καὶ τὴν φθορὰν ἀΐδιον τὴν τῶν αἰσθη τῶν ἀναγκαῖον, "generating and destructing all the intelligent entities are forever", which hints at the wheel of reincarnation. The phrase "nothing of nothing" (οὐῦέν οὐῦενός) hints at the concept of *śūnyatā*. So, the Early Buddhism influences on this text are not excluded. Then this text is the earliest eclectic document of Greco-Buddhism.

Till now we have no archaeological evidence of a direct Greek influence on Nyāya, but there are some direct proofs of their impact on astrology and geometry in India. So, in Buddhist inscriptions, written in the Kharoṣṭhī script and in the Gāndhārī language, excavated in Gandhāra, and dated to from the 1st century A.D. to the 3rd century A.D., the following eight Greco-Macedonian months have been recently identified after decoding: Ἀρτεμίσιος, Δαίσιος, Πάνημος, Λώιος, Γορπιαῖος, Ἀπελλαῖος, Αὐδυναῖος, and Ξανδικός (Salomon, 2011). It means that at that time the Greco-Macedonian calendar and astrology was accepted by the Gandhāran Buddhists.

In Medieval Hindu astrology there were some loanwords from Greek: (i) an angular sign: κέντρον (Sanskrit: *kendra*; Syriac: *qanṭrōn*); (ii) Sun: ἥλιος (Sanskrit: *heli*); (iii) diameter: διάμετρον (Sanskrit: *jāmitra*; Syriac: *jaθra*), (iv) the planet Jupiter: Ζεύς (Sanskrit: *jyau*); (v) the planet Mars: Ἄρης (Sanskrit: *āra*), etc. More technical terms: (i) a succedent house: *epanaphora* (in Greek), *panaphara* (in Sanskrit); (ii) a cadent house: *apoklima* (in Greek), *apoklima* (in Sanskrit); (iii) a void of course Moon: *kenodromia* (in Greek), *kemadruma* (in Sanskrit); (iv) an application: *sunaphe* (in Greek), *sunapha* (in Sanskrit); (vi) the 10° segments of the ecliptic: *dekanos* (in Greek), *drekanas* (in Sanskrit), etc.

In the *Yavanajātaka* (one of the first books on Hindu astrology that was written by a Greek author, '*Yavana*'), reconstructed by David Pingree (1978), we can observe some Greek patterns of thinking, e.g., a mathematical calculation is called 'inference' (*anumāna*), although in the Hindu tradition mathematics was never based on logic:

> One should find that the number of (lapsed) *tithis* diminished by the number of lapsed *avamas* equals the number of (civil) days which have passed in the *yuga*. There is a seven-fold measure of the planetary week days; in seeking the answer to this, one desires (the use of) inference (*anumāna*) (Pingree, 1978: vol. 1).

The *Yavanajātaka* was a translation from a book written first in Greek and devoted to astrology, and it is considered that this translation was made in the Western Kṣatrapa Empire in the 2nd century A.D. by Yavaneśvara and versified by Sphujidhvaja in the 3rd century A.D. (Pingree, 1981).

First of all the Greek culture had an effect on the social organization of communities in Bactria and Gandhāra. Therefore, in the Gāndhārī language there were some administrative terms loaned from Greek, e.g., (i) *stratega* 'general', 'commander' (στρατηγός); (ii) *meridarkha* 'meridarch' (εριδαρχη), etc.

As I said above, the Western Kṣatrapas and the Kuṣāṇas continued the Greek Hellenization of India (Jongeward et al., 2015; Shrava, 1993). Nevertheless, they stopped using the Greek language officially in the 1st–2nd century A.D. but continued to use the Greek alphabet: the Western Kṣatrapas for a Prakrit and the Kuṣāṇas for Bactrian. The Rabatak inscription of 127 A.D. was found and it contains a very important statement concerning the change of the official language in the Empire from the Greek language to the Bactrian one. So, Kaniṣka the Great (Greco-Bactrian: Κανηþκε) (his accession to the throne is estimated between ca. 90–140 A.D.) was the first who replaced the use of Greek by the 'Aryan' language after the 400-year history of the Greek and Greco-Scythian communities in the north-west of India. In fact, this 'Aryan' language was Bactrian—one of the Middle Iranian dialects with many loan words from Greek. The fragment of this Edict:

> [— — —]νο βωγο στοργο Κανηþκε κοþανο ραþτογο λαδειγο χοαζαοαργο βαγο
> εζνογο κιδι ασ[ο] Νανα οδο ασο οισποανο μι βαγανο ι Ραοδανι αβορδο κιδι ιωγο χþονο
> νοβαστο σαγωνδι βαγανο σινδαδο οτηια ι **ιωναγγο** οασο οζοαστο ταδηια **αριαο** ωσ-
> ταδο αβο ιωγο χþονο αβο [ι] Ιυνδο φροαγδαζο αβο þατριαγγε þαορε αγιτα κοο-
> αδηανο οδο ι ωζοπο οδο [ι Ζ]αγηδο οδο ι Κωζαμβο οδο ι Παλαβοτρο οιδρα αδα αβο ι Ζιριτ-
> (Sims-Williams, 1996).

> 1–3 The year one of Kaniṣka, the great deliverer, the righteous, the just, the autocrat, the god, worthy of worship, who has obtained the kingship from Nana and from all the gods, who has laid down (i.e., established) the year one as the gods pleased.
> 3–4 And it was he who laid out (i.e., discontinued the use of) the **Ionian** speech and then placed the **Arya** (or Aryan) speech (i.e., replaced the use of Greek by the Aryan or Bactrian language).
> 4–6 In the year one, it has been proclaimed unto India, unto the whole realm of the governing class including Koonadeano (Kaundinya) and the city of Ozeno (Ozene) and the city of Zageda (Saketa) and the city of Kozambo (Kausambi) and the city of Palabotro (Pataliputra) and so long unto (i.e., as far as) the city of Ziri-tambo (Śri-Campa) (Mukherjee, 1995).

Since that Kaniṣka edict, the Bactrian language in the Greek script has been used as official for many centuries, even at the Hephthalite time there were some inscriptions in Bactrian. So, many legal documents in Bactrian are found, including (i) the judgments concerned with the sale of agricultural lands, deeds of manumission, and undertakings to solve conflicts; (ii) the land-sale contracts written in two copies, an upper and a lower copy, usually on the same sheet; (iii) the tax receipts presented by storekeepers and millers to the people who had brought in goods. These documents were composed on skin in the way the Greeks had done before (Sims-Williams, 2001). Also, there are some Buddhist texts in Bactrian in the Greek script, made on skin which is untypical for the Indian civilization as such (Sims-Williams, 2007).

The main feature of Indian culture is that there is a huge gap and contrast between the traditional Sanskrit texts and the Indian archaeology. On the one hand, many descriptions, such as the *Mahābhārata*, are not confirmed by archaeology at all. On the other hand, archaeological data reconstructed after excavations have no parallel in Sanskrit texts. For instance, it is unclear why the Buddha was often accompanied by Cupids. What was the *tantra* for these images?

However, in the case of the Pāli tradition the phenomenon of Greco-Buddhism in Gandhāra is confirmed textually. This fact supports the authenticity of the *Milindapañha*—in the meaning that this text was created in Gandhāra indeed. In Sanskrit, there are no phrases on Yavanas in respect to their Buddhist faith. In contrast, the *Mahāvaṃsa* or the *Great Chronicle of Sri Lanka*, composed in Pāli in the late 5th or early 6th century A.D., many times mentions '*Yona*s' (the Greeks) as Buddhists. In particular, in this book, there is a description how Aśoka (ca. 268 B.C.– 232 B.C.), the great Indian emperor of the Maurya Dynasty, supported Moggaliputta-Tissa (ca. 327 B.C.–247 B.C.), his advisor and spiritual teacher, in organizing the Third Buddhist Council and sending *thera*s to the following countries, among which there are mentioned *Yona* (Greco-Bactria), *Kasmīra* (Kashmir), and *Gandhāra*, where later the Greco-Buddhism was founded:

> When the *thera* Moggaliputta, the illuminator of the religion of the Conqueror, had brought the (third) council to an end and when, looking into the future, he had beheld the founding of the religion in adjacent countries, (then) in the month Kattika he sent forth *thera*s, one here and one there. The *thera* Majjhantika he sent to **Kasmīra and Gandhāra**, the *thera*, Mahādeva he sent to **Mahisamaṇḍala**. To **Vanavāsa** he sent the *thera* named Rakkhita, and to **Aparantaka** the Yona named Dhammarakkhita; to **Mahāraṭṭha** (he sent) the *thera* named Mahādhammarakkhita, but the *thera* Mahārakkhita he sent into the country of the **Yona**. He sent the *thera* Majjhima to the **Himalaya country**, and to **Suvaṇṇabhūmi** he sent the two *thera*s Soṇa and Uttara. The great *thera* Mahinda, the *thera*s Iṭṭhiya, Uttiya, Sambala and Bhaddasāla his disciples, these five *thera*s he sent forth with the charge: "Ye shall found in the lovely **island of Laṅkā** the lovely religion of the Conqueror" (Geiger, 1912/1996: 82).

This quote is especially interesting, as the "*Yona* named Dhammarakkhita", i.e., a Greek, is mentioned among the important Buddhist leaders and teachers at the time of Aśoka (3rd century B.C.). This is quite early, because the Greeks had come to the region of India rather since the Indian campaign of Alexander the Great, starting in 326 B.C.

It is said further that the mission of Maharakkhita was successful among the Greeks of Greco-Bactria:

> The wise Maharakkhita who went to the country of the **Yona** delivered in themidst of the people the *Kalakarama-suttanta* [*A.Sch.*: the *Kāḷakārāmasutta, Aṅguttara-nikāya* IV: 24]. A hundred and seventy thousand living beings attained to the reward of the path (of salvation); ten thousand received the *pabbajja* (Geiger, 1912/1996: 85).

The next significant evidence in the *Mahāvaṃsa* about the Greco-Buddhists and their influence and spiritual power among all the Buddhist communities is as follows. One Sinhalese king of Sri Lanka, called Duṭṭhagāmaṇī or Gāmaṇī Abhaya ("fearless Gamini"), who reigned from 101 B.C. to 77 B.C., decided once to build up the Great Stūpa (Pāli: *thūpa*) that is known now as the Ruwanweliseya and Swarnamalee Chetiya and in order to celebrate the festival devoted to opening the Thūpa he invited many hundred thousand representatives of Buddhist communities from different places, including *Kasmīra* (Kashmir), *Alasanda* (the Greek city of Alexandria, a capital of Bactria), *Pallavabhogga* (Wilhelm Geiger means that it is Persia, but it is, most likely, Margiana in today's Afghanistan):

> From various (foreign) countries also did many *bhikkhu*s come hither; what need to speak of the coming of the brotherhood living here upon the island? With eighty thousand *bhikkhu*s from the region of **Rājagaha** came the *thera* Indagutta, the head of a great school. From **Isipatana** came the great *thera* Dhammasena with twelve thousand *bhikkhu*s to the place of the *cetiya*.
> With sixty thousand *bhikkhus* came hither the great *thera* Piyadassi from the **Jetārāma-vihāra**. From the **Mahāvana** (monastery) in Vesālī came the *thera* Urubuddharakkhita with eighteen thousand *bhikkhu*s. From the **Ghositārāma** in Kosambī came the *thera* Urudbammarakkhita with thirty thousand *bhikkhu*s. From the **Dakkhiṇāgiri** in Ujjenī came the *thera* Urusaṃgharakkhita with forty thousand ascetics.
> With a hundred and sixty thousand *bhikkhu*s came the *thera* named Mittiṇṇa from the **Asokārāma** in Pupphapura. From the **Kasmīra** country came the *thera* Utiṇṇa bringing with him two hundred and eighty thousand *bhikkhu*s. The wise Mahādeva came from **Pallavabhogga** with four hundred and sixty thousand *bhikkhu*s, and from **Alasanda** the city of the Yonas came the *thera* Yonamahādhammarakkhita with thirty thousand *bhikkhu*s [*A.Sch.*: *yonanagarā'lasandāso, yona mahādhammarakkhito; thero tiṃsa sahassāni, bhikkhū ādāya āgamā*].
> From his dwelling by the road through the **Viñjhā** forest mountains, came the *thera* Uttara with sixty thousand *bhikkhu*s.
> The great *thera* Cittagutta came hither from the **Bodhimaṇḍa-vihāra** with thirty thousand *bhikkhu*s. The great *thera* Candagutta came hither from the **Vanavāsa** country with eighty thousand ascetics. The great *thera*

> Suriyagutta came from the great **Kelāsa-vihāra** with ninety-six thousand *bhikkhus*. As for the number of the *bhikkhus* dwelling in the island who met together from every side, no strict account has been handed down by the ancients. Among all these *bhikkhus* who were met in that assembly those alone who had overcome the *āsavas*, as it is told, were ninety-six *koṭis*.
> These *bhikkhus* stood according to their rank around the place of the Great Thūpa, leaving in the midst an open space for the king (Geiger, 1912/1996: 193–194).

So, in this narration we are reading how the 30,000 Greco-Buddhists came from Bactria to take part in the festival in Sri Lanka. We should pay attention that Duṭṭhagāmaṇī who organized this festival was almost a contemporary of the same Menander, the King of Greco-Indian Empire with the capital in Gandhāra who became one of the two main characters of the *Milindapañha.* Due to coins and other archaeological facts, we also know that Menander as well as the majority of Hellenized elite of Greco-Bactria at that time or even earlier converted to Buddhism.

If we trust the *Mahāvaṃsa,* then the Greco-Buddhists can have influenced the Pāli Buddhists (Theravādins) since the 3rd century B.C.

Therefore, the *Milindapañha* can be a result of direct influences from Gandhāra; indeed there are some old narrations in Pāli about close contacts among the Pāli *theras* and the Bactrian-Gandhāran Greco-Buddhists at the time of Menander.

The sacred language of the *theras* of Sri Lanka was Pāli, while the sacred language of the Greco-Buddhists was Gāndhārī. In the *Mahāvaṃsa,* it is stated that the Pāli Canon was first written down at the time of the ruler of Sri Lanka, Vaṭṭagāmaṇi Abhaya (reigned from 29 B.C. to 17 B.C.), and due to his direct support:

> He (the king) built the cells of the *vihara* so that a greater number were joined together, for he reflected: "In this way it will be possible to restore them."
> The text of the three *piṭakas* and the *aṭṭhakathā* thereon did the most wise *bhikkhus* hand down in former times orally, but since they saw that the people were falling away (from religion) the *bhikkhus* came together, and in order that the true doctrine might endure, they wrote them down in books.
> Thus did the king Vaṭṭagāmaṇi Abhaya reign twelve years, and, at the beginning, five months beside (Geiger, 1912/1996: 237).

This event is traditionally interpreted as the so-called Fourth Buddhist Council that was held in Sri Lanka in the late 1st century B.C. However, it is known that there was another concurrent Fourth Buddhist Council that was held in Kashmir (Sanskrit: *Kaśmīr*) from the late 1st century A.D. to the early 2nd century A.D. and this council was supported by the emperor of the Kuṣāṇa dynasty, Kaniṣka the Great:

> Lately the king, Kaniṣka, with the honourable Pārśvika, summoning a council of five hundred saints and sages in the country of **Kaśmīr**, they drew up the Vibāshā Śāstra. These were the five hundred bats who formerly dwelt in that decayed tree (Beal, 1906: 117).

The centre of the Kuṣāṇa Empire was in Gandhāra, and the political elite of this empire remained Hellenized (Banerjee, 1961). At the sites of Bactria and Gandhāra, many fragments of the Buddhist manuscripts dated from the 1st to the 5th century A.D. were excavated, and these texts were written mainly in Gāndhārī and Bactrian. Consequently, we can assume that during the Fourth Buddhist Council in Kashmir the Gāndhārī Canon was founded as a parallel to the existing Pāli Canon. It is evidence that the Greco-Buddhism was influential at that time and the *Milindapañha* could have appeared due to some borrowings from the Gandhāran Buddhism in fact.

There is else another fact supporting the authenticity of the *Milindapañha*. In ca. 400 A.D., all the Hellenized Śaka dynasties (the Western Kṣatrapas and the Kuṣāṇas) fell. But they had sponsored Buddhism, and after their fall Buddhism in India faced many troubles: a lot of monasteries were closed and first of all many Buddhist temples were transformed into Hindu temples, mainly into Shaivite ones. Archaeologically, it can be readily seen that Buddhist images started to be replaced by Hindu deities since 400 A.D. and Shaivism became especially popular in the Buddhist regions, such as Kashmir. At the same time, Shaivism extended through the Gupta Empire and became popular among the Tamils.

Since then, all mentions of the Greeks (Yavanas) and Indo-Scythians (Śakas) and all attitudes towards them in Sanskrit were extremely negative. The Yavanas and Śakas are characterized as Kṣatriyas who have fallen to the level of Śūdras or even should be regarded as outcaste:

> *śanakais tu kriyālopād imāḥ kṣatriyajātayaḥ |*
> *vṛṣalatvaṁ gatā loke brāhmaṇādarśanena ca ||*
> *pauṇḍrakāś coḍadraviḍāḥ kāmbojā* ***yavanāḥ śakāḥ*** *|*
> *pāradāḥ pahlavāś cīnāḥ kirātā daradās tathā ||*
> (*Mānavadharmaśāstra* X: 43–44; Shastri, 1983).

> By neglecting rites and by failing to visit Brahmins, however, these men of Kṣatriya birth have gradually reached in the world the level of Śūdras—Puṇḍrakas, Coḍas, Draviḍas, Kāmbojas, **Yavanas**, **Śakas**, Pāradas, Pahlavas, Cīnas, Kirātas, and Daradas (Olivelle, 2006).

From this it follows that the *Milindapañha* could not have been written after 400 A.D.—in an atmosphere of hatred in relation to the Yavanas and Śakas (Karttunen, 2015). Most probably, considering the historical context reviewed above, this text should have appeared between 130 B.C. and 120 A.D., i.e., between the time of Menander and the time of Kaniṣka the Great.

Prior to 400 A.D., the Greeks were considered a prestigious caste within the Kṣatriyas who patronized Buddhism. So, the word *Yavana* or *Yona* often occurs among names of donators at Junnar, Karla, Nasik, and Junagadh caves (the territory controlled by the Western Kṣatrapas). For instance, the Karla cave:

> *dhenukākaṭā* ***yavanasa*** *sihadhayāna thaṁbho dānaṁ.*
>
> (This) pillar (is) the gift of the **Yavana** Sihadhaya from Dhenukākaṭa (Senart, 1905–1906).

Another example from the same cave:

> 1. *umēhanākaṭā* ***yavanasa*** 2. *Viṭasa[m*]gatānaṁ dānaṁ thabho.*
>
> (This) pillar (is) the gift of the **Yavana** Viṭasaṁgata from Umēhanākaṭa (Vats, 1925–1926).

The Nashik cave:

> 1. *sidhaṁ otarāhasa dātāmitiyakasa* ***yoṇakasa*** *Dhaṁmadevaputasa Īdrāgnidatasa dhaṁmātmanā* 2. *imaṁ leṇaṁ pavate tiraṁṇhumhi khānitaṁ abhaṁtaraṁ ca leṇasa cetiyagharo poḍhiyo ca mātāpi* 3. *taro udisa ima leṇa kāritaṁ savabudha-pujāya cātudiśasa bhikhūsaṁghasa niyātitaṁ sa* 4. *ha putena Dhaṁmarakhitena.*
>
> Success! (The gift) of Indrāgnidatta, son of Dhammadeva, the **Yavana**, a northerner from Dattāmitrī. By him, inspired by true religion, this cave has been caused to be excavated in mount Tiraṇhu, and inside the cave a Caityagṛha and cisterns. This cave made for the sake of his father and mother has been, in order to honour all Buddhas, bestowed on the universal Saṁgha of monks, together with his son Dhammarakhita (Senart, 1905–1906).

Let us pay attention that a Greek donator is designated in these inscriptions in the following two forms: either *yoṇakasa* (the plural Prakrit form) or *yavanasa* (the plural rather Sanskrit form). In these caves, only the class of Greek men has a plural form to designate its singular representative. There is only one other class of people, who is mentioned in the plural form, too, it is a class of religious leaders, such as *thera*s. Hence, the plural form is used rather to express a deep respect to a man. Therefore, we can assume that the word *yavanasa* (or *yoṇakasa*) was used in these inscriptions to indicate a prestigious caste of Greeks. Since 400 A.D. it would be impossible absolutely. It is worth noting that in the *Milindapañha*, it is assumed that *yonaka* (i.e., *yoṇakasa* or *yavanasa*) is the highest *varṇa*. This statement is drawn from the following order of *varṇa*s: Yavanas (*yonaka*), Kṣatriyas (*khattiya*), Brāhmaṇas (*brāhmaṇa*), Vaiśyas (*gahapati*), please see:

> ***yonaka****sukhumāliniyopi* ***khattiya****sukhumāliniyopi* ***brāhmaṇa****sukhumāliniyopi* ***gahapati****sukhumāliniyopi* (*Milindapañha* III: 4, 6).

Thus, the *Milindapañha* is an authentic book, indeed: first, Menander (one of its two main characters) was real; second, its narrative satisfies all the historical contexts, e.g., to be *yavana* before the 2nd century A.D. was honorable for Buddhist communities in fact.

4.2.6 Proto-Nyāya Doctrine of Pramāṇa in the Milindapañha

Let us now show that the *Milindapañha* or the *Questions of King Milinda* contains some obvious contextual references to the *pramāṇa* teaching. The monk bearing the name of Nāgasena talked to the Greek King Menander I Soter, the ruler of Indo-Greek Empire, and explained to him all the aspects of Buddhism by using different syllogisms.

One of the main differences of the *Milindapañha* and its syllogisms from the *Kathāvatthu* and its syllogisms is that each major premise used for drawing conclusions and mentioned in the *Milindapañha* has or presupposes an illustration to give verification. Let us provide an example from this text:

> The King said: "Revered Nāgasena, have you seen the Buddha?"
> "No, sire."
> "Then have your teachers seen the Buddha?"
> "No, sire."
> "Well then, revered Nāgasena, there is no Buddha."
> "But have you, sire, seen the river Ūhā in the Himalayas?"
> "No, revered sir."
> "Then has your father seen it?"
> "No, revered sir."
> "Well then, sire, there is no river Ūhā."
> (Horner, 1969, vol. 1: 95).

In this fragment, Menander affirms that the Buddha does not exist, because he has not been seen by Nāgasena or his teachers, i.e., he is not given by our perceptions in the meaning of *pratyakṣa* ("underlying things", 'evidence') of the Nyāya philosophy. However, Nāgasena exemplifies by illustration that there is a knowledge that is obtained by inferences in the meaning of *anumāna* of Nyāya. For instance, as he said further, if somebody sees a pleasant city, well planned out, he knows just by sign-inference (σημείωσις), see Subsection 1.2.4, how great the founder was, but he has never seen this founder. Hence, the knowledge that the Buddha exists (existed) is given by inference, not perception. And Nāgasena appeals to this term, *anumāna*, to prove the Buddha's existence.

It is worth noting that according to Dignāga there are only the following two real sources of knowledge: *pratyakṣa* ('evidence'; Pāli: *paccakkha*) and *anumāna* ('inference'). This thesis is assumed in the *Milindapañha*, also. For instance, the existence of the Buddha is deduced and in this inference we do not need recourse to *śabda* ('authority') or perception. So, the author of the *Milindapañha* is focused on the problem of the Buddha's existence to show that there are the two sources of our knowledge: perception and inference, e.g., the Buddha is not given to our perception, but he is deduced logically from signs/tokens (σημεῖον or *liṅga*), see Subsection 1.2.4.

The term of *paccakkha* occurs quite often in the Pāli Canon, but never in the logical meaning of the Nyāya philosophy. In most cases, its meaning is "this is evident" or "this is known". For instance:

> *sāvatthinidānaṃ. "sabbaṃ vo, bhikkhave, desessāmi. taṃ suṇātha. kiñca, bhikkhave, sabbaṃ? cakkhuñceva rūpā ca, sotañca saddā ca, ghānañca gandhā ca, jivhā ca rasā ca, kāyo ca phoṭṭhabbā ca, mano ca dhammā ca—idaṃ vuccati, bhikkhave, sabbaṃ. yo, bhikkhave, evaṃ vadeyya: 'ahametaṃ sabbaṃ* ***paccakkhāya*** *aññaṃ sabbaṃ paññāpessāmī'ti, tassa vācāvatthukamevassa; puṭṭho ca na sampāyeyya, uttariñca vighātaṃ āpajjeyya. taṃ kissa hetu? yathā taṃ, bhikkhave, avisayasmin"ti* (*Saṃyuttanikāya*, 1 *Saḷāyatanasaṃyuttaṃ* 23 [*Sabbasuttaṃ*]).

At Savatthi. "*Bhikkhu*s, I will teach you the 'all'. Listen to that..."
"And what, *bhikkhu*s, is 'the all'? The eye and forms, the ear and sounds, the nose and odours, the tongue and tastes, the body and tactile objects, the mind and mental phenomena. This is called the all."
"If anyone, *bhikkhu*s, should speak thus: "Having rejected this all, I **shall make known** [*paccakkhāya*] another all"—that would be a mere empty boast on his part. If he were questioned, he would not be able to reply and, further, he would meet with vexation. For what reason? Because, *bhikkhu*s, that would not be within his domain" (Horner, 1969, vol. 2: 1140).

In the *Kathāvatthu*, premises involved in reasoning are not evident (in the common meaning as well as in the meaning of the *nyāya* word *pratyakṣa*), i.e., their semantics remains unclear for us. In both the *nyāya* logic and in the Buddhist logic, well-expressed by Dignāga and Dharmakīrti, there is a requirement to illustrate reasoning by an example. This requirement to give an illustration is called *udāhārana* in Sanskrit. The comparison between the axiom, *upanayana* (observed phenomenon), and the provided example is called *upamā*. In turn, the comparison allows us to generalize the reason or condition for the observed phenomenon that is called *hetu* and to draw a conclusion that is called *nigamana*, by which we have applied the condition to the observed phenomenon to explain the latter. Let us consider the following classical example of the Hindu logic, also found in the Mahāyānist logic:

1) The *udāhārana*:

   yo yo aggimā so so dhūmāva. —Whatever is fiery, is smoky.
2) the *upamā* (metaphor) may be introduced:

   Smoky like a hearth.
3) The *upanayana*:

   ayam pabbato dhūmāva. —This hill is smoky.
4) The *niggamana*:

   tasmādayam aggima. —Therefore, it is fiery.
5) Smoky in (1), (2), and (3) is the *hetu* (condition or reason).

This reasoning differs a lot from all the syllogisms of the *Kathāvatthu*, because it assumes a verification procedure by *udāhārana* and *upamā*, i.e., by an example and an illustration, respectively.

Let us notice that the term *udāhārana* as a requirement to give an example in proofs occurs in the *Milindapañha* among the standard terms *naya* and *hetu* denoting logical reasoning as such:

ye pana te, mahārāja, bhikkhū navaṅgaṃ buddhavacanaṃ atthato ca byañjanato ca ***nayato*** *ca* ***kāraṇato*** *ca* ***hetuto*** *ca* ***udāharaṇato*** *ca vācenti anuvācenti bhāsanti anubhāsanti, evarūpā kho, mahārāja, bhikkhū bhagavato dhammanagare 'dhammāpaṇikā'ti vuccanti* (*Milindapañha* V: 4, 1).

> And those monks, sire, who teach and recite, speak, and repeat the nine-limbed speech of the Buddha in its literal senses and developed meanings, with its **methods** and **reasons** and **causes** and **examples**—monks such as these, sire, are called sellers of Dhamma in the Lord's City of Dhamma (Horner, 1969, vol. 2: 197).

However, instead of *udāhārana* and *upamā* mentioned in the introductory words of the *Milindapañha*, in the main body of the book the author just uses one term denoting a requirement to give illustrations and examples—*opamma* that is exposed in the very beginning of the text:

> Plunging into Further-Dhamma and Vinaya, deliberating the net of the Suttas,
> Nāgasena's talk was varied with **similes** and in the **method**.
> Aspiring to knowledge herein while gladdening the mind,
> Hearken to the abstruse questions, dissipating occasions for doubt (Horner, 1969, vol. 1: 1).
> *abhidhammavinayogāḷhā,*
> *suttajālasamattitā;*
> *nāgasenakathā citrā,*
> ***opammehi nayehi*** *ca.*
> *tattha ñāṇaṃ paṇidhāya,*
> *hāsayitvāna mānasaṃ;*
> *suṇātha nipuṇe pañhe,*
> *kaṅkhāṭṭhānavidālaneti*
> (*Milindapañha* I: 1, 1).

Also, it is important to point out that 'fiery' is regarded as a sample for *hetu* in the *Milindapañha*:

> "Suppose, sire, there were no (lower) piece of wood for making fire, no twirling-stick, no cord for the twirling-stick, no upper piece of wood for making fire, no little piece of cloth (for tinder), and no proper effort on the part of a man—would a fire be produced?"
> "O no, revered sir."
> "But if, sire, there were a (lower) piece of wood for making fire, a twirling-stick, a cord for the twirling-stick, an upper piece of wood for making fire, a little piece of cloth, and proper effort on the part of a man—would that fire be produced?" "Yes, revered sir, it would be produced" (Horner 1969, vol. 1: 74–75).

Later this 'fiery' became classical for the Hindu logic. It turned into the most popular example for *hetu* in the Indian thought.

Hence, the *Milindapañha* is the only book of the whole Pāli Canon where we can find some references to a true proto-Nyāya logic with a kind of semantics. Nevertheless, this 'Hindu' logic of the *Milindapañha* is much more archaic than the logic of the *Nyāya-sūtra* because it only contains hints of the following two *pramāṇa*: *paccakkha* ('evidence') and *anumāna* ('inference'), and instead of the two ways of verification called *udāhārana* and *upamā* there is just one way for verification called *opamma*. The existence of the Buddha is the main example for *anumāna* provided in the *Milindapañha*. It is quite surprising, as his existence is not established by authority, which is what might be expected, but by sign-inference, just logically.

In the *Milindapañha*, as well as it being in accordance with Dignāga's doctrine, all the first data are being collected by the *pratyakṣa* or *paccakkha*—states of affairs. In the Pāli text, this mechanism of collecting facts is quite sophisticated and it corresponds to the Pāli *abhidhamma* that may be considered the Buddhist way to epistemology. There are the following five means of sense: 'eye-door' (*cakkhu-dvāra*); 'ear-door' (*sota-dvāra*); 'nose-door' (*ghāna-dvāra*); 'tongue-door' (*jivhā-dvāra*); 'body-door' (*kāyadvāra*). These five groups of the characteristic marks of individuality (*upādānakkhandhā*) are united in a form of object (*rupakkhandhā*). This form can give rise to sensory contact, leading to feeling (*vedanā*). Due to this, we can develop an idea of object (*saññā*). From this, some mental potentialities, or conditions (*saṅkhāra*) can grow up. At the end, a consciousness (*viññāṇa*) takes all this in:

> And again, sire, the cat seeks after its food only in what is near; even so, sire, the yogin, the earnest student of yoga must dwell beholding the rise and fall among the five groups of grasping, thinking: "This is material shape, this the arising of material shape, this the going down of material shape; this is feeling, this the arising of feeling, this the going down of feeling; this is perception, this the arising of perception, this the going down of perception; these are the habitual tendencies, this the arising of the habitual tendencies, this the going down of the habitual tendencies; this is consciousness, this the arising of consciousness, this the going down of consciousness." This, sire, is the second quality of the cat that must be adopted. And this, sire, was said by the Lord, the *deva* above *deva*s:
> "One should not be far from here (or) how will one produce the Acme of Becoming?
> In the actual present, know your own body" (Horner, 1969, vol. 2: 269).

This epistemology is very similar to the transcendental-phenomenological reduction proposed by Edmond Husserl (1859–1938) so much later: we should start with analyzing pure phenomena and, as a result, we move to ourselves. "*Zu den Sachen selbst*" (go back to pure items as we go home to ourselves) was an appropriate famous philosophical motto in German.

Thus, based on the textual analysis of the Pāli Canon, we can draw the following conclusions (Schumann, 2018a, 2019c, 2020b):

1. Until the 1st century A.D., i.e., at the time when the Pāli Canon was established in its present form, the *Nyāya-sūtra* was not written yet and the Nyāya school of Hindu philosophy did not exist. Reasons: (i) there are no references to this school at all; (ii) the term *nyāya* (*ñāya*) is used in the meaning of the method of Buddhists distinguishing them from non-Buddhists; (iii) in the *Milindapañha*, the most logical book of early Pāli literature there are logical ideas which are more archaic than the ideas of the *Nyāya-sūtra* (the requirement of illustration

is simpler in the *Milindapañha* and there are only two sources of knowledge: *paccakkha* ('evidence') and *anumāna* ('inference')).

2. At the time of the early *sutta*s of the Pāli Canon, such as the *Kālāma-sutta*, the *Sutta-nipāta*, the *Jātaka*, and the *Mahāvagga* there was an art of debates (the so-called proto-logic) with the following two aspects of logical reasoning which were preserved later in Hindu logic, as well: (i) *naya* or defining premises for inferring; (ii) *hetu* or defining reasons or conditions for inferring.
3. In the early *sutta*s of the Pāli Canon all the attitudes towards this art of debates are rather sceptical and negative.
4. For the first time, some well-done logical syllogisms occurred in the *Kathāvatthu*. But this treatise also contains evident sophisms and there is no semantics for logical reasoning, i.e., there are no illustrations (*udāhāranas*) and examples (*upamās*) for verifying propositions. The *Kathāvatthu* includes debates with possible Sautrāntikas and Vaibhāśikas (the Buddhists from Gandhāra and Kashmir). This means that this text could have appeared quite late, e.g., it can be dated to the 1st century A.D., and it can be written under the influence of discussions with the northern (Gandhāran) Buddhists.
5. The only book of early Pāli literature that is logical indeed in all meanings is the *Milindapañha*. It is a unique document, because it represents a dialogue with Menander, the king of the large empire in the North and the political leader of all Bactrian-Gandhāran Buddhists. Considering that this book is entirely different from any other book from the time of the Pāli Canon, we can safely claim that it was written under a direct influence of the Greco-Buddhists or Bactrian-Gandhāran Buddhists. Reasons: (i) we can detect a strange dynamic in attitudes towards logic from negative ones in early texts to a neutral attitude in the *Kathāvatthu* and even to a positive attitude in the *Milindapañha*; (ii) the *Milindapañha* demonstrates a real logic that cannot be deduced from the other Pāli books even terminologically and has no analogues with the classical *nyāya*. So, the *Milindapañha* can have a non-Indian influence defined by us as the Greco-Buddhist one.

As we see, the Buddhist logic of the *Milindapañha* with the two sources of knowledge, *paccakkha* and *anumāna*, and with *opamma* as one way for verifying propositions that came from the Greco-Buddhist syncretic culture that flourished in Gandhāra.

4.2.7 Conclusions

As a result of the structuralist analysis of logical competence in early Pāli literature, the following statements can be inferred (Schumann, 2018a, 2019c, 2020b):

1. In the Pāli Canon there was a tradition of Buddhist logic, but this tradition was weak, and the (proto-)logic we can reconstruct on the basis of the Pāli texts by means of the historical reconstructive hermeneutics, which can be evaluated as a predecessor of the *nyāya* and *yogācāra* logic.

2. At the time of the Pāli Canon there did not exist the *nyāya* philosophy known by the *Nyāya-sūtra.*
3. The *Milindapañha*, the best logical source of the Pāli Canon, could have been written under a direct influence of the Greco-Buddhists.
4. From the viewpoint of the Pāli Canon, the origin of Indian logic is connected to the community that the author of the *Milindapañha* belonged to, and this community was Greco-Buddhist. Therefore, we can claim that the first correct application of inference rules in the early Indian logic may be explained by a Gandhāran influence.
5. In the *Milindapañha*, we find the earliest set of logemes of Indian logic.

4.3 Two Systems of Modal Logic in Antiquity: Aristotelian and Stoic, and the Possible Influence of the Stoic Modal Logic on the Schools of Madhyamaka, Yogācāra, and Nyāya

4.3.1 Problem Setting

The development of prose as a new literary genre (especially historical prose), which arose for the first time in Ancient Greece from the 6th century B.C., was possible in the context of logical intuition about the existence of random events. An event *A* is considered in logic to be contingent if the judgment "Possibly *A* and possibly not-*A*" is true. The Old Greek historical prose became the cultural context for the Aristotelian system of modal logic presented in his treatise *On Interpretation* (Περί ερμηνείας/*De Interpretatione*). This logic is formalized (axiomatized) now in the form of a system **D** and its extension **T**. In this logic, the existence of random (contingent) events is allowed.

The philosophers of Early Stoa (from the late 4th century B.C. to the mid-2nd century B.C.) were inspired by another cultural context—not historical prose, but the Babylonian art of divination (see Subsection 4.1.5), and they denied the existence of contingent events. As a result, for the first time they formulated the idea of logical determinism in an explicit form, namely, they began to consider the modal judgment "Either necessarily *A* or necessarily not-*A*" always true. This modal logic is now axiomatized in the form of the system **CD** and its extension **K=**. Note that in the system **T** the judgment "If *A* is necessary, then *A*" is always true, while in the system **K=** the judgment "If *A*, then *A* is necessary" is always true.

Both systems, both **D** and **CD**, are extensions of the same basic modal logic **K**, but additionally they include different modal judgments as always true. So, there is no similarity relation between the structures of the logemes of Aristotle's modal reasoning and of the modal reasoning of the philosophers of the Early Stoa. These are alternative versions of modal logic.

Basic philosophical assumptions of Stoic logic: (S_1) nothing happens without a reason; (S_2) every sentence *A* is either true or false—the judgment "either *A* or not-*A*" is always true; (S_3) every sentence *A* either must be true or must be false—the modal judgment "It is either necessarily *A* or necessarily not-*A*" is always true; (S_4) each (past, present, future) event *A* is actual, and therefore necessary—the modal proposition "If *A*, then *A* is necessary" is always true; (S_5) everything happens according to destiny.

This section shows that these assumptions from (S_1) to (S_5) give a logeme structure (explicable in the logics **CD** and **K=**), which is similar to the logeme structure induced by the philosophical assumptions of the *yogācāra* logic: (Y_1) the principle of *arthakriyātva* (causal efficiency) operates as a criterion for being real; (Y_2) each statement (*sādhya*) that the property *P* (*dharma*) belongs to the subject *S* (*dharmin*) either takes place or does not take place: either "*S* is *P*" or not that "*S* is *P*"; (Y_3) each statement (*sādhya*) that the property *P* (*dharma*) belongs to the subject *S* (*dharmin*) either necessarily expresses the connection (*svabhāva*) between *S* and *P*, or necessarily express its negation: it is necessary that "*S* is *P*" or it is necessary that it is not that "*S* is *P*"; (Y_4) each (past, present, future) event *A* is real (momentary, instantaneous, *kṣaṇika*), and therefore necessary—"If *A*, then *A* is necessary"; (Y_5) everything happens momentarily (*kṣaṇikatva*).

The structure of logemes (formalized in the logics **CD** and **K=**) is also induced in the philosophical reasoning of the Nyāya school. The difference from Yogācāra consists only in a different philosophical foundation of logical determinism. In Yogācāra, for each event *A*, there is a condition: "*A* is necessary, or not-*A* is necessary", but this is explained by the instantaneousness (*kṣaṇikatva*)—at one moment the cause, its action, and their observer meet each other. In Nyāya, the same condition "*A* is necessary, or not-*A* is necessary" for each *A* is explained by the fact that God (*Īśvara*) is the supreme observer of all non-empirical objects (*adṛṣṭa*) that are involved in explaining causality.

Logical determinism (and hence the logics **CD** and **K=**) are also implied in the reasoning of the Madhyamaka school with only one proviso—logical determinism takes place at the level of ordinary reality (*saṁvṛti-satya*), but not at the level of absolute reality (*paramārtha-satya*), at which logical laws cease to operate.

So, this section demonstrates that Aristotle implicitly used the logic **D** and its extension **T** in his modal reasoning, the philosophers of the Ancient Stoa, as well as the philosophers of the Indian schools of Madyamaka, Yogācāra, and Nyāya implicitly used the logic **CD** and its extension **K=**. There is a strong similarity between the structures of the logemes of modal reasoning of the early Stoics, on the one hand, and the Indian schools of logic, on the other hand.

Let us remember that the treatise *On Interpretation* (Περί ερμηνείας/*De Interpretatione*) written by Aristotle is the first source where well-defined logical relationships among different modal propositions are found, such as "It is possible that *A*" (symbolically from the point of view of modern modal logic: $\Diamond A$) or "It is necessary that *A*" (symbolically: $\Box A$). Recently, these relationships are formalized by the mathematicians within modal logic **D** and its extension **T**, see (Garson, 2006). Hence, we can state that Aristotle knew modal logic in the version of **T**, indeed (see Subsections 4.3.3 and 4.3.5). He was the first who understood it well and was able to describe its main properties, but without its formalization and axiomatization made just recently. It can be readily shown by many quotations from the Aristotelian text (see Subsections 4.3.3 and 4.3.5).

There was also a Stoic version of modal logic, but it has not been formalized as yet because there are a few Stoic fragments on modal reasoning to be formalized unequivocally. Nevertheless, we see that their modal logic was non-Aristotelian surely and it can be treated within the modal logic **CD** and its extension **K=**, see (Garson, 2006)—as I show in this section (see Subsections 4.3.4 and 4.3.6).

The difference is that in **T** the statement "If it is necessary that *A*, then *A*" ($\Box A \Rightarrow A$) is one of the main axioms, while in **K=** an appropriate converse "If *A*, then it is necessary that *A*" ($A \Rightarrow \Box A$) is a key axiom. The first statement assumes that there is a contingency—something that appears occasionally. The second statement supposes a logical determinism—if something exists, it exists necessarily. Aristotle follows the first modal axiom, while the Stoics share the second one (Subsection 4.3.6).

In this section, I show that the early Madhyamaka and Yogācāra support a kind of logical determinism, too, therefore their reasoning can be formalized within the modal logic **K=**, as well (see Subsection 4.3.6). So, it can be done in the way of Stoics and versus the Aristotelian intuition on modal logic.

Traditionally, Nyāya and Mīmāṃsā, the Indian logic and hermeneutics, have been considered a tradition beyond any modal logic, because we do not have clear direct Sanskrit analogies with expressions "It is possible that *A*" or "It is necessary that *A*" in the meaning of **T** or **K=**. However, as I show in this section, there are many texts of early Madhyamaka and Yogācāra with clear modal reasoning, but not in the way of Aristotle (i.e., not in the way of logic **T**, but within the logic **K=**). For instance, this reasoning is well presented in Nāgārjuna's *Īśvara-kartṛtva-nirākṛtiḥ-viṣṇoḥ-ekakartṛtva-nirākaraṇa*—its text and translation are contained in (Stcherbatsky, 1969). This short treatise is to propose a kind of critique of monotheism (the concept of *Īśvara*) by appealing to some modal relationships which can be formalized even mathematically within **K=** (see Subsection 4.3.7). Later these became traditional for Madhyamaka and Yogācāra and there are many other texts on the same subject that can be symbolically reconstructed within the modal logic **K=**. This logic is of interest because there are some logical and terminological similarities to Stoic modal reasoning. For example, Nāgārjuna's term *siddha* can be understood as a Stoic "to be resulted", "to be fated", or "to be necessitated".

In the *Īśvara-kartṛtva-nirākṛtiḥ-viṣṇoḥ-ekakartṛtva-nirākaraṇa*, Nāgārjuna's examples for the modality "It is impossible that it is *A*" (or "It is necessary that it is not *A*": "oil [crushed] out of sand, which is known to us as nonexistent; wool on a tortoise, which is known to us as non-existent" (Stcherbatsky, 1969: 9). Using these examples, Nāgārjuna shows that if the Creator (*Īśvara*) exists, He creates something from nothing, such as "wool on a tortoise". It means that He can make 'possible' from 'impossible'. The assumption that 'possibility' and 'impossibility' exist simultaneously gives a contradiction in **K=** and then it should be rejected. Therefore, the Creator does not exist, too. The assumption of His existence should be rejected, also, as entailing a contradiction between the modalities 'possible' and 'impossible'.

This short Nāgārjuna's argumentation is correct from the point of view of logic **K=**. All existent items are necessary (*siddha*): "If *A*, then it is necessary that *A*". Among non-existent items there are items *B* which are possible: "It is possible that *B*". From the statement "If *A*, then it is necessary that *A*" it follows logically that "If it is possible that it is not *A*, then non-*A*". These possible items cannot be created, because the conjunction "it is possible that non-*A* **and** it is necessary that *A*" is always false in **K=**. Thus, we can claim that Nāgārjuna knew some relationships of modal logic **K=**, but without its formalization and axiomatization made just recently.

He became the father of Indian rational thinking, indeed. Usually, in the secondary Western literature, Nāgārjuna is interpreted as a skeptic philosopher who avoids logic. But he does it only for an especial philosophical topic, not in general. He always starts with logical reasoning. This misunderstanding causes some mistakes in his translations. Nāgārjuna has stood at the gates of the beginning of logical discourse in India, while Aristotle has been at the forefront of the same in Greece. For instance, in a categorical proposition "*S* is *P*", Aristotle was one of the first in Greece who defined and used subject (τὸ ὑποκείμενον) *S* and predicate (τὸ κατηγορούμενον) *P* logically correctly and Nāgārjuna was one of the first in India who defined and used subject (*dharmin*) *S* and predicate (*dharma*) *P* perfectly from the point of view of logic to the same extent: (i) the subject *S* of categorical proposition "*S* is *P*" should be expressed by a non-empty name (for some *A*, *A* is *S*); (ii) the subject *S* of categorical proposition "*S* is *P*" should be a subset of *P* (for any *A*, if *A* is *S*, then *A* is *P*). Both properties may be shown by Venn diagrams:

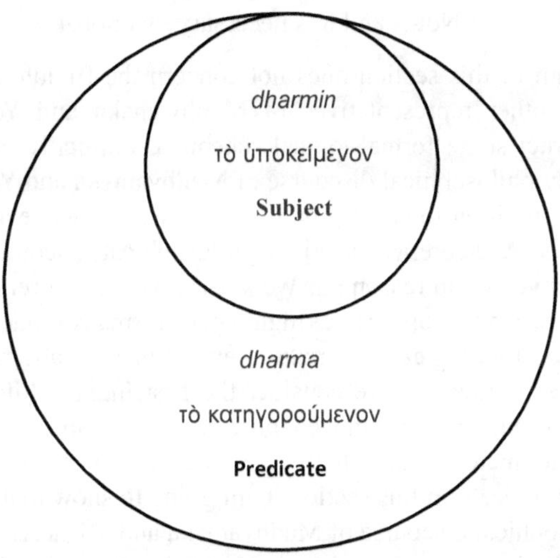

Nāgārjuna possessed a perfect logical competence, and his treatises should be translated with explicating his logical knowledge, too. For example, the *Mūlamadhyamakakārikā* contains many categorical statements in the form "*S* is *P*", see (Tachikawa, 1981: 11). So, the statement from this treatise (*Mūlamadhyamakakārikā* VII: 20), see de La Vallée Poussin (1960):

> *sataś ca tāvad utpattir asataś ca na yujyate* |
> *na sataś cāsataś ceti....* ||.

was understood by Musashi Tachikawa as a conjunction of the following three categorical propositions: (i) *sato notpattiḥ* ["There is no arising of that which exists";

sat (that which exists) is not *utpatti* (arising); it is false that if *sat*, then *utpatti*]; (ii) *asato notpattiḥ* ["There is no arising of that which does not exist"; *asat* (that which does not exist) is not *utpatti* (arising); it is false that if *asat*, then *utpatti*]; (iii) *sadasato notpattiḥ* ["There is no arising of that which both exists and does not exist"; *sadasat* (that which both exists and does not exist) is not *utpatti* (arising); it is false that if *sadasat*, then *utpatti*]. In this passage, Nāgārjuna claims that *utpatti* is not a property of *sat*, it is not a property of *asat*, and it is not a property of *sadasat*. Hence, *utpatti* cannot be inferred from *sat*, *asat*, and *sadasat*. If we understand *utpatti* of A modal-logically as "it is possible that A", *sat* of A as 'A' and *asat* of A as 'not-A', then Nāgārjuna's reasoning is formalized as follows:

1) $\neg (A \Rightarrow \Diamond A)$ It is false that if A, then it is possible that A
2) $\neg (\neg A \Rightarrow \Diamond A)$ It is false that if not-A, then it is possible that A

Proposition (1) is always false in the modal logic **T**, because $A \Rightarrow \Diamond A$ is an axiom of that logic, but (1) can be true in the modal logic **K=**. Statement (2) can be true in **K=**, also. Thus, we see that Nāgārjuna has a non-Aristotelian intuition on modal relationships. In **K=**, from (1) we can logically infer (3) and from (2) we can infer (4):

3) $A \,\&\, \Box\neg A$ A and it is necessary that not-A
4) $\neg A \,\&\, \Box\neg A$ Not-A and it is necessary that not-A

Hence, the aim of this section does not concern the Buddhist doctrine taught by Nāgārjuna or other representatives of Madhyamaka and Yogācāra, but the aim is to reconstruct some formal-logical schemata, implicitly or even explicitly applied in the early philosophical discourse of Madhyamaka and Yogācāra, to show that they have an intuition towards a modal logic with strict relationships among modal propositions. A theoretical level of philosophical discourse depends upon logical schemata involved in reasoning. We know that the theoretical level of post-Aristotelian Old Greek philosophers was high. There are many modern formalizations of Aristotelian and Stoic logical schemata to emphasize the abstract level of these ancient philosophers. However, we consider the first Indian philosophers such as Nāgārjuna as mystic authors still and we ignore the fact that many of them possessed a perfect logical competence, and they were able to engage very difficult logical schemata correctly indeed. In this section, I am going to show that since Nāgārjuna in the early philosophical discourse of Madhyamaka and Yogācāra, there has been a well-expressed intuition on the modal logic **K=**. This proves the claim that the first Indian philosophers (of the 2nd century A.D.—the time of Nāgārjuna, if we follow the early Chinese Buddhist chronics, see (Beal 1884/1981: 302–303)) had a theoretical level that was not worse than the level of post-Aristotelian ancient philosophers. The difference in respect to modal reasoning was that in India the modal logic **K=** was preferable in thinking, but in Greece the modal logic **T** (Schumann, 2018b, 2021c).

4.3.2 Madhyamaka and Logic

There were some recent attempts of analyzing the Nāgārjunian logical discourse worth mentioning: (i) explicating his logical schema called *catuṣkoṭi* (enumerating four alternatives: "something holds, it doesn't hold, it both holds and fails to hold, it neither holds nor fails to hold") (Westerhoff, 2005); (ii) explicating his argument "neither one nor many" (Tillemans, 1984); (iii) reconstructing his semantic views including his approach to the problem of empty names and the relation between language and momentariness (Kantor and Salvini, 2019; Westerhoff, 2019). These discoveries show that, in fact, logical reasoning played a significant role in the Nāgārjunian philosophy.

Let us quote the following statement from *Mūlamadhyamakakārikā* XXIV: 10, see (de La Vallée Poussin, 1960) showing that according to Nāgārjuna, logic is, indeed, necessary for understanding the Buddhist doctrine:

> *vyavahāram anāśritya paramārtho na deśyate* |
> *paramārtham anāgamya nirvāṇaṃ nādhigamyate* ||.

> Without relying on words [*vyavahāra*], the ultimate [*paramārtha*] is not taught. Without comprehending the ultimate [*paramārtha*], *nirvāṇa* is not attained.

The same verse in Tibetan, see (Ui et al., 1934), N 3824:

> *tha snyad la ni ma brten par* ||
> *dam pa'i don ni bstan mi nus* ||
> *dam pa'i don ni ma rtogs par* ||
> *mya ngan 'das pa thob mi 'gyur* ||

In the commentary on the *Mūlamadhyamakakārikā* entitled the *Prasannapadā* (de La Vallée Poussin, 1960) and written in the 7th century by the Buddhist master Candrakīrti, the term "words [*vyavahāra*]" is treated as the "level of conventional reality [*saṁvṛiti-satya*]":

> *tasmān nirvāṇādhigamopāyatvād avaśyam eva yathāvasthitā saṁvṛtir ādāv evābhyupeyā bhājanam iva salilārthineti* ||10||.

> So, if it is the means (*upāya*) to attain *nirvāṇa*, then the conventional [*saṁvṛti*] should be necessarily used from the beginning, as it happens to be; just like a container is used by someone who wants water (*Prasannapadā* 24: 10).

It is worth noting that this phrase of Candrakīrti is composed as a logical reasoning supported by *upamā* (metaphor): "like a container is used by someone who wants water"—this *upamā* is a necessary step in inferring, according to the Yogācāra and Nyāya logic (Vattanky, 2003). It corresponds to the requirement of *opamma* in the *Milindapañha*.

Hence, by this commentary, Candrakīrti postulates following Nāgārjuna that to reach the "level of ultimate reality [*paramārtha-satya*]" we should go first through the "level of conventional reality [*saṁvṛti-satya*]". At the end, the *paramārtha-satya* shall be revealed as the emptiness of all phenomena—as the *śūnyatā* of *vyavahāra* or *saṁvṛti* (Sebastian, 2016: 62). But in the beginning, we should investigate all the phenomena through logic—that is, through a "valid cognition of empirical reality [*vyāvahārika-pramāṇa*]", because this cognition is "a cause of the correct cognition of the *paramārtha*" (Seyfort Ruegg, 2000: 280). The idea that the "level of conventional reality [*saṁvṛti-satya*]" concerns only words and their correct usage is well expressed by Buddhapālita, another early commentator on the Nāgārjunian *Mūlamadhayamakakārikā*, please see his *upamā* (metaphor):

> As two villagers were passing through a city on business, they entered a temple to take in the sights. As they began examining the paintings, one remarked, "The one holding the trident is Narāyāna; the one with the discus is Maheśvara." The other answered, "You have it wrong. Maheśvara holds the trident and Narāyāna has the discus." As they argued, they came upon a nearby wandering sage. They paid their respects, and each explained his opinion. To one of them the sage replied, "What you say is true," and to the other he said, "What you say is not true." (The Cowherds, 2011: 7).

This metaphor (called *opamma* in the *Milindapañha*) was to demonstrate that, according to the conventions (*vyavahāra*), Maheśvara holds the trident and Narāyāna holds the discus, but not vice versa, although they do not exist as such to the same extent. Therefore, something is either true or false only due to some conventions sharing by us and logic is the best way for checking our correctness. Thus, the Nāgārjunian treatise *Īśvara-kartṛtva-nirākṛtiḥ-viṣṇoḥ-ekakartṛtva-nirākaraṇa* (Stcherbatsky, 1969) is devoted to the "level of conventional reality [*saṁvṛiti-satya*]" and how we can consider our conventions in respect to gods due to logic as well as how we can show that some conventions about gods are wrong in fact, since they contradict themselves. As we see, the *Īśvara-kartṛtva-nirākṛtiḥ-viṣṇoḥ-ekakartṛtva-nirākaraṇa* is well placed in a cultural and philosophical context of the early Madhyamaka and Yogācāra texts, indeed; therefore, it can be dated to the 2nd–4th centuries A.D.

The founder of the Gelug school of Tibetan Buddhism, Je Tsongkhapa (1357–1419), followed the Nāgārjunian-Candrakīrtian treatment of differences between *saṁvṛti-satya* and *paramārtha-satya* and the belief that the *saṁvṛti-satya* is a prolegomenon to the *paramārtha-satya.* As a result, it is no wonder that he organized the monastery education in the following manner: logic is taught first followed by the Madhyamaka philosophy. So, in the Gelug tradition, the Buddhist doctrine is taught at the following five successive levels:

1. The first subject (course) of monastery education is called "valid cognition (logic)" (*tshad ma*). The main text (*gzhung*) of this course is written by the Yogācāra philosopher Dharmakīrti (*chos kyi grags pa*, ca. 6th or 7th century), it is entitled the *Commentary on Valid Cognition* (*Pramāṇavārttika, tshad ma rnam 'grel*) (Ui et al., 1934), N 3713. Therefore, the Buddhist education in Gelug is started with studying the logic of Yogācāra.
2. The second subject (course) is called "perfection of understanding" (*phar phyin*). Its main text (*gzhung*) is presented by the *Treatise Containing the Commentary to* [the *Sūtra* of] *Perfection of Understanding, Called the "Ornament of Clear Realization"* (*Abhisamayālaṃkāra, shes rab kyi pha rol tu phyin pa'i man ngag gi bstan bcos mngon par rtogs pa'i rgyan ces nya ba*) (Ui et al., 1934), N 3786. The authorship of this book is ascribed to Maitreya (*byams pa*).

3. The third subject (course) is called "middle way" (*madhyamaka, dbu ma*). The *Introduction to Madhyamaka* (*Prasannapadā, dbu ma la 'jug pa*) by Candrakīrti (*zla ba grags pa*) (Ui et al., 1934), N 3861, is its main text (*gzhung*).
4. The fourth subject (course) is called "knowledge about *dharma*" (*abhidharma, chos mngon*). The *Treasury of Abhidharma* (*Abhidharmakośa, chos mngon pa'i mdzod*) authored by Vasubandhu (*dbyig gnyen*, ca. 4th or 5th century) (Ui et al., 1934), N 4089, is the main text (*gzhung*) of the course.
5. The last, fifth subject (course) is called "monastic discipline", its main text (*gzhung*) is presented by the *Sūtras on Monastic Disciplines* (*'dul ba'i mtha' dpyod*) (Ui et al., 1934), N 4117.

The Yogācāra logic has been examined as a foundation of Buddhist knowledge in the Gelug school until now. Representatives of this school have written many books on the "valid cognition" (logic) since Je Tsongkhapa who wrote his well-known *Introduction to the Seven Treatises of the Hetuvidyā by Dharmakīrti* (*sde bdun la 'jug pa'i sgo don gnyer yid kyi mun sel*) (Yoshimizu et al., 1989: 213), N 1589. In the Gelug monastery education and in the education of other schools of Tibetan Buddhism, the learning process has two dimensions assuming a perfect logical knowledge: (i) studying some written commentaries (*yig cha*) and oral commentaries (*khrid*) composed as a set of logical reasoning; (ii) participating in debates about these commentaries (*rtags gsal, rtsod pa*). In the debates, argumentation is formed either as a Yogācāra syllogism (*sbyor ba*) or as *reductio ad absurdum* (*thal 'gyur*). About the Tibetan Buddhist debate logic please see Tillemans (1989).

To sum up, in Gelug as in a living tradition of Madhyamaka, the logical discourse is a necessary step for comprehending the ultimate (*paramārtha*). In this section, I assume that modern symbolic logic can be applied as a reliable tool for the valid cognition at the level of conventional reality (*saṁvṛti-satya*) and for the explicating our conventions about gods in the way of how Nāgārjuna did it in his *Īśvara-kartṛtva-nirākṛtiḥ-viṣṇoḥ-ekakartṛtva-nirākaraṇa* (Stcherbatsky, 1969).

4.3.3 Some General Remarks on Modal Logic

Philosophers of language since John Langshaw Austin (1911–1960) and John Rogers Searle (born 1932) have distinguished performative propositions designating and expressing our behavior from informative propositions denoting facts. While informative propositions describe states of affairs (objective reality), performative propositions express our emotional and cognitive valuations in respect to states of affairs to commit common interactions. So, meanings of informative propositions do not depend upon us, but meanings of performative propositions depend on contexts of our interactions with other people. For example, the proposition "this apple is red" is informative, while the proposition "I insist that I am right" is performative.

Our evaluations in performative propositions are made by using performative verbs: 'think', 'like', 'order', 'ask', etc. These verbs express a kind of modality. In the meanwhile, they can show a weaker or stronger aspect of the same modality. For instance, 'asking' is weaker, than 'ordering', and 'ordering' is stronger, than 'asking', although they demonstrate the same modality expressing to do something for us.

Another example is that 'assuming' is weaker, than 'believing', and 'believing' is stronger, than 'assuming', although they express the same modality of our subjective acceptance for something. Hence, for each performative verb we can find its couple which appeals to the same modality, but in a weaker or stronger form.

The following two are among the different performative verbs: "be necessarily" and "be possibly". They express the same modality showing that something is caused, but "be necessarily" is stronger and at the same time "be possibly" is weaker. For the first time, Aristotle in his treatise *On Interpretation* (Περί ερμηνείας/*De Interpretatione*) paid attention that we can construct a kind of symbolic logic for all relationships among propositions with these two forms of the same modality expressing that something is caused.

In his logic, he defined the following basic propositions (see Subsection 4.3.5):

- *A*, any informative proposition;
- □*A*, to read: "*A* is necessarily";
- ◊*A*, to read: "*A* is possibly";
- □¬*A*, to read: "Not-*A* is necessarily";
- ◊¬*A*, to read: "Not-*A* is possibly".

All other propositions are obtained as their logical compositions. Aristotle demonstrated some logical relationships among these basic propositions and these relationships were presented by Scholastic and modern philosophers graphically as the following three squares of oppositions (see Subsection 4.3.5):

Square 4.3.1.

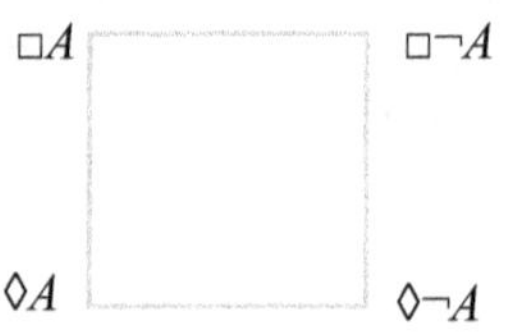

In this square, □*A* and ◊¬*A* are contradictories, □¬*A* and ◊*A* are contradictories, □*A* and □¬*A* are contraries, ◊*A* and ◊¬*A* are subcontraries, ◊*A* is a subaltern of □*A*, ◊¬*A* is a subaltern of □¬*A*.

Square 4.3.2.

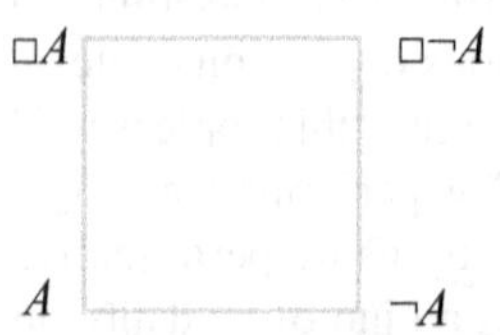

In this square, $\Box A$ and $\neg A$ are contraries, $\Box\neg A$ and A are contraries, $\Box A$ and $\Box\neg A$ are contraries, A and $\neg A$ are contradictories, A is a subaltern of $\Box A$, $\neg A$ is a subaltern of $\Box\neg A$.

Square 4.3.3.

In this square, A and $\Diamond\neg A$ are subcontraries, $\neg A$ and $\Diamond A$ are subcontraries, A and $\neg A$ are contradictories, $\Diamond A$ and $\Diamond\neg A$ are subcontraries, $\Diamond A$ is a subaltern of A, $\Diamond\neg A$ is a subaltern of $\neg A$.

Let us remember that two propositions are contradictory if and only if they cannot both be true, and they cannot both be false. Two propositions are contraries if and only if they cannot both be true but both can be false. Two propositions are subcontraries if and only if they cannot both be false but both can be true. A proposition is a subaltern of another if and only if it must be true if its superaltern is true, and the superaltern must be false if the subaltern is false.

Squares 4.3.1, 4.3.2, and 4.3.3 satisfy the modal logic **T** (see Subsection 4.3.5). Aristotle considered two modal (performative) verbs: "be necessarily" and "be possibly", but the modal logic introduced by him describes all the relationships for all couples of stronger and weaker performances with the same modality. For example, let $\Box A$ mean "I insist to do A", $\Diamond A$ mean "I ask to do A", $\Box\neg A$ mean "I insist not to do A", $\Diamond\neg A$ mean "I ask not to do A". Then Squares 4.3.1, 4.3.2, and 4.3.3 hold true, as well.

Hence, the basic modal logic (such as **T**) is not a study of necessity and possibility as such, but it is a study of weaker and stronger performances with the same modality. Therefore, if we are going to discover a modal-logical reasoning in ancient Indian philosophy, we do not have to find words 'possible' and 'necessary', because the Indian modal logic can be founded on other performative verbs. For example, the term *utpatti* ('arising') from the Nāgārjunian treatise *Mūlamadhyamakakārikā* is studied in its logical relations to the terms *sat* ('being') and *asat* ("not being"). This *utpatti* expresses a weak performance within the modality showing that something is caused, while *sat* and *asat* express a neutral performance of the same modality. Therefore, we can understand *utpatti* as $\Diamond A$, *sat* as A, *asat* as $\neg A$. Then we can check whether the Nāgārjunian reasoning satisfies a modal logic.

4.3.4 Some General Remarks on Translations of Logical Terms

The main modal-logical intuition of Aristotle is summarized by Squares 4.3.1, 4.3.2, and 4.3.3. It is based on the idea that any weak performance follows from an appropriate strong performance of the same modality. For example, "I ask you" follows from "I order you" or "It is possible" follows from "It is necessary."

In the Nāgārjunian treatise *Īśvara-kartṛtva-nirākṛtiḥ-viṣṇoḥ-ekakartṛtva-nirākaraṇa* (Stcherbatsky, 1969), we find the following two performative terms: *siddha* and its negation *asiddha*. The literal meaning of *siddha* has two aspects: (i) something ontic expressing that it is resulted (produced or established) and (ii) something epistemic expressing that it is proved (i.e., resulted also, but epistemicly). So, this word shows a strong performance for the modality demonstrating causal relations. It is worth noting that in the Sroic logic, performative words for expressing a strong performance of the same modality have the close literal meaning "be resulted". So, even in Greek philosophy different thinkers used different performative verbs to denote the modal operator $\Box$.

Something existent is a neutral form of this modality. According to Aristotle (see Square 4.3.2), it means that something existent (A) follows from the strong performance ($\Box A$). Nevertheless, it is wrong from the Nāgārjunian point of view. For him, the strong performance or *siddha* ($\Box A$) follows from something existent or *sat* (A). Thus, instead of Square 4.3.2 we can reconstruct the following square in his treatise (see Subsection 4.3.7):

Nāgārjunian Square.

In this square, $\Box A$ and $\neg A$ are subcontraries, $\Box\neg A$ and A are subcontraries, $\Box A$ and $\Box\neg A$ are subcontraries, A and $\neg A$ are contradictories, $\Box A$ is a subaltern of A, $\Box\neg A$ is a subaltern of $\neg A$.

In the Stoic fragments, we can reconstruct the following squares (see Subsection 4.3.6):

Square 4.3.4.

In this square, $\Box A$ and $\Diamond\neg A$ are contradictories, $\Box\neg A$ and $\Diamond A$ are contradictories, $\Box A$ and $\Box\neg A$ are subcontraries, $\Diamond A$ and $\Diamond\neg A$ are contraries, $\Box A$ is a subaltern of $\Diamond A$, $\Box\neg A$ is a subaltern of $\Diamond\neg A$.

Square 4.3.5.

In this square, $\Box A$ and $\neg A$ are subcontraries, $\Box\neg A$ and A are subcontraries, $\Box A$ and $\Box\neg A$ are subcontraries, A and $\neg A$ are contradictories, $\Box A$ is a subaltern of A, $\Box\neg A$ is a subaltern of $\neg A$.

Square 4.3.6.

In this square, $\Diamond A$ and $\neg A$ are contraries, $\Diamond\neg A$ and A are contraries, $\Diamond A$ and $\Diamond\neg A$ are contraries, A and $\neg A$ are contradictories, $\Diamond A$ is a subaltern of A, $\Diamond\neg A$ is a subaltern of $\neg A$.

As we see, the Nāgārjunian Square corresponds to Square 4.3.5 of the Stoics. In turn, Squares 4.3.4, 4.3.5, and 4.3.6 satisfy the modal logic **K=** (see Subsection 4.3.6). In these squares, we deal with another inequality between the weaker and stronger forms of the same modality expressing that something is caused: any strong performance follows from an appropriate weak performance.

The same performances can be denoted by different performative verbs, as we said. In modal logics **T** and **K=** we treat $\Box$ as "be necessarily" and $\Diamond$ as "be possibly" just technically. The point is that both performative terms came from Scholastic philosophy: *necessarius* (necessary) and *possibilis* (possible), therefore we follow tradition of Scholastic modal logic there. The Stoic modal operator $\Box$ is translated "be necessitated" as well as the Nāgārjunian $\Box$ (i.e., *siddha*) is translated "be necessitated", too, only to distinguish them from the Aristotelian one.

It is common knowledge that philosophical terms cannot be translated literally. The same applies to logic. So, both terms of Indian logic *dharma* and *dharmin* denoting logical predicate and logical subject according to their functions do not have these meanings literally. The same is with *siddha*. This term denotes a strong performance literally, but functionally it can be translated "be necessitated" within an appropriate formalization of modal reasoning.

4.3.5 Aristotelian Modal Logic

Let us recall that any system of modal logic consists of the following vocabulary (see Subsection 4.1.2):

- $p_0, p_1, \ldots$ – propositional letters (atoms), *Prop*;
- $\neg$, $\vee$, &, $\Rightarrow$, $\Leftrightarrow$ – propositional connectives: negation ("not…"), disjunction ("… or …"), conjunction ("… and …"), implication ("if…, then…"), equivalence ("… if and only if…"), respectively;
- $\Box$, $\Diamond$ – modal operators: the symbol $\Box$ is used for 'necessity' ("... is necessarily") or for any other strong performance and the symbol $\Diamond$ is for 'possibility' ("… is possibly") or for any other weak performance of the same modality.

A well-formed formula of modal logic is defined as follows:

- Each propositional letter from *Prop* is a formula;
- If A and B are formulas, then $\neg A$, $\neg B$, $A \vee B$, $A \,\&\, B$, $A \Rightarrow B$, $A \Leftrightarrow B$, $\Box A$, $\Box B$, $\Diamond A$, $\Diamond B$ are formulas, as well.

The weakest modal logic, denoted by **K** in honor of Soul Kripke, consists of the following axioms/theorems (formulas which are ever proved):

- All propositional axioms/theorems such as $A \vee \neg A$ (excluded middle or *tertium non datur*), $A \Rightarrow A$ (reflexivity of implication), $(A \,\&\, B) \Rightarrow B$ (elimination of conjunction), etc.;
- All instances of the Kripke schema: $\Box(A \Rightarrow B) \Rightarrow (\Box A \Rightarrow \Box B)$.

The set of these axioms/theorems is closed under the following two inference rules:

- *modus ponens*: from $A \Rightarrow B$ and A it follows that B;
- *necessitation rule*: if A is an axiom/theorem, then $\Box A$ is an axiom/theorem, too.

To obtain a system of modal logic we can add different additional modal axiom schemes to the set of propositional theorems. The most important of these additional schemes are as follows (Garson, 2006):

(K) $\Box(A \Rightarrow B) \Rightarrow (\Box A \Rightarrow \Box B)$
(D) $\Box A \Rightarrow \Diamond A$
(T) $\Box A \Rightarrow A$
(=) $A \Rightarrow \Box A$
(CD) $\Diamond A \Rightarrow \Box A$

The modal logic, denoted by **D**, has additional schemes (K) and (D). The modal logic, denoted by **T**, has additional schemes (K) and (T).

It is worth noting that Aristotle might be appraised as a father of modal logic, because he considered some modal propositions as axioms which are axioms of logic **T** in fact. Hence, Aristotle had an inspiration in respect to logical reasoning formalized now within **T**. Let us show it on some statements from his book Περί ερμηνείας (*De Interpretatione*).

First, it is evident for him that if a proposition A is a tautology (it is ever true), then $\Box A$ is a tautology, too:

$A \vee \neg A$ is a tautology, then $\Box(A \vee \neg A)$ is a tautology, as well.

[…] ἀνάγκη τὴν κατάφασιν ἢ τὴν ἀπόφασιν ἀληθῆ ἢ ψευδῆ εἶναι (*De Interpretatione* 9: 18a). […] ὥστ' ἀνάγκη τὴν κατάφασιν ἢ τὴν ἀπόφασιν ἀληθῆ εἶναι (*De Interpretatione* 9: 18b). […] Εἶναι μὲν ἢ μὴ εἶναι ἅπαν ἀνάγκη, καὶ ἔσεσθαί γε ἢ μή• οὐ μέντοι διελόντα γε εἰπεῖν θάτερον ἀναγκαῖον (*De Interpretatione* 9: 19a).

[…] Necesse est affirmationem vel negationem veram vel falsam esse (*De Interpretatione* 9: 18a). […] Quare necesse est aut affirmationem aut negationem veram esse (*De Interpretatione* 9: 18b). […] Esse quidem vel non esse omne necesse est et futurum esse vel non; non tamen dividentem dicere alterum necessario (*De Interpretatione* 9: 19a).

[…] Propositions, whether positive or negative, are true or false necessarily [*A.Sch.*: according to the rule of *tertium non datur*] […] It may therefore be claimed that it is necessary that affirmations or negations must be either true or false. […] Everything must either be or not be, whether in the present or in the future, but it is not always possible to distinguish and state determinately which of these alternatives must necessarily come about.

This statement is a basic theorem even for logic **K** (it is inferred directly from $A \vee \neg A$ due to the necessitation rule and an introduction of implication).

In the meanwhile, Aristotle warned us that we cannot argue $\Box A \vee \Box B$ from $A \vee B$:

It is false that if $A \vee B$ is a tautology, then $\Box A \vee \Box B$ is a tautology, as well.

Δὲ οἷον ἀνάγκη μὲν ἔσεσθαι ναυμαχίαν αὔριον ἢ μὴ ἔσεσθαι, οὐ μέντοι γενέσθαι αὔριον ναυμαχίαν ἀναγκαῖον οὐδὲ μὴ γενέσθαι• γενέσθαι μέντοι ἢ μὴ γενέσθαι ἀναγκαῖον (*De Interpretatione* 9: 19a).

Necesse est quidem futurum esse bellum navale cras vel non esse futurum sed non futurum esse cras bellum navale necesse est vel non futurum esse, futurum autem esse vel non esse necesse est (*De Interpretatione* 9: 19a).

A sea-fight must either take place tomorrow or not, but it is not necessary that it should take place tomorrow, neither is it necessary that it should not take place, yet it is necessary that it either should or should not take place tomorrow.

Otherwise, we would accept that $\Diamond A \Rightarrow A$ is ever false and $A \Rightarrow \Box A$ is ever true. Nevertheless, we cannot do it in any way:

Οὐδὲν ἄρα οὔτε ἔστιν οὔτε γίγνεται οὔτε ἀπὸ **τύχης** οὔθ' ὁπότερ' ἔτυχεν, οὐδ' ἔσται ἢ οὐκ ἔσται, **ἀλλ' ἐξ ἀνάγκης ἅπαντα** καὶ οὐχ ὁπότερ' ἔτυχεν (*De Interpretatione* 9: 18b). […] ἅπαντα οὖν τὰ ἐσόμενα ἀναγκαῖον γενέσθαι (*De Interpretatione* 9: 18b).

> Nihil igitur neque est neque fit nec a casu nec **utrumlibet**, nec erit nec non erit sed **ex necessitate omnia** et non utrumlibet (*De Interpretatione* 9: 18b). [...] Omnia ergo quae futura sunt necesse est fiery (*De Interpretatione* 9: 18b).

> Then nothing is or takes place **occasionally**, either in the present or in the future, and there are no real alternatives; **everything takes place of necessity** and not occasionally [...]. [...] Then all that is to be must necessarily take place in the future.

According to Aristotle, the proposition $A \Rightarrow \Box A$ is always true if and only if A is always true. This takes place only under the following condition: A is a universal affirmation such as "Each human being is intelligent" that is true indeed at any time. In this case, A does not depend on time, and it can be stated truly in any tense: past, present, and future. However, if A is an individual affirmation such as "Socrates is white", we cannot accept $A \Rightarrow \Box A$. The main reason is that A is time dependent. Now it can be true, but tomorrow it can be false. Hence, accepting $A \Rightarrow \Box A$ for an individual A gives rise to inconsistencies:

> ἔτι εἰ ἔστι λευκὸν νῦν, ἀληθὲς ἦν εἰπεῖν πρότερον ὅτι ἔσται λευκόν, ὥστε ἀεὶ ἀληθὲς ἦν εἰπεῖν ὁτιοῦν τῶν γενομένων ὅτι ἔσται• εἰ δ' ἀεὶ ἀληθὲς ἦν εἰπεῖν ὅτι ἔστιν ἢ ἔσται, οὐχ οἷόν τε τοῦτο μὴ εἶναι οὐδὲ μὴ ἔσεσθαι (*De Interpretatione* 9: 18b).

> Amplius, si est album nunc, verum erat dicere primo quoniam erit album, quare semper verum fuit dicere quodlibet illud eorum quae facta sunt quoniam erit; quod si semper verum est dicere quoniam est vel erit, non potest hoc non esse nec non futurum esse (*De Interpretatione* 9: 18b).

> So, if **a thing is white now**, it was true before to say that **it will be white**, so that of anything that has taken place it was always true to say "it is" or "it will be". But if it was ever true to say that a thing is or will be, it is not possible that it should not be or not be about to be.

As we see, Aristotle denied modal axiom (=).

In his Περί ερμηνείας, he implicitly used a new modal operator *to be contingent* (*admissible*) (ἐνδεχόμενον) for some propositions such as "A sea-fight will be tomorrow": $\Diamond A \,\&\, \Diamond\neg A$ (ἐν οἷς ἄμφω ἐνδέχεται καὶ τὸ εἶναι καὶ τὸ μὴ εἶναι; *in quibus utrumque contingit et esse et non esse*). It can be read thus: "It is possible that A and it is possible that not-A". It means, "A is contingent". So, it is possible that a sea-fight will happen tomorrow and simultaneously it is possible that a sea-fight will not happen tomorrow. This event is just contingent.

Some theorems of **T** (and of **D**), which were examined as true propositions by Aristotle:

$$(\Diamond A \,\&\, \Diamond\neg A) \Rightarrow \neg\neg\Diamond A$$
$$(\Diamond A \,\&\, \Diamond\neg A) \Rightarrow \neg\Box A$$
$$(\Diamond\neg A \,\&\, \Diamond\neg\neg A) \Rightarrow \neg\Box\neg A$$
$$(\Diamond\neg A \,\&\, \Diamond\neg\neg A) \Rightarrow \neg\neg\Diamond\neg A$$
$$(\neg\Diamond A \,\&\, \neg\Diamond\neg A) \Rightarrow \Box\neg A$$
$$(\neg\Diamond A \,\&\, \neg\Diamond\neg A) \Rightarrow \neg\Diamond A$$

$(\neg\Diamond\neg A \;\&\; \neg\Diamond\neg\neg A) \Rightarrow \Box A$
$(\neg\Diamond\neg A \;\&\; \neg\Diamond\neg\neg A) \Rightarrow \neg\,\Diamond\neg A$
$\neg\Diamond A \Leftrightarrow \Box\neg A$
$\neg\Diamond\neg A \Leftrightarrow \Box A$
$\Box A \Rightarrow \Diamond A$

Καὶ αἱ ἀκολουθήσεις δὲ κατὰ λόγον γίγνονται οὕτω τιθε μένοις• τῷ μὲν γὰρ δυνατῷ εἶναι τὸ ἐνδέχεσθαι εἶναι, καὶ τοῦτο ἐκείνῳ ἀντιστρέφει, καὶ τὸ μὴ ἀδύνατον εἶναι καὶ τὸ μὴ ἀναγκαῖον εἶναι• τῷ δὲ δυνατῷ μὴ εἶναι καὶ ἐνδεχομένῳ μὴ εἶναι τό τε μὴ ἀναγκαῖον μὴ εἶναι καὶ οὐκ ἀδύνατον μὴ εἶναι, τῷ δὲ μὴ δυνατῷ εἶναι καὶ μὴ ἐνδεχομένῳ εἶναι τὸ ἀναγ καῖον μὴ εἶναι καὶ τὸ ἀδύνατον εἶναι, τῷ δὲ μὴ δυνατῷ μὴ εἶναι καὶ μὴ ἐνδεχομένῳ μὴ εἶναι τὸ ἀναγκαῖον εἶναι καὶ τὸ ἀδύνατον μὴ εἶναι (*De Interpretatione* 13: 22a). [...] ἀποδίδοται, τὸ αὐτὸ δυνάμενον• εἰ γὰρ ἀδύνατον εἶναι, ἀναγκαῖον τοῦτο οὐχὶ εἶναι ἀλλὰ μὴ εἶναι• εἰ δὲ ἀδύνατον μὴ εἶναι, τοῦτο ἀνάγκη εἶναι• ὥστ' εἰ ἐκεῖνα ὁμοίως τῷ δυνατῷ καὶ μή, ταῦτα ἐξ ἐναντίας, ἐπεὶ σημαίνει γε ταὐτὸν τό τε ἀναγκαῖον καὶ τὸ ἀδύνατον, ἀλλ' ὥσπερ εἴρηται, ἀντ εστραμμένως (*De Interpretatione* 13: 22b). [...] Τὸ μὲν γὰρ ἀναγκαῖον εἶναι δυνατὸν εἶναι (*De Interpretatione* 13: 22b).

Et consequentiae vero secundum ordinem fiunt ita ponentibus: illi enim quae est "possibile esse" illa quae est "contingit esse", et hoc illi convertit, et "non impossibile esse" et "non necessarium esse"; illi vero quae est "possibile non esse" et "contingere non esse" ea quae est "non necessarium non esse" et "non impossibile non esse"; illi vero quae est "non possibile esse" et "non contingens esse" illa quae est "necessarium non esse" et "impossibile esse"; illi vero quae est "non possibile non esse" et "non contingens non esse" illa quae est "necesse esse" et "impossibile non esse" (*De Interpretatione* 13: 22a). [...] Nam si impossibile est esse, necesse est hoc non esse sed non-esse; si vero impossibile non esse, hoc necessarium est esse; quare, si illa similiter possibile et non, haec e contrario, nam idem significat 'necessarium' et 'impossibile' sed, quemadmodum dictum est, contrarie (*De Interpretatione* 13: 22a). [...] Nam quod est necessarium esse, possibile est esse (*De Interpretatione* 13: 22b).

Implications follow in due course when we have arranged the propositions thus. From the proposition "contingent to be" it follows that "admissible to be", and the relation is reciprocal. It follows also "not impossible to be" [*A.Sch.*: $(\Diamond A \;\&\; \Diamond\neg A) \Rightarrow \neg\,\neg\Diamond A$] and "not necessary to be" [*A.Sch.*: $(\Diamond A \;\&\; \Diamond\neg A) \Rightarrow \neg\,\Box A$]. From "contingent not to be" and "admissible not to be" follow both "'not necessary not to be" [*A.Sch.*: $(\Diamond\neg A \;\&\; \Diamond\neg\neg A) \Rightarrow \neg\Box\neg A$] and "not impossible not to be" [*A.Sch.*: $(\Diamond\neg A \;\&\; \Diamond\neg\neg A) \Rightarrow \neg\neg\Diamond\neg A$]. From "not contingent to be" and "not admissible to be" follow both "necessary not to be" [*A.Sch.*: $(\neg\Diamond A \;\&\; \neg\Diamond\neg A) \Rightarrow \Box\neg\, A$] and "impossible to be" [*A.Sch.*: $(\neg\Diamond A \;\&\; \neg\Diamond\neg A) \Rightarrow \neg\Diamond A$]. From "not contingent not to be" and "not admissible not to be" follow "necessary to be" [*A.Sch.*: $(\neg\Diamond\neg A \;\&\; \neg\Diamond\neg\neg A) \Rightarrow \Box A$] and "impossible not to be" [*A.Sch.*: $(\neg\Diamond\neg A \;\&\; \neg\Diamond\neg\neg A) \Rightarrow \neg\,\Diamond\neg A$].

> [...] For "impossible to be" it is necessary for this (not, to be, but) not to be [*A.Sch.*: $\neg\Diamond A \Leftrightarrow \Box\neg A$]; and for "impossible not to be" it is necessary for this to be [*A.Sch.*: $\neg\Diamond\neg A \Leftrightarrow \Box A$]. Thus, if those follow from 'possible' and "not possible" in the same way, these follow in a contrary way, since 'necessary' and 'impossible' do signify the same but (as we said) when applied conversely. [...] For the necessary to be is possible to be [*A.Sch.*: $\Box A \Rightarrow \Diamond A$, but it is not the case that $\Box A \Rightarrow (\Diamond A \;\&\; \Diamond\neg A)$].

Let us now remember a Kripke model $M = (X, a, R, V)$ for modal propositions. This model consists of the following components (see Subsection 4.1.2):

- a set X of indices;
- a distinguished index a;
- a binary relation R on the indices;
- a function V assigning a valuation $V(x)$ to each index x and a truth-value $V(x)(A)$ to each index x and atom A.

The distinguished index a is to represent an actual time. The relation R is to be said a possibility since a. Let us take x from X such that aRx. Then this aRx shows us a possibility since a at an index x.

A true valuation of formulas at an index x in a model M is defined as follows:

1) for atomic A, A is true at x in M	Iff	$V(x)(A) = \text{T}$, where T means truth
2) $\neg A$ is true at x in M	Iff	A is not true at x in M
3) $A \;\&\; B$ is true at x in M	Iff	A is true at x in M and B is true at x in M
4) $A \vee B$ is true at x in M	Iff	A is true at x in M or B is true at x in M
5) $A \Rightarrow B$ is true at x in M	Iff	if A is true at x in M, then B is true at x in M
6) $A \Leftrightarrow B$ is true at x in M	Iff	A is true at x in M if and only if B is true at x in M
7) $\Box A$ is true at x in M	Iff	for all y with xRy, A is true at y in M
8) $\Diamond A$ is true at x in M	Iff	for some y with xRy, A is true at y in M

A formula A is valid if A is true in all models, and A is satisfiable if A is true in some model. Let us note that A is valid if and only if $\neg A$ is not satisfiable.

To demonstrate an intuitive meaning of R in M, we may consider the following example given by Aristotle:

> Οἷον ὅτι τουτὶ τὸ ἱμάτιον δυνατόν ἐστι διατμηθῆναι καὶ οὐ διατμηθήσεται, ἀλλ' ἔμπροσθεν κατατριβήσεται· ὁμοίως δὲ καὶ τὸ μὴ διατμη θῆναι δυνατόν· οὐ γὰρ ἂν ὑπῆρχε τὸ ἔμπροσθεν αὐτὸ κατατριβῆναι, εἴγε μὴ δυνατὸν ἦν τὸ μὴ διατμηθῆναι (*De Interpretatione* 9: 19a).

Ut quoniam hanc uestem possibile est, incidi et non inciditur sed prius exteritur; similiter autem et non incidi possibile est; non enim esset eam prius exteri nisi esset possibile non incidi (*De Interpretatione* 9: 19a).

It is possible that this coat may be cut in half, but it will not be cut in half, because it will be worn out first. In the same way, it is possible that it is not cut in half; since it would not be possible to wear it out first because of impossibility to cut it in half.

Let x from X be a future point when the proposition A := "This coat is cut in half" is true and let y from X be a future point when the proposition B := "This coat is worn out" is true. Then for all z with xRz, B is not true at z in M. It means that $\Box\neg B$ is true at z in M. In the same way, it is readily seen that for all z with yRz, A is not true at z in M. It means that $\Box\neg A$ is true at z in M. In other words, xRz gives all the points z for which it is necessary that this coat is *not* worn out and yRz gives all the points z for which it is necessary that this coat is *not* cut in half. Thus, R is to represent a sequence of events validating a modal proposition. If there was the event x that this coat was cut in half, then for all subsequent events the proposition "It is necessary that this coat is *not* worn out" is true; and if there was the event y that this coat was worn out, then for all subsequent events the proposition "It is necessary that this coat is *not* cut in half" is true.

This R can be different for different axioms (Garson, 2006):

(K)	$\Box(A \Rightarrow B) \Rightarrow (\Box A \Rightarrow \Box B)$	R can be any relation
(D)	$\Box A \Rightarrow \Diamond A$	R is serial: there exist y such that xRy
(T)	$\Box A \Rightarrow A$	R is reflexive: xRx
(=)	$A \Rightarrow \Box A$	R is bisimilar: $xRy \Rightarrow x = y$
(CD)	$\Diamond A \Rightarrow \Box A$	R is unique: $(zRx \,\&\, zRy) \Rightarrow x = y$

For instance, the theorems of **D** correspond to the class of Kripke models in which the relation R is *serial*. If a formula is a theorem in **D**, then it is true in all serial Kripke models (this property is called a soundness of **D**); and if a formula is true in all serial Kripke models, then it is provable in **D** (this property is called a completeness of **D**). The truth of $\Box A$ at a amounts to the truth of A at all x such that aRx, and the seriality of R guarantees that at a there exist y such that aRy, so that A will be true at some y, too. Therefore, if $\Box A$ is true at a, then $\Diamond A$ is true at a. As a result, $\Box A \Rightarrow \Diamond A$ is true.

Assume that R is *reflexive*, meaning that xRx for all x. Then all the theorems of **T** are true in all M with the reflexive R (soundness) and all valid formulas in these M are provable in **T** (completeness). The truth of $\Box A$ at a means the truth of A at all x such that aRx. The reflexivity of R guarantees that a itself will be among these x. Therefore, if $\Box A$ is true at a, then A is true at a, as well. Thus, $\Box A \Rightarrow A$ is true.

It is worth noting that if R is reflexive, then R is serial. From this it follows that the class of Kripke models in which the relation R is reflexive includes the class of Kripke models in which the relation R is serial. So, from axiom (T) we can infer axiom (D). It means that **T** is an extension of **D**.

In his works, Aristotle considered many modal propositions which are provable in **T** and therewith true in Kripke models in which the relation R is reflexive. Later these propositions were analyzed in medieval scholastic philosophy. We must be honest and frank about saying that these propositions of **T** do not occur in classical Indian philosophical treatises at all. Nevertheless, we can face many correct modal propositions corresponding to axioms of the modal logic, denoted by **CD**, with additional modal schemes (K) and (CD). And for the first time, these propositions can be reconstructed on Nāgārjuna's texts.

4.3.6 Logical Determinism and Hetutā

The main claim in the modal logic reconstructed on the text of Περί ερμηνείας is that there exists contingency: $\Diamond A \mathbin{\&} \Diamond\neg A$. Contingent events were called by Aristotle 'luck', τύχη (*De Interpretatione* 9: 18b)—by name of Tyche (Roman equivalent: Fortuna), the goddess of luck and destiny. These events are possible only in modal logic **T** (and **D**). The negation of contingency gives to rise a *logical determinism*: $\neg(\Diamond A \mathbin{\&} \Diamond\neg A) = \neg\Diamond A \vee \neg\Diamond\neg A = \Box\neg A \vee \Box A = \Box A \vee \Box\neg A$ so that if there is no contingency, then each event is necessary to be, or it is necessary not to be: $\Box A \vee \Box\neg A$.

Let us notice that besides **T** and **D** there are modal logics **K=** and **CD**. The logic, denoted by **K=**, has additional modal schemes (K) and (=). The modal logic, denoted by **CD**, has additional modal schemes (K) and (CD).

Suppose that R is *bisimilar*: $xRy \Rightarrow x = y$. Then all the theorems of **K=** are true in all M with the bisimilar R (soundness) and all valid formulas in these M are provable in **K=** (completeness). Let us assume the truth of A at a. The truth of $\Box A$ at a would mean the truth of A at all x such that aRx, but for all aRx we have that $a = x$ according to the property of R. Then $A \Rightarrow \Box A$ is true. Now let R be *unique*: $(zRx \mathbin{\&} zRy) \Rightarrow x = y$. In this case, all the theorems of **CD** are true in all M with the unique relation, R (soundness) and all valid formulas in these M are provable in **CD** (completeness). The truth of $\Diamond A$ at a would mean the truth of A at some x such that aRx, and the truth of $\Box A$ at x would mean the truth of A at all y such that xRy. Because of the property of R, we have for these (some) x: $x = y$. Then $\Diamond A \Rightarrow \Box A$ is true.

In both **K=** and **CD**, the following propositions are theorems:

(CD) $\Diamond A \Rightarrow \Box A$
(CD1) $\Box(A \vee B) \Rightarrow (\Box A \vee \Box B)$
(CD2) $(\Box A \vee \Box B) \Rightarrow \Box(A \vee B)$
(CD3) $\Box(A \Rightarrow B) \Rightarrow (\Box A \Rightarrow \Box B)$
(CD4) $(\Box A \Rightarrow \Box B) \Rightarrow \Box(A \Rightarrow B)$
(CD5) $\Box(A \mathbin{\&} B) \Rightarrow (\Box A \mathbin{\&} \Box B)$
(CD6) $(\Box A \mathbin{\&} \Box B) \Rightarrow \Box(A \mathbin{\&} B)$

For more details, see Garson (2006). Hence, in logics **K=** and **CD** we can deduce that the proposition $\Box A \vee \Box\neg A$ is a theorem. Indeed, $A \vee \neg A$ is a propositional axiom/theorem. Then $\Box(A \vee \neg A)$ is a theorem, too, by the necessitation rule. According to (CD1), the proposition $\Box(A \vee \neg A) \Rightarrow (\Box A \vee \Box\neg A)$ is a theorem. By *modus ponens* we conclude that $\Box A \vee \Box\neg A$ is a theorem of **K=** and **CD**. Hence, in logics **K=** and **CD** no contingent events are possible. The proposition $\Diamond A \mathbin{\&} \Diamond\neg A$ is always false.

Let us show that if R is bisimilar, then R is unique, i.e., the class of Kripke models in which the relation R is bisimilar includes the class of Kripke models in which the relation R is unique. Let us take the bisimilar R for zRx and zRy. Then $z = x$ and $z = y$. From this it follows that $x = y$. Thus, from axiom (=) we can infer axiom (CD). In other words, **K=** is an extension of **CD**.

Among the Greek philosophers the majority accepted the existence of contingency, ἐνδεχόμενον (an intuition for modal logics **T** and **D**), but there were followers of logical determinism, also (an intuition for modal logics **K=** and **CD**). The Stoics are the most prominent among these followers.

The main modal postulate proposed by Chrysippus, who became the most prominent Stoic logician, is as follows:

(S_1) *Nihil enim fieri sine causa potest* (Cicero, *De Divinatione* II: 61).

Nothing happens without a cause.

Motum nullum esse sine causa (Cicero, *De Fato* 23).

No motion is without a cause.

To understand what it means, we should note that for the Greek philosophers, each individual proposition A is time dependent. Hence, the semantic statement "A is true" (A ἀληθές ἐστι) always means "A is true *now*". For instance, "Aspasia is walking" is true if Aspasia is now walking and it takes place due to a cause that might be explored. All the future propositions such as "Aspasia will walk tomorrow" or "A sea-fight will take place tomorrow" can be evaluated only now, whether recently they are true owing to some reasons. Hence, they are true *now* if and only if now there are some causes determining these events tomorrow.

An emergence of cause implies an effect. If this effect is described by an individual proposition (ἀξιώματα), the latter is called an actualization (ὑπάρχειν) of predicate, see (Bobzien, 1998). Examples of predicates are as follows: "to walk", "to be alive", "to be cut in half", and "to be worn out". If we observe some events such as walking, being alive, being cut in half, and being worn out, respectively, as effects of some causes, the predicates "to walk", "to be alive", "to be cut in half", and "to be worn out" are actualized now and we can express them in individual propositions truly. For instance, "to walk" is actualized if Aspasia is walking now for some reasons. These actualized predicates are then called attributes (συμβεβηκότα) (see Stobaeus, *Eclogarum physicarum et ethicarum* I: 106, 20–23). So, due to (S_1) at any one time each individual proposition can contain an actualized and deactualized predicate to be either true or false:

(S_2) *Each individual proposition* (ἀξιώματα) *A* (*even about future*) *is either true or false*: $A \vee \neg A$.

From (S_2) the Stoics infer as follows:

(S_3) *Each individual proposition* (*even about future*) *A is necessarily true, or it is necessarily false*: $\Box A \vee \Box\neg A$.

Thus, according to the Stoics, modal logics **K=** and **CD** are more natural than logics **T** and **D**. Some Stoic synonyms for the word 'necessity' (ἀνάγκη): 'inexorable' (ἀπαράβατος), 'inflexible' (ἄτρεπτός), 'invincible' (ἀνίκητος), 'unconquerable' (ἀνεκβίαστος), 'unpreventable' (ἀκώλυτος), 'immutable' (ἀμετάβλητος), and 'unchangeable' (ἀμετάθετος) (Bobzien, 1998). All of them are attributes of fate which rules over the world. Some Stoic synonymous terms for 'fate' (εἱμαρμένη): 'destiny' (πεπρωμένη), 'pneuma' (πνεῦμα), 'cause' (αιτία), 'reason' (λόγος), Zeus, "will (βούλησις) of Zeus". Due to the fate, each event has its cause. So, each event should be regarded as actual as if it happens in the present tense.

(S_4) *Each (past, present, future) event A is actual and then necessary*: $A \Rightarrow \Box A$.

Let us take the bisimilar relation R: $xRy \Rightarrow x = y$. The event a represents an actual time. So, if the proposition A is true at a, then A is necessary for the bisimilar R: $\Box A$ is true at a. Furthermore, the R makes all events actual.

The same epiphany of bisimilar R came to the Stoics—as we see. According to them, making all events actual and all actual events necessary, fate can be called "the greatest necessity" (κνριωτάτη ἀνάγκη; Plutarch, *De Stoicorum Repugnantiis* 1055e), "the greatest Cause" (ἡ μίγίστη αἰτία; *Ibid* 1055e), "the Cause of all things" (πάντων αἰτία; *Ibid*., 1056b), "invincible and unpreventable and unchangeable Cause" (αἰτίαν ἀνίκητον καὶ ἀκώλυτον και ἄτρεπτον; *Ibid*., 1056c). Hence, the Stoics claim:

(S_5) Τὰ πάντα καθ' εἱμαρμένην γίνεται.

Everything happens in accordance with fate.

First, it means that the universe is a causal network. This network is called by Chrysippus "interweaving [of causes]" (ἐπιπλοκή) and "chain [of causes]" (είρμός). Within the network, every event is 'necessitated' (καταναγκάζεσθαι) or "happened in accordance with necessity" (κατ' ἀνάγκην) as well as 'fated' (καθειμάρθαι) or "happened in accordance with fate" (καθ' εἱμαρμένην) (Bobzien, 1998).

For Chrysippus as well as other Stoics, causes (ποιοῦν, ενεργοῦν) are expressed by some 'predicates' (κατηγορήματα) saying about relations (πρός τι) between two corporeal bodies: the first one that affects and the second one that is caused. Causes can be used for explanations (logical inferences): "that because of which" (δι ' ὅ). Formally, causes can be represented as relations $r_{tp}(x, y)$ at a time t and at a place p among corporeal bodies x and y with an antecedent x affecting on a body of y at t and p and a succedent y being a result of this affecting. Each body from $r_{tp}(x, y)$ can be described as a quantitative state (κινήσεις) or qualitative state (σχέσεις). So, each state of body as well as a relation between bodies is 'sayable' (λεκτά)—it is worth noting that this "to be sayable" is the only incorporeal thing of our universe for the Stoics. Let us quote how causes are treated in the Stoic fragments:

> [...] Στωικοι μεν παν αίτιον σώμα φασι σώματι άσευμάτου τίνος αίτιον γίνεσθαι, οίον σώμα μεν τδ σμιλίον, σώματι δε τη σαρκί, άσωμάτου δε τοῦ τεμνεσθαι κατηγορήματος (Sextus Empiricus, *Adversus Mathematicos* IX: 211).

> [...] the Stoics say that every cause is a body which becomes a cause, to a body, of something incorporeal; as for instance the scalpel, which is a body, becomes a cause, to the flesh, which is a body, of the incorporeal predicate "being cut".

Hence, (=) as an axiom of **K=** is understood as a tautology by the Stoics, too, see (S_4) and (S_5). The statement about possibilities with the same logical meaning is as follows (**K=** can be obtained by adding (K) and (=2) to the propositional axioms):

(=2) $\Diamond A \Rightarrow A$

According to the Stoics, a proposition A is possible now, i.e., $\Diamond A$, if and only if nothing external hinders it from happening at some time from now on, see (Bobzien, 1998). For instance, the proposition "this coat will be cut in half" is possible from now on if and only if there exists the coat that was not worn out yet. In this case "being worn out" is an effect of something external that hinders the event "being cut in half" from happening. Let us pay attention that we have the following two general conditions: (i) an event is actual (A) if and only if there is a cause now that affected on the happening of this event; (ii) an event is possible ($\Diamond A$) if and only if there are no causes now which hinder this event from happening. As we see, the condition (i) is more general than (ii): (ii) ⇒ (i), i.e., we deal with (=2), but in the Stoic terms:

> Τὸ δὲ λέγειν μὴ ἀναιρεῖσθαι πάντων γινομένων καθ'εἱμαρμένην τὸ δυνατόν τε καὶ ἐνδεχόμενον τῷ δυνατὸν μὲν εἶναι γενέσθαι τοῦτο ὃ ὑπ' οὐδενός κωλύεται γενέσθαι, κἂν μὴ γένηται, τῶν δὲ καθ' εἱμαρμένην γινομένων οὐ κεκωλῦσθαι τὰ ἀντικείμενα γενέσθαι- διὸ καίτοι μὴ γινόμενα ὅμως ἐστὶ δυνατά, καὶ τοῦ μὴ κεκωλῦσθαι γενέσθαι αὐτὰ ἀπόδειξιν φέρειν τὸ ἡμῖν τὰ κωλύοντα αὐτά [ἂν] ἄγνωστα εἶναι πάντως μέν τινα ὄντα ἃ γάρ ἐστιν αἴτια τοῦ γίνεσθαι τὰ ἀντικείμενα αὐτοῖς καθ' εἱμαρμένην, ταῦτα καὶ τοῦ μὴ γίνεσθαι τούτοις αἴτια, εἴ γε ὥς φασιν ἀδύνατον τῶν αὐτῶν περιεστώτων γίνεσθαι τὰ ἀντικείμενα ἀλλ' ὅτι μὴ ἡμῖν ἐστι γνώριμά τινα ἅ ἐστι, διά τοῦτο ἀκώλυτον αὐτῶν τὸ [μὴ] γίνεσθαι λέγουσιν (Alexander of Aphrodisias, *De Fato* 176: 14–23).

> But to say that the possible and the contingent are not avoided if everything happens in accordance with fate [*A.Sch.*: see (S_4) and (S_5)], since that is possible to happen which is hindered by nothing from happening even if it does not happen, and that the contradictories of the things that happen in accordance with fate are not hindered from happening <and to say that> for this reason those things that do not happen are equally possible [*A.Sch.*: $\Diamond A \vee \Diamond \neg A$], and to put forward as proof of their not being prevented the fact that the things that hinder them are unknown to us, although there certainly are some, namely the causes of their contradictories happening in accordance with fate, which are also the causes of their not happening (since, as they say, it is impossible that in the same circumstances contradictory things happen); but since it is not known to us what they are, because of this, they say, the things that do not happen are not hindered from happening (Bobzien, 1998).

Thus, the Stoics have accepted the following two modal axioms: (=) and (=2), i.e., the logic **K=**. It means they have followed the logical determinism and avoided any contingency. The same logical determinism is observed in the Yogācāra texts, but it is put forward by means of other postulates. So, instead of the concept of motion proposed by the Stoics in (S_1), the Yogācārins have introduced the new term *arthakriyātva*:

(Y_1) *There is* arthakriyātva (*a causal efficacy as a criterion to be real*).

> *hetupratyayānām eva tarhi sāmarthyaṃ paśyāmaḥ, sati sāmagrye bhāvād asati cābhavān na jāter iti hetupratyayā eva janakāḥ santi*
> (Vasubandhu, *Abhidharmakośabhāṣya* 79: 19–21), see (Pradhan, 1967).

> So, we see the causal efficiency only of causes and conditions. [It is the case] of origination, because when the [causal] complex is given, there is existence, and when not there is non-existence. Therefore, only the causes and conditions are bringing forth.

> *arthakriyāsamarthaṃ yat tad atra paramārthasat |*
> *anyat saṃvṛtisat proktaṃ te svasāmānyalakṣaṇe ||*
> (Dharmakīrti, *Nyayabindu* 14–15), see (Malvania, 1971).

> To have a real existence (*paramārthasat*) an entity should have some causal powers (*arthakriyāsamartha*) there. Other entities are declared to be customarily existent (*saṃvṛtisat*). These two [i.e., the real notion and the customary notion] are [respectively] particulars and generals.

According to this *arthakriyātva*, everything that exists is fully subordinated to causality (*hetutā*): each real entity should have a cause, on the one hand, and an effect, on the other hand. As we see, the Yogācāra assumption of (Y_1) is even stronger in principle than the Stoic assumption of (S_1). In the meanwhile, by the Yogācārins there are two logical terms: the ultimate particular (*svalakṣaṇa*) that expresses real entities and the general (*sāmānyalakṣaṇa*) which is unreal and used for our inferences to characterize and generalize an appropriate 'particular'.

So, due to (Y_1), each event can be considered "being a cause (*upajīvyatva*)" or "being an effect (*upajīvakatva*)". Between them, i.e., between a cause A and an effect B, there is a necessary connection (*sambandha*) $A \Rightarrow B$ which can be articulated in our desire to know which is the reason of treating inference after perception of B (*anumānajñānaṃ me bhavatu*, "let me have knowledge of inference"). In turn, the knowledge of inference produces what is desirable for me (*anumānajñānaṃ madiṣṭasādhananī*) (Vattanky, 2003). As a result, my claim (*sādhya*) about the causal connection $A \Rightarrow B$ is expressed in the form that at a locus (*pakṣa*) x there is the property (*dharma*) A of a subject (*dharmin*) x, because at this locus x there is the sign /token (*liṅga*) B:

1) The *udāhārana/udāharaṇa*: $A \Rightarrow B$

   "Whatever is fiery (A), is smoky (B)."

2) The *upanayana/upanaya*:

   "This hill is smoky"—at the locus x ('hill') there is the sign/token (*liṅga* or *hetu*) B ('smoky').

3) The *niggamana/nigamana*:

   "Therefore, it is fiery"—at the locus x there is the property (*dharma*) A ('fiery').

The causal connection between 'fiery' and 'smoky' is represented by the implication: "If fiery, then smoky", but our inference is based on a sign of 'smoky' and it has the form of converse: "If smoky there, then fiery there". Hence, the implication for the Yogācārins, first, can connect events *A* and *B* and then we have: $A \Rightarrow B$, and second, the implication can connect signs *A* and *B* and, as a result, there is a converse: $B \Rightarrow A$.

The same situation of two possible implications with the same content was known by the Stoics. Their examples of reasoning based on signs or tokens: "If this one has milk in her breast, then she has had a child", "If sweat flows through the surface, then there are imperceptible pores", "If this one has a scar, then he has had a wound" (cf. Sextus Empiricus, *Pyrrhoniae Hypotyposes* II: 106; Sextus Empiricus, *Adversus Logicos* II: 252, 254–255, 309). In all these cases, the causal relationships are expressed by converses: "If this one gave birth, then she has milk in her breast", "If there are imperceptible pores, then sweat flows through the surface", "If this one has had a wound, then he has a scar".

For the Yogācārins, each claim (*sādhya*) about a causal relationship is either acceptable or not to the same extent as for the Stoics—see (S_2):

(Y_2) *Each claim* (*sādhya*) *that a property P* (*dharma*) *belongs to a subject S* (*dharmin*) *is either acceptable or not*: "*S* is *P*" ∨ ¬ "*S* is *P*".

This statement is equivalent to the following:

¬ ("*S* is *P*" & ¬ "*S* is *P*")
¬ ("*S* is *P*" & "*S* is ¬ *P*")

viruddhayor ekadharmiṇy ayogād astu bādhanam |
viruddhaikāntike nātra tadvad asti virodhitā ||
(Tillemans, 2000: 92).

Since two contradictory [properties (*dharma*)] [*A.Sch.*: *P* and ¬ *P*] cannot occur in one subject (*dharmin*) [*A.Sch.*: *S*], let us grant that there is invalidation of the antinomic [reason] (*viruddhaikāntika*). [But] there is no contradiction like that here (Tillemans, 2000: 92).

Furthermore, the claim (*sādhya*) about a causal relationship is either necessarily true or it is necessarily false in the same way as for the Stoics—see (S_3):

(Y_3) *Each claim* (*sādhya*) *that a property P* (*dharma*) *belongs to a subject S* (*dharmin*) *is either necessarily true* [*it expresses a necessarily connection, svabhāva, between S and P*] *or it is necessarily false*: □ "*S* is *P*" ∨ □¬ "*S* is *P*".

liṅgaṃ svabhāvaḥ kāryaṃ vā dṛśyādarśanam eva vā |
sambaddhaṃ vastutaḥ siddhaṃ tad asiddhaṃ kim ātmanaḥ ||
(Tillemans, 2000: 20).

A [valid] reason, which is necessarily connected [with the property to be proved], can only be an essential property (*svabhāva*) or an effect (*kārya*) or a nonperception of something perceptible (*dṛśyādarśana*). Why is this [reason, viz., *utpattimattvādi*,] which is established in reality, not established for oneself [i.e., for the Sāṃkhya]? (Tillemans, 2000: 20).

According to the Sarvāstivādins, each existing item (corporeal body) goes through the three moments (*kṣaṇa*) of all its life: origination, duration, and destruction.

> *uppādaṭṭhitibhaṅgavasena khaṇattayaṃ ekacittakkhaṇaṃ nāma. tāni pana sattarasa cittakkhaṇāni rūpadhammānam āyu* (Anuruddha, *Abhidhammatthasaṅgaha* IV: 3) (Rospatt, 1995: 39).

> The three moments [required] on account of origination, duration and destruction are called "the moment of (i.e., taken by) one mental event" (*citta*). The lifetime of material entities is seventeen of these moments of a *citta* (Rospatt, 1995: 39).

Only such items (i.e., momentary events, *kṣaṇa*, with origination, duration, and destruction) can be causally efficient—they are caused by one item and has affects on other items. All the permanent (non-momentary) entities do not exist and stay out of causality. Let us consider Dharmakīrti's example of permanent entities: "Rabbit's horns (*śaśaviṣāṇa*)". This entity does not exist, as far as we know: $\neg A$, where A := "rabbit's horns". As consequence, it has no cause and cannot affect other things. So, it stays out of causality, indeed, so that there is no cause making the rabbit's horns existing and then we can assert: "It is impossible that there are rabbit's horns" ($\neg\Diamond A$, where A := "rabbit's horns"). Hence, from $\neg A$ Dharmakīrti implicitly infers $\neg\Diamond A$:

(=3) $\neg A \Rightarrow \neg\Diamond A$

In the basic modal logic **K**, from (=3) there are inferred (=) and (=2). So, adding (=3) to **K** gives for us the system **K=** in the same way as adding (=) or (=2) to **K** implies obtaining the logic **K=** for the logical determinism.

For the Sarvāstivādins all conditioned entities (*saṃskṛta, saṃskāra*) are present for a moment, but for the Yogācārins they exist for a moment (Rospatt, 1995: 39). At this moment they manage to function as efficient causes—in the words of Yogācārins: "they realize their own fruit". If there is an uninterrupted flow of causally connected momentary entities of the same kind, they constitute *santāna* or *citta-santāna*—a stream or mind-stream in the meaning close to the stream of phenomena defined by Edmond Husserl in the 20th century A.D.

We can easily note that the Yogācārins have examined "being momentary" as "being actual" or "being just now". Due to this momentariness, each conditioned entity was actual in the past or it is actual in the present or it will be actual in the future. This implies that each entity has own causes and since that it is necessitated in fact—please compare to (S_4):

(Y_4) *Each* (*past, present, future*) *event A is actual* (*momentary*) *and then necessary*: $A \Rightarrow \Box A$.

The momentariness of all real beings is explained by the Yogācārins by the momentariness of our mind (*citta*) as an ultimate observer of all appropriate moments:

> *kṣaṇikaṃ hi cittaṃ prasiddhaṃ, tasya cānye saṃskārāś cakṣūrūpādayo hetutaḥ. tasmāt te 'pi kṣaṇikā iti siddhaṃ. na tv akṣaṇikāt kṣaṇikaṃ bhavitum arhati, yathā nityād anityam iti*
> (Vasubandhu, *Mahāyānasūtrālaṃkārabhāṣya* 151: 1–3), see Lévi (1907).

The mind (*citta*) is commonly known to be momentary (*kṣaṇika*), as well as the other conditioned factors (*saṃskāra*): the eye, the visible and so on as its cause. Therefore, it is proved that they also are momentary. Nevertheless, it is not possible that something momentary [such as mind] originates out of something non-momentary, as it is said (*iti*) [to be impossible that] something impermanent [originates] out of something eternal.

Quite similar:

If the soul were unchangeable, then even at the moment of the rise of a cognition, the soul would remain the same as in its [previous] state of being a noncognizer (*apramātṛ*), and it could not be a cognizer (*pramātṛ*) (Hattori, 1968: 69).

So, first, our mind (*citta*) is momentary (*kṣaṇika*) at all steps of our observations, and therefore all corporal bodies are momentary, too. Hence, for the Yogācārins, momentariness (*kṣaṇikatva*) means that at one moment (*kṣaṇa*) a cause, its effect, and their observer meet each other.

The Stoics, for justifying the logical determinism of axiom (S_4), have appealed to fate—an ultimate personified or depersonified observer who contemplates all causal relationships, see (S_5), while the Yogācārins have supposed a momentary mind scattered everywhere and have put forward the following statement:

(Y_5) *Everything is momentary.*

As we see, the Yogācārins have accepted two modal axioms: first, (=) in the form of (Y_4); and second, (=2) in the form of (=3), i.e., the modal logic **K=**, as well as the Stoics have done the same. Nevertheless, an appropriate philosophy of logical determinism proposed by the Yogācārins was even much more theoretical than the Stoic philosophy. In any case, this Yogācāra discourse has been at the root of establishing logic (*nyāya*) in India. In the *Nyāyasūtra*, one of the first logical treatises, we find a lot of quotations from different texts of Madhyamaka and Yogācāra schools, see some examples in (Vidyabhusana, 1921: 46–47). These quotations prove that the *Nyāyasūtra* was written in the 2nd century A.D. in the Kuṣāṇa Empire—at the time and place of writing the early Madhyamaka and Yogācāra texts. In this treatise, we see even the Madhyamaka manner of refutation of the God existence probably taken from the Nāgārjuna's *Īśvara-kartṛtva-nirākṛtiḥ-viṣṇoḥ-ekakartṛtva-nirākaraṇa*:

īśvaraḥ kāraṇam, puruṣakarmāphalyadarśanāt ||19||
na, puruṣakarmābhāve phalāniṣpatteḥ ||20||
tatkāritatvād ahetuḥ ||21||
(*Nyāyasūtra* IV: 1, 19–21), see Gautama (1936–1944).

Īśvara is a reason for observing that human actions are fruitless. Wrong, because without human actions there is no 'fruiting'. Not an argument—due to the conditionality of the latter.

"If the Lord (*īśvara*) is a cause (A), then human actions are without fruits (B). Nevertheless, they are ever with fruits ($\neg B$). Then the Lord is not a cause ($\neg A$)":

$$\frac{(A \Rightarrow B); \neg B}{\neg A.}$$

The Mādhyamikas and Yogācārins were atheists. Therefore, the Yogācārins have refuted any personified way of describing the logical determinism—they have not used the Stoic terms like fate, destiny, God. The Yogācāra term of momentariness (*kṣaṇika*) denotes a moment of meeting a cause, its effect, and their observer. So, we do not need any assumption of the God existence as a guarantee of causality. Sometimes, the Stoics have tried to describe the logical determinism in a depersonalized manner too by means of the term of τόνος ('tension'), a momentary movement of pneuma (πνεῦμα) at all things causing their attraction and repulsion, on the one hand, and an observation of appearing a causal relationship then and there, on the other hand.

Later, the Naiyāyikas (such as Udayana in his *Nyāyakusumāñjali*) have supposed that for the validity of logical determinism we should accept the existence of God. According to Udayana, there is always an antecedent fact existing prior to an event which determines when this event should begin to be (*niravadhitve aniyaṭāvadhikatve vā kādācitkatvavyāghātāt*) (Bhattacharyya, 1961). If there is no effect, it means that at that moment there is no cause. So, we have two implications: "If a cause (A), then an effect (B)"; "If no effect ($\neg B$), then no cause ($\neg A$)" ($A \Rightarrow B; \neg B \Rightarrow \neg A$). At each locus called *pakṣa,* we can observe an appropriate effect B to draw a suitable causal inference: "If there is the cause A at the locus (*pakṣa*), then there is the effect B at the same place and at the same time." This B is a determinate concomitant (*vyāpya*) of A. So, this B at the locus allows us to conclude that there is the cause A of B (*sādhya*), also. But in this inference, we deal with an instance (*dṛṣṭānta*) that in the locus there is the causality: "If A, then B" ($A \Rightarrow B$). Hence, we deal always with specific instances at different loci, but we infer if they implement a general case of causality of that type (*sāmānyataḥ sādhyasiddhi*). In other words, our inferences concern loci and then they are particular in its character (*viśeṣanumāna*), but we ever suppose that they are general in character (*sāmānjānumāna*). According to the Naiyāyikas, God is a supervisor and guarantor allowing any observer to see general inferences in particular ones. He is the founding causality as such at each moment, and He is a supervisor of all forms of causal connections among things in the same way as "the wood-cutter is seen to supervise the axe in the act of cutting" (Bhattacharyya, 1961). Therefore, God is an intelligent author of causality as a general law (which is *adṛṣṭa,* immaterial), but we see only its examples among material things. This reasoning of Naiyāyikas gives a personification of logical determinism in the spirit of Stoics with their terms such as 'fate' (εἱμαρμένη): 'destiny' (πεπρωμένη), Zeus, "will (βούλησις) of Zeus".

Thus, the Stoics have based the understanding the logical determinism on different moods: personified and depersonified, but the Yogācārins have preferred only the depersonified (atheistic) mood, while the Naiyāyikas have followed only the personified (theistic) mood, see Table 4.3.1.

Table 4.3.1. Comparisons of logic of Stoics, Yogācārins, and Naiyāyikas.

	Stoics	**Yogācārins**	**Naiyāyikas**
Depersonification of logical determinism, *atheistic* approach	$\Box A \lor \Box\neg A$ because of 'pneuma' (πνεῦμα), 'cause' (αιτία), 'reason' (λόγος)	$\Box A \lor \Box\neg A$ because of momentariness (*kṣaṇikatva*)—at one moment a cause, its effect and their observer meet each other	—
Personification of logical determinism, *theistic* approach	$\Box A \lor \Box\neg A$ because of 'fate' (εἱμαρμένη): 'destiny' (πεπρωμένη), Zeus, "will (βούλησις) of Zeus"	—	$\Box A \lor \Box\neg A$ because of God who is the highest supervisor of all non-empirical objects (*adṛṣṭa*) which are being involved into explanations by causality

4.3.7 Nāgārjuna's Intuition towards Modal Logic

The atheistic approach of Yogācārins can be explained, first by their modal intuition that the logical determinism holds true everywhere—this position shared by them was strongly opposed to the Hinduism of the first *purāṇa* texts. The point is that at the time of the early Madhyamaka and Yogācāra, the religion based on the ideas of different *purāṇa* texts became quite popular among common people and, according to these ideas, in our world there are ever possible divine miracles, because some mighty deities can be born among us, such as Kṛṣṇa or Rāma who can perform miracles at any moment. These deities can work wonders for our protection or just for their own sake. The Mādhyamikas and Yogācārins were pitting themselves against the *purāṇ*ic faith in miracles and magic. The existence of miracles would mean the existence of contingency, but according to the logical determinism there is no place for contingency in the world.

In the *Īśvara-kartṛtva-nirākṛtiḥ-viṣṇoḥ-ekakartṛtva-nirākaraṇa* (Stcherbatsky, 1969), Nāgārjuna criticizes the *purāṇ*ic faith from the point of view of modal logic **K=**. It is one of the earliest philosophical texts in India assuming a modal logic. Then the ideas of this short treatise were well developed by the Yogācārins presenting the logical determinism even more explicitly.

Nāgārjuna begins with the following postulate:

(N_1) Everything is either *siddha* or *asiddha*.

Th. Stcherbatsky (1969) translates *siddha* as 'existent' and *asiddha* as 'non-existent'. Nevertheless, *siddha* (as well as its negation, *asiddha*) has an additional modal aspect: *siddha* means something that was resulted from a committed act or something that is reliable because of evidence through our thought. This modal aspect is quite close to the following Stoic logical terms: "to be fated" (τὸ εἱμαρμένον) and "to be necessitated" (τὸ κατηναγκασμένον). The Stoics distinguish the term "to be necessitated" from the term "that which is necessary" (τὸ ἀναγκαῖον). We can say that "to be necessitated" is an *a posteriori* necessity, while "that which is necessary" is an *a priori* necessity. The Stoic claim of (S_3) that each individual proposition (even about future) A is necessarily true, or it is necessarily false: $\Box A \vee \Box\neg A$, concerns only the *a posteriori* necessity. It means that *de facto* we have that $\Box A \vee \Box\neg A$, not in advance. Claim (S_3) of the Stoics holds true only because of causality—see (S_1).

The Nāgārjuna's term *siddha* is not *sat* ('existent'), but something that was caused—see the Yogācāra postulate of (Y_1). Therefore claim (N_1) should be understood as follows:

> (N_2) Each event A happens necessarily (*siddha*) or it does not happen necessarily (*asiddha*): $\Box A \vee \Box\neg A$.

Indeed, Nāgārjuna states that between both terms, *siddha* and *asiddha*, there is a mutual contradiction (*paraspara-virodha*):

> *yaḥ siddhaḥ saḥ siddhaḥ eva. yaḥ tu asiddhaḥ saḥ eva asiddhaḥ. evaṃ tadanayoḥ paraspara-virodhaḥ syād eva. yathā ca āloka-andhakārayoḥ jīvana-maraṇayoḥ iva. atha yatra ālokaḥ vidyate tatra andhakāraḥ na asti. yatra andhakāraḥ tatra ālokaḥ na asti eva. yaḥ hi jīvati saḥ jīvati eva, yaḥ mṛtaḥ mṛtaḥ eva* (Stcherbatsky, 1969: 11).

> Something which is necessitated to be (*siddha*) remains something which is necessitated to be. And something which is necessitated not to be (*asiddha*) remains something which is necessitated not to be. Thus, between both [concepts], there is inevitable mutual contradiction (*paraspara-virodha*)—in the same way as between light and darkness, between life and death. In fact, where there is light there is no darkness and where there is darkness there is no light. Who is alive, is alive, and who is dead, is dead.

It is assumed that *siddha* is subordinated to causality (*hetutā*) and *asiddha* stays out of causality. Now, let us suppose that there exists God (*Īśvara*) as an actor (*karta*)—this idea is expressed in the *purāṇic* texts well. From this it follows that

> *kim asau siddhaṃ karoti atha asiddhaṃ vā* (Stcherbatsky, 1969: 11).

> He [*A.Sch.*: *Īśvara*] can create something which we know as what was caused (*siddha*) or which we know as something what is beyond causality (*asiddha*).

Let us consider the first possibility, namely that a deity creates something causal (*siddha*). But it is impossible, as "for example, we know that man exists. Creating him further cannot be an act of creation; because his existence is already established [i.e., before this alleged creation by God]" (Stcherbatsky, 1969: 9). Now, we can examine the second possibility that God creates something staying outside causality (*asiddha*) such as "oil [crushed] out of sand, which is known to us as non-existent; wool on a tortoise, which is known to us as non-existent" and so on. However, such things are impossible—they are necessitated not to be. Hence, they cannot be created as well.

We can assume also that God "makes non-existent existent" (Stcherbatsky, 1969: 9), i.e., He creates something that is *siddha* from something that is *asiddha*. But it is impossible, too. The point is that both concepts are mutually exclusive, see (N_2).

Thus, God cannot create (i) *siddha* (it exists because of causality, not because of God); (ii) *asiddha* (it stays out of causality and cannot exist at all even by the will of God); (iii) *siddha* from *asiddha* (both concepts are mutually exclusive and cannot be connected).

After that Nāgārjuna asks the question whether God, to be an actor (*karta*), should be born or unborn. Let us regard the first possibility that He can act being unborn. But it is impossible, "because He is Himself something unborn, like 'the son of a barren woman', who being unborn, cannot perform any action like the digging of the earth" (Stcherbatsky, 1969: 9). As we see, the term 'unborn' is exemplified by something that is *asiddha* (necessitated not to be): "the son of a barren woman." In other words, if God is unborn, then it means that He is *asiddha* and cannot act in any way—He is outside causality. Thus, to be active, God should be born—He should be *siddha* Himself, i.e., to be Himself subordinated to causality.

Now, there are the next two possibilities: (i) God was born from something or someone else and (ii) God was born from Himself. Let us start from the first possibility and assume that He was born from another being. Then it contradicts to His definition as someone who has no origin (*anādi*). Consequently, in this case He is *asiddha* as containing the following contradiction: "having no beginning" and "born from another being" at once. So, He does not exist. Now, let us consider the second possibility that God was born from Himself. Nevertheless, it is impossible, too, "since one's own actions cannot relate to one's own self. The blade of a sword, howsoever sharp it may be, cannot cut itself. Even the most expert dancer, howsoever skilful he may be, cannot dance standing on his own shoulders. Besides, it is never observed that one and the same object is the produced (*janya*) and the producer (*janaka*). A person, who is the father is himself also the son—such an assertion is quite unknown in common discourse" (Stcherbatsky, 1969: 9). As we see, according to Nāgārjuna, "being born from Himself" is *asiddha*, i.e., this notion denotes something unreal, because it contradicts to the notion of causality.

Hence, if we posit the logical determinism (Nāgārjuna's claims (N_1) and (N_2)), then *purāṇic* deities are impossible in principle, because they are *asiddha*. This Nāgārjuna's atheistic approach has been inherited by the Yogācārins. So, their atheism is rather explained by the intuitive application of modal logic **K=** in their reasoning, than by a religious believe to avoid the *purāṇic* gods.

4.3.8 Conclusions

There exists the basic modal logic **K** which has two different extensions (Garson, 2006): the logic **T** assuming contingency and the logic **K=** assuming logical determinism. The Aristotelian treatise *On Interpretation* (Περί ερμηνείας/*De Interpretatione*) has introduced some modal-logical relationships which correspond to **T** (Schumann, 2018b, 2021c). In other words, Aristotle has supposed that there are contingent events A such that $\Diamond A \ \& \ \Diamond\neg A$ is true. The Nāgārjunian treatise *Īśvara-kartṛtva-nirākṛtiḥ-viṣṇoḥ-ekakartṛtva-nirākaraṇa* has introduced some modal-logical relationships which correspond to **K=**, namely Nāgārjuna has assumed that any event A is either *siddha* or *asiddha*: $\Box A \vee \Box\neg A$ (Schumann, 2018b, 2021c). His approach has been inherited by the Yogācārins who have developed, first, the doctrine of causality of all real entities (*arthakriyātva*), see (Y_1), and second, the doctrine of momentariness of all real entities (*kṣaṇikavāda*), see (Y_5). The Nāgārjunian ideas of logical determinism can be observed also in the Stoic logical fragments. The Stoics have created, first, the doctrine of causality of all real entities (ἀνάγκη), see (S_1), and second, the doctrine that everything happens in accordance with fate (καθ' εἱμαρμένην), see (S_5). Nevertheless, the Stoic philosophy differs a lot from the Yogācāra philosophy, although both philosophies support the same logical determinism—they contain the same set of logemes (Schumann, 2018b, 2021c).

Thus, after analyzing the earliest logeme structures, we can conclude that the idea of logical determinism first came to mind of the Babylonians in their art of divination (ca. 1800 B.C.). Then this art of composing omens via algebraic techniques found a logical reflection in the Stoic works on logic (3rd century B.C.). Then we observe the same logeme structure, presenting the logical determinism, but in the Buddhist-Indian logic (2nd century A.D.).

References

Abraham, M., Gabbay, D., Hazut, G., Maruvka, Y.E. and Schild, U. 2010a. *The Textual Inference Rules Klal uPrat How the Talmud Defines Sets*. London: College Publications.

Abraham, M., Gabbay, D. and Schild, U. 2010b. *Non-deductive Inferences in the Talmud*. London: College Publications.

Abraham, M., Gabbay, D. and Schild, U. 2013. *Principles of Talmudic Logic*. London: College Publications.

Abraham, M., Belfer, I., Gabbay, D. and Schild, U. 2015. *Fuzzy Logic, and Quantum States in Talmudic Reasoning*. London: College Publications.

Aung, Sh.Z. and Davids, Rh. (trs.). 1915. *Points of Controversy or Subjects of Discourse Being a Translation of the Kathā-vatthu*. Oxford University Press, London.

Banerjee, G.N. 1961. *Hellenism in Ancient India*. Delhi: Munshi Ram Manohar Lal, Delhi.

Beal, S. (tr.). 1884/1981. *Si-Yu-Ki. Buddhist Records of the Western World*. Two vols. Vol. I. London Kegan: Paul, Trench, Trübner & Co.

Beal, S. (tr.). 1906. *Si-Yu-Ki. Buddhist Records of the Western World*. Two vols. Vol. II. Paul, Trench, Trübner & Co., London.

Bhattacharyya, G. 1961. *Studies in Nyāya-Vaiśeṣika Theism*. Calcutta: The Principal, Sanskrit College.

Bobzien, S. 1998. *Determinism and Freedom in Stoic Philosophy*. Oxford: Clarendon Press.

Bodhi, Bh. (ed.). 2000. *The Connected Discourses of the Buddha*. Vol. 1–2. Wisdom Publications, USA.

Bopearachchi, O. 1993. *Indo-Greek, Indo-Scythian, and Indo-Parthian Coins in the Smithsonian Institution*. Washington D.C.

Bull, R. and Segerberg, K. 1984. Basic modal logic. pp. 1–81. *In*: Gabbay, D. and Guenthner, F. (eds.). *Handbook of Philosophical Logic*. Synthese Library (Studies in Epistemology, Logic, Methodology, and Philosophy of Science). Vol. 165. Dordrecht: Springer.

Falconer, W.A. (tr.). 1923. *Cicero: De Senectute De Amicitia De Divinatione*. Cambridge: Harvard University Press.

Falk, H. and Karashima, S. 2012. A first-century Prajñāpāramitā manuscript from Gandhāra—Parivarta 1 (Texts from the Split Collection 1). *Annual Report of the International Research Institute for Advanced Buddhology at Soka University* 15: 19–61.

Freedman, S. 1998. *If a City is Set on a Height: the Akkadian Omen Series "Šumma Ālu Ina Mēlê Šakin"*, Vol.1, Tablets 1–21 (OPKF 17). Philadelphia: The University of Pennsylvania Museum.

Freedman, S. 2017. *If a City Is Set on a Height: The Akkadian Omen Series "Šumma Alu Ina Mēlê Šakin"*, Vol.3; Tablets 41–63. Eisenbrauns.

Fyfe, W.H. (tr.). 1932. *Aristotle in 23 Volumes*, Vol. 23. Cambridge, MA: Harvard University Press; London: William Heinemann Ltd.

Gabbay, U. 2016. *The Exegetical Terminology of Akkadian Commentaries*. Series: Culture and History of the Ancient Near East. Vol. 8. Leiden, Boston, Brill.

Garson, J.W. 2006. *Modal Logic for Philosophers*. Cambridge: Cambridge University Press.

Gautama. 1936–1944. *Nyayasutra with Vatsyayana's Nyayabhasya Based on the Edition by Taranatha Nyaya-Tarkatirtha*. (Calcutta Sanskrit Series, 18 and 19). Calcutta: Metropolitan Printing & Publ.

Geiger, W. (tr.). 1912/1996. *The Mahavamsa, The Great Chronicle of Sri Lanka. Composed in the Late 5th or Early 6th Century CE, by Ven. Mahanama Thera*. Asian Educational Services.

Godley, A.D. (tr.). 1920. *Herodotus*. Cambridge: Harvard University Press.

Hattori, M. (tr.). 1968. *Dignāga, On Perception, Being the Pratyakṣapariccheda of Dignāga's Pramāṇasamuccaya from the Sanskrit Fragments and the Tibetan Versions*. Cambridge, Massachusetts: Harvard University Press.

Hoover, O.D. 2013. *Handbook of Coins of Baktria and Ancient India*. London: Classical Numismatic Group.

Horner, I.B. (tr.). 1969. *Milinda's Questions*. Vols. 1–2. London: Luzac & Co., Ltd.

Jayatilleke, K.N. 1963. *Early Buddhist Theory of Knowledge*. London: George Allen & Unwin Ltd.

Jeyes, U. 1989. *Old Babylonian Extispicy. Omens Texts in the British Museum*. Nederlands historisch-arhaeologisch instituut te Istanbul.

Jongeward, D., Cribb, J. and Donovan, P. 2015. *Kuṣāṇa, Kuṣāṇao-Sasanian, and Kidarite Coins: A Catalogue of Coins from the American Numismatic Society*. New York: The American Numismatic Society.

Kantor, H. and Salvini, M. 2019. Words, concepts, and the middle way: language in the traditions of Madhyamaka thought. *Journal of Indian Philosophy* 47(4): 603–611.

Karttunen, K. 2015. *Yonas and Yavanas in Indian Literature*. Studia Orientalia., Vol. 116, Finnish Oriental Society.

Kienast, B. and Volk, K. 1995. *Die Sumerischen und Akkadischen Briefe des III Jahrtausends aus der Zeit vor der III Dynastie von Ur*. Freiburg: F. Steiner Verlag.

Koch(-Westenholz), U. 2015. *Mesopotamian Divination Texts: Conversing with the Gods: Sources from the First Millennium BCE*. Munster: Ugarit-Verlag.

La Vallée Poussin, L. de (tr.). 1977. *Mūlamadhyamakakārikās* (as found in the *Prasannapadā Madhyamakavṛttiḥ* by Candrakīrti). Bibliotheca Buddhica iv, St. Petersburg, 1903–1913; reprinted with alterations by P.L. Vaidya, Madhyamakaśāstra of Nāgārjuna, Buddhist Sanskrit Texts no. 10, Darbhanga, 1960. New edition (using a fourth manuscript of the Kārikās only) by J.W. de Jong, Adyar.

Lahiri, A.N. 1965. *Corpus of Indo-Greek Coins*. Calcutta: Poddar Publications.

Levi, S. 1896. Notes sur les indo-scythes, *Journal Asialique*, pp. 444–484. Levi's article is translated in: Philips, W.R. 1903. "Notes on the Indo-Scythians". *Indian Antiquary* 32: 381–389.

Lévi, S. (tr.). 1907. *Mahāyānasūtrālaṃkāra, Exposé de la doctrine du Grand Véhicule*. Vol. I. Paris.

Malvania, D. (tr.). 1971. *Dharmottarapradipa* (subcommentary on Dharmottara's *Nyayabindutika* on Dharmakīrti's *Nyayabindu*). Patna: Kashiprasad Jayaswal Research Institute (Tibetan Sanskrit Works, 2).

Maul, S.M. 2007. Divination culture and the handling of the future. pp. 361–372. *In*: Leick, G. (ed.). *The Babylonian World*. Oxford: Routledge.

Mill, L.Kh. (tr.). 2015. *Sutta Nipāta, Smaller Discourse on Quarrelling*. Sutta Central.

Mitchiner, M. 1975–1976. *Indo-Greek and Indo-Scythian Coinage*. Sanderstead, London: Hawkins Publications.

Mukherjee, B.N. 1995. *The Great Kuṣāṇaa Testament*. Calcutta: Indian Museum Bulletin.

Nārada, U. and Kumārābhivaṁsa, U. (tr.). 1998. *The Sixth Book of The Abhidhamma Piṭaka. The Book on Pairs* (*Yamaka*). Vol. 1. Penang: Burmese Buddhist Temple.

Olivelle, P. (tr.). 2006. *Manu's Code of Law*. Oxford University Press.

Pingree, D. 1978. *The Yavanajātaka of Sphujidhvaja*. 2 vols. Harvard Oriental Series 48.

Pingree, D. 1981. Jyotiḥśāstra. *In*: Jan Gonda (ed.). *A History of Indian Literature*. Vol. VI, Fasc. 4. Wiesbaden: Otto Harrassowitz.

Potter, K.H. 1977. *Encyclopedia of Indian Philosophies*. Vol. II: Indian Metaphysics and Epistemology: The Tradition of Nyāya-Vaiśesika up to Gangeśa. Princeton: Princeton University Press.

Pradhan, P. (ed.). 1967. *Abhidharma-koshabhāṣyaṃ of Vasubandhu*. Patna: K.P. Jayaswal Research Institute, TSWS.

Rapin, C. 1992. *La Trésorerie du palais hellénistique d'Aï Khanoum. L'apogée et la chute du royaume grec de Bactriane* (*Fouilles d'Aï Khanoum VIII, Mémoires de la Délégation archéologique française en Afghanistan 33*), Paris, De Boccard. Chapitre XII sur les textes littéraires: pp. 115–130. Sur le papyrus philosophique: pp. 115–121, dessins pl. 52, photographie couleur pl. 125. Bibliographie sur les textes littéraires d'Aï Khanoum, p. 387.

Rea, J.R., Senior, R.C. and Hollis, A.S. 1994. A tax receipt from Hellenistic Bactria. *Zeitschrift für Papyrologie und Epigraphik* 104: 261–280.

Reiner, E. and Pingree, D. 1998. *Babylonian Planetary Omens*. Part Three. Cuneiform Monograph 11. Groningen: Styx.

Ronkin, N. 2005. *Early Buddhist Metaphysics. The Making of a Philosophical Tradition*. Routledge Curzon.

Rospatt, A. von. 1995. *The Buddhist Doctrine of Momentariness. A Survey of the Origins and Early Phase of This Doctrine up to Vasubandhu*. Stuttgart: Franz Steiner Verlag.

Salomon, R. 2011. The Macedonian Month Xandikos in Gandhāran Inscriptions. *Studia Orientalia* 110: 165–170.

Schlosser, A. 2016. On the Bodhisattva Path in Gandhāra Edition of Fragment 4 and 11 from the Bajaur Collection of Kharoṣṭhī Manuscripts. *In*: *Zur Erlangung des Doktorgrades eingereicht am Fachbereich Geschichts- und Kulturwissenschaften der Freien Universität Berlin im Dezember 2013*. Überarbeitete Fassung.

Schumann, A. 2008. Non-well-foundedness in Judaic logic. *Studies in Logic, Grammar, and Rhetoric* 13(26): 41–60.

Schumann, A. (ed.). 2010a. *Judaic Logic*. Gorgias Press.

Schumann, A. 2010b. Preface, Introduction. pp. 7–8, 11–33. *In*: Schumann, A. (ed.). *Judaic Logic*. Book series: Judaism in Context, 8. Gorgias Press.

Schumann, A. 2010c. Judaic Syllogistics: The Baba Qama from the logical point of view. pp. 239–254. *In*: Schumann, A. (ed.). *Judaic Logic*. Book series: Judaism in Context, 8. Gorgias Press.

Schumann, A. 2010d. Preface: Religious logic as a part of philosophical logic. pp. 10–19. *In*: Schumann, A. (ed.). *Logic in Religious Discourse*. Ontos Verlag.

Schumann, A. 2011a. Preface. *History and Philosophy of Logic* 32(1): 1–8.

Schumann, A. 2011b. Qal wa-Homer and Theory of Massive-Parallel Proofs. *History and Philosophy of Logic* 32(1): 71–83.

Schumann, A. 2012a. *Talmudic Logic*. London: College Publications.

Schumann, A. 2012b. Kabbalistic-Leibnizian automata for simulating the universe. pp. 259–280. *In*: Evgeny Katz (ed.). *Molecular and Supramolecular Information Processing. From Molecular Switches to Logic Systems*. Weinheim, Germany: Wiley-VCH.

Schumann, A. 2012c. Cellular Automaton of the Lord. *Polish Journal of Philosophy* 6(2): 103–110.

Schumann, A. 2013a. Logical cornestones of judaic argumentation theory. *Argumentation* 27(3): 305–326.

Schumann, A. (ed.). 2013b. *Pragmatic Studies in Judaism*. Gorgias Press.

Schumann, A. 2013c. Preface, Introduction. pp. 11–12, 15–24. *In*: Schumann, A. (ed.). *Pragmatic Studies in Judaism*. Book series: Judaism in Context, 14. USA: Gorgias Press.

Schumann, A. 2013d. Pragmatic limits of Halakhic logic. pp. 67–98. *In*: Schumann, A. (ed.). *Pragmatic Studies in Judaism*. Book series: Judaism in Context, 14. USA: Gorgias Press.

Schumann, A. 2015. *Logika Talmudu*. Kraków: Copernicus Center Press.

Schumann, A. 2016. Physarum Polycephalum syllogistic L-systems and Judaic roots of unconventional computing. *Studies in Logic, Grammar, and Rhetoric* 44(57): 181–201.

Schumann, A. (ed.). 2017a. *Philosophy and History of Talmudic Logic*. London: College Publications.

Schumann, A. 2017b. On the Babylonian origin of Symbolic logic. *Studia Humana* 6(2): 126–154.

Schumann, A. and Kuznetsov, A.V. 2017. Talmudic foundations of mathematics. pp. 67–74. *BICT 2017 – 10th EAI International Conference on Bio-Inspired Information and Communications Technologies*.

Schumann, A. 2018a. On a Logical Reconstruction of the Kathavatthu. pp. 1509–1512. *In*: *Proceedings of the 3rd International Conference on Contemporary Education, Social Sciences and Humanities* (ICCESSH 2018).

Schumann, A. 2018b. Did Nāgārjuna now Modal Logic? *Логико-философские штудии* 16(1–2): 95.

Schumann, A. 2019a. Did the Neo-Babylonians construct a symbolic logic for legal proceedings? *Journal of Applied Logics – IfCoLog Journal of Logics and Their Applications* 6(1): 31–82.

Schumann, A. 2019b. Do ant colonies obey the Talmud? pp. 261–272. *In*: Adamatzky, A., Akl, S. and Sirakoulis, G.Ch. (eds.). *From Parallel to Emergent Computing*, CRC Press.

Schumann, A. 2019c. On the origin of Indian Logic from the viewpoint of the Pāli Canon. *Logica Universalis* 13: 347–393.

Schumann, A. 2020a. Legal argumentation in Mesopotamia since Ur III. *Journal of Argumentation in Context* 9(2): 126–154.

Schumann, A. 2020b. The Milindapañha in the context of History of Indian Civilization. *RUDN Journal of Philosophy* 24(4): 544–569.

Schumann, A. 2021a. Logical reasoning for forecasting in Mesopotamia. *Studies in Logic, Grammar, and Rhetoric* (accepted).

Schumann, A. 2021b. On the Origin of logical determinism in Babylonia. *Logica Universalis* 15: 331–357.

Schumann, A. 2021c. Nāgārjunian-Yogācārian Modal Logic versus Aristotelian Modal Logic. *Journal of Indian Philosophy*, https://doi.org/10.1007/s10781-021-09470-5.

Schumann, A. and Sazonov, V. 2021. Omens as Logic. *NABU: Nouvelles Assyriologiques Brèves et Utilitaires* 3: 192–195.

Sear, D. 1979. *Greek Coins and Their Values*, Vol. 2, Asia and Africa. London.

Sebastian, C.D. 2016. *The Cloud of Nothingness. The Negative Way in Nagarjuna and John of the Cross.* Sophia Studies in Cross-cultural Philosophy of Traditions and Cultures. India: Springer.

Senart, É. 1905–1906. The Inscriptions in the caves at Nasik. *Epigraphia Indica and Record of the Archaeological Survey of India* 8: 59–96.

Seyfort Ruegg, D. 2000. *Three Studies in the History of Indian and Tibetan Madhyamaka Philosophy. Studies in Indian and Tibetan Madhyamaka thought.* Part 1. Wiener Studien zur Tibetologie und Buddhismuskunde; Heft 50. Wien: Arbeitskreis für Tibetische und Buddhistische Studien, Universität Wien.

Sharma, P. (tr.). 2014. *Caraka-saṃhitā: Agniveśa's Treatise Refined and Annotated by Caraka and Redacted by Dṛḍhabala: Text with English Translation.* Vol. 1. Varanasi: Chaukhambha Orientalia.

Shastri, J.L. (ed.). 1983. *Shastri Mānava-Dharmaśāstra. Manusmṛti, With the Sanskrit Commentary Manvartha-Muktāvalī of Kullūka Bhaṭṭa.* Delhi: Ministry of Human Resources.

Shrava, S. 1993. *The Dated Kushāṇa Inscriptions.* New Delhi: Pranava Prakashan.

Sims-Williams, N. and Cribb, J. 1996. A new Bactrian inscription of Kanishka the Great. *Silk Road Art and Archaeology* 4: 75–142.

Sims-Williams, N. 2001. *Bactrian Documents from Northern Afghanistan I: Legal and Economic Documents.* Oxford: Oxford University Press.

Sims-Williams, N. 2007. *Bactrian Documents from Northern Afghanistan II: Letters and Buddhist Texts.* London: Nour Foundation.

Stcherbatsky, Th. 1969. A Buddhist Philosopher on Monotheism. pp. 1–11. *In: Papers of Th. Stcherbatsky* (Soviet Indology series, no. 2). Calcutta: Indian Studies: Past & Present.

Swamy, M.C. (tr.). 1874. *The Daṭhavansa or, The History of the Tooth-Relic Gotama Buddha.* London: Trübner & Co.

Tachikawa, M. 1981. *The Structure of the World in Udayana's Realism: A Study of the Lakṣaṇāvalī and the Kiraṇāvalī (Studies of Classical India)* (Vol. 4). Dordrecht, Boston, London: D. Reidel Publishing Company.

Tarn, W.W. 1951. *The Greeks in Bactria and India.* Cambridge: Cambridge University Press.

The Cowherds. 2011. *Moonshadows: Conventional Truth in Buddhist Philosophy.* Oxford: Oxford University Press.

Thera, Ñ. (tr.). 2013. *Salha Sutta: To Salha.* Kandy, Sri Lanka: Buddhist Publication Society.

Thera, S. (tr.). 1981. *Kālāma Sutta. The Buddha's Charter of Free Inquiry.* The Wheel Publication No. 8. Kandy, Sri Lanka: Buddhist Publication Society.

Thucydides. 1910. *The Peloponnesian War.* London, New York.

Tillemans T.F. 1984. Two Tibetan texts on the "neither one nor many" argument for śūnyatā. *Journal of Indian Philosophy* 12(4): 357–388.

Tillemans, T.F. 1989. Formal and semantic aspects of Tibetan Buddhist debate logic. *Journal of Indian Philosophy* 17(3): 265–297.

Tillemans, T.F. 2000. *Dharmakīrti's Pramāṇavārttika: An Annotated Translation of the Fourth Chapter* (parārthānumāna): K.1-148. Wien, Germany: Österreichische Akademie der Wissenschaften.

Tversky, A. and Kahneman, D. 1974. Judgments under uncertainty: Heuristics and biases. *Science* 185(4157): 1124–1131.

Ui, H., Suzuki, M., Kanakura, Y. and Tada Sendai, T. (eds.). 1934. *A Complete Catalogue of the Tibetan Buddhist Canons* (Bkaḥ-ḥgyur and Bstan-ḥgyur). Tokyo: Tōhoku Imperial University.

Vats, M.S. 1925–1926. Unpublished votive inscriptions in the Chaitya Cave at Karle. *Epigraphia Indica and Record of the Archaeological Survey of India* 18: 325–329.

Vattanky, J. 2003. *A System of Indian Logic. The Nyāya Theory of Inference*. Analysis, Text, Translation, and Interpretation of the *anumāna* section of *Kārikāvalī, Muktāvalī* and *Dinakarī*. NJ, New York: Routledge Curzon.

Vidyabhusana, S.Ch. 1921. *A History of Indian Logic: Ancient, Mediaeval, and Modern Schools*. Motilal Banarsidass.

Watanabe, F. 1983. *Philosophy and its Development in the Nikayas and Abhidhamma*. Delhi, Varanasi, Patna: Motilal Banarsidass.

Westerhoff, J. 2005. Nāgārjuna's Catuṣkoṭi. *Journal of Indian Philosophy* 34(4): 367–395.

Westerhoff, J. 2019. Nāgārjuna and the Philosophy of Language. *Journal of Indian Philosophy* 47(4): 779–793.

Winitzer, A. 2017. *Early Mesopotamian divination literature: Its Organizational Framework and Generative and Paradigmatic Characteristics*. Leiden; Boston: Brill.

Yoshimizu, Ch., Sithar, K. and Gyaltsen Gonta, S. (eds.). 1989. *Descriptive Catalogue of the Naritasan Institute Collection of Tibetan Works*. Vol. I. Narita: Naritasan Shinshoji.

Zorzi, N. De. 2009. Bird divination in Mesopotamia, new evidence from BM 108874. *KASKAL Rivista di storia, ambienti e culture del Vicino Oriente Antico*. Vol. 6.

Zorzi, N. De. 2011. The omen series *Šumma Izbu* – internal structure and hermeneutic strategies. *KASKAL. Rivista di storia, ambienti e culture del Vicino Oriente Antico*. Vol. 8.

Chapter 5

Some Prospects for the Practical Application of the Theory of Logical Competence in the Modern Legal Culture

Legal argumentation is now considered sometimes within the framework of the so-called informal logic as a rhetorical process that cannot be formalized by means of symbolic or mathematical logic. However, as we have seen, already at the dawn of its appearance, the legal discourse in Mesopotamia presupposed an impeccable logical competence, which means a comprehensive possibility of logical formalization by means of modern symbolic logic. This competence can be developed further, and the chapter discusses the promising task of building expert systems for the automation or semi-automation of legal decisions. Obviously, these systems use a rich arsenal of formal logic to build an integral argumentation model with the subsequent possibility of a computer implementation of this model. The novelty of this task is associated with an opportunity of building an expert system of an electronic lawyer to strengthen the role of logical competence, already represented by artificial intelligence, in the legal practice. This chapter shows (Section 5.1) that the strategies used in a court are dynamic and their mathematical formalization is possible within a new game theory, in which we can model the dynamics of strategies depending on changing contexts and new reactions of the opponent. The trial is thus a reflective (reflexive) game in which the opponents seek to predict each other's strategies. This process can also proceed as a conscious control when it is possible to impose on the opponent the interpretation of events through the modification of motives or modification of information.

In the formalization of legal argumentation for systems of an electronic lawyer, models of preference can be divided into criminal and civil cases (Section 5.1). For the first model, the Aristotelian syllogistic is applicable, and for the second model the non-Aristotelian one.

Thus, in the chapter it is affirmed that a creation of working system of an electronic lawyer is probable based on some new logical methods for modelling the legal argumentation from the point of view of reflective games. Hence, some theoretical fundamentals are prepared for further practical research in the field of electronic justice and for the creation of an expert system of an electronic lawyer (Section 5.2).

5.1 Some Ways to Formalize Litigation Using Logic

5.1.1 Problem Setting

The formation of informal logic took place in line with the general trend of avoiding the apriorism in modern philosophy. In symbolic logic, an anti-apriorism was expressed in creating non-classical logics, particularly intuitionistic and probabilistic logics.

Currently, informal logic is the most popular branch of logical knowledge used in the analysis of legal argumentation. As a result, the legal argumentation continues to be traditionally viewed more as a rhetorical process, and the success in a court case is associated with the oratorical persuasiveness of the arguments presented.

However, recently, some attempts to logically formalize the trial became relevant in respect to: (a) the possibility of electronic document management (the electronic filing of a statement of claim, the automatic verification of the completeness of documents, the automatic calculation of opportunities for an outcome of a court case), and (b) the possibility of building an expert system of an electronic lawyer, which implements the process of automatic decision-making based on existing facts and legal documents.

Traditionally the legal argumentation has been considered as a rhetorical process that cannot be formalized by means of symbolic logic. Nevertheless, the informal approach to the legal argumentation cannot be used in expert systems to automate or semi-automate legal decisions, since this approach is intuitive and contextual. Thus, we are still forced to look for means for the logical formalization of legal argumentation—this is the only way to create an architecture for the automation or semi-automation of legal decisions. In this study, it is possible to examine a game-theoretic approach for this formalization.

Within the game-theoretic approach, the semantics of legal argumentation can be logically formalized within the situation theory of meanings created by Kenneth John Barwise (1942–2000) and John Richard Perry (b. 1943), taking into account a specific judicial practice, giving a set of 'situations'. As a result, the situation semantics (Barwise and Perry, 1980, 1983) makes it possible to create a formal database of a possible court case with a specific 'situation' and a specific 'reference'. In this database, we can create a court case library and propose a decision model in the way of Schumann and Woleński (2015), and Schumann and Szelc (2014, 2015).

In this section, it is shown that each trial can be presented as a complex concurrent game between the main participants in the trial. Meanwhile, the game scenarios of criminal and civil proceedings are different, which is determined by the different nature of the participants in the trial and different rules for their participation. Litigation strategies, in this case, are formed by the prosecutor and the defense attorney in the framework of a criminal case, as well as by the plaintiff and the defendant (or their lawyers) in the framework of a civil claim. In both the cases, we have two opposing sides.

So, the trial can be modeled using the game theory. In this case, the protagonists and the antagonists of the trial (such as judge, prosecutor, defender, accused, and witness) are players, and their strategies of proof and argumentation are, in fact, game strategies. The section proposes a formalization of the trial in the non-linear game theory.

5.1.2 Litigation as a Game Situation

Each criminal case can be presented in a game-theoretic form as a complex parallel game between the main participants in a trial: (1) the presiding judge; (2) other participants of the trial (for example, judges or jurors), who make a judgment; (3) the prosecutor (he or she makes an accusation, then lists all the facts and arguments confirming the accused's guilt); (4) the defense counsel (he or she finds facts and arguments justifying the accused); (5) the accused (whose guilt or innocence must be established in the course of the trial); (6) the prosecution witnesses (they are involved by the prosecutor); (7) the defense witnesses (they are invited by the defense lawyer). There are also auxiliary participants in the criminal trial: (8) the secretary of the court session; (9) the bailiffs, etc.

The criminal process begins with the presiding judge giving the floor to the prosecutor. After that, the prosecution witnesses are called, from whom the lawyer can ask questions, or simply listen to their testimony. Then the defense counsel comes forward and calls the defense witnesses, who, in turn, may be asked additional questions by the prosecutor. After that, the presiding judge gives the floor to the accused. After all the statements, the facts presented to the court are announced.

Consequently, the court must (1) understand the corpus delicti by hearing the prosecutor, (2) find out the corpus delicti on the basis of the testimony of the prosecution witnesses, (3) get acquainted with the defendant's position after hearing the lawyer and defense witnesses, (4) hear the testimony of the accused, and (5) make a fair decision. At the same time, the scope of the trial is limited only by the charge brought forward and is carried out only against the accused on the facts that are imputed to him or her.

To the same extent, a civil litigation can be presented in a game-theoretic form as a complex concurrent game between the main participants in a civil process: (1) the presiding judge, (2) the plaintiff (he or she formulates a claim against the defendant), (3) the lawyer or defender of the interests of the plaintiff (he or she finds facts, arguments in defense of the claim), (4) the defendant (he or she disputes the validity of the claim), and (5) the attorney or defender of the interests of the defendant (he or she finds facts, arguments against the claim). There are also ancillary participants in the civil litigation such as (6) the court secretary, etc.

The civil litigation begins with the presiding officer giving the floor to the plaintiff or his/her lawyer. Then the defendant or his/her lawyer speaks. After that, the parties begin the debate on the materials of the claim. The presiding judge must

- fully implement the principles of civil procedural law,
- establish factual circumstances relevant to the case,
- define the rights, obligations, and legitimate interests of two parties,
- make a court decision on the merits.

In both criminal and civil cases, the presiding judge must

- hear all the explanations of the parties, including third parties, testimonies of witnesses, expert opinions,
- examine written as well as material evidence,
- listen to audio recordings concerning the matter, and
- watch appropriate video recordings.

During the court session, a protocol must be kept, for which the court secretary is responsible. The trial record should reflect the entire course of the process.

As we can see, game scenarios of criminal and civil litigation are different, which is determined by the different nature of the participants in the trial and the different rules of their participation.

Before the trial, each of the two parties creates a suitable plan for (1) the opening speech, (2) initial questioning of the opposing witness, (3) cross-examination, (4) re-examination, and (5) closing argument. This plan is a game strategy of each opposing side. However, the plan is unclear, because it can be interrupted, rejected, or not implemented in one form or another at each step of the trial. For example, the chairperson might ask a question out of turn. The witness may be confused. In this regard, it sometimes makes sense to improvise in these new, unforeseen conditions. And then analyze where and why the plan failed to improve our strategy for the next trial.

So, the peculiarity of the trial is that in the theoretical game formalization, we have two communicating parallel processes: (1) the prosecutor or the plaintiff, his/her vision of the problem and his/her arguments, and (2) the defender or the defendant, his/her vision of the problem and his/her arguments. The chosen strategies, however, are mobile and change over time. Hence, in both the criminal and civil litigation we have several players who intercommunicate among themselves and can change their strategies at each stage—so, their games are concurrent and have an especial mathematics (Abramsky and Mellies, 1999; Brenguier, 2013; Schumann et al., 2014, 2015; Pancerz and Schumann, 2018). In other words, their game is not classical, where players, their strategies and payoffs should be fixed in advance and cannot change.

5.1.3 Payoff Cellular Automata

In some modern studies in game theory outside of the classical approach, it was found that we often deal with the dynamically communicating processes for which there may be no equilibria, for example, equilibria in the sense of Nash (Ackerman, 2002; Berger, 2009; Crawford and Iriberri, 2007). In these games, the strategies themselves change. And if we deal with repeated or infinite games (Brandenburger, 2008; Brandenburger and Keisler, 2006; Brandenburger et al., 2012; Capretta, 2007; Mailath and Samuelson, 2006), we should use co-algebraic or other non-classical tools. In Schumann (2014b), a mathematical formalization of such games was proposed, in which the intermediate solutions, obtained during the game, affect the strategies of other players in the framework of changing contexts. The game is defined by (1) the players, who are represented in an appropriate transition rule of cellular automata, (2) the strategies of the players, which are given as combinations of all possible payments for each outcome of the strategy known before the game and presented as states of cells of cellular automata, and (3) the neighborhood of cellular automata, which makes some strategies actual and others non-actual depending on the context (that is, it activates the most important strategies in a given context at time t).

Such a formalization can simulate the trial as a game not only logically, but also contextually. Therefore, the players make decisions not only in the conditions of these payments, which correspond to each possible outcome of the strategy, but also in the conditions of various other circumstances, for example, how the strategies can be applied in this context, how they can be changed by this context, how the past contexts have influenced the existing ones, whether some statements during the trial are genuinely false, etc.

So, instead of payoff matrices, we use payoff cellular automata (Schumann, 2014b), which consider the limitation of contexts. The symbolic logic used in payoff matrices becomes unconventional in payoff cellular automata (such as dynamic, temporal, and massively parallel logics) (see Schumann, 2012).

5.1.4 Litigation as a Reflective Game

The opposing parties in a lawsuit seek to predict the opponent's strategies and hide the predispositions of their own strategies for a while. Such games are called reflective. The rules in reflective games depend on the following parameters (Novikov and Chkhartishvili, 2005):

- the number of agents (a winner is presented as a player who is able to be at the level of reflection n or $n + 1$, while all other interlocutors remain at the level of reflection n, and for each pair of agents the number n may be different);
- the agents' preferences (different goal functions and a dependence of their payments on actions, for example, when we know that each agent is interested in maximizing payments and for this purpose, he or she implements the minimum set of steps, and for different agents this set may be different);
- the set of admissible actions for the agents (there are actions that are unacceptable for everyone in the group, and there are actions that are expected or not expected by other agents, but these actions are allowed);
- the knowledge of agents (at the time of making a decision, all preferences of other agents must be reported, possibly falsely);
- the order of steps (i.e., the sequence of selection of actions in the group of agents).

The more agents there are in a group, the more difficult it is to become a winner in it (i.e., to achieve victory in the reflective game in the trial).

In reflective games, we are dealing with an unlimited hierarchy of cognitive pictures. Consider a bimatrix game with agents i and j. Each of them may have their own picture of event A. Let us designate these pictures K_iA and K_jA, respectively. A reflection of the first order (thoughts about the enemy's pictures) is expressed through pictures of the second order, which are defined as K_jK_iA and K_iK_jA, where the pictures K_jK_iA are representations of agent j about the pictures of agent i, the pictures K_iK_jA are representations of agent i about the pictures of agent j. A reflection of the second order determines which pictures of the proponent relate to pictures of the opponent. At this level, the third-order reflection patterns $K_iK_jK_iA$ and $K_jK_iK_jA$ appear. And, so on *ad infinitum*. A collection of all pictures K_iA, K_jA, K_jK_iA, K_iK_jA, $K_iK_jK_iA$, $K_jK_iK_jA$, etc. gives an endless hierarchy. Let $K_jK_iA = K_j^2A$, $K_iK_jA = K_i^2A$, $K_iK_jK_iA = K_i^3A$, $K_jK_iK_jA = K_j^3A$, etc.

We have the following properties:

- both $K_i^{n+1}A$ and $K_j^{n+1}A$ are implemented by both agents simultaneously—this means that agents i and j are at the n-th reflection level at the same time;
- only $K_i^{n+1}A$ is realized in the game (then we can take $K_j^{n+1}A = K_j^nA$)—this means that agent i stays at the n-th level of reflection, and agent j remains at the $(n-1)$-th level of reflection;
- only $K_j^{n+1}A$ is realized in the game (then we can take $K_i^{n+1}A = K_i^nA$)—this means that agent j is at the n-th level of reflection, and agent i remains at the $(n-1)$-th level of reflection.

For any reflective game at the n-th level of reflection, we can create a tree of diagrams. The nodes of the tree correspond to real or virtual agents participating in the reflective game. The branches of the tree model the mutual knowledge of agents at the level of reflection n: if there is a path from (real or virtual) agent i to agent j, then agent j is correctly aware of agent i. In this case, $K_j^{n+1}A$ is a realizable strategy. If both $K_i^{n+1}A$ and $K_j^{n+1}A$ are realizable strategies of the same game, then the corresponding tree has a cycle.

Let us consider how the reflective game is presented in a lawsuit. Let i be a plaintiff/accuser and j be a defendant/accused. By A we denote a claim/charge within the framework of a civil/criminal court case. Then K_iA means the explanation of facts A by agent i for the benefit of i, and K_jA means the explanation of facts A by agent j for the benefit of j. Accordingly, K_jK_iA means a successful explanation of the picture K_iA by agent j for his or her own benefit, and K_iK_jA means a successful explanation of the picture K_jA by agent i to his or her advantage. Further, $K_iK_jK_iA$ means a successful explanation of the picture K_jK_iA by agent i for his or her own benefit, and $K_jK_iK_jA$ means a successful explanation of the picture K_iK_jA by agent j for his or her own benefit, etc. To win the lawsuit, agent i must successfully explain the picture from agent j, and agent j must successfully explain the picture from agent i. As a result, at level n for agent j, it is fulfilled that K_i^nA is a subset of $K_j^{n+1}A$, which means that agent j has really corresponded to level n and there he or she has successfully explained the picture of K_i^nA from agent i. Correctly determining the level of reflection of opponent i implies victory in the game by agent j.

Any everyday dialogue can be considered a reflective game. Each person, saying certain things, is trying to learn something from us or even get from us. We are always trying to understand the motives of our interlocutor. How does he or she want to influence us? How exclusive is his or her message that is uttered? Will we know more after the conversation? Is he or she sincere?

Emotions, which are expressed in illocutions, are one of the main forms of reflection. The exchange of emotions is always a reflective game, i.e., a way to manipulate others.

The insufficient knowledge (the lack of $K_i^{n+1}A$) for agent i at the level of reflection n leads to an actual vector of actions at level n, which may differ from the vector expected by agent j. To achieve the implementation of our strategy, it is advisable to observe the following provisions (Schumann, 2014c, 2014d):

- The reflective game involves a finite set of possible contexts that change strategies.
- The equally informed agents choose identical actions according to the level of reflection n.
- The rational behavior of agents is that each of them seeks to maximize the goal function by choosing appropriate measures, predicting which actions are chosen by other agents who are rational in terms of knowing the level of reflection n about other agents.

5.1.5 Specificity of Reflective Games

So, we can try to simulate the trial using the game theory. In this case, the protagonists and the antagonists of the trial (judge, prosecutor, defender, accused, witness, etc.) are players, and their argumentation strategies are game strategies. The main problem in this modeling is to determine how acceptable the classical game theory is in the game-theoretic formalization of the trial.

In the classical game theory, the human behavior is understood as an interaction of decision-makers for the sake of their own gains. It is assumed that each player has a specific goal, called a payoff, and takes steps to consciously achieve this goal. For this purpose, each player takes into account the knowledge or mathematical expectations of other decision makers, because the payoffs of different players may conflict. So, the main element of game theory is the player, who can be interpreted as a person or as a group of people. A player is anyone who influences the payments of others.

Consequently, in the classical game theory, the set of possible actions of individual players is considered atomic (i.e., actions have no common elements). Therefore, we are dealing with the set Ω_G of all strategies (actions available to each player) in the game G in such a way that Ω_G is a division into mutually disjoint subsets E_1, E_2 ..., where each E_i contains the actions of player i, $i = 1, 2,...$ Each element $\langle e_1, e_2,\ldots\rangle$ of the set $E_1 \times E_2 \times ...$ is a result of the game, and this is associated with payments $\langle a_1, a_2,\ldots\rangle$, where a_i is a payoff of player i after using the strategy e_i, $i = 1, 2,...$ So, each player has his or her own strategies, and combinations with the strategies of others give payoffs. Thus, the number of players is established and known to all parties. This is the first assumption of classical game theory. We have mutually disjoint subsets E_1, E_2,... from Ω_G and each player knows this division. In other words, each player chooses among two or more possible strategies and knows how each strategy chosen by him (her) or another player determines the whole game.

However, in reflective interactions, very often we find ourselves in situations where we do not know all the players (for example, lobbyists in political games can be hidden), thus, we do not know an appropriate division of Ω_G, and perhaps we do not know all the strategies of Ω_G as those. In this case, the possible actions of individual players cannot be considered atomic.

The second assumption of classical game theory is that all players are considered initially rational. This is understood as follows (Schumann and Szelc, 2014, 2015):

i) The players know all the rules of the game, but it may happen that some of the rules change during the game. In reflective interactions, everything depends on the players—we can always move from one level of reflection to the next one.
ii) The players assume that the other parties are completely rational. This means that all players have a zero level of reflection: they know each other and know everything about each other, including all strategies. Obviously, this assumption does not hold for the reflective interactions. I can cheat to hide my true motives and misinterpret facts. Therefore, I may, justifiably, distrust others.
iii) All players try to maximize their utility. As a result, all players strive for the highest payoffs. However, in the reflective interactions, I can give up the highest payoffs for some symbolic reasons—I can, for example, act altruistically and sacrifice something for someone. For example, take the blame of others.
iv) All players have resistance points, that is, they can only make decisions that match their security levels. In reflective interactions, I can avoid these points, also, for altruistic reasons.
v) All players know the usefulness of certain actions and the preferences of other players and choose one action from many preferred options. The preference remains constant throughout the game. But in reflective interactions, players can lie and hide their true preferences or change their preferences through the interaction.
vi) Pareto's efficiency holds for any game. All players can make the most effective decisions that maximize their own winnings. It can be recalled that a distribution of utility A is called Pareto superior over another distribution B if from state B there is a possible redistribution of utility to A such that at least one player has the better payoff in A than in B and no player has the worse payoffs. In situations of reflective interactions, where preferences can change during the game, there is no Pareto efficiency in general.

From all the given assumptions of the classical game theory mentioned above, it follows that there are always general solutions for a game, giving an endogenously stable or balanced state. These decisions are called equilibria. This concept is extrapolated from physics, where it means a steady state in which all causes, internal to the system, balance each other, if this has not changed because of the intervention of some external force. Thus, classical game theorists view economic systems as a mutual equilibrium of causal relationships, just as they do it in physical systems. This balance can only be found through mathematical calculations of payoffs.

If we use reflective games (Chkhartishvili, 2003, 2010, 2012; Chkhartishvili and Novikov, 2004; Novikov and Chkhartishvili, 2005), we can turn to other game-theoretic assumptions (Schumann, 2014; Schumann and Szelc, 2014, 2015):

i') Every game can be endless or repeated, because its strategies can change, see (Abreu et al., 1990).

ii') Players may have different levels of reflection: one player may know everything about the other, but the other may only know pictures that have deliberately misled him or her by the former.

iii') Some payments may have symbolic meanings, i.e., instead of payments A, we are dealing with pictures or interpretations of $K_i^{n+1}A$ and $K_j^{n+1}A$ regarding the same payments A. A higher symbolism of payments means a higher level of reflection of the respective players. At the zero level of reflection, payments have no symbolic meanings at all. For agreement, players are looking for joint symbolic meanings—common pictures regarding A.

iv') Resistance points for players can only be found for zero reflection payoffs.

v') Joint symbolic meanings (joint pictures $K_i^{n+1}A$ or $K_j^{n+1}A$) can change during the game if the player increases his level of reflection.

vi') For any game, there is a performative efficiency when all symbolic values of one player are shared by other players (the picture $K_i^{n+1}A$ or $K_j^{n+1}A$ is accepted by all).

Considering these new assumptions, we can compute some aspects of symbolic interactions with probabilistic tools on non-Archimedean numbers (Schumann, 2009, 2014c, 2014d). The Aumann Agreement Theorem does not apply to these assumptions (Schumann, 2014c, 2014d).

Aumann's Agreement Theorem (Aumann, 1976, 1989) says that two agents acting rationally (according to Bayesian formulas) and with a common knowledge of each other cannot come to disagreement. More precisely, if two people share a common premise and have a common knowledge of the current probability distributions (their total for a given case A is a common knowledge), then they should have equal probability distributions (the results should be the same). This is one of the most important theorems in game theory and epistemic logic. For example, according to this statement, every rational player must behave in the same manner under the same circumstances. Rational players always have a common knowledge, they know all the parameters of the game and must make sure that their opponents know that they know the parameters of the game that they know what they know, and so on *ad infinitum*.

However, we know that every rational agent is capable of deliberate deception, and especially in courts. For the first time, assumptions which contradict the ideas of Aumann were formulated by Vladimir Lefebvre (1936–2020) in his concept of "reflexive (reflective) games" (Lefebvre, 1977, 2010). A game is called reflexive (reflective) if, when choosing an action, the agent must simulate (predict) the actions of his/her opponents, for example, he or she can try to control them or deceive them. The early Lefebvre formulated reflective games that have many levels. At level zero, I ignore the beliefs of opponents; at the first level, I consider their beliefs; at the second level, I scrutinize that they are trying to predict my beliefs; at the third level, I foresee their beliefs in which my beliefs are predicted by them, etc. Recall that game-theoretic mathematics for the ideas of the early Lefebvre was developed by D.A. Novikov and A.G. Chkhartishvili (Chkhartishvili, 2003, 2010, 2012; Chkhartishvili and Novikov, 2004; Novikov and Chkhartishvili, 2005).

The later Lefebvre attempted to model decision-making in reflective games by means of Boolean functions (Lefebvre, 1982). The main disadvantage of the approach proposed by the later Lefebvre is that reflective levels are ignored, and agents are represented as automata. However, a self-esteem in reflection is a variable value. Reflection levels change depending on speech situations and the emotionality of our interlocutors. In this sense, the dynamism of the levels of reflection fully corresponds to the well-known paradox of Chevalley and Belzung (2002), which is formulated as follows: the emotional reaction of the same person in the same situation can be different at different points in time. Thus, modeling decision-making in reflective games by means of Boolean functions is too idealized and cannot help in the analysis of everyday situations. Within this approach, we assume that the reactions and assessments of the same agent remain the same forever. However, they do not. This approach can only be useful in explaining some of the main characteristics of reflective control.

5.1.6 Litigation as a Reflective Management

If a reflective game is realized by means of a targeted modification of some information components, then such a modification for a guaranteed victory in a reflective game is called a reflective control or reflective management (Mints et al., 2020; Schumann, 2015a). The main types of reflective control are as follows:

- the target control (the modification of admissible sets of actions of all groups of agents)—e.g., it can be implemented by the presiding officer;
- the motivational management (the modification of the functions of the goal of specific agents)—e.g., it can be implemented both by the prosecutor or the plaintiff, and by the lawyer or the defendant;
- the information management (the modification of information that agents use in making a decision)—e.g., it can be implemented by all participants in the trial.

The information management has the following types:

- the information regulation (the targeted influence on information about events);
- the experienced management (the targeted influence on information about decision-making models).

The reflective control problem is formulated as follows: one agent i creates the knowledge structure of other agents $K_j^{n+1}A$ in such a way that the implemented strategy of j in the framework of $K_j^{n+1}A$ would be maximally favorable for i, not j.

Thus:

1. In both the criminal and civil litigation, there are two opposing parties who, even before the trial, develop a strategy to maximize their possible gains, however, these strategies may change within the framework of the trial itself. This means that there can be no fixed strategies in litigation.
2. For dynamic strategies, which are used in the court, a mathematical formalization is possible, considering the levels of reflection. Thanks to this formalization, we can model the dynamics of strategies depending on changing contexts, as well as the mutual influence of opponent's strategies.
3. The litigation is a reflective game in which opponents try to correctly predict each other's strategies and successfully hide their own predispositions. The winner is an opponent who moves to a higher level of reflection $n + 1$, when the second opponent becomes completely transparent to the first opponent and, at the same time, this first opponent remains not transparent to the second.
4. The trial can also proceed as a reflective control, when it is possible to impose on the opponent the interpretation of events through the modification of motives or modification of information.
5. Hence, the game scenarios of the trial are considered by us within the framework of mathematical formalization as a reflective game with a possible reflective control.

5.1.7 Preference Relations

In modeling the legal argumentation with the aim to simulate it, we can use the group of mixed methods: qualitative (content analysis, case study, etc.), quantitative (statistics) and logical. Usually, in the modeling of legal argumentation, there were two types of reductionism: (i) to logic (the reduction of legal argumentation to logical schemes of reasoning from propositional logic—the way made first in the Sumerian-Akkadian legal culture), and (ii) to argumentation (the reduction of legal argumentation to rhetoric with *topoi*, tropes and other standard rules of persuading interlocutors). But we can combine both approaches to simulate even psychological aspects of legal argumentation. In this way we can formalize preferences of different participants in the litigation, causing their attitudes and estimations.

The preference relation R_i (a, b), which reads as follows: "agent i prefers alternative a to alternative b" is one of the basic concepts in decision-making theory. Any choice is possible due to a preference relationship. There are two possibilities in determining preferences: (i) evaluating each alternative individually; (ii) evaluating each pair of alternatives. The second approach has been formulated as Thurston's Law of Comparative Decision (Thurstone, 1994). This method was then extended as a pairwise comparison method (Kułakowski, 2015a, 2015b), which is used in the reconstruction of the analytical hierarchy process (Saaty, 2008) and in the analysis of complex decision-making (Chu et al., 1979; Koczkodaj et al., 1997, 2020).

Comparisons can be formulated in many ways: "player j beats player i" or "a director accepts the report of j instead of i", etc. In any case, it is assumed that the decision-maker can compare any two objects. This assumption is called comparability. This is a very strong assumption, because comparisons can sometimes yield not only a preferred attitude, but also an attitude of indifference or an attitude of incomparability.

Comparison is one of the most fundamental human cognitive actions in decision-making (David, 1988). For the legal argumentation, we can offer two logical comparison models with four relations ("*S* is at least as permissible as *P*", "*S* is not at least as permissible as *P*", "*S* is at least as avoidable as *P*", and "*S* is not at least as avoidable as *P*"): (1) a preference model in civil cases, and (2) a preference model in criminal cases. We can call the first model of the preference logic the compromise logic, and the second model the uncompromising one.

We can show that the logical model of uncompromising preferences can be formalized as an Aristotelian system, while the logical model of compromise preferences can be formalized as a non-Aristotelian system. In the Aristotelian system, the preference relation is quantitative (numerically expressible), and in the non-Aristotelian system, the preference relation is qualitative (without a numerical equivalent).

Let us first consider the syllogistic with the logic of preferences for the criminal procedure. In this syllogistic, all the positions of the opposing parties in respect to the case are indicated by the corresponding syllogistic variables, which are mentioned or not in the litigation. A position *S* is considered empty if and only if the relevant fact is not mentioned in the lawsuit. The syllogistic lines (words) *SP* have the following interpretation: "*S* and *P* are positively comparable", and with the following meaning: *SP* is true if and only if *S* and *P* are not empty and appear together within a given set of arguments or facts, otherwise *SP* is false. According to this definition of syllogistic lines, we can define atomic syllogistic judgments as follows:

- In formal syllogistic: "for agent *i*, alternative *S* is at least as permissible as alternative *P*", formally: there is an *A* with *AS* and for any *A*, if *AS*, then *AP*. In the model of criminal cases: there is an argumentative block with *A* such that *A* and *S* are linked in this block and for any block with *A*, if *A* and *S* are linked in this block, then *A* and *P* are linked in this block, too.
- In formal syllogistic: "for agent *i*, alternative *S* is not at least as avoidable as alternative *P*", formally: there is an *A* such that *AS* and *AP*. In the criminal case model: there is an argumentative block with *A* such that *A* and *S* are linked in this block, and *A* and *P* are linked in this block, too.
- In formal syllogistic: "for agent *i*, alternative *S* is at least as avoidable as alternative *P*", formally: for any *A*, *AS* is false or *AP* is false. In the criminal case model: for any argumentative block with *A*, *A* and *S* are not linked in this block, or *A* and *P* are not linked in this block.
- In formal syllogistic: "for agent *i*, alternative *S* is not at least as permissible as alternative *P*", formally: for any *A*, *AS* is false or there is an *A* such that *AS* is true, and *AP* is false. In the model of criminal cases: for any argumentative block with *A*, *A* and *S* are not connected in this block or there is a position *A* in a block such that *AS* is true, and *AP* is false.

According to these definitions, it is easy to see that we are dealing with the syllogistic of Aristotle, formalized within the ontology of Leśniewski (Schumann, 2011, 2013a). The semantics for this system is represented by the announcement of materials and evidence in the framework of the criminal process.

This system is uncompromising, because (i) between the two judgments "for agent *i*, alternative *S* is at least as permissible as alternative *P*" and "for agent *i*, alternative *S* is at least as avoidable as alternative *P*" there is a contrary relationship, and (ii) between the two judgments "for agent *i*, alternative *S* is not at least as permissible as alternative *P*" and "for agent *i*, alternative *S* is not at least as avoidable as alternative *P*" there is a subcontrary relationship.

Now let us look at the syllogistic for the logic of preferences in the civil litigation. In the same way as in the case of Aristotelian syllogistic for the criminal procedure, we will begin to interpret all positions that can be pronounced as syllogistic variables. A variable is considered empty if and only if an appropriate position has not been announced. Let us define the syllogistic lines *SP* with the following interpretation: "*S* and *P* are positively comparable", and with the following meaning: *SP* is true if and only if *S* and *P* are reachable for each other within the framework of a common citation, otherwise *SP* is false. Using this definition of syllogistic lines, we can define atomic syllogistic propositions as follows:

- In formal syllogistic: "for agent *i*, alternative *S* is at least as permissible as alternative *P*", formally: there is an *A* such that for any *A*, *AS* is true, and *AP* is true. In the civil case model: there is an argumentative block with *A* such that *A* and *S* are linked in this block of announcement and for any block with the position *A*, *A* and *S* are linked in this block, and *A* and *P* are linked in this block.
- In formal syllogistic: "for agent *i*, alternative *S* is not at least as avoidable as alternative *P*", formally: for any *A*, *AS* is false, and *AP* is false. In the civil case model: for any announcement with *A*, *A* and *S* are not linked in this announcement block, and *A* and *P* are not linked there as well.
- In formal syllogistic: "for agent *i*, alternative *S* is at least as avoidable as alternative *P*", formally: there is an *A* such that if *AS* is false, then *AP* is true. In the civil case model: there is an announcement with *A* that if *A* and *S* are not linked in this block, then *A* and *P* are linked there.
- In formal syllogistic: "for agent *i*, alternative *S* is not at least as permissible as alternative *P*", formally: for any *A*, *AS* is false or there is an *A* such that *AS* is false, or *AP* is false. In the civil case model: for any argumentative block with *A*, *A* and *S* are not connected in this block or there is a block with *A* such that *AS* is false, or *AP* is false.

The formal properties of this non-Aristotelian syllogistic are discussed in Schumann (2011, 2013a); Schumann and Adamatzky (2015); Schumann and Akimova (2013, 2015); Schumann and Woleński (2016).

This system is the compromise one, because (i) from the judgment "for agent *i*, alternative *S* is at least as permissible as alternative *P*" the following judgment follows: "for agent *i*, alternative *S* is at least as avoidable as alternative *P*", and (ii) from the judgment "for agent *i*, alternative *S* is not at least as avoidable as alternative *P*" the following judgment follows: "for agent *i*, alternative *S* is not at least as permissible as alternative *P*". It is Mozi's square of opposition, see Subsection 1.2.3.

So, we can differ the preference model for criminal cases from the preference model for civil cases. For the first model, the Aristotelian syllogistic can be applied, for the second model the non-Aristotelian one.

5.1.8 Conclusions

The e-lawyer software supposes the development of logical foundations for the design of expert systems for automating or semi-automating legal decisions. To build legal catalogs and create a database of e-lawyer, we can use situation semantics (Barwise and Perry, 1980, 1983), which will consider for each legal article also the judicial practice of its application with a set of typical 'situations'. Instead of algorithms for decision-making based on inductive trees as a logical model of legal argumentation, it is more efficient to use co-algorithms (Capretta, 2007; Schumann and Woleński, 2015) and creative decision-making on infinite trees, because the strategies applied in courts are variable and dynamic. In this way, a mathematical formalization of these strategies is possible within the framework of cellular payoff automata (see Subsection 5.1.3). Thanks to these automata, we can simulate the dynamics of strategies depending on changing contexts, as well as the mutual influence of opponents' strategies. The litigation is a reflective game in which opponents seek to predict each other's strategies to move to a higher level of reflection (Subsections 5.1.4, 5.1.5). The judicial process can also proceed as a reflective control when it is possible to impose on the opponent the interpretation of events through the modification of motives or the modification of information (Subsection 5.1.6). Thus, the game scenarios of the trial may be considered within the framework of the mathematical formalization in the form of cellular payoff automata as a reflective game with a possible reflective control.

In this formalization of legal reasoning for the e-lawyer system, it is better to distinguish between preference models for criminal and civil cases. For the first model, I have proposed the Aristotelian syllogistic, for the second model, the non-Aristotelian one (Subsection 5.1.7).

So, the strategies used in a court are variable and dynamic. In this section, I have shown that the mathematical formalization of these strategies is possible within the framework of the theory of reflective games. Thanks to these games, we can simulate the dynamics of strategies. Any litigation is a game in which opponents seek to predict each other's strategies, and I have proposed the game theory in which the opponent who goes to a higher level of reflection wins.

This approach to formalizing legal processes is very different from the approach created by the Babylonians. The main features are as follows:

1a. The Babylonian approach was normative—there were strict rules (some constituted norms) for resolving conflicts by propositional logic which must be followed by all the participants in courts. These rules/norms were transmitted as obligatory within the legal hermeneutics and in this way, they must be understood and shared by all the members of the trial. So, the high self-reflection for different social roles in conflict resolutions was widely assumed.

1b. In the meanwhile, the modern approach does not need normativity as such because all possible actions of all agents are formalizable. So, the self-reflection of the participants in obeying propositional logic can be low here. The e-lawyer system can check and correct logical fallacies.

2a. The Babylonian approach was based on classical propositional logic.

2b. The modern approach involves various non-classical logics, unconventional (for example, reflective) game theory, and some other advanced mathematical tools (Schumann and Zenil, 2020).

3a. The Babylonian legal process was mechanized through logical rules of legal hermeneutics.

3b. The modern legal process can be automated through the development of automatic or semi-automatic expert systems (the so-called electronic lawyer or e-lawyer).

5.2 Some Prospects for the e-Lawyer Project

5.2.1 Problem Setting

The project of e-lawyer as automating or semi-automating legal decisions is a significant part of the electronic justice or e-justice which consists of different applications of information technologies in the maintenance of judicial system to increase the transparency of the judicial process, improve the functioning of existing justice, simplify the procedures and reduce the costs (Contini and Lanzara, 2009, 2014; Fabri and Contini, 2001). As a result of this use, the access to justice is increased and improved, legal powers are effectively allocated, and the efficiency of the justice system is enhanced. So, the main result is underlined by the improvement of citizens' access to justice, which is a measure of a more effective judicial process. Thus, the e-lawyer or, more generally, the e-justice presupposes a modernization of the judicial system and the widest eventual use of information technologies and artificial intelligence in litigations (Reiling, 2010; Velicogna, 2008; Velicogna and Errera, 2011).

In the European Union, the European Commission has announced the electronic justice in civil and commercial law since 2003. Later the European Commission in May 2008 approved a development strategy entitled "Towards a European e-Justice Strategy". The European Union has already implemented the E-CODEX system, a project funded by the EU Commission and the Ministry of Justice of the German state of North Rhine-Westphalia.

The e-justice itself is a part of e-government, in which information technologies are involved across the entire spectrum of case management (Contini and Lanzara, 2009, 2014; Fabri and Contini, 2001). The most important element of e-government used in the e-justice is the introduction of an electronic signature (e-signature) and electronic identification (e-identity)—an opportunity for individuals and legal entities to guarantee the authenticity of their actions in the electronic document flow.

In this section, I am going to discuss some prospects of e-lawyer software, namely: (i) some logical methods which can be used in implementing the e-justice (Subsection 5.2.2) and (ii) some advantages of e-justice (Subsection 5.2.3).

5.2.2 Some Logical Methods in the Implementation of e-Justice

The court process includes many players acting concurrently. Therefore, for the electronic formalization of litigation within e-justice (the so-called e-lawyer project), the most optimal are the unconventional logical tools (Adamatzky et al., 2017; Schumann, 2013b, 2014a, 2017a, 2017b, 2017c, 2018) that allow analyzing a wide range of parallel and competing solutions, in particular, a theory of parallel proofs is needed—in which mutually exclusive decisions have independent parallel proof processes (Schumann, 2010; Schumann and Kuznetsov, 2018) and some proofs can have loops (Brotherston, 2005). This theory allows us to develop new technologies for the design of legal databases and implement new systems of supporting judicial decisions. Some algorithms and logical tools that are obtained in the theory of concurrent processes (Schumann, 2015b, 2015c), particularly in the theory of concurrent games, could be used in the implementation of parallel computing technologies in various decision-making areas (for example, in business process management or in litigation management). In addition, based on a new logical approach to decision-making, it would be possible to design an appropriate architecture of a judicial decision support system that can be used to automate the judicial process in the framework of e-justice.

The parallel proof theory uses an integral logical approach based on non-well-founded mathematics (Barwise and Etchemendy, 1987; Heifetz, 1996; Schumann, 2009, 2010, 2012; Schumann and Kuznetsov, 2018). This approach can allow us to analyze and predict complex game strategies of different participants in the trial, to support decision-making in cases of massively parallel decisions. This can also be applied in the development of an appropriate architecture for decision support systems for courts, such as the e-lawyer system.

Some directions in using the parallel proof theory in e-justice include:

- Explaining the intellectual behavior of different participants in the trial by means of unconventional logics (such as behavioral logic, coalgebra, modal logic). This would be a further practical development of applications of non-well-founded mathematics, as well as parallel and massively parallel computational paradigm (Barwise and Etchemendy, 1987; Heifetz, 1996; Schumann, 2010, 2012; Schumann and Kuznetsov, 2018). These applications are already heavily used in behavioral economics (Knudsen, 1993; Lescanne and Perrinel, 2012).
- Developing a comprehensive logical theory of judicial decision in the framework of e-justice. This means automating the decision-making mechanism based on rules (legal databases, court libraries, archives of important or innovative decisions), court data (evidence, documents), game strategies of court parties (expectations and statements). Thus, it is expected to develop logical tools in making the most transparent judicial decisions.

So, in the implementation of the e-justice system, the use of unconventional logics is expected: illocutionary logic, non-well-founded mathematics, parallel and massively parallel computations. These logics can contribute to explaining the rational behavior of the agents of the judicial process and in the analysis of their decision-making. Thus, the theory of parallel proofs can form the basis for designing the architecture of systems for supporting a judicial decision at all levels of the court and, therefore, for developing the e-lawyer system.

From the point of view of philosophy, the theory of parallel proofs means the rejection of Platonism in mathematics—the doctrine asserting the existence of *a priori* truths, in particular axioms. Thanks to Platonism, the axiomatic method arose. According to this doctrine, there are entities with an absolute status of existence (primary objects of mathematics, for example, the notion of point or line in geometry) and there are sentences with the highest degree of persuasion (axioms). In fact, Plato's teachings are the cornerstone of all classical mathematics.

The axiomatic method seeks to obtain the maximum information from the minimum number of premises. David Hilbert (1862–1943), in his program of foundations of mathematics, considered the axiomatic method as the main way of mathematical understanding of reality. He proposed to transform all existing mathematics into a formal theory and prove the consistency of this formal theory (that is, to prove that no proposition can be proved and refuted at the same time). He regarded the elementary number theory as a privileged part of mathematics. The formal theory obtained by Hilbert was built based on two types of axioms: (a) logical axioms, which establish the rules of reasoning, and (b) non-logical axioms, by which the primary objects of mathematics are defined.

However, there can be no axioms in the theory of parallel proofs (Schumann, 2010, 2012; Schumann and Kuznetsov, 2018). We deal with contextual decisions based on many random factors—legal rules (legal databases, court libraries, archives of past decisions), court data, game strategies of court parties. When we simulate this by means of parallel computing, the possibility of outcomes of such contextual court decisions is infinite—as the possibility of outcomes of an online game like the World of Tanks.

So, the introduction of e-justice is not possible without the development of new logical methods, such as the theory of parallel proofs, and other applied experimentation in the field of automating decision-making mechanisms in all spheres of society. These methods can model reality, including economic phenomena or litigation as a computational system, which allows us to better predict the dynamics of both the system as a whole and its parts. New methods show that, due to modern logical tools, each phenomenon can be considered a computational process modeled by an abstract machine. Preliminary results show that, for example, a lawsuit can be modeled using this unconventional logic as well and this can be implemented in the e-lawyer system.

To sum up, applying the parallel proof theory to the applied litigation research will enable us to develop more effective architecture design methods for legal support systems.

The parallel proof theory assumes the development of the following tools:

- Several new computational methods of behavioral logic which are used to formalize the decision-making process.
- A new mathematical metalanguage which is used to formalize a trial within reflective games of its participants.
- Some new computational tools within the paradigm of parallelism and non-Platonic mathematics (Gabbay and Schumann, 2014; Schumann, 2013c).
- Some new programming tools which can be created to program judicial decision-making, for example, for the e-lawyer system (an important step in the further development and implementation of e-government—digital interactions between the government and citizens, the government and firms, the government and employees, and also between a government and governments).

The e-justice system can provide greater transparency in court decisions and significantly simplify our access to justice. One of the elements of e-justice can be presented as a system of electronic lawyer—a database for the automation or semi-automation of legal decisions.

It is assumed that the electronic lawyer system can open the access to legal information not only for lawyers, but also for ordinary users. Such a system can be effective if it is deployed as a new social network. At the same time, the effectiveness of an electronic lawyer can be achieved only in the case of including legal acts of a particular country, data from its judicial practice, as well as data from its law enforcement practice in the form of a game-theoretic ontology. In this ontology, each article of the codes should have several labels (possible consequences, conditions, events), which allow the articles to be grouped in different ways depending on the user's request. An open base of facts with methods of establishing, checking, and confirming them should be linked to each article. The set of articles and facts must be matched with the exclusionary strategies of the plaintiff/prosecutor or the defendant/defender in the prosecution and defense.

As we can see, this ontology will imply a smart search for articles of codes at the user's request. A game-theoretic interface may be built over the ontology as a home laboratory that allows us to fill an electronic statement of claim, generating it automatically depending on the problem posed. With the formation of the statement of claim, the prosecutor/plaintiff's strategy will unfold, and the ontology with a set of articles and facts will contribute to the correct preparation for the court session.

So, thanks to several logical methods, it is possible to create an effective e-lawyer system with an extensive use of logical competence. At the audience level, using the theory of random graphs, it is possible to outline the main block of necessary contacts to keep users in communication within the given system. At the single user level, process calculus (Milner, 1993, 1999) can be used to define a home legal lab that tracks all legal aspects of a user's activity. And at the level of user requests, it is possible to create a system of ontology of legal documents, which will make it possible to form the strategy of the plaintiff or the defendant based on the legal acts and documentation of actual states.

Developing the e-lawyer software assumes an integral model of legal argumentation with the possibility of its computer implementation, and therefore with the possibility of its most comprehensive formalization.

In practice, solving this problem means introducing a whole range of new logical theories into the arsenal of logical science:

1. *Dynamic game theory*. In both the criminal and civil litigation, there are two opposing parties who, even before the trial, develop a strategy to maximize their possible payoffs; however, these strategies may change within the framework of the trial. This means that there can be no fixed strategies in litigations and the game is constantly changing.
2. *Theory of reflective games*. In the litigation, the parties initially try to hide their own true predispositions, on the one hand, and try to predict the strategies of their opponents, on the other hand.
3. *Logical ways of setting legal catalogs*. Legal directories are constantly changing, even within the same trial. Due to their dynamism, it is necessary to use logical methods that will allow the processing of dynamic legal directories.
4. *Logical network*. It is necessary to design a logical network within which the gameplay of many trials can be built due to dynamic legal directories.

5.2.3 Some Advantages of e-Justice

The main advantages of e-justice are as follows:

1. The possibility of implementing joint actions and efforts on the part of various social forces to increase the transparency of justice. For example, this concerns the provision of information online, regarding legal systems and legislations, as well as the implementation of some systems for the exchange of information electronically between various parties and the court (and in some cases a complete electronic procedure for the legal process).
2. Increasing the access to justice for individuals and legal entities. This guarantees a common access to information on judicial systems and procedures and makes electronic claims to a court possible. For instance, now an electronic payment for legal costs is always possible and easy.
3. Improving the efficiency of court proceedings. First, the creation of electronic tools contributes to accepting balanced and logically correct court decisions. It is based on the development of electronic legal databases, court libraries, archives of important or innovative decisions. Legal databases, having become publicly available on the Internet, influence the public perception of courts—there is a growing public demand for transparency in decisions. Second, it is based on an implementation of technologies which facilitate the very communication in the litigation, for example, it assumes an installation of equipment in courtrooms, including sound amplification, surveillance cameras, testimony monitors, electronic conference tables, monitors intended for the audience, as well as other audio and video equipment for conference calls.

4. Orientation of judicial system to the client. It is an ability of the system to manage the entire flow of electronic interactions with clients individually—from electronic registration to electronic document management but also considering electronic judicial archives.

Thus, one of the main trends of the information society consists of introducing the e-government system and its most important part which is presented by the e-justice (Reiling, 2010).

The electronic justice makes the entire judicial system as transparent as possible for society, thereby it highly minimizes the possibility of judicial errors, including logical fallacies. Meanwhile, such a system is more efficient and minimizes the costs of retaining the judicial system.

5.2.4 Conclusions

1. The e-justice system is an integral part of the new wave of democratization in the information society—this wave is called e-democracy and e-government.
2. The electronic justice is the higher form of justice—it maximally guarantees the transparency and consistency of court decisions.
3. For the introduction of e-justice in the automation or semi-automation of judicial decisions, the theory of parallel proofs is needed, which is based on parallel and massively parallel computations.
4. The theory of parallel proofs is based on unconventional logics. As a result, there is a rejection of the paradigm of Platonism in mathematics.

As we see, the development of electronic justice (including electronic lawyer) is based on the progress of artificial intelligence, i.e., on the growth of logical competence. For the first known time, since Ur-Nammu (the founder of the III dynasty of Ur) the judicial system has started to be based on logical competence, represented by the knowledge of some inference rules of propositional logic, i.e., it has taken place since ca. 4,000 years ago. The modern system is just a further cultivation of this legal tradition. Recently, some advanced mathematical tools, including non-classical logic, have been used in the development of legal expert systems such as the electronic lawyer, while the Babylonian legal hermeneutics involved logical tools which are now characterized as classical, for example, the laws of excluded middle and non-contradiction were assumed here.

In this chapter, I have discussed some prospects in applying artificial intelligence for the development of modern justice, just to show that logic is its important part. The same high role of logic in law and justice is well reconstructed on the materials of early states due to a structuralist genealogical analysis called the "archaeology of logic".

References

Abramsky, S. and Mellies, P.A. 1999. Concurrent games, and full completeness. pp. 431–442. *In*: *Proceedings of the 14th Symposium on Logic in Computer Science*.

Abreu, D., Pearce, D. and Stacchetti, E. 1990. Toward a theory of discounted repeated games with imperfect monitoring. *Econometrica* 58(5): 1041–1063.

Ackerman, F. 2002. Still dead after all these years: Interpreting the failure of general equilibrium theory. *Journal of Economic Methodology* 9(2): 119–139.

Adamatzky, A., Selim, Akl, Mark Burgin, Cristian S. Calude, José Félix Costa, Mohammad Mahdi Dehshibi, Yukio-Pegio Gunji, Zoran Konkoli, Bruce MacLennan, Bruno Marchal, Maurice Margenstern, Genaro J. Martínez, Richard Mayne, Kenichi Morita, Andrew Schumann, Yaroslav D. Sergeyev, Georgios Ch. Sirakoulis, Susan Stepney, Karl Svozil and Hector Zenil. 2017. East-West paths to unconventional computing. *Progress in Biophysics & Molecular Biology* 131: 469–493.

Aumann, R.J. 1976. Agreeing to disagree. *The Annals of Statistics* 4(6): 1236–1239.

Aumann, R.J. 1989. *Notes on Interactive Epistemology*. Jerusalem: Mimeo, Hebrew University of Jerusalem.

Barwise, J. and Perry, J. 1980. The situation underground. pp. 1–55. *In*: Barwise, J. and Sag, I. (eds.). *Stanford Working Papers in Semantics*, Vol. 1. Stanford Cognitive Science Group, Section D.

Barwise, J. and Perry, J. 1983. *Situations and Attitudes*. Bradford Books, MIT Press.

Barwise, J. and Etchemendy, J. 1987. *The Liar: An Essay on Truth and Circularity*. Oxford University Press.

Berger, S. 2009. *The Foundations of Non-Equilibrium Economics: The Principle of Circular Cumulative Causation*. Routledge Chapman & Hall.

Brandenburger, A. and Keisler, H.J. 2006. An impossibility theorem on beliefs in games. *Studia Logica* 84(2): 211–240.

Brandenburger, A. 2008. Epistemic game theory: An overview. *In*: Macmillan Publishers Ltd. (eds.) *The New Palgrave Dictionary of Economics*. London: Palgrave Macmillan.

Brandenburger, A., Friedenberg, A. and Keisler, H.J. 2012. Fixed points in epistemic game theory. *In*: Samson Abramsky and Michael Mislove (eds.). *Mathematical Foundations of Information Flow*. American Mathematical Society.

Brenguier, R. 2013. PRALINE: A tool for computing nash equilibria in concurrent games. pp. 890–895. *In*: Sharygina, N. and Veith, H. (eds.). *Computer Aided Verification*. CAV 2013. Lecture Notes in Computer Science, Vol. 8044. Berlin, Heidelberg: Springer.

Brotherston, J. 2005. Cyclic proofs for first-order logic with inductive definitions. pp. 78–92. *In*: Beckert, B. (ed.). ABLEAUX 2005, Vol. 3702 of LNAI. Springer-Verlag.

Capretta, V. 2007. Common knowledge as a coinductive modality. *In*: Barendsen, E., Geuvers, H., Capretta, V. and Niqui, M. (eds.). *Reflections on Type Theory, Lambda Calculus, and the Mind*. Essays Dedicated to Henk Barendregt on the Occasion of His 60th Birthday. Radboud University, Nijmegen.

Chevalley, C. and Belzung, C. 2002. Emotional behaviour as the result of stochastic interactions: A process crucial for cognition. *Behavioral Processes* 60(2): 115–132.

Chkhartishvili, A.G. 2003. Bayes-Nash Equilibrium: Infinite-depth point information structures. *Automation and Remote Control* 64(12): 1922–1927.

Chkhartishvili, A.G. and Novikov, D.A. 2004. Models of reflexive decision-making. *Systems Science* 30(2): 45–59.

Chkhartishvili, A.G. 2010. Reflexive games: Transformation of awareness structure. *Automation and Remote Control* 71(6): 1208–1216.

Chkhartishvili, A.G. 2012. Concordant informational control. *Automation and Remote Control* 73(8): 1401–1409.

Chu, A.T.W., Kalaba, R.E. and Spingarn, K. 1979. A comparison of two methods for determining the weight belonging to fuzzy sets. *Journal of Optimization Theory and Applications* 4: 531–538.

Contini, F. and Lanzara, G.F. (eds.). 2009. *ICT and Innovation in the Public Sector: European Studies in the Making of e-government*. Palgrave Macmillan.

Contini, F. and Lanzara, G.F. (eds.). 2014. *The Circulation of Agency in e-Justice*. Springer.

Crawford, V.P. and Iriberri, N. 2007. Level-k Auctions: Can a nonequilibrium model of strategic thinking explain the winner's curse and overbidding in private-value auctions? *Econometrica* 75(6): 1721–1770.

David, H.A. *The Method of Paired Comparisons*. New York: Oxford University Press, 1988.

Fabri, M., Contini F. *Justice and Technology in Europe: How ICT is changing the judicial business*. Kluwer, 2001.

Gabbay, D. and Schumann, A. 2014. The logic system is the way you do logic. *Studia Humana* 3(4): 41–44.

Heifetz, A. 1996. Non-well-founded-type spaces. *Games and Economic Behaviour* 16(2): 202–217.

Knudsen, Ch. 1993. Equilibrium, perfect rationality, and the problem of self-reference in economics. *In*: Uskali Maki (ed.). *Rationality, Institutions and Economic Methodology*. Routledge Chapman & Hall.

Koczkodaj, W.W., Herman, M.W. and Orlowski, M. 1997. Using consistency-driven pairwise comparisons in knowledge-based systems. pp. 91–96. *Proceedings of the Sixth International Conference on Information and Knowledge Management*. ACM Press.

Koczkodaj, W.W., Liu, F., Marek, V.W., Mazurek, J., Mazurek, M., Mikhailov, L., Özel, C., Pedrycz, W., Przelaskowski, A., Schumann, A., Smarzewski, R., Strzalka, D., Szybowski, J. and Yayli, Y. 2020. On the use of group theory to generalize elements of pairwise comparisons matrix: A cautionary note. *International Journal of Approximate Reasoning* 124: 59–65.

Kułakowski, K. 2015a. A heuristic rating estimation algorithm for the pairwise comparisons method. *Central European Journal of Operations Research* 23(1): 187–203.

Kułakowski, K. 2015b. On the properties of the priority deriving procedure in the pairwise comparisons method. *Fundamentae Informaticae* 139(4): 403–419.

Lefebvre, V. 1977. *The Structure of Awareness: Toward a Symbolic Language of Human Reflexion*. New York: Sage Publications.

Lefebvre, V.A. 1982. *Algebra of Conscience*. D. Reidel.

Lefebvre, V.A. 2010. *Lectures on Reflexive Game Theory*. Los Angeles: Leaf & Oaks.

Lescanne, P. and Perrinel, M. 2012. Backward coinduction, Nash equilibrium and the rationality of escalation. *Acta Informatica* 49(3): 117–137.

Mailath, G.J. and Samuelson, L. 2006. *Repeated Games and Reputations: Long-run relationships*. Oxford University Press.

Milner, R. 1993. The Polyadic π-Calculus: A tutorial. *In*: Hamer, F.L., Brauer, W. and Schwichtenberg, H. (eds.). *Logic and Algebra of Specification*. Springer-Verlag.

Milner, R. 1999. *Communicating and Mobile Systems: The π-calculus*. Cambridge: Cambridge University Press.

Mints, A., Schumann, A. and Kamyshnykova, E. 2020. Stakeholders' rank of reflexion diagnostics in a corporate social responsibility system. *Economic Annals-XXI* 181(1–2): 92–104.

Novikov, D.A. and Chkhartishvili, A.G. 2005. Stability of information equilibrium in reflexive games. *Automation and Remote Control* 66(3): 441–448.

Pancerz, K. and Schumann, A. 2018. Slime mould games based on rough set theory. *International Journal of Applied Mathematics and Computer Science* 28(3): 531–544.

Reiling, D. 2010. *Technology for Justice. How Information Technology can Support Judicial Reform*. Leiden University Press.

Saaty, T.L. 2008. Relative measurement and its generalization in decision making. Why pairwise comparisons are central in mathematics for the measurement of intangible factors. The analytic hierarchy/network process. *Statistics and Operations Research (RACSAM)* 102: 251–318.

Schumann, A. 2009. Non-well-founded probabilities and coinductive probability logic. pp. 54–57. *In*: *Proceedings of the 10th International Symposium on Symbolic and Numeric Algorithms for Scientific Computing*.

Schumann, A. 2010. Towards theory of massive-parallel proofs. Cellular automata approach. *Bulletin of the Section of Logic* 39(3/4): 133–145.

Schumann, A. 2011. Two squares of opposition: For analytic and synthetic propositions. *Bulletin of the Section of Logic* 40(3/4): 165–178.

Schumann, A. 2012. Proof-theoretic cellular automata as Logic of unconventional computing. *International Journal of Unconventional Computing* 8(3): 263–280.

Schumann, A. 2013a. On two squares of opposition: The Leśniewski's style formalization of synthetic propositions. *Acta Analytica* 28: 71–93.

Schumann, A. 2013b. Unconventional Logic for massively parallel reasoning. pp. 298–305. *In*: Paja, W.A. and Wilamowski, B.M. (eds.). *6th International Conference on Human System Interactions* (HSI), Book Series: Conference on Human System Interaction.

Schumann, A. 2013c. Anti-Platonic logic and mathematics. *Journal of Multiple-Valued Logic and Soft Computing* 21(1–2): 53–88.

Schumann, A. and Akimova, L. 2013. Simulating of Schistosomatidae (Trematoda: Digenea) Behavior by Physarum spatial logic. pp. 225–230. *In*: Ganzha, M., Maciaszek, L. and Paprzycki, M. (eds.). *Federated Conference on Computer Science and Information Systems* (*FedCSIS*).

Schumann, A. and Szelc, A. 2014. Towards new probabilistic assumptions in Business Intelligence. *Studia Humana* 3(4): 11–21.

Schumann, A., Pancerz, K., Adamatzky, A. and Grube, M. 2014. Bio-inspired Game Theory: The case of Physarum Polycephalum. *BICT 2014*.

Schumann, A. 2014a. *Unconventional Logics in Decision Making*. Rzeszow: WSIiZ.

Schumann, A. 2014b. Payoff cellular automata and reflexive games. *Journal of Cellular Automata* 9(4): 287–313.

Schumann, A. 2014c. Probabilities on streams and reflexive games. *Operations Research and Decisions* 24(1): 71–96.

Schumann, A. 2014d. Reflexive games and non-archimedean probabilities. *p-Adic Numbers, Ultrametric Analysis and Applications* 6(1): 66–79.

Schumann, A., Pancerz, K. and Szelc, A. 2015. The swarm computing approach to business intelligence. *Studia Humana* 4(3): 41–50.

Schumann, A. and Szelc, A. 2015. Nowe granice symulacji logicznej w inteligencji biznesowej. pp. 173–201. *In*: Andrzej Dąbrowski, Andrew Schumann and Jan Woleński (red.). *Podejmowanie decyzji. Pojęcia, teorie, kontrowersje*. Kraków: Copernicus Center Press.

Schumann, A. and Woleński, J. 2015. Decisions involving databases, fuzzy databases and codatabases. *Operations Research and Decisions* 25(3): 59–72.

Schumann, A. and Adamatzky, A. 2015. Physarum Polycephalum diagrams for syllogistic systems. *IfCoLog Journal of Logics and Their Applications* 2(1): 35–68.

Schumann, A. and Akimova, L. 2015. Syllogistic System for the propagation of parasites. The case of Schistosomatidae (Trematoda: Digenea). *Studies in Logic, Grammar, and Rhetoric* 40(53): 303–319.

Schumann, A. 2015a. Reflexive games in e-University. pp. 770–775. *IEEE 2015 Science and Information Conference* (SAI).

Schumann, A. 2015b. Towards context-based concurrent formal theories. *Parallel Processing Letters* 25(1): 1540008.

Schumann, A. 2015c. Podejmowanie decyzji na zbiorach nie-ufundowanych. pp. 155–172. w: Andrzej Dąbrowski, Andrew Schumann, Jan Woleński (red.). *Podejmowanie decyzji. Pojęcia, teorie, kontrowersje*. Kraków: Copernicus Center Press.

Schumann, A. and Woleński, J. 2016. Two squares of oppositions and their applications in pairwise comparisons analysis. *Fundamenta Informaticae* 144(3–4): 241–254.

Schumann, A. 2017a. Conventional and unconventional approaches to Swarm Logic. pp. 711–734. *In*: *Advances in Unconventional Computing*, Vol. 1: *Theory*. Book Series: Emergence Complexity and Computation. Vol. 22. New York: Springer International Publishing.

Schumann, A. 2017b. Conventional and unconventional reversible logic gates on Physarum polycephalum. *International Journal of Parallel Emergent and Distributed Systems* 32(2): 218–231.

Schumann, A. 2017c. Group theory and p-adic valued models of swarm behavior. *Mathematical Methods in the Applied Sciences* 40(18): 7438–7452.

Schumann, A. 2018. Logics for unconventional computing. *International Journal of Parallel, Emergent and Distributed Systems* 33(3): 233–236.

Schumann, A. and Kuznetsov, A.V. 2018. Foundations of Mathematics under Neuroscience conditions of lateral inhibition and lateral activation. *International Journal of Parallel, Emergent and Distributed Systems* 33(3): 237–256.

Schumann, A. and Zenil, H. 2020. Non-classical logics in unconventional computing. *International Journal of Unconventional Computing* 15(4): 237–244.

Thurstone, L.L. 1994. A law of comparative judgment, reprint of an original work published in 1927. *Psychological Review* 101(2): 266–270.

Velicogna, M. 2008. Use of information and communication technologies (ICT) in European judicial systems. CEPEJ Studies No. 7. Council of Europe.

Velicogna, M., Errera, A. and Derlange, S. 2011. e-Justice in France: The e-Barreau experience. *Utrecht Law Review* 7(1): 163–187.

Conclusion

1. In this work, the concept of logical competence has been presented as an ability to draw logically correct conclusions. This ability is not innate and in its natural manifestations is inevitably accompanied by the so-called cognitive biases—by the impact of affective states on decision-making. Logical competence arose as a social cognitive mechanism that can be cleared of natural cognitive distortions and biases. This mechanism was transmitted through training and was improved in the process of practice. In its perfect forms, logical competence was embodied in various artificial intelligence systems, which were first created in the 20th century, but their idea was formulated much earlier—for example, in the works of Gottfried Wilhelm Leibniz (1646–1716).

   The development of the concept of logical competence made it possible to present the dynamics of the use of logical reasoning as a long-term process of forming the skills, abilities, and competences of applying logically correct conclusions. This process began in Mesopotamia from the 22nd century B.C. Logical competence began to develop from the simplest forms to more and more sophisticated ones within the framework of legal discourse. During the Hellenistic and then Roman periods, the following four independent directions of reflective comprehension of logical competence appeared: (1) the ancient Greek logic created by Aristotle and Chrysippus (from the 4th to the 3rd centuries B.C.); (2) the Judaic logical hermeneutics, the foundations of which were laid by Hillel, Rabbi Ishmael, and Rabbi Akiva (from the 1st to the 2nd centuries A.D.); (3) the Indian-Buddhist logic of the Yogācāra and Nyāya schools (from the 2nd to the 4th centuries A.D.); (4) the Chinese proto-logic in the form of situation semantics, proposed by Mozi (from the 5th to the 3rd centuries B.C.).

2. The new structuralist methodology has been developed, which allows comparing logical constructs within the semantic structures of different cultural traditions and finding similarities and differences in these structures. Within the framework of the new methodology, the project of the so-called "archaeology of logic" has been proposed, which makes it possible to identify the logical competence in its historical forms. The main idea of the "archaeology of logic" is to evaluate the different traditions of logic, even those that were interrupted and survived only fragmentarily in some texts.

   The structuralist approach to the study of dynamics of logical competence, proposed in the book, made it possible to identify the early forms of logically correct reasoning. The peculiarity of this approach is in using formalizations of logeme structures by means of modern symbolic logic—where each logeme is understood as an elementary step of logically correct reasoning. Logeme structures may or may not have similarities with each other (an isomorphism with a preservation of the logeme order). A similarity occurs if in a text the same logical schemes produce conclusions with the same logical inferences by using the same inference rules (such as *modus ponens* and *modus tollens*). In the case of philosophical texts, the similarity of logeme structures takes place if the same judgments are considered as logical axioms (for example, if the judgment of logical determinism "Either *A* is necessary or not-*A* is necessary" is always considered equally true).

At the same time, some other methodological strategies have been applied in the context of the structuralist approach to the early history of logic, such as (a) historical reconstruction based on archaeological data; (b) reconstruction presented in historical sociology; (c) comparative textual studies and hermeneutics, studying different corpora of texts of the same or different traditions. To reconstruct a social context of logical reasoning, the structural-functional approach has been used in order to identify the normative-value (illocutionary) aspects of written documents and explicate their social background.

So, the use of the systematic approach, combining several methods of related humanities (history, logic, cognitive science), made it possible to substantiate the philosophical concept of "logical competence" on the basis of documents from different cultures of the ancient world of the historical period—from the Sumerian-Akkadian to the ancient Indian one, and to trace the dynamics of the development of logical competence in the early stages.

3. The study has constituted the dynamics of logical competence as an independent scientific direction of logical-philosophical and historical-logical research. It has been shown that logical competence is not an innate property, but a social mechanism of cognition, which is transmitted through training; therefore, logical competence has its own historical forms and develops from elementary methods of logical thinking to more and more complex ones. Being a social mechanism of cognition, logical competence is inseparable from the corresponding social practices, and the work reveals specific ways of functioning of logical competence in social practices, namely, it shows the role and place of logical competence in the legal support of administrative and economic affairs within the developed legal system. Thus, the connection between logical competence and the special legal culture of Mesopotamia has been established. It is in this culture that logical competence was first discovered. With the spread and diffusion of the legal culture of Mesopotamia throughout the Middle East, and then through the Achaemenid Empire and the Hellenistic or Hellenized kingdoms of some regions of Eurasia, logical competence also transferred. During the Hellenistic and then Roman periods, this competence had a thorough reflective understanding within the framework of four independent traditions that have survived to this day: (1) the ancient Greek logic (from the 4th to the

3rd centuries B.C.); (2) the Judaic logical hermeneutics (from the 1st to the 2nd centuries A.D.); (3) the Indian-Buddhist logic (from the 2nd to the 4th centuries A.D.); (4) the traditional Chinese hermeneutics as a proto-logic (4th century B.C.).

4. Thus, the work has revealed the early historical forms of logical competence, which became a basic need for humanity in the future—the methods of constructing logically correct reasoning within the framework of legal discourse. As a result, the theory of logical argumentation was reconstructed, which appeared for the first time in the Sumerian and Akkadian languages and was used in lawsuits and business correspondence. Here, the logical rules of inference, used implicitly in this theory, have received their explication, and it has also been shown that propositional logic of the early Stoic philosophers is a reflective interpretation of this tradition of legal argumentation, which was formed for the first time in Mesopotamia.

   In Babylonia, the idea of logical determinism was also formulated for the first time (namely, the assumption that the judgment "Either *A* is necessary or not-*A* is necessary" is always true). This idea was presented through an unspoken requirement, reflected in the Assyrian and Babylonian art of divination (fortune-telling), to compile, by all means, complete lists of all possible logical combinations of the available omens. This art of divination with the intuition of logical determinism influenced the development of the modal logic of the early Stoics, in which logical determinism was first formulated explicitly. Thus, the work explicates the idea of logical determinism, which first appeared in the texts on divination in Akkadian and later became the quintessence of the modal logic of the Stoics. Subsequently, the idea of logical determinism was also basic in the logical studies of the schools of Indian philosophy: Madhyamaka, Yogācāra, and Nyāya.

5. The main specific regularity of the development of logical competence is its resistance to all historical changes due to the demand for social practices of finding a consensus, cleared of cognitive distortions. These practices were born in the legal culture of Mesopotamia (Babylonia) and have gained a worldwide significance over time. So, the work establishes specific patterns of the dynamics of logical competence associated with its inherent effective way of regulating litigation and looking for intersubjective grounds for resolving conflicts. It was found that with the growth of international trade, starting with the developed states of Mesopotamia and, especially, during the Hellenistic and then Roman periods (the time of the creation of the Silk Trade route connecting the whole of Eurasia), there was an urgent need for a unified legal support for trade, which would include logical competence as an integral part of legal hermeneutics to ensure a consensus free of all cognitive distortions. In this social context, logical competence has been highly developed and, during our times, has already reached in artificial intelligence systems.

6. The analysis of the features of the dynamics of logical competence made it possible to formulate not only specific, but also general patterns of its functioning

and development. As a result, the model of the formation of logical competence in antiquity has been proposed, which shows the following general patterns of manifestation of logical competence in the context of various religious and philosophical traditions: (a) logical competence, in the process of its application, improves and becomes more complicated over time; (b) logical competence, in the process of reflection on it, acquires more and more sophisticated forms; (c) logical research eventually becomes self-sufficient and leaves its original cultural context, represented at the beginning by legal practice and divination.

This process of reflective comprehension of logical competence begins at the stage of the emergence of three new literary genres of philosophical reflection, which are (1) a philosophical prose in Ancient Greece (from the 4th century B.C.); (2) collections of Talmudic thematic dialogues (from the 1st century A.D.); (3) thematic dialogues or philosophical *sūtra*s in Ancient India (from the 2nd century A.D.). Within the framework of three new genres, the very first logical works appeared in the strict sense of the word: (1) in the 4th–3rd centuries B.C. in Ancient Greece; (2) in the 1st–2nd centuries A.D. in Ancient Judea; (3) in the 2nd–4th centuries A.D. in Ancient India. All three traditions have survived to this day. In all the three traditions, a key role was played by the logical-philosophical reflection on logical determinism, which most schools supported (Stoics, authors of *'aggādāh*, Yogācārins, and Naiyāyikas). Only Aristotle became the creator of an alternative modal logic, which already admits the denial of logical determinism—the statement about the existence of random (contingent) events *A*, for which it is always true to say "Both *A* is possible and not-*A* is possible".

7. As a practical application of the obtained theoretical results, a package of recommendations on the application of the theory of logical competence in the modern legal culture is proposed. Namely, the emphasis is placed on the prospects of the idea of an electronic lawyer, in whose system it is possible to simulate a trial using game theory, when the main characters of the trial (judge, prosecutor, defender, accused, witness) are represented by players with certain strategies of proof and argumentation.

Thus, the results obtained concerning the regularities of the dynamics of logical competence have an explanatory, ideological, prognostic, and socio-managerial significance.

Recommendations for the Practical Use of the Results

The scientific conclusions obtained in the book lay the theoretical foundations for further researches of the dynamics of logical competence and create the prerequisites for building a modern theory of logic. The materials of scientific work are of practical importance for the analysis of historical documents in order to identify the dynamics of not only logical competence, but also various semantic structures. The work shows the possibilities of reconstructing historical processes using structuralism enriched with logical methods. This methodology can be applied to the analysis of various forms of discourse, including legal argumentation. In particular, the book shows the possibilities of using logic in the design of e-justice and e-democracy systems. Also, the results can be used to teach university courses: “History of Philosophy” (sections “Philosophy of the Ancient World”, “Philosophy of Antiquity”), “Logic”, as well as a special course “Methods of Structuralism in the Humanities”.

The materials of scientific work can be of practical importance for making effective management decisions in the field of e-democracy and e-government (in the context of increasing the role of logical competence in making socially significant decisions).

The results of the work were used in the following projects: (1) “Model of innovative development and socialization of science in the Republic of Belarus” (Belarus, BRFFR No. G06-320 from 01.04.2006 to 31.03.2008); (2) “Philosophical and logical analysis of complex self-organizing systems in modern science and culture” (Belarus, BRFFR No. G07MS-048 from 01.04.2007 to 31.03.2009); (3) “Transformation of the norms of rationality in post-industrial culture” (Belarus, RHSF-BRFFR No. G09MS-027 from 15.04.2009 to 31.03.2011); (4) “Development of communicative models of Belarusian science management” (Belarus, BRFFR No. G11OB-107 from 15.04.2011 to 31.03.2013); (5) “New logical and mathematical methods in applied sciences and modeling of decision-making processes” (Poland, UMO-2012/07/B/HS1/00263 from 12.07.2013 to 11.07.2015); (6) “Physarum Chip: Growing Computers from Slime Mold” (Council of Europe, FP7-ICT-2011-8 from 01.03.2013 to 29.02.2016); (7) “Formal models and cognitive mechanisms of argumentation in law” (Belarus, RHSF-BRFFR No. G15R-025 from 04.05.2015 to 30.04.2017); (8) “Massive-parallel games and information war simulation using Web Mining and Big Data analysis” (Poland, F3_40 from 03.12.2019 to 31.05.2020); (9) “Dictionary of Polish Logicians 1900–1939” (Poland, 0411/NPRH7/H30/86/2019 from 02.10.2019 to 01.10.2022).

Thus, the results of the work allow for the use of logical methods for the structuralist analysis of various phenomena with a complex structure and Big Data in order to identify general and specific patterns of dynamics.

Index

B

E

F

J

K

M

N

Q

R

T

U

V

W

X

Y

Z

For Product Safety Concerns and Information please contact our EU representative GPSR@taylorandfrancis.com
Taylor & Francis Verlag GmbH, Kaufingerstraße 24, 80331 München, Germany

www.ingramcontent.com/pod-product-compliance
Lightning Source LLC
LaVergne TN
LVHW011105300926
849160LV00008B/285

* 9 7 8 1 0 3 2 3 2 7 3 7 2 *